Studies in Systems, Decision and Control

Volume 69

Series editor

Janusz Kacprzyk, Polish Academy of Sciences, Warsaw, Poland
e-mail: kacprzyk@ibspan.waw.pl

About this Series

The series "Studies in Systems, Decision and Control" (SSDC) covers both new developments and advances, as well as the state of the art, in the various areas of broadly perceived systems, decision making and control- quickly, up to date and with a high quality. The intent is to cover the theory, applications, and perspectives on the state of the art and future developments relevant to systems, decision making, control, complex processes and related areas, as embedded in the fields of engineering, computer science, physics, economics, social and life sciences, as well as the paradigms and methodologies behind them. The series contains monographs, textbooks, lecture notes and edited volumes in systems, decision making and control spanning the areas of Cyber-Physical Systems, Autonomous Systems, Sensor Networks, Control Systems, Energy Systems, Automotive Systems, Biological Systems, Vehicular Networking and Connected Vehicles, Aerospace Systems, Automation, Manufacturing, Smart Grids, Nonlinear Systems, Power Systems, Robotics, Social Systems, Economic Systems and other. Of particular value to both the contributors and the readership are the short publication timeframe and the world-wide distribution and exposure which enable both a wide and rapid dissemination of research output.

More information about this series at http://www.springer.com/series/13304

Victor A. Sadovnichiy · Mikhail Z. Zgurovsky
Editors

Advances in Dynamical Systems and Control

Springer

Editors
Victor A. Sadovnichiy
Lomonosov Moscow State University
Moscow
Russia

Mikhail Z. Zgurovsky
National Technical University of Ukraine
"Kyiv Polytechnic Institute"
Kyiv
Ukraine

ISSN 2198-4182 ISSN 2198-4190 (electronic)
Studies in Systems, Decision and Control
ISBN 978-3-319-40672-5 ISBN 978-3-319-40673-2 (eBook)
DOI 10.1007/978-3-319-40673-2

Library of Congress Control Number: 2016942510

Printed on acid-free paper

This Springer imprint is published by Springer Nature
The registered company is Springer International Publishing AG Switzerland

Preface

Given collected articles have been organized as a result of open joint academic panels of research workers from Faculty of Mechanics and Mathematics of Lomonosov Moscow State University and Institute for Applied Systems Analysis of the National Technical University of Ukraine "Kyiv Polytechnic Institute," devoted to applied problems of mathematics, mechanics, and engineering, which attracted attention of researchers from leading scientific schools of Brazil, France, Germany, Poland, Russian Federation, Spain, Mexico, Ukraine, USA, and other countries. Modern technological applications require development and synthesis of fundamental and applied scientific areas, with a view to reducing the gap that may still exist between theoretical basis used for solving complicated technical problems and implementation of obtained innovations. To solve these problems, mathematicians, mechanics, and engineers from wide research and scientific centers have been worked together. Results of their joint efforts, including applied methods of modern algebra and analysis, fundamental and computational mechanics, nonautonomous and stochastic dynamical systems, optimization, control and decision sciences for continuum mechanics problems, are partially presented here. In fact, serial publication of such collected papers to similar seminars is planned.

This is the sequel of earlier two volumes "Continuous and Distributed Systems: Theory and Applications." In this volume, we are focusing on recent advances in dynamical systems and control (theoretical bases as well as various applications):

(1) we benefit from the presentation of modern mathematical modeling methods for the qualitative and numerical analysis of solutions for complicated engineering problems in physics, mechanics, biochemistry, geophysics, biology, and climatology;
(2) we try to close the gap between mathematical approaches and practical applications (international team of experienced authors closes the gap between abstract mathematical approaches, such as applied methods of modern analysis, algebra, fundamental and computational mechanics, nonautonomous and

stochastic dynamical systems, on the one hand, and practical applications in nonlinear mechanics, optimization, decision-making theory, and control theory on the other); and

(3) we hope that this compilations will be of interest to mathematicians and engineers working at the interface of these fields.

Moscow
Kyiv
April 2016

Victor A. Sadovnichiy
Mikhail Z. Zgurovsky

International Editorial Board of This Volume

Acknowledgements

We express our gratitude to editors of the “Springer” Publishing House who worked with collection and everybody who took part in the preparation of the manuscript.

We want to express the special gratitude to Olena L. Poptsova for a technical support of our collection.

Contents

Contributors

Yusuke Asai Department of Hygiene, Graduate School of Medicine, Hokkaido University, Sapporo, Japan

Irina N. Balaba Tula State Lev Tolstoy Pedagogical University, Tula, Russia

Francisco Balibrea Facultad de Matemáticas, Campus de Espinardo, Universidad de Murcia, Murcia, Spain

Lesia V. Baranovska Institute for Applied System Analysis, National Technical University of Ukraine "Kyiv Polytechnic Institute", Kyiv, Ukraine

Tomás Caraballo Departamento de Ecuaciones Diferenciales y Análisis Numérico, Universidad de Sevilla, Sevilla, Spain

Laurence Cherfils Université de La Rochelle, Laboratoire Mathématiques, Image et Applications, La Rochelle Cedex, France

Jose S. Cánovas Departamento de Matemática Aplicada y Estadística, Universidad Politécnica de Cartagena, Cartagena, Spain

Nikolai N. Dobrovolsky Tula State Lev Tolstoy Pedagogical University, Tula, Russia

Nikolai M. Dobrovol'skii Tula State Lev Tolstoy Pedagogical University, Tula, Russia

Gleb V. Fedorov Mechanics and Mathematics Faculty, Moscow State University, Moscow, Russia; Research Institute of System Development, Russian Academy of Sciences, Moscow, Russia

Victoria V. Fokicheva Lomonosov Moscow State University, Moscow, Russia

Anatoly T. Fomenko Lomonosov Moscow State University, Moscow, Russia

Vladimir V. Galatenko Lomonosov Moscow State University, Moscow, Russian Federation

María J. Garrido-Atienza Departamento de Ecuaciones Diferenciales y Análisis Numérico, Universidad de Sevilla, Sevilla, Spain

Dimitri V. Georgievskii Moscow State University, Moscow, Russia

Mark O. Gluzman Department of Applied Physics and Applied Mathematics, Columbia University, New York, NY, USA

Iryna M. Gorban Institute of Hydromechanics, National Academy of Sciences of Ukraine, Kyiv, Ukraine

Nataliia V. Gorban Institute for Applied System Analysis, National Technical University of Ukraine "Kyiv Polytechnic Institute", Kyiv, Ukraine

Xiaoying Han Department of Mathematics and Statistics, Auburn University, Auburn, AL, USA

Alexander O. Ivanov Mechanical and Mathematical Faculty, Lomonosov Moscow State University, Moscow, Russian Federation; Bauman Moscow Technical University, Moscow, Russia

Olena A. Kapustian Taras Shevchenko National University of Kyiv, Kyiv, Ukraine; Institute for Applied System Analysis, National Technical University of Ukraine "Kyiv Polytechnic Institute", Kyiv, Ukraine

Volodymyr O. Kapustyan National Technical University of Ukraine "Kyiv Polytechnic Institute", Kyiv, Ukraine

Oleksiy V. Kapustyan Taras Shevchenko National University of Kyiv, Kyiv, Ukraine; Institute for Applied System Analysis, National Technical University of Ukraine "Kyiv Polytechnic Institute", Kyiv, Ukraine

Pavlo O. Kasyanov Institute for Applied System Analysis, National Technical University of Ukraine "Kyiv Polytechnic Institute", Kyiv, Ukraine

Olha V. Khomenko Institute for Applied System Analysis, National Technical University of Ukraine "Kyiv Polytechnic Institute", Kyiv, Ukraine

Peter E. Kloeden School of Mathematics and Statistics, Huazhong University of Science & Technology, Wuhan, China; Felix-Klein-Zentrum Für Mathematik, TU Kaiserslautern, Kaiserslautern, Germany

Peter I. Kogut Department of Differential Equations, Dnipropetrovsk National University, Dnipropetrovsk, Ukraine

Olha P. Kupenko Department of System Analysis and Control, National Mining University, Dnipro, Ukraine; Institute for Applied and System Analysis of National Technical University of Ukraine "Kyiv Polytechnic Institute", Kyiv, Ukraine

Taras P. Lukashenko Lomonosov Moscow State University, Moscow, Russian Federation

Javier López-de-la-Cruz Departamento de Ecuaciones Diferenciales y Análisis Numérico, Universidad de Sevilla, Sevilla, Spain

Oleg K. Mazur National University of Food Technologies, Kyiv, Ukraine

Yuriy L. Milyavsky Institute for Applied System Analysis, National Technical University of Ukraine "Kyiv Polytechnic Institute", Kyiv, Ukraine

Alain Miranville Université de Poitiers, Laboratoire de Mathématiques et Applications, UMR CNRS 7348 - SP2MI, Chasseneuil Futuroscope Cedex, France

Nadezhda I. Nedashkovskaya Institute for Applied System Analysis, National Technical University of Ukraine "Kyiv Polytechnic Institute", Kyiv, Ukraine

Liliia S. Paliichuk Institute for Applied System Analysis, National Technical University of Ukraine "Kyiv Polytechnic Institute", Kyiv, Ukraine

Nataliya D. Pankratova Institute for Applied System Analysis, National Technical University of Ukraine "Kyiv Polytechnic Institute", Kyiv, Ukraine

Shuiran Peng Université de Poitiers, Laboratoire de Mathématiques et Applications, UMR CNRS 7348 - SP2MI, Chasseneuil Futuroscope Cedex, France

Boris E. Pobedria Moscow State University, Moscow, Russia

Ivan O. Pyshnograiev National Technical University of Ukraine "Kyiv Polytechnic Institute", Kyiv, Ukraine

Irina Yu. Rebrova Tula State Lev Tolstoy Pedagogical University, Tula, Russia

Victor D. Romanenko Institute for Applied System Analysis, National Technical University of Ukraine "Kyiv Polytechnic Institute", Kyiv, Ukraine

Iryna V. Romaniuk Taras Shevchenko National University of Kyiv, Kyiv, Ukraine

Alina V. Rusina Taras Shevchenko National University of Kyiv, Kyiv, Ukraine

Victor A. Sadovnichiy Lomonosov Moscow State University, Moscow, Russian Federation

Jacson Simsen Instituto de Matemática e Computação, Universidade Federal de Itajubá, Itajubá, MG, Brazil; Fakultät für Mathematik, Universität Duisburg-Essen, Essen, Germany

Dmitrii K. Sobolev Moscow State Pedagogical University, Moscow, Russian Federation

Valentina N. Soboleva Moscow State Pedagogical University, Moscow, Russian Federation

Alla M. Tkachuk Faculty of Automation and Computer Systems, National University of Food Technologies, Kyiv, Ukraine

Alexey A. Tuzhilin Mechanical and Mathematical Faculty, Lomonosov Moscow State University, Moscow, Russian Federation

José Valero Centro de Investigación Operativa, Universidad Miguel Hernández de Elche, Elche (Alicante), Spain

Mikhail Z. Zgurovsky National Technical University of Ukraine "Kyiv Polytechnic Institute", Kyiv, Ukraine

Part I
Applied Methods of Modern Algebra and Analysis

Chapter 1
Convergence Almost Everywhere of Orthorecursive Expansions in Functional Systems

Vladimir V. Galatenko, Taras P. Lukashenko and Victor A. Sadovnichiy

Abstract Along with the convergence in L^2-norm, convergence almost everywhere of expansions in functional systems is a property of interest for both theoretical studies and applications. In this paper we present results on convergence almost everywhere for orthorecursive expansions which are a natural generalization of classical expansions in orthogonal systems. As a corollary of a more general result, we obtain a condition on coefficients of an expansion that guarantees convergence almost everywhere. We also show that this condition cannot be relaxed.

1.1 Introduction

Orthorecursive expansion [8] is a natural generalization of orthogonal expansion. In case of an orthogonal system, these types of expansions give the same result, but orthorecursive expansions can be utilized for a much broader class of systems, and for redundant systems, they provide an absolute stability with respect to errors in coefficient computation [2].

Let us recall the definition of orthorecursive expansions in a system of elements. Let H be a Hilbert space (here we consider spaces over $\mathbb{R}$; however, the case of spaces over $\mathbb{C}$ is similar), and let $\{e_n\}_{n=1}^{\infty}$ be an arbitrary sequence of nonzero elements from H. For $f \in H$, we define a sequence of remainders $\{r_n(f)\}_{n=0}^{\infty}$ and a sequence of coefficients $\{\hat{f}_n\}_{n=1}^{\infty}$:

$$r_0(f) = f;$$

V.V. Galatenko (✉) · T.P. Lukashenko · V.A. Sadovnichiy
Lomonosov Moscow State University, GSP-1, Leninskie Gory 1, 119991 Moscow, Russian Federation
e-mail: vgalat@imscs.msu.ru

T.P. Lukashenko
e-mail: lukashenko@mail.ru

V.A. Sadovnichiy
e-mail: info@rector.msu.ru

V.A. Sadovnichiy and M.Z. Zgurovsky (eds.), *Advances in Dynamical Systems and Control*, Studies in Systems, Decision and Control 69,
DOI 10.1007/978-3-319-40673-2_1

$$\hat{f}_{n+1} = \frac{(r_n(f), e_{n+1})}{(e_{n+1}, e_{n+1})}, \quad r_{n+1}(f) = r_n(f) - \hat{f}_{n+1} e_{n+1} \quad (n = 0,\ 1,\ 2,\ \ldots).$$

Definition 1.1 The series $\sum_{n=1}^{\infty} \hat{f}_n e_n$ is called *an orthorecursive expansion* of f in a system $\{e_n\}_{n=1}^{\infty}$.

Orthorecursive expansions share with orthogonal expansions such properties as Bessel's identity

$$\left\| f - \sum_{n=1}^{N} \hat{f}_n e_n \right\|^2 = \|r_N(f)\|^2 = \|f\|^2 - \sum_{n=1}^{N} \hat{f}_n^2 \|e_n\|^2 \quad (N = 0,\ 1,\ 2,\ \ldots),$$

Bessel's inequality

$$\sum_{n=1}^{\infty} \hat{f}_n^2 \|e_n\|^2 \leq \|f\|^2,$$

the equivalence of convergence to the expanded element and Parseval's identity

$$\|f\|^2 = \sum_{n=1}^{\infty} \hat{f}_n^2 \|e_n\|^2$$

(see [8]).

Similarly to orthogonal expansions, there are at least two main types of results concerning orthorecursive expansions. The first one deals with the general properties of these expansions (e.g., [2, 10]), while the second one is focused on the properties of expansions in given functional systems or classes of functional systems (e.g., [1, 6, 7]).

Most of the results of both types concern the convergence with respect to a norm induced by a scalar product (for functional systems, it is L^2-norm). However, for some functional systems, results on pointwise convergence were also obtained (e.g., [1, 8]). At the same time, no general results on the pointwise convergence of orthorecursive expansions were known.

In this paper, we obtain a condition on coefficients of an orthorecursive expansion that guarantees convergence almost everywhere. The condition is obtained as a corollary of a more general result. We also show that this condition cannot be relaxed.

In order to simplify formulas without loss of generality, we consider normed systems, i.e., we suppose that $\|e_n\| = 1$ for all n.

1.2 Main Results

We start with a very simple positive result on the pointwise convergence. In fact, it is a result on Weyl multipliers [3, Chap. VIII, Sect. 1] for general (not necessarily orthogonal) systems.

Theorem 1.1 *Let* $\{e_n(x)\}_{n=1}^{\infty}$ *be a normed functional system in* $L^2(\Omega)$*, and let* $\{\lambda_n\}_{n=1}^{\infty} \subset [1, +\infty)$ *be a sequence for which*

$$\sum_{n=1}^{\infty} \frac{1}{\lambda_n} = \Lambda < \infty.$$

Then, for every sequence $\{a_n\}_{n=1}^{\infty} \subset \mathbb{R}$ *which satisfies the condition*

$$\sum_{n=1}^{\infty} a_n^2 \cdot \lambda_n = L < \infty$$

the functional series $\sum\limits_{n=1}^{\infty} a_n e_n(x)$ *absolutely converges almost everywhere on* Ω *and absolutely converges in* $L^2(\Omega)$*, and*

$$\left\| \sum_{n=1}^{\infty} a_n e_n(x) \right\| \leq \sqrt{L\Lambda}. \tag{1.1}$$

Note that in case of normed orthogonal systems, the classical Menshov's result [9] implies that an exact Weyl multiplier for the almost everywhere convergence is $\{\log^2 n\}$.

Corollary 1.1 *If there exists such a sequence* $\{\lambda_n\}_{n=1}^{\infty} \subset [1, +\infty)$ *with*

$$\sum_{n=1}^{\infty} \frac{1}{\lambda_n} < \infty$$

that coefficients of an orthorecursive expansion of a function $f \in L^2(\Omega)$ *in a normed functional system* $\{e_n(x)\}_{n=1}^{\infty}$ *satisfy the condition*

$$\sum_{n=1}^{\infty} \hat{f}_n^2 \cdot \lambda_n < \infty,$$

then the orthorecursive expansion $\sum\limits_{n=1}^{\infty} \hat{f}_n e_n(x)$ *absolutely converges almost everywhere on* Ω.

For example, if $|\hat{f}_n|$ do not exceed $\frac{C}{n^{1+\alpha}}$ for all n, where C and α are arbitrary positive constants, then convergence almost everywhere can be guaranteed for the orthorecursive expansion as $\{n^{1+\alpha}\}_{n=1}^{\infty}$ can be taken as $\{\lambda_n\}_{n=1}^{\infty}$.

Note that in this case, the orthorecursive expansion also absolutely converges in $L^2(\Omega)$. However, the limit does not necessarily coincide with f.

The next result shows that in spite of its simplicity, the condition in Theorem 1.1 cannot be relaxed.

Theorem 1.2 *Let $L^2(\Omega)$ be a separable space, and let $\{\lambda_n\}_{n=1}^{\infty} \subset [1, \infty)$ be an arbitrary sequence for which*

$$\sum_{n=1}^{\infty} \frac{1}{\lambda_n} = \infty.$$

Then, for every function f from $L^2(\Omega)$ with $\|f\| > 0$, there exists a normed functional system $\{e_n(x)\}_{n=1}^{\infty} \subset L^2(\Omega)$ such that the orthorecursive expansion of f in this system diverges almost everywhere and diverges in $L^2(\Omega)$, while

$$\sum_{n=1}^{\infty} \hat{f}_n^2 \cdot \lambda_n < \infty.$$

Note that if the condition of convergence of an orthorecursive expansion in L^2 is additionally imposed, then almost everywhere convergence can be guaranteed by a softer condition on $\{\lambda_n\}_{n=1}^{\infty}$ in comparison with the condition from Theorem 1.1. The details will be given in subsequent publications.

1.3 Proofs

We start with the proof of Theorem 1.1 which is quite simple and straightforward.

For an arbitrary measurable set $E \subset \Omega$ with measure $\mu E < \infty$ due to the Cauchy–Schwarz inequality

$$\int_E |a_n e_n(x)|\, d\mu \leq \left(\int_E d\mu\right)^{1/2} \cdot \left(\int_\Omega |a_n e_n(x)|^2\, d\mu\right)^{1/2} = \sqrt{\mu E} \cdot |a_n|.$$

At the same time,

$$\sum_{n=1}^{\infty} |a_n| = \sum_{n=1}^{\infty} \frac{1}{\sqrt{\lambda_n}} \cdot \left(|a_n|\sqrt{\lambda_n}\right) \leq \left(\sum_{n=1}^{\infty} \frac{1}{\lambda_n}\right)^{1/2} \cdot \left(\sum_{n=1}^{\infty} a_n^2 \lambda_n\right)^{1/2} = \sqrt{L\Lambda} < \infty. \tag{1.2}$$

Thus,

$$\sum_{n=1}^{\infty} \int_E |a_n e_n(x)| \, d\mu < \infty$$

and Beppo Levi's theorem implies that the series $\sum_{n=1}^{\infty} |a_n e_n(x)|$ converges almost everywhere on E. As E is an arbitrary measurable subset of Ω with a finite measure, the series converges almost everywhere on Ω as well.

Absolute convergence in $L^2(\Omega)$ and the estimate (1.1) directly follow from (1.2) as the system $\{e_n(x)\}_{n=1}^{\infty}$ is normed.

The proof of Theorem 1.2 is less straightforward. We start it with a number of technical lemmas.

Lemma 1.1 *Let d and h be vectors from $\mathbb{R}^2$ with $\|d\| = \delta > 0$, $\|h\| = 1$, and an angle between these vectors equal to γ $(0 \le \gamma \le \pi)$. Then, for every finite set $\{\alpha_n\}_{n=1}^{N}$ of positive numbers with*

$$\sum_{n=1}^{N} \alpha_n \ge \gamma \quad \textit{and} \quad \sum_{n=1}^{N} \alpha_n^2 < 1$$

there exists a finite system of normed vectors $\{e_n\}_{n=1}^{N} \subset \mathbb{R}^2$, such that the finite orthorecursive expansion $\sum_{n=1}^{N} \hat{d}_n e_n$ of d in this system satisfies the conditions

$$\left|\hat{d}_n\right| \le \delta \alpha_n \ (n = 1, 2, \ldots, N), \quad r_N = \beta h,$$

where $r_N = d - \sum_{n=1}^{N} \hat{d}_n e_n$, and

$$\delta \left(1 - \sum_{n=1}^{N} \alpha_n^2\right)^{1/2} \le \beta \le \delta.$$

In order to prove this lemma, we first consider a trivial case $\gamma = 0$ (i.e., $d = \delta h$). In this case, we find a unit vector e orthogonal to d and set all e_n to e. We get $\hat{d}_n = 0$ $(n = 1, 2, \ldots, N)$, $r_N = d$, and $\beta = \delta$.

In case of $\gamma > 0$, we divide the angle between d and h into N angles with angle measure γ_n, where $0 < \gamma_n \le \alpha_n$ for all $n = 1, 2, \ldots, N$. Then, for each n in the angle between vectors d and h, we find a unit vector v_n such that the angle between v_n and d is $\sum_{k=1}^{n} \gamma_k$. One can easily see that v_N coincides with h. Finally, for each $n = 1, 2, \ldots, N$, we find a unit vector orthogonal to v_n and take it as e_n (an arbitrary unit vector orthogonal to v_n is taken). For this selection, we have

$$|\hat{d}_1| = |(d, e_1)| = \|d\| \sin\gamma_1 \le \delta\gamma_1, \quad r_1 = \beta_1 v_1 \quad (\beta_1 > 0),$$

$$\delta^2 \ge \|r_1\|^2 = \beta_1^2 = \|d\|^2 - \hat{d}_1^2 \ge \delta^2\left(1 - \gamma_1^2\right),$$

$$|\hat{d}_2| = |(r_1, e_2)| = \|r_1\| \sin\gamma_2 \le \delta\gamma_2, \quad r_2 = \beta_2 v_2 \quad (\beta_2 > 0),$$

$$\delta^2 \ge \|r_2\|^2 = \beta_2^2 = \|r_1\|^2 - \hat{d}_2^2 \ge \delta^2\left(1 - \gamma_1^2 - \gamma_2^2\right),$$

$$\dots\dots\dots\dots\dots\dots\dots\dots\dots\dots\dots\dots\dots\dots,$$

$$|\hat{d}_N| = |(r_{N-1}, e_N)| = \|r_{N-1}\| \sin\gamma_N \le \delta\gamma_N, \quad r_N = \beta_N v_N \quad (\beta_N > 0),$$

$$\delta^2 \ge \|r_N\|^2 = \beta_N^2 = \|r_{N-1}\|^2 - \hat{d}_N^2 \ge \delta^2\left(1 - \sum_{n=1}^{N} \gamma_n^2\right).$$

Thus, for all $n = 1, 2, \dots, N$, we have $\left|\hat{d}_n\right| \le \delta\gamma_n \le \delta\alpha_n$, $r_N = \beta_N v_N = \beta h$, and

$$\delta\left(1 - \sum_{n=1}^{N} \alpha_n^2\right)^{1/2} \le \delta\left(1 - \sum_{n=1}^{N} \gamma_n^2\right)^{1/2} \le \beta \le \delta,$$

so all the required conditions are satisfied.

Lemma 1.2 *Let f be an arbitrary unit vector of a Hilbert space H, and let $\{\alpha_n\}_{n=1}^{\infty}$ be a sequence of positive numbers such that*

$$\sum_{n=1}^{\infty} \alpha_n = \infty, \quad \sum_{n=1}^{\infty} \alpha_n^2 = \nu \in (0, 1).$$

Let $\{h_k\}_{k=1}^{\infty}$ be an arbitrary sequence of unit vectors from H. Then, there exists a normed system $\{e_n\}_{n=1}^{\infty} \subset H$ such that coefficients of the orthorecursive expansion of f in this system satisfy the estimate $|\hat{f}_n| \le \alpha_n$ $(n = 1, 2, 3, \dots)$ and there exists an increasing sequence of indices $\{n_k\}_{k=1}^{\infty}$ for which

$$r_{n_k}(f) = r_{n_k} = \beta_k h_k, \quad \beta_k > 0, \quad \beta_k^2 = 1 - \sum_{j=1}^{n_k} \hat{f}_j^2 \ge 1 - \sum_{j=1}^{n_k} \alpha_j^2$$

(so for all $k = 1, 2, 3, \dots$ the remainder $r_{n_k}(f)$ is collinear to h_k and $\|r_{n_k}(f)\|^2$ exceeds $1 - \nu$).

Lemma 1.2 can be proved simply using Lemma 1.1 and induction.

Let γ_1 be the angle between f and h_1. We find such n_1 that $\sum_{j=1}^{n_1} \alpha_j \geq \gamma_1$ and applying Lemma 1.1 to f and h_1, we construct vectors $\{e_n\}_{n=1}^{n_1}$ which give $|\hat{f}_n| \leq \alpha_n$ $(n = 1, 2, \ldots, n_1)$, $r_{n_1}(f) = \beta_1 h_1$, $\beta_1 > 0$ (and due to Bessel's identity, we immediately have the equality $\beta_1^2 = 1 - \sum_{j=1}^{n_1} \hat{f}_j^2$).

Assume that vectors $\{e_n\}_{n=1}^{n_m}$ are already constructed and $r_{n_m}(f) = \beta_m h_m$ with $\beta_m > 0$. Let γ_{m+1} be the angle between r_{n_m} and h_{m+1}. We find such $n_{m+1} > n_m$ that $\sum_{j=n_m+1}^{n_{m+1}} \alpha_j \geq \gamma_{m+1}$ and applying Lemma 1.1 to r_{n_m} and h_{m+1}, we construct the system $\{e_n\}_{n=n_m+1}^{n_{m+1}}$. Taking into consideration that coefficients and remainders of the orthorecursive expansion of r_{n_m} in this system with indices $k = 1, 2, \ldots, n_{m+1} - n_m$ coincide with coefficients and remainders of the orthorecursive expansion of f in the system $\{e_n\}_{n=1}^{n_{m+1}}$ with indices $n_m + k$, we have $|\hat{f}_n| \leq \alpha_n$ $(k = n_m + 1, n_m + 2, \ldots, n_{m+1})$, $r_{n_{m+1}}(f) = \beta_{m+1} h_{m+1}$, $\beta_{m+1} > 0$ and

$$\beta_{m+1}^2 = \beta_m^2 - \sum_{j=n_m+1}^{n_{m+1}} \hat{f}_j^2 = 1 - \sum_{j=1}^{n_{m+1}} \hat{f}_j^2.$$

Lemma 1.3 *Let f be a unit vector of a separable Hilbert space H, and let $\{\alpha_n\}_{n=1}^{\infty}$ be a sequence of positive numbers such that*

$$\sum_{n=1}^{\infty} \alpha_n = \infty, \quad \sum_{n=1}^{\infty} \alpha_n^2 \in \left(0, \frac{3}{4}\right).$$

Then, there exists a normed system of elements $\{e_n\}_{n=1}^{\infty} \subset H$ such that coefficients of the orthorecursive expansion of f in this system satisfy the estimate $|\hat{f}_n| \leq \alpha_n$ (and hence all remainders $r_n(f)$ have norms exceeding $\frac{1}{2}$) and the sequence of remainders normed $\left\{\frac{r_n}{\|r_n\|}\right\}_{n=0}^{\infty}$ is everywhere dense in a unit sphere $S = \{x \in H : \|x\| = 1\}$.

Lemma 1.3 directly follows from the monotony of the sequence of norms $\{\|r_n\|\}_{n=0}^{\infty}$ (which is a corollary of Bessel's identity) and Lemma 1.2 as we can apply this lemma to a sequence of unit vectors $\{h_k\}_{k=1}^{\infty}$ that is everywhere dense in S. In this case, $\frac{r_{n_k}}{\|r_{n_k}\|} = h_k$ and $\|r_{n_k}\|^2 > 1 - \frac{3}{4} = \frac{1}{4}$ for all k.

Remark 1.1 Note that the expansion of f from Lemma 1.3 diverges. Moreover, its orthogonal projection on an arbitrary closed non-trivial (i.e., different from $\{0\}$) subspace also diverges. It follows from the fact that there exists an infinite subsequence of remainders which has the following properties: all remainders in the subsequence have norms exceeding $\frac{1}{2}$; after norming, these remainders form an everywhere dense subset of the unit sphere of the subspace.

Now, we proceed directly to the proof of Theorem 1.2. Let Λ_n denote the partial sum $\sum_{k=1}^{n} \frac{1}{\lambda_k}$. As the series $\sum_{n=1}^{\infty} \frac{1}{\lambda_n}$ diverges, due to Abel–Dini theorem [4, Chap. IX, Sect. 39], the series $\sum_{n=1}^{\infty} \frac{1}{\lambda_n \Lambda_n}$ also diverges, but the series $\sum_{n=1}^{\infty} \frac{1}{\lambda_n \Lambda_n^2}$ converges.

We take a number sequence $\alpha_n = \frac{\delta}{\lambda_n \Lambda_n}$, where a positive δ is selected in such a way that

$$\sum_{n=1}^{\infty} \alpha_n^2 \lambda_n = \sum_{n=1}^{\infty} \frac{\delta^2}{\lambda_n \Lambda_n^2} < \frac{3}{4}.$$

Note that $\sum_{n=1}^{\infty} \alpha_n$ diverges.

Let f be an arbitrary function from $L^2(\Omega)$ with $\|f\| = 1$ (the case of f with another positive norm is brought to this case by norming). Lemma 1.3 guarantees that there exists such a normed system $\{e_n(x)\}_{n=1}^{\infty} \subset L^2(\Omega)$ that the orthorecursive expansion of f in this system has the following properties: $|\hat{f}_n| \leq \alpha_n$ for all n and hence,

$$\sum_{n=1}^{\infty} \hat{f}_n^2 \lambda_n \leq \sum_{n=1}^{\infty} \alpha_n^2 \lambda_n < \infty,$$

all remainders $r_n(f)$ have norms exceeding $\frac{1}{2}$, and the sequence of normed remainders $\left\{ \frac{r_n(f)}{\|r_n(f)\|} \right\}_{n=0}^{\infty}$ is everywhere dense in a unit sphere of $L^2(\Omega)$.

According to Remark 1.1, the orthorecursive expansion of f in this system diverges in $L^2(\Omega)$ and its orthogonal projection on an arbitrary closed non-trivial subset of $L^2(\Omega)$ also diverges. If we assume that the orthorecursive expansion of f in the constructed system $\{e_n(x)\}_{n=1}^{\infty}$ converges pointwise on a set E with a positive measure, then due to Egorov's theorem [5, Chap. 8, Sect. 28.5], it would uniformly converge on a set $E_0 \subset E$ with $\mu E_0 > 0$. Hence, the orthogonal projection of the orthorecursive expansion on the subspace $L^2(E_0) \subset L^2(\Omega)$ would converge in L^2-norm. This contradiction completes the proof.

1.4 Conclusion

The results of the paper are the first non-trivial general results on Weyl multipliers for orthorecursive expansions. In the subsequent papers, we plan to state and prove similar results in case of additional assumption of expansion convergence in L^2.

We hope that these results would attract attention to problems of pointwise convergence of orthorecursive expansions and stir up the studies both for a general case and for cases of specific functional systems, including non-orthogonal wavelets [6, 7].

Acknowledgments The authors thank Dr. Alexey Galatenko for valuable comments and discussions.

The research was supported by the Russian Foundation for Basic Research (project 14–01–00417) and the President grant for the support of the leading scientific schools of the Russian Federation (grant NSh–7461.2016.1).

References

1. Galatenko, V.V.: On the orthorecursive expansion with respect to a certain function system with computational errors in the coefficients. Mat. Sb. **195**(7), 935–949 (2004)
2. Galatenko, V.V.: On orthorecursive expansions with errors in the calculation of coefficients. Izv. Math. **69**(1), 1–14 (2005)
3. Kashin, B.S., Saakyan, A.A.: Orthogonal Series. American Mathematical Society, Providence (1989)
4. Knopp, K.: Theory and Applications of Infinite Series. Dover Publications Inc., New York (1990)
5. Kolmogorov, A.N., Fomin, S.V.: Introductory Real Analysis. Dover Publications Inc., New York (1975)
6. Kudryavtsev, A.Yu.: On the convergence of orthorecursive expansions in nonorthogonal wavelets. Math. Notes. **92**(5), 643–656 (2012)
7. Kudryavtsev, A.Yu.: On the rate of convergence of orthorecursive expansions over non-orthogonal wavelets. Izv. Math. **76**(4), 688–701 (2012)
8. Lukashenko, T.P.: Properties of orthorecursive expansions in nonorthogonal systems. Moscow Univ. Math. Bull. **56**(1), 5–9 (2001)
9. Men'shov, D.E.: Sur les series de fonctions orthogonalen I. Fund. Math. **4**, 82–105 (1923)
10. Politov, A.V.: A convergence criterion for orthorecursive expansions in Euclidean spaces. Math. Notes. **93**(3), 636–640 (2013)

Chapter 2
Billiard Systems as the Models for the Rigid Body Dynamics

Victoria V. Fokicheva and Anatoly T. Fomenko

Abstract Description of the rigid body dynamics is a complex problem, which goes back to Euler and Lagrange. These systems are described in the six-dimensional phase space and have two integrals the energy integral and the momentum integral. Of particular interest are the cases of rigid body dynamics, where there exists the additional integral, and where the Liouville integrability can be established. Because many of such a systems are difficult to describe, the next step in their analysis is the calculation of invariants for integrable systems, namely, the so called Fomenko–Zieschang molecules, which allow us to describe such a systems in the simple terms, and also allow us to set the Liouville equivalence between different integrable systems. Billiard systems describe the motion of the material point on a plane domain, bounded by a closed curve. The phase space is the four-dimensional manifold. Billiard systems can be integrable for a suitable choice of the boundary, for example, when the boundary consists of the arcs of the confocal ellipses, hyperbolas and parabolas. Since such a billiard systems are Liouville integrable, they are classified by the Fomenko–Zieschang invariants. In this article, we simulate many cases of motion of a rigid body in 3-space by more simple billiard systems. Namely, we set the Liouville equivalence between different systems by comparing the Fomenko–Zieschang invariants for the rigid body dynamics and for the billiard systems. For example, the Euler case can be simulated by the billiards for all values of energy integral. For many values of energy, such billard simulation is done for the systems of the Lagrange top and Kovalevskaya top, then for the Zhukovskii gyrostat, for the systems by Goryachev–Chaplygin–Sretenskii, Clebsch, Sokolov, as well as expanding the classical Kovalevskaya top Kovalevskaya–Yahia case.

V.V. Fokicheva (✉) · A.T. Fomenko (✉)
Lomonosov Moscow State University, Leninskie Gory, 1, Moscow, Russia
e-mail: arinir@yandex.ru

A.T. Fomenko
e-mail: atfomenko@mail.ru

V.A. Sadovnichiy and M.Z. Zgurovsky (eds.), *Advances in Dynamical Systems and Control*, Studies in Systems, Decision and Control 69,
DOI 10.1007/978-3-319-40673-2_2

2.1 Introduction

Definition 2.1 A symplectic structure on a smooth manifold M is a differential 2-form ω satisfying the following two properties:

(1) ω is closed, i.e., $d\omega = 0$;
(2) ω is non-degenerate at each point of the manifold, i.e., in local coordinates, $\det\Omega(x) \neq 0$, where $\Omega(x) = (\omega_{ij}(x))$ is the matrix of this form.

The manifold endowed with a symplectic structure is called symplectic.

Let H be a smooth function on a symplectic manifold M. We define the vector of skew-symmetric gradient $\operatorname{sgrad} H$ for this function by using the following identity:

$$\omega(v, \operatorname{sgrad} H) = v(H),$$

where v is an arbitrary tangent vector v. In local coordinates $x^1, \dots, x^n$, we obtain the following expression:

$$(\operatorname{sgrad} H)^i = \sum \omega^{ij} \frac{\partial H}{\partial x^j},$$

where ω^{ij} are components of the inverse matrix to the matrix Ω.

Definition 2.2 The vector field $\operatorname{sgrad} H$ is called a Hamiltonian vector field. The function H is called the Hamiltonian of the vector field $\operatorname{sgrad} H$.

One of the main properties of Hamiltonian vector fields is that they preserve the symplectic structure ω.

Definition 2.3 Dynamical system $\dot{x} = v$ on the smooth manifold M is called Hamiltonian if and only if on the manifold M we can find symplectic structure ω and the function H such that system can be wrote as $v = \operatorname{sgrad} H$.

Definition 2.4 Let f and g be two smooth functions on a symplectic manifold M. By definition, we set $\{f, g\} = \omega(\operatorname{sgrad} f, \operatorname{sgrad} g) = (\operatorname{sgrad} f)(g)$ This operation $\{\cdot, \cdot\} : C^\infty \times C^\infty \to C^\infty$ on the space of smooth functions on M is called the *Poisson bracket.*

Let M^{2n} be a smooth symplectic manifold, and let $v = \operatorname{sgrad} H$ be a Hamiltonian dynamical system with a smooth Hamiltonian H.

Definition 2.5 The Hamiltonian system is called *Liouville integrable* if there exists a set of smooth functions $f_1, \dots, f_n$ such that

(1) $f_1, \dots, f_n$ are integrals of v,
(2) they are functionally independent on M, i.e., their gradients are linearly independent on M almost everywhere.

(3) $\{f_i, f_j\} = 0$ for any i and j,
(4) the vector fields sgrad f_i are complete, i.e., the natural parameter on their integral trajectories is defined on the whole real axis.

Definition 2.6 The decomposition of the manifold M^{2n} into connected components of common level surfaces of the integrals $f_1, \dots, f_n$ is called the *Liouville foliation* corresponding to the integrable system $v = \operatorname{sgrad} H$.

Since $f_1, \dots, f_n$ are preserved by the flow v, every leaf of the Liouville foliation is an invariant surface. The Liouville foliation consists of regular leaves (filling M almost in the whole) and singular ones (filling a set of zero measure). The Liouville theorem formulated below describes the structure of the Liouville foliation near regular leaves.

Consider a common regular level T_ξ for the functions $f_1, \dots, f_n$, that is $T_\xi = \{x \in M | f_i(x) = \xi_i, i = 1, \dots, n\}$. The regularity means that all 1-forms df_i are linearly independent on T_ξ.

Theorem 2.1 (J. Liouville) *Let* $v = \operatorname{sgrad} H$ *be a Liouville integrable Hamiltonian system on* M^{2n}*, and let* T_ξ *be a regular level surface of the integrals* $f_1, \dots, f_n$*. Then*

(1) T_ξ *is a smooth Lagrangian submanifold that is invariant with respect to the flow* $v = \operatorname{sgrad} H$ *and* $\operatorname{sgrad} f_1, \dots, \operatorname{sgrad} f_n$.
(2) if T_ξ *is connected and compact, then* T_ξ *is diffeomorphic to the n-dimensional torus* T^n *(this torus is called the Liouville torus);*
(3) the Liouville foliation is trivial in some neighborhood of the Liouville torus, that is, a neighborhood U of the torus T_ξ *is the direct product of the torus* T_ξ *and the disc* D^n*;*
(4) in the neighborhood $U = T^n \times D^n$ *there exists a coordinate system* $s_1, \dots, s_n$, $\varphi_1, \dots, \varphi_n$, *(which is called the action-angle variables), where* $s_1, \dots, s_n$ *are coordinates on the disc* D^n *and* $\varphi_1, \dots, \varphi_n$ *are standard angle coordinates on the torus, such that*

- $\omega = \Sigma d\varphi_i \wedge ds_i$, *are functions of the integrals,*
- *the action variables* s_i *are functions of the integrals* $f_1, \dots, f_n$,
- *in the action-angle variables* $s_1, \dots, s_n, \varphi_1, \dots, \varphi_n$*, the Hamiltonian flow v is straightened on each of the Liouville tori in the neighborhood U, that is,*

$$\dot{s}_i = 0, \ \dot{\varphi}_i = q_i(s_1, \dots, s_n), i = 1, 2, \dots, n.$$

(this means that the flow v determines the conditionally periodic motion that generates a rational or irrational rectilinear winding on each of the tori).

The problems of the rigid body dynamics can be described on the six-dimensional phase manifold, which in some cases is the Poisson manifold. In integrable case we can restrict the system to a submanifold M^4, where it is possible to introduce a symplectic structure. As a result we assume the existence of such four-dimensional symplectic manifold. Thus, the Liouville tori are two-dimensional tori.

Liouville foliation provides a lot of information about the solutions of the system. In fact, according to the Liouville theorem, the solutions on each torus, are its rectilinear windings. The manifold of the parameters of the integrals, where the rectilinear winding is rational (the case of the so-called resonant torus) has measure zero. Thus, for almost all values of the additional integral the closure of the solution forms the Liouville torus. If you change the initial data the change entails the change the Liouville torus, which makes it possible to describe the behavior of the solutions of the system. This weakening of the orbital equivalence is called the Liouville equivalence, see below.

Definition 2.7 Let $(M_1^4, \omega_1, f_1, g_1)$ and $(M_2^4, \omega_2, f_2, g_2)$ be two Liouville integrable systems on symplectic manifolds M_1^4 and M_2^4. Consider the isoenergy surfaces $Q_1^3 = \{x \in M_1^4 : f_1(x) = c_1\}$ $Q_2^3 = \{x \in M_2^4 : f_2(x) = c_2\}$, endowed with the Liouville foliations. Integrable systems on these 3-manifolds are said to be Liouville equivalent if there exists a leafwise diffeomorphism $Q_1^3 \to Q_2^3$, preserving the orientation of the 3-manifolds Q_1^3 and Q_2^3 and of all critical circles.

Let (M^4, ω, f_1, f_2) be Liouville integrable system on symplectic manifolds M^4. The manifold $Q^3 = \{x \in M^4 : f_1(x) = c_1\}$ is foliated into tori and singular leaves. Consider the base of the Liouville foliation on Q^3. This is a one-dimensional graph W called the Kronecker–Reeb graph of the function $f_2|_{Q^3}$. The structure of a foliation in a small neighborhood of the singular leaf corresponding to a vertex of the graph is described by a combinatorial object, called atom. A graph each of whose vertices is assigned an atom is called a Fomenko invariant (rough molecule). At the vertices of "atoms" are placed; they describe the corresponding bifurcations of the Liouville tori.

We now describe the atoms we need.

The minimax 3-atom A. Topologically, this 3-atom is presented as a solid torus foliated into concentric tori, shrinking into the axis of the solid torus. In other words, the 3-atom A is the direct product of a circle and a disc foliated into concentric circles (see Fig. 2.1). From the viewpoint of the corresponding dynamical system, A is a neighborhood of a stable periodic orbit.

The saddle 3-atoms without stars. Consider an arbitrary 2-atom without stars, i.e., a two-dimensional oriented compact surface P with a Morse function $f : P \to \mathbb{R}$ having just one critical value. The corresponding 3-atom is the direct product $U = P \times S^1$. An example is shown in Fig. 2.1: this is the simple 3-atom B.

The simple 3-atom A^* with star is presented in Fig. 2.1.

The molecule W contains a lot of essential information on the structure of the Liouville foliation on Q^3. However, this information is not quite complete. We have to add some additional information to the molecule W, namely, the rules that clarify how to glue the isoenergy surface Q^3 from individual 3-atoms.

To this end, cut every edge of the molecule in the middle. The molecule will be divided into individual atoms. From the point of view of the manifold Q^3 this operation means that we cut it along some Liouville tori into 3-atoms. Imagine that we want to make the backward gluing. The molecule W tells us which pairs of boundary atoms we have to glue together. To realize how exactly they should be glued, for every edge of W, we have to define the gluing matrix C, which determines the isomorphism between the fundamental groups of the two glued tori. To write down this matrix, we have to fix some coordinate systems on the tori. As usual, by a coordinate system on the torus, we mean a pair of independent oriented cycles (λ, μ) that are generators of the fundamental group $\pi_1(T^2) = \mathbb{Z} \oplus \mathbb{Z}$ (or, what is the same in this case, of the one-dimensional homology group). Geometrically, this simply means that the cycles λ and μ are both nontrivial and are intersected transversely at a single point. According to the fixed rules for each type of 3-atom we must choose a special coordinate system on the boundary tori of the atom (see [1]) which will be called admissible.

To the gluing matrix $C_i = \begin{pmatrix} \alpha_i & \beta_i \\ \gamma_i & \delta_i \end{pmatrix}$ on the edge e_i we assign two following numerical marks.

Definition 2.8 The mark r_i on the edge e_i of the molecule W is:

$$r_i = \begin{cases} \frac{\alpha_i}{\beta_i} \mod 1 \in \mathbb{Q}/\mathbb{Z}, & \text{if } \beta_i \neq 0, \\ \text{symbol } \infty, & \text{if } \beta_i = 0. \end{cases}$$

Definition 2.9 The mark ε_i on the edge e_i of the molecule W is:

$$\varepsilon_i = \begin{cases} \operatorname{sign} \beta_i, & \text{if } \beta_i \neq 0, \\ \operatorname{sign} \alpha_i, & \text{if } \beta_i = 0. \end{cases}$$

First, we need some preliminary construction. An edge of the molecule with mark r_i equal to ∞ is said to be it an infinite edge. The other edges are called *finite*. Let

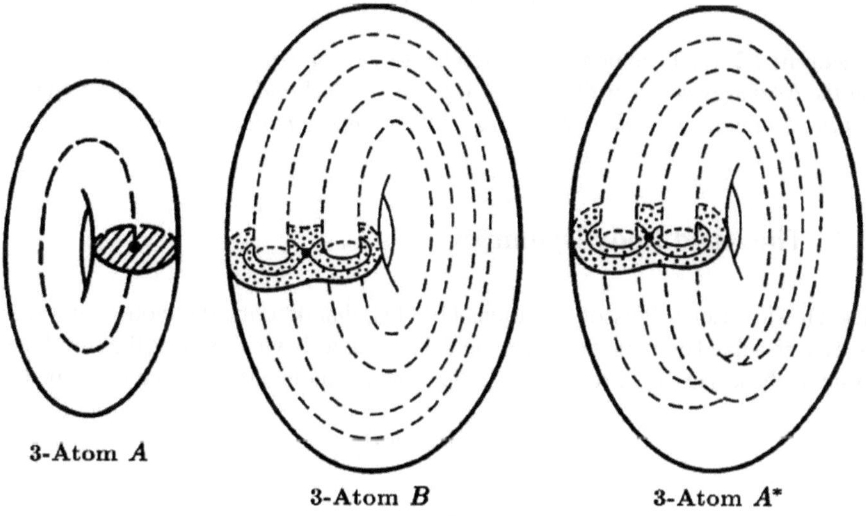

Fig. 2.1 The simple 3-atoms A, B and A^*.

us cut the molecule along all the finite edges. As a result, the molecule splits into several connected pieces.

Definition 2.10 Those pieces which do nor contain atoms of type A are said to be families. For example, if all the edges of a molecule are finite, then each of its saddle atoms is a family by definition.

Consider a single family $U = U_k$. All its edges can be divided into three classes: incoming, outgoing, and interior.

Definition 2.11 To each of these edges e_i, we assign an integer number Θ_i by the following rule:

$$\Theta_i = \begin{cases} [\frac{\alpha_i}{\beta_i}], & \text{if } e_i - \text{outgoing edge,} \\ [-\frac{\delta_i}{\beta_i}], & \text{if } e_i - \text{incoming edge,} \\ [-\frac{\gamma_i}{\alpha_i}], & \text{if } e_i - \text{interior edge.} \end{cases}$$

For every family U_k, we define an integer number n_k by setting

$$n_k = \sum \Theta_i,$$

where the sum is taken over all edges of the given family, and k is the number of the family.

Definition 2.12 The molecule W endowed with the marks r_i, ε_i and n_k is called a *marked molecule*. We denote it by

$$W^* = (W, r_i, \varepsilon_i, n_k).$$

Theorem 2.2 (A.T. Fomenko, X. Zieschang) *Two integrable Hamiltonian systems on the isoenergy surfaces* $Q_1^3 = \{x \in M_1^4 : f_1(x) = c_1\}$ *and* $Q_2^3 = \{x \in M_2^4 : f_2(x) = c_2\}$ *are Liouville equivalent if and only if their marked molecules coincide.*

2.2 The Rigid Body Dynamics

The classical Euler–Poisson equations [10, 11], that describe the motion of a rigid body with a fixed point in the gravity field, have the following form (in the coordinate system whose axes are directed along the principal moments of inertia of the body):

$$A\dot{\omega} = A\omega \times \omega - Pr \times \nu,$$

$$\dot{\nu} = \nu \times \omega. \tag{2.1}$$

Here ω and ν are phase variables of the system, where ω is the angular velocity vector, ν is the unit vector for the vertical line. The parameters of (2.1) are the diagonal matrix

$A = diag(A_1, A_2, A_3)$ that determines the tensor of inertia of the body, the vector r joining the fixed point with the center of mass, and the weight P of the body. Notation $a \times b$ is used for the vector product in $\mathbb{R}^3$. The vector $A\omega$ has the meaning of the angular momentum of the rigid body with respect to the fixed point.

N.E. Zhukovskii studied the problem on the motion of a rigid body having cavities entirely filled by an ideal incompressible fluid performing irrotational motion [12]. In this case, the angular momentum is equal to $A\omega + \lambda$, where λ is a constant vector characterizing the cyclic motion of the fluid in cavities. The angular momentum has a similar form in the case when a flywheel is fixed in the body such that its axis is directed along the vector λ. Such a mechanical system is called a gyrostat. The motion of a gyrostat in the gravity field, as well as some other problems in mechanics (see, for instance, [13]), are described by the system of equations

$$A\dot{\omega} = (A\omega + \lambda) \times \omega - Pr \times \nu,$$
$$\dot{\nu} = \nu \times \omega, \tag{2.2}$$

whose particular case for $\lambda = 0$ is system (2.1).

Another generalization of Eq. (2.1) can be obtained by replacing the homogeneous gravity field with a more complicated one. The equations of motion of a rigid body with a fixed point in an arbitrary potential force field were obtained by Lagrange. If this field has an axis of symmetry, then this axis can be assumed to be vertical, and the equations become

$$A\dot{\omega} = A\omega \times \omega + \nu \times \frac{\partial U}{\partial \nu},$$
$$\dot{\nu} = \nu \times \omega, \tag{2.3}$$

where $U(\nu)$ is the potential function, and $\frac{\partial U}{\partial \nu}$ denotes the vector with coordinates $(\frac{\partial U}{\partial \nu_1}, \frac{\partial U}{\partial \nu_2}, \frac{\partial U}{\partial \nu_3})$. For $U = P\langle r, \nu\rangle$ we obtain system (2.1). By $\langle a, b\rangle$ we denote the standard Euclidean inner product in $\mathbb{R}^3$.

The generalized Eqs. (2.2) and (2.3) can be combined by considering, the motion of a gyrostat in an axially symmetric force field. The most general equations that describe various problems in rigid body dynamics have the following form (see, for example, Kharlamov's book [14]):

$$A\dot{\omega} = (A\omega + \kappa) \times \omega + \nu \times \frac{\partial U}{\partial \nu},$$
$$\dot{\nu} = \nu \times \omega, \tag{2.4}$$

where $\kappa(\nu)$—is the vector function whose components are the coefficients of a certain closed 2-form on the rotation group $SO(3)$, the so-called form of gyroscopic forces. Moreover, $\kappa(\nu)$ is not arbitrary, but has the form

$$\kappa = \lambda + (\Lambda - \mathrm{div}\lambda \cdot E)\nu, \tag{2.5}$$

where $\lambda(\nu)$—is an arbitrary vector function, $\mathrm{div}\lambda = \frac{\partial \lambda_1}{\partial \nu_1} + \frac{\partial \lambda_2}{\partial \nu_2} + \frac{\partial \lambda_3}{\partial \nu_3}$, and $\Lambda = \left(\frac{\partial \lambda_i}{\partial \nu_j}\right)^T$ is the transposed Jacobi matrix. Obviously, systems (2.1)–(2.3) are particular cases of (2.4).

System (2.4) always possesses the geometrical integral

$$F = \langle \nu, \nu \rangle = 1$$

and the energy integral

$$E = \frac{1}{2}\langle A\omega, \omega \rangle + U(\nu).$$

If the vector function $\kappa(\nu)$ has the form (2.5) then there exists another integral the so-called area integral

$$G = \langle A\omega + \lambda, \nu \rangle.$$

It can be shown that Eqs. (2.4), (2.5) are Hamiltonian on common four-dimensional levels of the geometrical and area integrals. Moreover, (2.4) and (2.5) can be represented as the Euler equations for the six-dimensional Lie algebra $e(3)$ of the group of isometrical transformations (motions) of three-dimensional Euclidean space.

On the dual space $e(3)^*$, there is the standard Lie-Poisson bracket defined for arbitrary smooth functions f and g:

$$\{f, g\}(x) = x([d_x f, d_x g]),$$

where $x \in e(3)^*$, [,] denotes the commutator in the Lie algebra $e(3)$, and $d_x f$ and $d_x g$—are the differentials of f and g at the point x. These differentials in fact belong to the Lie algebra $e(3)$ after standard identification of $e(3)^{**}$ with $e(3)$. In terms of the natural coordinates

$$S_1, S_2, S_3, R_1, R_2, R_3$$

on the space $e(3)^*$ this bracket takes the form:

$$\{S_i, S_j\} = \varepsilon_{ijk} S_k, \ \{R_i, S_j\} = \varepsilon_{ijk} R_k, \ \{R_i, R_j\} = 0, \tag{2.6}$$

where $\{i, j, k\} = \{1, 2, 3\}$, and $\varepsilon_{ijk} = \frac{1}{2}(i-j)(j-k)(k-i)$.

A Hamiltonian system on $e(3)^*$ relative to the bracket (2.6), i.e. the so-called Euler equations, by definition has the form:

$$\dot{S}_i = \{S_i, H\}, \ \dot{R}_i = \{R_i, H\},$$

where H is a function on $e(3)^*$ called the Hamiltonian of the system. By introducing the vectors

$$S = (S_1, S_2, S_3) \text{ and } R = (R_1, R_2, R_3),$$

these equations can be rewritten in the form of the generalized Kirchhoff equations:

$$\dot{S} = \left(\frac{\partial H}{\partial S}\right) \times S + \left(\frac{\partial H}{\partial R}\right) \times R, \ \dot{R} = \left(\frac{\partial H}{\partial S}\right) \times R. \tag{2.7}$$

Proposition 2.1 *The mapping $\varphi : \mathbb{R}^6(\omega, \nu) \to \mathbb{R}^6(S, R)$, given by the formulas*

$$S = -(A\omega + \lambda), R = \nu, \tag{2.8}$$

establishes an isomorphism between system (2.4), (2.5) and system (2.7) with the Hamiltonian

$$H = \frac{(S_1 + \lambda_1)^2}{2A_1} + \frac{(S_2 + \lambda_2)^2}{2A_2} + \frac{(S_3 + \lambda_3)^2}{2A_3} + U, \tag{2.9}$$

where the parameters A_1, A_2, A_3 and the function $\lambda_1, \lambda_2, \lambda_3, U$ are taken from (2.4), (2.5), but the functions are defined not on the space $\mathbb{R}^3(\nu)$, but on $\mathbb{R}^3(R)$.

Corollary 2.1 *Condition (2.5) imposed on the vector function $\kappa(\nu)$ is equivalent to the fact that (2.4) is isomorphic to the Euler equations (2.7) on $e(3)^*$ with the quadratic (in variables S) Hamiltonian of the form*

$$H = \langle CS, S\rangle + \langle W, S\rangle + V, \tag{2.10}$$

where is a constant symmetric 3×3-matrix, $W(R)$ is a vector function, and $V(R)$ is a smooth scalar function.

Under mapping (2.8), the integrals $F = \langle \nu, \nu\rangle$ and $G = \langle A\omega + \lambda, \nu\rangle$ transform into the invariants of the Lie algebra $e(3)$

$$f_1 = R_1^2 + R_2^2 + R_3^2, \ f_2 = S_1R_1 + S_2R_2 + S_3R_3,$$

and the energy integral $E = \frac{1}{2}\langle A\omega, \omega\rangle + U(\nu)$ transforms into Hamiltonian (2.9). System (2.7) is Hamiltonian on common four-dimensional level surfaces of the two invariants f_1 and f_2:

$$M^4_{c,g} = \{f_1 = R_1^2 + R_2^2 + R_3^2 = c, \ f_2 = S_1R_1 + S_2R_2 + S_3R_3 = g\}. \tag{2.11}$$

For almost all values of c and g, these common levels are non-singular smooth submanifolds in $e(3)^*$. In what follows, we shall assume that c and g are such regular values.

It is easily seen that these symplectic 4-manifolds $M^4_{c,g}$ are diffeomorphic (for $c > 0$) to the cotangent bundle TS^2 of the 2-sphere S^2. The symplectic structure on $M^4_{c,g}$ is given by the restriction of the Lie-Poisson bracket onto $TS^2 = M^4_{c,g}$ from the ambient six-dimensional space $e(3)^*$. Since the linear transformation $S' = S$, $R' = \gamma R$, where $\gamma = \text{const}$, preserves bracket (2.6), we shall assume in what follows that $c = 1$.

Thus, from now on, we shall consider Eq. (2.7) with Hamiltonian (2.9) on symplectic four-dimensional manifolds $M^4_{1,g} = \{f_1 = 1, f_2 = g\}$ in the six- dimensional space $e(3)^*$. In each specific problem, the phase variables and parameters of the system obtain a concrete physical meaning.

Now we give the list of main integrable cases of Eqs. (2.7), (2.9) with necessary comments. For each case we indicate explicitly the Hamiltonian H and the additional integral K independent of H. Note that sometimes the additional integral K may exist only for exceptional values of the area constant g.

The *Euler case* (1750). The motion of a rigid body about a fixed point that coincides with its center of mass.

$$H = \frac{S_1^2}{2A_1} + \frac{S_2^2}{2A_2} + \frac{S_3^2}{2A_3}, \quad K = S_1^2 + S_2^2 + S_3^2. \tag{2.12}$$

The *Lagrange case* (1788). The motion of an axially symmetric rigid body about a fixed point located at the symmetry axis.

$$H = \frac{S_1^2}{2A} + \frac{S_2^2}{2A} + \frac{S_3^2}{2B} + aR_3, \quad K = S_3. \tag{2.13}$$

The *Kovalevskaya case* (1899). The motion of a rigid body about a fixed point with the special symmetry conditions indicated below.

$$\begin{aligned} H &= \frac{S_1^2}{2A} + \frac{S_2^2}{2A} + \frac{S_3^2}{A} + a_1R_1 + a_2R^2, \\ K &= \left(\frac{S_1^2 - S_2^2}{2A} + a_2R_2 - a_1R_1\right)^2 + \left(\frac{S_1S_2}{A} - a_1R_2 - a_2R_1\right)^2. \end{aligned} \tag{2.14}$$

The integral has degree 4. In this case, $A_1 = A_2 = 2A_3$ (in particular, the body is axially symmetric), and the center of mass is located in the equatorial plane related to the coinciding axes of the inertia ellipsoid.

The *Goryachev–Chaplygin* case (1899). The motion of a rigid body about a fixed point with the special symmetry conditions indicated below.

$$\begin{aligned} H &= \frac{S_1^2}{2A} + \frac{S_2^2}{2A} + \frac{2S_3^2}{A} + a_1R_1 + a_2R^2, \\ K &= S_3(S_1^2 + S_2^2) - AR_3(a_1S_1 + a_2S_2). \end{aligned} \tag{2.15}$$

The integral has degree 3. In this case, $A_1 = A_2 = 4A_3$, and the center of mass is located in the equatorial plane related to the coinciding axes of the inertia ellipsoid.

In this case, the Poisson bracket of H and is

$$\{H, K\} = (S_1R_1 + S_2R_2 + S_3R_3)(a_2S_1 - a_1S_2).$$

Hence the functions H and K do not commute on all the manifolds $M^4_{1,g}$. Therefore, the system is integrable only on the single special manifold $M^4_{1,0} = \{f_1 = 1, f_2 = 0\}$. This is a case of partial integrability corresponding to the zero value of the area constant f_2.

Each of these four cases admits an integrable generalization the the case of gyroscopic forces.

The *Zhukovskii case* (1885). The motion of a gyrostat in the gravity field when the body is fixed at its center of mass.

$$\begin{aligned} H &= \frac{(S_1+\lambda_1)^2}{2A_1} + \frac{(S_2+\lambda_2)^2}{2A_2} + \frac{(S_3+\lambda_3)^2}{2A_3}, \\ K &= S_1^2 + S_2^2 + S_3^2. \end{aligned} \tag{2.16}$$

This case is a generalization of the classical Euler case (the Euler case is obtained for $\lambda_1 = \lambda_2 = \lambda_3 = 0$).

The *Kovalevskaya–Yahia* case (1986). The Kovalevskaya case with gyrostat.

$$\begin{aligned} H =& \frac{S_1^2}{2A} + \frac{S_2^2}{2A} + \frac{(S_3+\lambda)^2}{A} + a_1R_1 + a_2R^2, \\ K =& \left(\frac{S_1^2 - S_2^2}{2A} + a_2R_2 - a_1R_1\right)^2 + \left(\frac{S_1S_2}{A} - a_1R_2 - a_2R_1\right)^2 \\ & - \frac{2\lambda}{A^2}(S_3 + 2\lambda)(S_1^2 + S_2^2) + \frac{4\lambda R_3}{A}(a_1S_1 + a_2S_2). \end{aligned} \tag{2.17}$$

The classical Kovalevskaya case is obtained for $\lambda = 0$.

The *Sretenskii case* (1963). The Goryachev–Chaplygin case with gyrostat.

$$\begin{aligned} H &= \frac{S_1^2}{2A} + \frac{S_2^2}{2A} + \frac{2(S_3+\lambda)^2}{A} + a_1R_1 + a_2R^2, \\ K &= (S_3 + 2\lambda)(S_1^2 + S_2^2) - AR_3(a_1S_1 + a_2S_2). \end{aligned} \tag{2.18}$$

If $\lambda = 0$, then we obtain the classical Goryachev–Chaplygin case. This system is integrable on the zero level of the area integral.

The *Clebsch case* (1871). Motion of a rigid body in a fluid.

$$\begin{aligned} H &= \frac{S_1^2}{2A_1} + \frac{S_2^2}{2A_2} + \frac{S_3^2}{2A_3} + \frac{\varepsilon}{2}(A_1R_1^2 + A_2R_2^2 + A_3R_3^2), \\ K &= \frac{1}{2}(S_1^2 + S_2^2 + S_3^2) - \frac{\varepsilon}{2}(A_2A_3R_1^2 + A_3A_1R_2^2 + A_1A_2R_3^2). \end{aligned} \tag{2.19}$$

The calculation of Fomenko–Zieschang invariants is an effective method for recognizing the Liouville equivalence of the systems. The bifurcations of Liouville tori, bifurcation diagrams, and molecules W for these cases were first calculated by M.P. Kharlamov [14] and A.A. Oshemkov [15–17]. Then the complete invariants of the Liouville foliations (marked molecules W^*) were computed in a series of papers by several authors (A.V. Bolsinov [10], P. Topalov [18], A.V. Bolsinov, A.T. Fomenko [7, 8], O.E. Orel [19], O.E. Orel, S. Takahashi [20]). As a result, a complete classification of the main integrable cases in rigid body dynamics has been obtained up to Liouville equivalence. P. Morozov proved the Liouville equivalence of the Clebsch case [21] and the Sokolov case [22] for certain values of the integrals. In [23], the Liouville equivalence invariants for the Kovalevskaya–Yehia case (this is a generalization of the classical Kovalevskaya top to the case of the problem on the motion of a heavy gyrostat) were calculated.

2.3 Billiard Motion

Let the domain Ω be the domain on the plane $\mathbb{R}^2$ such that its boundary is the piecewise smooth curve and the angle at the corner points equals $\frac{\pi}{2}$. Consider the billiard dynamical system in Ω that describes the motion of a point inside Ω with natural reflection at the boundary $P = \partial\Omega$. At those points where the boundary P is not smooth, the trajectory of the system is extended by continuity: hitting a corner vertex, a material point is reflected back along the same trajectory without losing the rate.

The phase space of the system is the manifold

$$M^4 := \{(x, v)|x \in \Omega, v \in T_x\mathbb{R}^2, |v > 0|\}/\sim$$

where the equivalence relation $\sim$ is defined by

$(x_1, v_1) \sim (x_2, v_2)$ if and only if $x_1 = x_2 \in P$, $|v_1| = |v_2|$ and $v_1 + v_2 \parallel T_xP$. Here, T_xP denotes the tangent to the domain Ω at the point x and $|v|$ is the Euclidean length of the vector v.

Billiard motion has the natural integral—the speed $|v|$ of the material point x. If $|v| > 0$ then we can restrict the system to the isoenergy surface $Q^3 := \{(x, v) \in M^4 : |v| = 1\}$. Such isoenergetic surfaces are homeomorphic to each other and in the subsequent discussion we put $|v| = 1$. Some restriction of the choice of the boundary allows to find the additional integral.

Theorem 2.3 (Jacobi, Chasles [24]) *Given a geodesic curve on a quadric in n-dimensional Euclidean space, tangent lines which are drawn at arbitrary points on the geodesic are tangent both to this quadric and to $n-2$ confocal quadrics, which are the same for all the points on the geodesic.*

Now fix the family of the confocal quadrics on the plane $\mathbb{R}^2$ and consider the equation

$$(b - \lambda)x^2 + (a - \lambda)y^2 = (a - \lambda)(b - \lambda), \lambda \leq a. \tag{2.20}$$

where $\infty \geq a \geq b > 0$ is the fixed pair of numbers, which describe the family of quadrics, λ is the parameter defining the quadric which belongs to the family.

Suppose that a domain Ω in the plane $\mathbb{R}^2$ is such that its boundary is the union of piecewise smooth curves consisting of arcs of the confocal quadric. This domain will be called *elementary*.

From the Jacobi Chasles theorem it follows that the tangent lines to a billiard trajectory at any point inside a plane two-dimensional domain are tangent to an ellipse or a hyperbola confocal with the family of quadrics forming the boundary of this domain [24].

This implies the integrability of the billiard in a plane domain bounded by arcs of confocal ellipses and hyperbolas. The functions $|v|$—the speed of the material point—and λ—the parameter of the confocal quadric—commutate inside the domain Ω. Thus, they commutate in the boundary P of the domain Ω because they are integrals of the system.

As a result the billiard system which if defined in the plane domain bounded by the arcs of the confocal quadrics has two independent (see [24]) integrals $|v|$ and λ. Function λ sets on the isoenergy surface Q^3 the Liouville foliation which can be described in terms of the Fomenko–Zieshang invariant.

To classify all the domains bounded by ellipses and hyperbolas it is convenient to take the equivalence relation, which would allow, smoothly changing the class of confocal quadrics of the border region, to preserve the Liouville foliation of the billiard motion in it.

Definition 2.13 Elementary domain Ω, bounded by arcs of the confocal family of quadrics (2.20), is called *equivalent* to the other elementary domain Ω', which is bounded by arcs of quadrics from the same family (2.20), if Ω' can be obtained from Ω by the following composition of transformations:

- sequential changing borders by continuous segments deformation in the class of quadrics (2.20), so that the value of the parameter λ of the variable segment of the border did not take the value b;
- symmetry with respect to the axis of the family (2.20).

As a result of such definition of equivalence all elementary domains can be divided into three classes:

- pieces of the ellipse: domain A_2 (bounded by ellipse), A_1 (right part of the A_2), A_0 (rectangle, limited by ellipse and two branches of a hyperbola) and its upper halfs A_2',A_1' and A_0';
- ring-domain C_2, bounded by two ellipses;
- simply connected domain-bands series B, which are parts of the ring-domain C_2.

We can extend the class of elementary domains, adding to them the flat domains that do not have an immersion into the plane. In this case, to have the above-described non-simply connected domains we need to add the domains C_{2n}–n−sheets coverings over the domain C_2 and results of C_n of the quotient by the group $\mathbb{Z}_2$. As for simply connected domain, we must add "prolongations" of the domains B, which are now subsets of relevant domains C_n.

The Fomenko–Zieschang invariants of these systems were calculated by M. Radnovic and V. Dragovic in [25] and V.V. Fokicheva in [26].

The generalized billiard system in a generalized locally flat domain is defined in a similar way as the billiard in domains glued together along common convex segments of their boundaries. In this case, if a point reaches such a segment, its trajectory passes from one elementary domain to another. If a pair of domains is glued together along the common corner (the case of *a conical point*), then, by continuity, the motion must be defined as follows: a point moving on a sheet and hitting the corner is reflected along the same trajectory on the same sheet.

The equivalence relation on the set of generalized domains if taken as a continuation of the equivalence relation on the set of elementary domains. Namely, the domains will be called equivalent if they can be obtained from each other by replacing their constituent elementary domains on their equivalent. All such domains were classified by V. Fokicheva [28].

Obviously, with such a definition phase manifold M^4 preserves integrability of the system, namely, retained additional integral λ—parameter of the confocal quadric, which concerns the billiard trajectory. This is due to the fact that the boundary of any elementary domain Ω_i, which is part of the generalized domain Δ, and in particular, all the gluing edges pass into the arc of the same family of confocal quadrics in the isometric immersion of the field Ω_i or double covering in the plane.

The Fomenko–Zieschang invariants of these systems were calculated by V.V. Fokicheva in [28].

For further convenience, we assume the following notation. By Ω will be denoted an elementary domain. Glued from several elementary domains Ω_i the generalized domain is denoted by Δ. For a fixed generalized domain Δ the unification of all the borders of its constituent regions Ω_i, which are not glue arcs will be called a free boundary. By Θ we will denote the domain bounded by arcs of confocal quadrics, without specifying whether it is a elementary (Ω) or generalized (Δ).

The generalized domain without conical points is denoted by Δ_α, with conical points by Δ_β. We distinguish three types of conical points: type x is formed by the intersection of the focal line ($\lambda = b$) and confocal ellipse ($\lambda < b$), type c—at the intersection of the focal line ($\lambda = b$) and confocal hyperbola ($b < \lambda < a$), type y—at the intersection of the confocal ellipse ($\lambda < b$) and confocal hyperbola ($\lambda > b$). In the notation of the generalized domain in parentheses we specify the types of domains that make up this region and generalized types of conical points, if they exist.

We describe several classes of generalized domains and calculate Fomenko–Zieschang invariants that describe the topology of the Liouville foliation of the billiard motion in them. More precisely, we describe the domains of the invariants of the billiard motion that occur in problems of rigid body dynamics.

Proposition 2.2 ([28]) *Let the domain Θ be that, first, the interior of each elementary domain in its composition does not include points of the focal line, and*

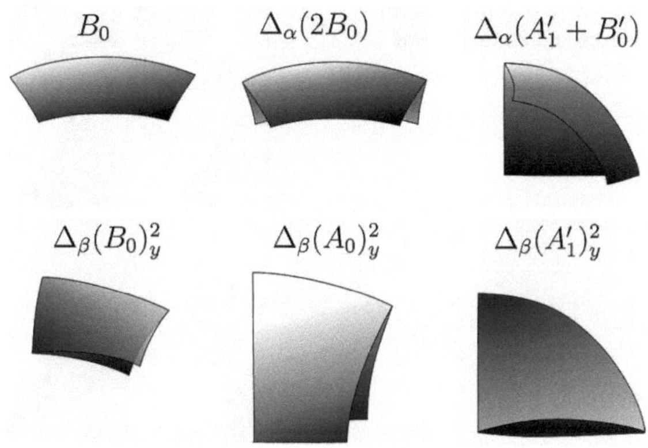

Fig. 2.2 In the *top row* there are domains without conical points at the *bottom*—with one conical point

secondly, any conical point is of the type y (see examples on the Fig. 2.2). Then Fomenko–Zieschang invariant describing the topology of Liouville foliation for the billiard motion in Θ is of the form:

- $A \xrightarrow{r=0,\varepsilon=1} A$, *if the domain Θ does not contain conical points;*
- $A \xrightarrow{r=\frac{1}{2},\varepsilon=1} A$, *if the domain Θ contains conical points.*

Remark 2.1 According to the classification of generalized domains [28] the domain, which has no common points with the focal line, and contains a conical point is arranged as follows: it contains exactly one conical point, with its free boundary homeomorphic to a circle.

Proposition 2.3 ([28]) *Suppose that domain Θ without conical points is such that each elementary domain Ω in its composition does not contain any focuses (see examples on the Fig. 2.3). Then Fomenko–Zieschang invariant describing the topology of Liouville foliation for the billiard motion in Θ is of the form:*

- $A \xrightarrow{r=\infty,\varepsilon=1} B \rightrightarrows \begin{matrix} A \\ A \end{matrix}$, *where marks on the right edges are $r = 0$, $\varepsilon = 1$, if domain Θ is equivalent to B_1, $\Delta_\alpha(2B_1)$,*
  A_0, $\Delta_\alpha(A_0 + B_0), \Delta_\alpha(A_0 + A'_0), \Delta_\alpha(B_0 + A_0 + B_0)$,
  $\Delta_\alpha(A'_0 + A_0 + B_0)$ or $\Delta_\alpha(A'_0 + A_0 + A'_0)$, i.e. domain Θ is homeomorphic to a disc and contains only one line segment of the focal line (either only one elementary domain Ω in its composition contains line segment of the focal line or these segments are glued into one along the arcs of the focal line);
- $A \xrightarrow{r=0,\varepsilon=1} B \rightrightarrows \begin{matrix} A \\ A \end{matrix}$, *where marks on the right edges are $r = \infty$, $\varepsilon = 1$, if domain Θ is equivalent to $\Delta_\alpha(A_0)^2$ or C_1, i.e. domain Θ is homeomorphic to a cartesian product $S^1 \times [0, 1]$ and contains only one line segment of the focal line;*

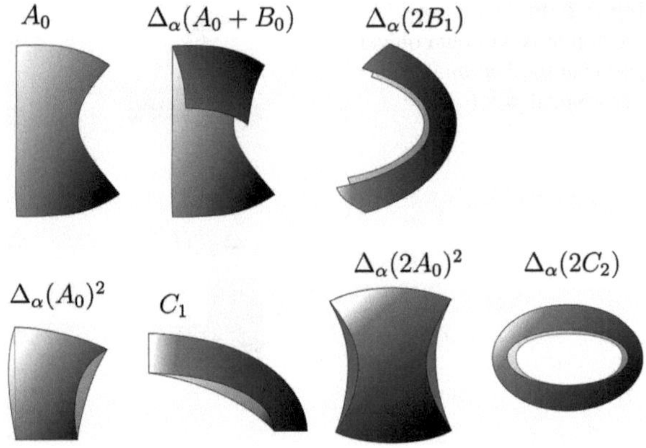

Fig. 2.3 Domains without focuses and conical points

- $\begin{matrix}A\\A\end{matrix} \rightrightarrows C_2 \rightrightarrows \begin{matrix}A\\A\end{matrix}$, *where marks on the left edges are* $r=\infty$, $\varepsilon=1$, *and on the right edges are* $r=0$, $\varepsilon=1$, *if domain* Θ *is equivalent to* $\Delta_\alpha(2A_0)^2$*or* $\Delta_\alpha(2C_2)$, *i.e. domain* Θ *is homeomorphic to a cartesian product* $S^1 \times [0,1]$ *and contains two line segments of the focal line.*

If the domain contains the focuses, all the edges of the molecule are finite, that makes compute mark n.

Proposition 2.4 ([28]) *Let the domain* Θ *be such that an elementary domain in its composition contains focuses of the confocal family of the domain's border (see examples on the Fig. 2.4). Then Fomenko–Zieschang invariant describing the topology of Liouville foliation for the billiard motion in* Θ *is of the form:*

- $\begin{matrix}A\\A\end{matrix} \rightrightarrows B \rightarrow A$, *marks on the all edges are* $r=0$, $\varepsilon=1$, *and mark* n *in the family is equal to* 1, *if domain* Θ *is equivalent to* A_2, $\Delta_\alpha(2A_1)$ *or* $\Delta_\alpha(A_2+C_2)$;
- $A \xrightarrow{r=0,\varepsilon=1} A^* \xrightarrow{r=0,\varepsilon=1} A$, *mark* n *in the family is equal to* 0, *if domain* Θ *is equivalent to* A_1 *or* $\Delta_\alpha(A_1+B_1)$;
- $\begin{matrix}A\\A\end{matrix} \rightrightarrows C_2 \rightrightarrows \begin{matrix}A\\A\end{matrix}$, *marks on the all edges are* $r=0$, $\varepsilon=1$, *and mark* n *in the family is equal to* 2, *if domain* Θ *is equivalent to* $\Delta_\alpha(2A_2)$;
- $\begin{matrix}A\\A\end{matrix} \rightrightarrows B \rightarrow A$, *marks on the all edges are* $r=0$, $\varepsilon=1$, *and mark* n *in the family is equal to* 2, *if domain* Θ *is equivalent to* $\Delta_\beta(A_1')^2_c, \Delta_\beta((A_1')^2_c+C_1)$ *or* $\Delta_\beta(A_1')^2_x$.

2.4 Main Results

The descriptions of all systems of the rigid body dynamics are fairly complex. It turns out that, in many cases, the Fomenko–Zieschang theorem makes it possible to establish the Liouville equivalence of these systems to certain simpler billiard systems on the four-dimensional phase space M^4.

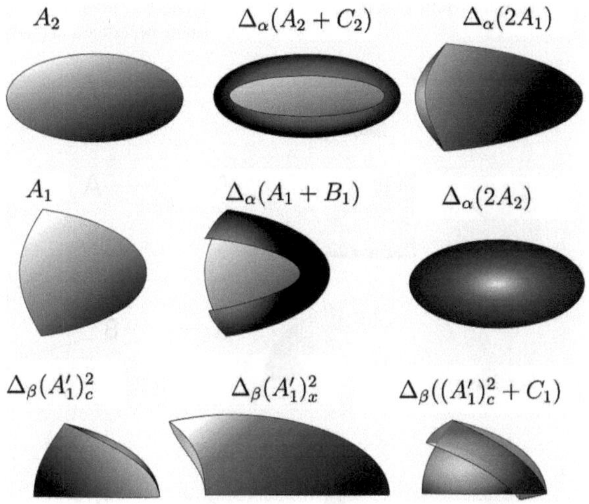

Fig. 2.4 Domains which contains the focuses

Theorem 2.4 ([27]) *The following cases of rigid body dynamics are modeled by (Liouville equivalent to) the following generalized billiards*

- *the Euler case (see [9]) is completly modeled by the billiards in the generalized domains shown in Figs. 2.5 and 2.6;*
- *the Zhukovskii case (see [15]) is modeled by the billiards in the generalized domains shown in Fig. 2.5b (energy zone 11, $Q^3 \simeq RP^3$), Fig. 2.5c (energy zone 2, $Q^3 \simeq S^1 \times S^2$), Fig. 2.5d (energy zone 8, $Q^3 \simeq S^3$), and Fig. 2.5f (energy zone 12, $Q^3 \simeq RP^3$);*
- *the Lagrange case (see [9]) is modeled by the billiards in the generalized domains shown in Fig. 2.5a (energy zone 2, $Q^3 \simeq S^3$) and Fig. 2.5b (energy zone 3, $Q^3 \simeq RP^3$);*
- *the Goryachev–Chaplygin–Sretenskii case (see [19]) is modeled by the billiards in the generalized domains shown in Fig. 2.5c (energy zone 4, $Q^3 \simeq S^1 \times S^2$) and Fig. 2.5g (energy zone 2, $Q^3 \simeq S^3$);*
- *the Kovalevskaya–Yehia case (see [23]) is modeled by the billiards in the generalized domains shown in Fig. 2.5c (energy zone h_{28}, $Q^3 \simeq S^1 \times S^2$) and Fig. 2.5e (energy zone h_{18}, $Q^3 \simeq S^3$);*
- *the Clebsch case (see [21]) is modeled by the billiards in the generalized domains shown in Fig. 2.5e (energy zone 2, $Q^3 \simeq S^3$), Fig. 2.5h (energy zones 10 and 12, $Q^3 \simeq S^1 \times S^2$), and Fig. 2.5i (energy zone 5, $Q^3 \simeq RP^3$);*
- *the Sokolov case (see [22]) is modeled by the billiards in the generalized domains shown in Fig. 2.5e (energy zone B, $Q^3 \simeq S^3$) and Fig. 2.5i (energy zone I, $Q^3 \simeq RP^3$).*

The Liouville equivalence of these billiard systems and integrable systems of the motion of a rigid body follows from the Fomenko–Zieschang theorem and the

	Generalized billiard domain	The Fomenko-Zieschang invariant describing the billiard in this domain	Equivalent known cases of integrability for a rigid body
a		A —($r = 0$, $\varepsilon = 1$)— A	Lagrange (1), Euler (1)
b	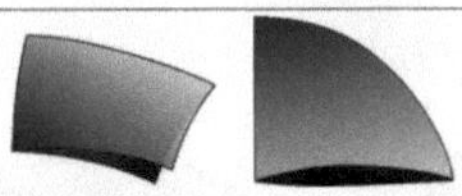	A —($r = 1/2$, $\varepsilon = 1$)— A	Lagrange (3), Zhukovskii (11)
c	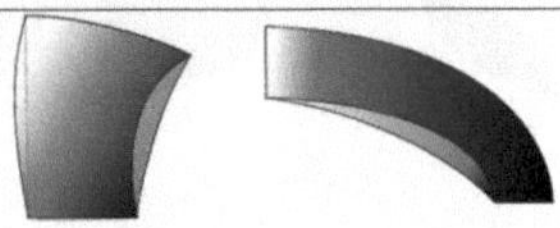	A —($r = 0$, $\varepsilon = 1$)— B; B —($r = \infty$, $\varepsilon = 1$)— A; B —($r = \infty$, $\varepsilon = 1$)— A	Kovalevskaya (5), Zhukovskii (2), Goryachev—Chaplygin—Sretenskii (4), Kovalevskaya—Yehia (h_{18})
d	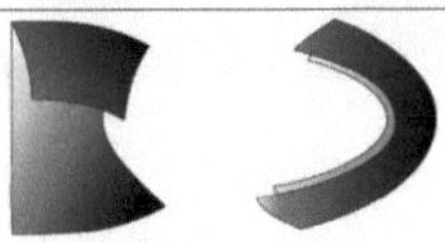	A —($r = \infty$, $\varepsilon = 1$)— B; B —($r = 0$, $\varepsilon = 1$)— A; B —($r = 0$, $\varepsilon = 1$)— A	Zhukovskii (8)
e	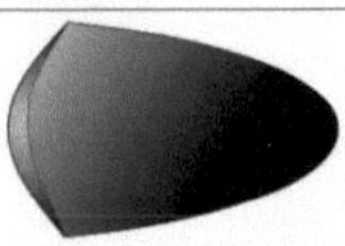	A —($r = 0$, $\varepsilon = 1$)— B ($n=1$); B —($r = 0$, $\varepsilon = 1$)— A; B —($r = 0$, $\varepsilon = 1$)— A	Clebsch (2), Sokolov (B), Kovalevskaya—Yehia (h_{18})
f	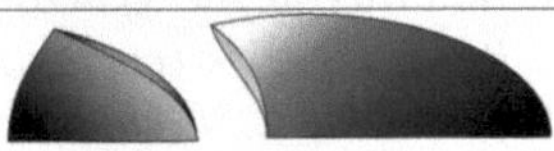	A —($r = 0$, $\varepsilon = 1$)— B ($n=2$); B —($r = 0$, $\varepsilon = 1$)— A; B —($r = 0$, $\varepsilon = 1$)— A	Zhukovskii (12)
g	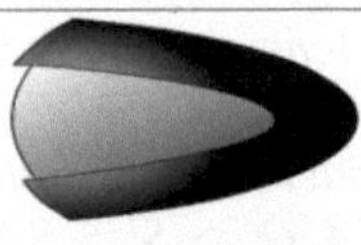	A —($r = 0$, $\varepsilon = 1$)— A* ($n=0$) —($r = 0$, $\varepsilon = 1$)— A	Goryachev—Chaplygin—Sretenskii (2)
h		A —($r = \infty$, $\varepsilon = 1$)— C_2; A —($r = \infty$, $\varepsilon = 1$)— C_2; C_2 —($r = 0$, $\varepsilon = 1$)— A; C_2 —($r = 0$, $\varepsilon = 1$)— A	Euler (2), Clebsch (10, 12)

Fig. 2.5 The left column shows the billiard domains, in the middle column – Fomenko-Zieschang invariants describing the topology of the billiard motion in them. The right column shows the cases of rigid body dynamics Fomenko-Zieschang which also have the form shown in the middle column (in parentheses are the numbers of the isoenergy surfaces in accordance with the numbering of the authors, the data to calculate the invariants)

comparison of the invariants of generalized billiards found by these authors with invariants calculated in the cited works of other authors.

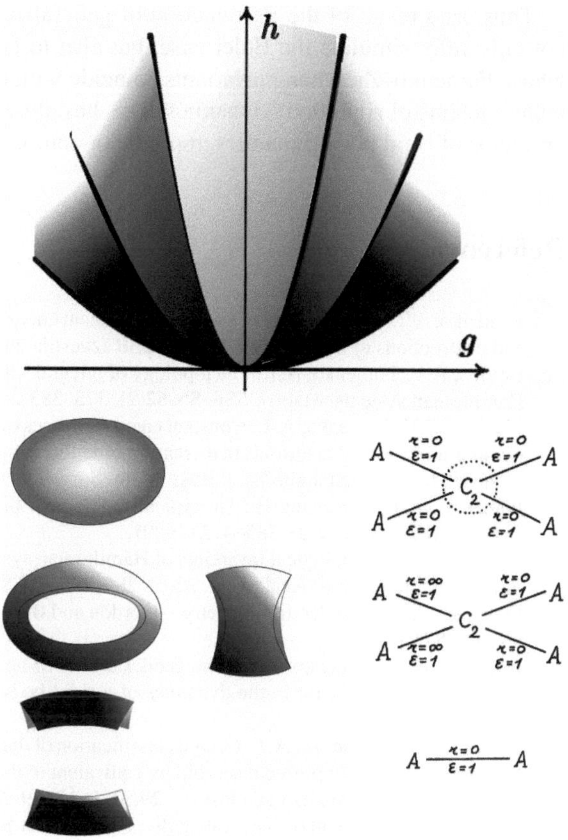

Fig. 2.6 Billiard system and the Euler case of the rigid body dynamics. The motion of a rigid body, the appropriate settings in the shaded gray area on the bifurcation diagram, modeled billiards in the domain shaded in the same shade

The Euler case is Liouville equivalent to the case of the geodesic flow on the ellipsoid [7]. This has been proven by the application of the theory of Fomenko–Zieschang—by calculating and comparing the invariants. On the other hand, the problem of the geodesic flow is closely connected with the integrable billiard problem in the domain bounded by arcs of confocal quadrics—by limiting to zero at the half-axis the ellipsoid becomes the flat ellipse, and geodesic lines on it become straight line segments. However, as can be seen, the billiard in an ellipse will not be Liouville equivalent to the geodesic flow.

The introduction of generalized billiards allowed to expand the class of classical billiard systems and successfully simulate not only the case of Euler fixed type isoenergy surfaces, but also to select for each constant-energy surface of a billiard a movement which will simulate the motion of a rigid body fixed at its center of mass.

It turns out that, in a sense, the billiard system is not so simple. However, its complexity lies in the complexity of a generalized billiard table—the more exotic the boundary the more complicated the topology of the Liouville foliation isoenergy surface Q^3.

Thus, as a result of the introduction of generalized billiards we have been able not only fully simulate the Euler case, but also to get a large number of systems, whose Fomenko–Zieschang invariants coincide with those calculated previously for many systems of rigid body dynamics. This has allowed to simulate a wide class of problems of rigid body dynamics, though not completely.

References

1. Fomenko, A.T.: The topology of surfaces of constant energy in integrable Hamiltonian systems, and obstructions to integrability. Math. USSR Izvestija **29**(3), 629–658 (1987)
2. Brailov, A.V., Fomenko, A.T.: The topology of integral submanifolds of completely integrable Hamiltonian systems. Math. USSR-Sb. **62**(2), 373–383 (1989)
3. Matveev, S.V., Fomenko, A.T.: Constant energy surfaces of Hamiltonian systems, enumeration of three-dimensional manifolds in increasing order of complexity, and computation of volumes of closed hyperbolic manifolds. -. Russ. Math. Surv. **43**(1), 3–24 (1988)
4. Fomenko, A.T., Zieschang H.: On typical topological properties of integrable Hamiltonian systems. USSR-Izv. **32**(2), 385–412 (1989)
5. Fomenko, A.T.: Topological invariants of Hamiltonian systems that are integrable in the sense of Liouville functional. Anal. Appl. **22**(4), 286–296 (1988)
6. Fomenko, A.T.: Symplectic Geometry. – Gordon and Breach. Advanced Studies in Mathematics. vol. 5 (1988)
7. Bolsinov, A.V., Fomenko, A.T.: The geodesic flow of an ellipsoid is orbit ally equivalent to the Euler integrable case in the dynamics of a rigid body. Dokl. Akad. Nauk SSSB. **339**(3), 293–296 (1994)
8. Bolsinov, A.V., Fomenko, A.T.: Orbital classification of the geodesic flows on two- dimensional ellipsoids. The Jacobi problem is orbitally equivalent to the integrable Euler case in rigid body dynamics. Funkts. Analiz i ego Prilozh. **29**(3), 1–15 (1995)
9. Bolsinov, A.V., Fomenko, A.T.: Integrable Hamiltonian Systems: Geometry, Topology, Classification, 1, 2. Regulyarnaya i Khaolichcskaya Dinamika, Izhevsk (1999). [in Russian]
10. Bolsinov, A.V.: Methods of calculation of the Fomenko – Ziesehang invariant. In: Topological Classification of Integrable Systems – Advances in Soviet Mathematics, vol 6, 147–183. AMS, Providence (1991)
11. Arkhangelskii, YuA: Analytical Dynamics of a Rigid Body. Nauka, Moscow (1977)
12. Zhukovskii, N.E.: On the motion of a rigid body having cavities filled with homogeneous liquid. Zh. Russk. Fiz-Khim. Obsch. **17**(6), 81–113; **7**, 145–149; **8**, 231–280 (1885)
13. Kharlamov, P.V.: Lectures on Rigid Body Dynamics. Novosibirsk State University, Novosibirsk (1965)
14. Kharlamov, M.P.: Topological Analysis of Integrable Problems in Rigid Body Dynamics. Leningrad University, Leningrad (1988)
15. Oshemkov, A.A.: Fomenko invariants for the main integrable cases of the rigid body motion equations. In: Topological Classification of Integrable Systems –Advances in Soviet Mathematics, vol. 6, 67–146. AMS, Providence (1991)
16. Oshemkov, A.A.: Topology of isoenergy surfaces and bifurcation diagrams for integrahle cases of rigid body dynamics on $SO(4)$. Uspekhi Mat. Nauk **42**(6), 199–200 (1987)
17. Oshemkov, A.A.: Description of isoenergetic surfaces of some integrable Hamiltonian systems with two degrees of freedom. In: Trudy Semin. po Vektor. Tenzor. Analizu, vol. 23, 122–132. Izdatel'stvo Moskovskogo Universiteta, Moscow (1988)
18. Topalov, P.I.: Calculation of the fine Fomenko-Ziesehang invariant for the main integrable cases in rigid body motion. Matem. Sbornik **187**(3), 143–160 (1996)

19. Orel, O.E.: The rotation function for integrable problems that are reducible to Abel equations. Trajectory classification of Goryachev–Chaplygin systems. Matem. Sbormk **186**(2): 105–128 (1995)
20. Orel, O.E., Takahashi, S.: Trajectory classification of integrable Lagrange and Goryachev Chaplygin problems by computer analysis methods. Mat em. Sbomik **187**(1), 95–112 (1996)
21. Morozov, P.V.: The Liouville classification of integrable systems of the Clebsch case. Sb. Math. **193**(10), 1507–1533 (2002)
22. Morozov, P.V.: Topology of Liouville foliations in the Steklov and the Sokolov integrable cases of Kirchhoff's equations. Sb. Math. **195**(3), 369–412 (2004)
23. Slavina, N.S.: Topological classification of systems of Kovalevskaya-Yehia type. Sb. Math. **205**(1), 101–155 (2014)
24. Kozlov, V.V., Treshchev, D.V.: (Russian) [Billiards] A genetic introduction to the dynamics of systems with impacts. Moskov Gos University, Moscow (1991)
25. Dragovich, V., Radnovich, M.: Integrable Billiards, Quadrics, and Multidymensional Poncelet Porisms. Regulyarnaya i Khaolichcskaya Dinamika, Izhevsk (2010). [in Russian]
26. Fokicheva, V.V.: Description of singularities for billiard system bounded by confocal ellipses and hyperbolas. Moscow Univ. Math. Bull. **69**(4), 148–158 (2014)
27. Fokicheva, V.V., Fomenko, A.T.: Integrable billiards model important integrable cases of rigid body dynamics. Doklady Math. **92**(3), 1–3 (2015). doi:10.7868/S0869565215320055
28. Fokicheva, V.V.: A topological classification of billiards in locally planar domains bounded by arcs of confocal quadrics. Sb. Math. **206**(10), 127–176 (2015). doi:10.1070/SM2015v206nl0ABEH004502

[illegible]

Chapter 3
Uniform Global Attractors for Nonautonomous Evolution Inclusions

Mikhail Z. Zgurovsky and Pavlo O. Kasyanov

Abstract In this note, we prove the existence and provide basic structure properties of compact (in the natural phase space) uniform global attractor for all global weak solutions of the general classes of nonautonomous evolution equations and inclusions that satisfy standard sign and polynomial growth conditions. The obtained results allow to reduce the problem of the complete qualitative investigation of various nonlinear systems into the "small" (compact) part of the natural phase space.

3.1 Introduction and Setting of the Problem

For evolution triple $(V_i; H; V_i^*)$[1] and multivalued map $A_i : \mathbb{R}_+ \times V \rightrightarrows V^*$, $i = 1, 2, \ldots, N$, $N = 1, 2, \ldots,$ we consider a problem of longtime behavior (in the natural phase space H) of all globally defined weak solutions for nonautonomous evolution inclusion

$$y'(t) + \sum_{i=1}^{N} A_i(t, y(t)) \ni \bar{0}, \tag{3.1}$$

as $t \to +\infty$. Let $\langle \cdot, \cdot \rangle_{V_i} : V_i^* \times V_i \to \mathbb{R}$ be the pairing in $V_i^* \times V_i$ that coincides on $H \times V_i$ with the inner product $(\cdot, \cdot)$ in the Hilbert space H.

[1] That is, V_i is a real reflexive separable Banach space continuously and densely embedded into a real Hilbert space H, H is identified with its topologically conjugated space H^*, V_i^* is a dual space to V_i. So, there is a chain of continuous and dense embeddings: $V_i \subset H \equiv H^* \subset V_i^*$ (see, e.g., Gajewski, Gröger, and Zacharias [1, Chap. I]).

M.Z. Zgurovsky
National Technical University of Ukraine "Kyiv Polytechnic Institute",
Peremogy Ave. 37, Build. 1, Kyiv 03056, Ukraine
e-mail: zgurovsm@hotmail.com

P.O. Kasyanov (✉)
Institute for Applied System Analysis, National Technical University of Ukraine "Kyiv Polytechnic Institute", Peremogy Ave. 37, Build. 35, Kyiv 03056, Ukraine
e-mail: kasyanov@i.ua

V.A. Sadovnichiy and M.Z. Zgurovsky (eds.), *Advances in Dynamical Systems and Control*, Studies in Systems, Decision and Control 69,
DOI 10.1007/978-3-319-40673-2_3

To introduce the assumptions on parameters of Problem (3.1) let us introduce additional constructions. A function $\varphi \in L_\gamma^{\mathrm{loc}}(\mathbb{R}_+)$, $\gamma > 1$, is called *translation bounded* in $L_\gamma^{\mathrm{loc}}(\mathbb{R}_+)$, if

$$\sup_{t\geq 0}\int_t^{t+1} |\varphi(s)|^\gamma ds < +\infty;$$

Chepyzhov and Vishik [2, p. 105]. A function $\varphi \in L_1^{\mathrm{loc}}(\mathbb{R}_+)$ is called *translation uniform integrable (t.u.i.)* in $L_1^{\mathrm{loc}}(\mathbb{R}_+)$, if

$$\lim_{K\to+\infty}\sup_{t\geq 0}\int_t^{t+1} |\varphi(s)|\mathbf{I}\{|\varphi(s)|\geq K\}ds = 0.$$

Note that Dunford–Pettis compactness criterion provides that a function $\varphi \in L_1^{\mathrm{loc}}(\mathbb{R}_+)$ is t.u.i. in $L_1^{\mathrm{loc}}(\mathbb{R}_+)$ if and only if for every sequence of elements $\{\tau_n\}_{n\geq 1} \subset \mathbb{R}_+$ the sequence $\{\varphi(\cdot + \tau_n)\}_{n\geq 1}$ contains a subsequence which converges weakly in $L_1^{\mathrm{loc}}(\mathbb{R}_+)$. Note that for any $\gamma > 1$ every translation bounded in $L_\gamma^{\mathrm{loc}}(\mathbb{R}_+)$ function is t.u.i. in $L_1^{\mathrm{loc}}(\mathbb{R}_+)$; Gorban et al. [3].

Throughout this paper, we suppose that the listed below assumptions hold:

Assumption 1 Let $p_i \geq 2$, $q_i > 1$ are such that $\frac{1}{p_i} + \frac{1}{q_i} = 1$, for each for $i = 1, 2, \ldots, N$, and the embedding $V_i \subset H$ is compact one, for some for $i = 1, 2, \ldots, N$.

Assumption 2 (*Growth Condition*) There exist a t.u.i. in $L_1^{\mathrm{loc}}(\mathbb{R}_+)$ function $c_1 : \mathbb{R}_+ \to \mathbb{R}_+$ and a constant $c_2 > 0$ such that

$$\max_{i=1}^{N} \|d_i\|_{V_i^*}^q \leq c_1(t) + c_2\sum_{i=1}^{N} \|u\|_{V_i}^p$$

for any $u \in V_i$, $d_i \in A_i(t, u)$, $i = 1, 2, \ldots, N$, and a.e. $t > 0$.

Assumption 3 (*Signed Assumption*) There exists a constant $\alpha > 0$ and a t.u.i. in $L_1^{\mathrm{loc}}(\mathbb{R}_+)$ function $\beta : \mathbb{R}_+ \to \mathbb{R}_+$ such that

$$\sum_{i=1}^{N}\langle d_i, u\rangle_{V_i} \geq \alpha\sum_{i=1}^{N} \|u\|_{V_i}^p - \beta(t)$$

for any $u \in V_i$, $d_i \in A_i(t, u)$, $i = 1, 2, \ldots, N$, and a.e. $t > 0$.

Assumption 4 (*Strong Measurability*) If $C \subseteq V_i^*$ is a closed set, then the set $\{(t, u) \in (0, +\infty) \times V_i : A_i(t, u) \cap C \neq \emptyset\}$ is a Borel subset in $(0, +\infty) \times V_i$.

Assumption 5 (*Pointwise Pseudomonotonicity*) Let for each $i = 1, 2, \ldots, N$ and a.e. $t > 0$, two assumptions hold:

(a) for every $u \in V_i$ the set $A_i(t, u)$ is nonempty, convex, and weakly compact one in V_i^*;
(b) if a sequence $\{u_n\}_{n\geq 1}$ converges weakly in V_i toward $u \in V_i$ as $n \to +\infty$, $d_n \in A_i(t, u_n)$ for any $n \geq 1$, and $\limsup_{n\to+\infty}\langle d_n, u_n - u\rangle_{V_i} \leq 0$, then for any $\omega \in V_i$ there exists $d(\omega) \in A_i(t, u)$ such that

$$\liminf_{n\to+\infty}\langle d_n, u_n - \omega\rangle_{V_i} \geq \langle d(\omega), u - \omega\rangle_{V_i}.$$

Let $0 \leq \tau < T < +\infty$. As a *weak solution* of evolution inclusion (3.1) on the interval $[\tau, T]$, we consider an element $u(\cdot)$ of the space $\cap_{i=1}^{N} L_{p_i}(\tau, T; V_i)$ such that for some $d_i(\cdot) \in L_{q_i}(\tau, T; V_i^*)$, $i = 1, 2, \ldots, N$, it is fulfilled:

$$-\int_{\tau}^{T}(\xi'(t), y(t))dt + \sum_{i=1}^{N}\int_{\tau}^{T}\langle d_i(t), \xi(t)\rangle_{V_i}dt = 0 \quad \forall \xi \in C_0^\infty([\tau, T]; V_i), \tag{3.2}$$

and $d_i(t) \in A_i(t, y(t))$ for each $i = 1, 2, \ldots, N$ and a.e. $t \in (\tau, T)$.

3.2 Preliminary Properties of Weak Solutions

Zgurovsky and Kasyanov [4, p. 225] provide the existence of a weak solution of Cauchy problem (3.1) with initial data $y(\tau) = y^{(\tau)}$ on the interval $[\tau, T]$, for any $y^{(\tau)} \in H$. For fixed τ and T, such that $0 \leq \tau < T < +\infty$, we denote

$$\mathscr{D}_{\tau,T}(y^{(\tau)}) = \{y(\cdot) \mid y \text{ is a weak solution of (3.1) on } [\tau, T],\ y(\tau) = y^{(\tau)}\}, \quad y^{(\tau)} \in H.$$

We remark that $\mathscr{D}_{\tau,T}(y^{(\tau)}) \neq \emptyset$, if $0 \leq \tau < T < +\infty$ and $y^{(\tau)} \in H$. Moreover, the concatenation of Problem (3.1) weak solutions is a weak solutions too, i.e., if $0 \leq \tau < t < T$, $y^{(\tau)} \in H$, $y(\cdot) \in \mathscr{D}_{\tau,t}(y^{(\tau)})$, and $v(\cdot) \in \mathscr{D}_{t,T}(y(t))$, then

$$z(s) = \begin{cases} y(s), & s \in [\tau, t], \\ v(s), & s \in [t, T], \end{cases}$$

belongs to $\mathscr{D}_{\tau,T}(y^{(\tau)})$; cf. Zgurovsky et al. [5, pp. 55–56].

Gronwall lemma provides that for any finite time interval $[\tau, T] \subset \mathbb{R}_+$ each weak solution y of Problem (3.1) on $[\tau, T]$ satisfies estimates

$$\|y(t)\|_H^2 - 2\int_0^t \beta(\xi)d\xi + 2\alpha \sum_{i=1}^N \int_s^t \|y(\xi)\|_{V_i}^p d\xi \le \|y(s)\|_H^2 - 2\int_0^s \beta(\xi)d\xi, \tag{3.3}$$

$$\|y(t)\|_H^2 \le \|y(s)\|_H^2 e^{-2\alpha\gamma(t-s)} + 2\int_s^t (\beta(\xi) + \alpha\gamma)e^{-2\alpha\gamma(t-\xi)}d\xi, \tag{3.4}$$

where $t, s \in [\tau, T]$, $t \ge s$; γ is a constant that does not depend on y, s, and t; see Zgurovsky and Kasyanov [4, p. 225]. Therefore, any weak solution y of Problem (3.1) on a finite time interval $[\tau, T] \subset \mathbb{R}_+$ can be extended to a global one, defined on $[\tau, +\infty)$.

For each $\tau \ge 0$ and $y^{(\tau)} \in H$ let $\mathscr{D}_\tau(y^{(\tau)})$ be the set of all weak solutions (defined on $[\tau, +\infty)$) of Problem (3.1) with initial data $y(\tau) = y^{(\tau)}$. Let us consider the family $\mathscr{K}_\tau^+ = \cup_{y^{(\tau)}\in H}\mathscr{D}_\tau(y^{(\tau)})$ of all weak solutions of Problem (3.1) defined on the semi-infinite time interval $[\tau, +\infty)$.

Consider the Fréchet space $C^{\mathrm{loc}}(\mathbb{R}_+; H)$. We remark that the sequence $\{f_n\}_{n\ge1}$ converges in $C^{\mathrm{loc}}(\mathbb{R}_+; H)$ toward $f \in C^{\mathrm{loc}}(\mathbb{R}_+; H)$ as $n \to +\infty$ iff the sequence $\{\Pi_{t_1,t_2}f_n\}_{n\ge1}$ converges in $C([t_1, t_2]; H)$ toward $\Pi_{t_1,t_2}f$ as $n \to +\infty$ for any finite interval $[t_1, t_2] \subset \mathbb{R}_+$, where Π_{t_1,t_2} is the restriction operator to the interval $[t_1, t_2]$; Chepyzhov and Vishik [6, p. 918]. We denote $T(h)y(\cdot) = y_h(\cdot)$, where $y_h(t) = y(t + h)$ for any $y \in C^{\mathrm{loc}}(\mathbb{R}_+; H)$ and $t, h \ge 0$.

Let us consider *united trajectory space* that includes all globally defined on any $[\tau, +\infty) \subseteq \mathbb{R}_+$ weak solutions of Problem (3.1) shifted to $\tau = 0$:

$$\mathscr{K}^+ = \mathrm{cl}_{C^{\mathrm{loc}}(\mathbb{R}_+;H)}\left[\bigcup_{\tau\ge0}\{y(\cdot + \tau) : y \in \mathscr{K}_\tau^+\}\right],$$

where $\mathrm{cl}_{C^{\mathrm{loc}}(\mathbb{R}_+;H)}[\,\cdot\,]$ is the closure in $C^{\mathrm{loc}}(\mathbb{R}_+; H)$. Note that $T(h)\{y(\cdot + \tau) : y \in \mathscr{K}_\tau^+\} \subseteq \{y(\cdot + \tau + h) : y \in \mathscr{K}_{\tau+h}^+\}$ for any $\tau, h \ge 0$. Moreover,

$$T(h)\mathscr{K}^+ \subseteq \mathscr{K}^+ \text{ for any } h \ge 0,$$

because

$$\rho_{C^{\mathrm{loc}}(\mathbb{R}_+;H)}(T(h)u, T(h)v) \le \rho_{C^{\mathrm{loc}}(\mathbb{R}_+;H)}(u, v) \text{ for any } u, v \in C^{\mathrm{loc}}(\mathbb{R}_+; H),$$

where $\rho_{C^{\mathrm{loc}}(\mathbb{R}_+;H)}$ is a standard metric on Fréchet space $C^{\mathrm{loc}}(\mathbb{R}_+; H)$; Zgurovsky and Kasyanov [4, p. 226].

The following Lemma 3.1 and Theorem 3.1 are keynote for the existence of compact (in the natural phase space H) uniform global attractor for all weak solutions of Problem (3.1).

Lemma 3.1 (Zgurovsky and Kasyanov [4]) *Let Assumptions (1)–(5) hold. Then, there exist positive constants c_3 and c_4 such that the following inequalities hold:*

$$\|y(t)\|_H^2 \leq \|y(s)\|_H^2 e^{-c_3(t-s)} + c_4,$$

for each $y \in \mathscr{K}^+$, $t \geq s \geq 0$.

Theorem 3.1 (Zgurovsky and Kasyanov [4]) *Let Assumptions (1)–(5) hold. Let* $\{y_n\}_{n\geq 1} \subset \mathscr{K}^+$ *be a bounded in* $L_\infty(\mathbb{R}_+; H)$ *sequence. Then, there exist a subsequence* $\{y_{n_k}\}_{k\geq 1} \subset \{y_n\}_{n\geq 1}$ *and an element* $y \in \mathscr{K}^+$ *such that*

$$\max_{t\in[\tau,T]} \|y_{n_k}(t) - y(t)\|_H \to 0, \quad k \to +\infty,$$

for any finite time interval $[\tau, T] \subset (0, +\infty)$.

3.3 Uniform Global Attractor for all Weak Solutions of Problem (3.1)

Let us define the multivalued semi-flow (*m-semi-flow*) $G : \mathbb{R}_+ \times H \to 2^H$:

$$G(t, y_0) := \{y(t) \, : \, y(\cdot) \in \mathscr{K}^+ \text{ and } y(0) = y_0\}, \quad t \geq 0, \; y_0 \in H. \tag{3.5}$$

For each $t \geq 0$ and $y_0 \in H$, the set $G(t, y_0)$ is nonempty. Moreover, the following two conditions hold:

(i) $G(0, \cdot) = I$ is the identity map;
(ii) $G(t_1 + t_2, y_0) \subseteq G(t_1, G(t_2, y_0))$, $\forall t_1, t_2 \in \mathbb{R}_+$, $\forall y_0 \in H$,

where $G(t, D) = \underset{y\in D}{\cup} G(t, y)$, $D \subseteq H$.

We denote by $\operatorname{dist}_H(C, D) = \sup_{c\in C} \inf_{d\in D} \rho(c, d)$ the *Hausdorff semi-distance* between nonempty subsets C and D of the Polish space H. Recall that the set $\mathscr{R} \subset H$ is a *global attractor* of the m-semi-flow G if it satisfies the following conditions:

(i) $\mathscr{R}$ attracts each bounded subset $B \subset H$, i.e.,

$$\operatorname{dist}_H(G(t, B), \mathscr{R}) \to 0, \quad t \to +\infty; \tag{3.6}$$

(ii) $\mathscr{R}$ is negatively semi-invariant set, i.e., $\mathscr{R} \subseteq G(t, \mathscr{R})$ for each $t \geq 0$;
(iii) $\mathscr{R}$ is the minimal set among all nonempty closed subsets $C \subseteq H$ that satisfy (3.6).

The main result of this paper has the following form.

Theorem 3.2 *Let Assumptions (1)–(5) hold. Then, the m-semi-flow G, defined in (3.5), has a compact global attractor* $\mathscr{R}$ *in the phase space H.*

3.4 Proof of Theorem 3.2

Lemma 3.1 and Theorem 3.1 imply the following properties for the m-semiflow G, defined in (3.5):

(a) for each $t \geq 0$, the mapping $G(t, \cdot) : H \to 2^H \setminus \{\emptyset\}$ has a closed graph;
(b) for each $t \geq 0$ and $y_0 \in H$, the set $G(t, y_0)$ is compact in H;
(c) the set $G(1, \tilde{C})$, where $\tilde{C} := \{z \in H \, : \, \|z\|_H^2 < c_4 + 1\}$, is precompact and attracts each bounded subset $C \subset H$.

Indeed, property (a) follows from Theorem 3.1; property (b) directly follows from (a) and Theorem 3.1; property (c) holds, because of Lemma 3.1 and since the set $G(1, \tilde{C})$ is precompact in H (Theorem 3.1).

According to properties (a)–(c), Mel'nik and Valero [7, Theorems 1, 2, Remark 2, Proposition 1] yields that the m-semi-flow G has a compact global attractor $\mathscr{R}$ in the phase space H.

3.5 Conclusions

For the class of nonautonomous differential-operator inclusions with pointwise pseudomonotone operators, the dynamics (as $t \to +\infty$) of all global weak solutions defined on $[0, +\infty)$ is examined. The existence of a compact global attractor in the natural phase space H is proved. The results obtained allow one to study the dynamics of solutions for new classes of evolution inclusions related to nonlinear mathematical models of geophysical and socioeconomic processes and for fields with interaction functions of pseudomonotone type satisfying the power growth and sign conditions. For applications, one can consider new classes of problems with degeneracy, feedback control problems, problems on manifolds, problems with delay, stochastic partial differential equations, etc. (see Balibrea et al. [8]; Hu and Papageorgiou [9]; Gasinski and Papageorgiou [10]; Kasyanov [11]; Kasyanov, Toscano, and Zadoianchuk [12]; Mel'nik and Valero [13]; Denkowski, Migórski, and Papageorgiou [14]; Gasinski and Papageorgiou [10]; Zgurovsky et al. [5]; etc., see, also, [16–31]) involving differential operators of pseudomonotone type and the corresponding choice of the phase spaces. This note is a continuation of Zgurovsky and Kasyanov [4, 15].

Acknowledgments This work was partially supported by the Ukrainian State Fund for Fundamental Researches under grant GP/F66/14921, and by the National Academy of Sciences of Ukraine under grant 2284.

References

1. Gajewski, H., Gröger, K., Zacharias, K.: Nichtlineare operatorgleichungen und operatordifferentialgleichungen. Akademie-Verlag, Berlin (1978)
2. Chepyzhov, V.V., Vishik, M.I.: Attractors for Equations of Mathematical Physics. American Mathematical Society, Providence (2002)
3. Gorban, N.V., Kapustyan, O.V., Kasyanov, P.O.: Uniform trajectory attractor for non-autonomous reaction-diffusion equations with Caratheodory's nonlinearity. Nonlinear Anal. Theory Methods Appl. **98**, 13–26 (2014). doi:10.1016/j.na.2013.12.004
4. Zgurovsky, M.Z., Kasyanov, P.O.: Uniform trajectory attractors for nonautonomous dissipative dynamical systems. Continuous and Distributed Systems II. Studies in Systems, Decision and Control Volume 30, pp. 221–232. Springer, New York (2015)
5. Zgurovsky, M.Z., Kasyanov, P.O., Kapustyan, O.V., Valero, J., Zadoianchuk, N.V.: Evolution inclusions and variation Inequalities for Earth data processing III. Springer, Berlin (2012)
6. Chepyzhov, V.V., Vishik, M.I.: Evolution equations and their trajectory attractors. J. Math. Pures Appl. **76**, 913–964 (1997)
7. Melnik, V.S., Valero, J.: On attractors of multivalued semi-flows and generalized differential equations. Set-Valued Anal. **6**(1), 83–111 (1998)
8. Balibrea, F., Caraballo, T., Kloeden, P.E., Valero, J.: Recent developments in dynamical systems: three perspectives. Int. J. Bifurc. Chaos (2010). doi:10.1142/S0218127410027246
9. Hu, S., Papageorgiou, N.S.: Handbook of Multivalued Analysis. Volume II: Applications. Kluwer, Dordrecht (2000)
10. Gasinski, L., Papageorgiou, N.S.: Nonlinear Analysis. Series in Mathematical Analysis and Applications 9. Chapman & Hall/CRC, Boca Raton (2005)
11. Kasyanov, P.O.: Multivalued dynamics of solutions of autonomous operator differential equations with pseudomonotone nonlinearity. Math. Notes **92**, 205–218 (2012)
12. Kasyanov, P.O., Toscano, L., Zadoianchuk, N.V.: Regularity of Weak Solutions and Their Attractors for a Parabolic Feedback Control Problem. Set-Valued Var. Anal. (2013). doi:10.1007/s11228-013-0233-8
13. Mel'nik, V.S., Valero, J.: On global attractors of multivalued semiprocesses and nonautonomous evolution inclusions. Set-Valued Anal. doi:10.1023/A:1026514727329
14. Denkowski, Z., Migórski, S., Papageorgiou, N.S.: An Introduction to Nonlinear Analysis: Applications. Kluwer Academic/Plenum Publishers, Boston (2003)
15. Zgurovsky, M.Z., Kasyanov, P.O.: Evolution inclusions in nonsmooth systems with applications for earth data processing. Advances in Global Optimization. Springer Proceedings in Mathematics & Statistics, vol. 95 (2014). doi:10.1007/978-3-319-08377-3_29
16. Babin, A.V., Vishik, M.I.: Attractors of Evolution Equations. Nauka, Moscow (1989). [in Russian]
17. Chepyzhov, V.V., Vishik, M.I.: Trajectory attractors for evolution equations. C. R. Acad. Sci. Paris. Ser. I **321**, 1309–1314 (1995)
18. Chepyzhov, V.V., Vishik, M.I.: Trajectory and global attractors for 3D Navier-Stokes system. Mat. Zametki. (2002). doi:10.1023/A:1014190629738
19. Chepyzhov, V.V., Vishik, M.I.: Trajectory attractor for reaction-diffusion system with diffusion coefficient vanishing in time. Discrete Contin. Dyn. Syst. **27**(4), 1498–1509 (2010)
20. Clarke, F.H.: Optimization and Nonsmooth Analysis. Wiley, New York (1983)
21. Hale, J.K.: Asymptotic behavior of dissipative systems. AMS, Providence (1988)
22. Kasyanov, P.O.: Multivalued dynamics of solutions of an autonomous differential-operator inclusion with pseudomonotone nonlinearity. Cybern. Syst. Anal. **47**, 800–811 (2011)
23. Kapustyan, O.V., Kasyanov, P.O., Valero, J.: Pullback attractors for a class of extremal solutions of the 3D Navier-Stokes equations. J. Math. Anal. Appl. (2011). doi:10.1016/j.jmaa.2010.07.040
24. Ladyzhenskaya, O.A.: Attractors for Semigroups and Evolution Equations. Cambridge University Press, Cambridge (1991)

25. Migórski, S., Ochal, A.: Optimal control of parabolic hemivariational inequalities. J. Glob Optim. **17**, 285–300 (2000)
26. Migórski, S.: Boundary hemivariational inequalities of hyperbolic type and applications. J. Glob. Optim. **31**(3), 505–533 (2005)
27. Panagiotopoulos, P.D.: Inequality Problems in Mechanics and Applications. Convex and Nonconvex Energy Functions. Birkhauser, Basel (1985)
28. Sell, G.R.: Global attractors for the three-dimensional Navier-Stokes equations. J. Dyn. Differ. Equ. **8**(12), 1–33 (1996)
29. Temam, R.: Infinite-Dimensional Dynamical Systems in Mechanics and Physics. Applied Mathematical Sciences, vol. 68. Springer, New York (1988)
30. Zgurovsky, M.Z., Mel'nik, V.S., Kasyanov, P.O.: Evolution Inclusions and Variation Inequalities for Earth Data Processing II. Springer, Berlin (2011)
31. Zgurovsky, M.Z., Kasyanov, P.O., Zadoianchuk (Zadoyanchuk), N.V.: Long-time behavior of solutions for quasilinear hyperbolic hemivariational inequalities with application to piezoelectricity problem. Appl. Math. Lett. **25**(10), 1569–1574 (2012)

Chapter 4
Minimal Networks: A Review

Alexander O. Ivanov and Alexey A. Tuzhilin

Abstract Minimal Networks Theory is a branch of mathematics that goes back to 17th century and unites ideas and methods of metric, differential, and combinatorial geometry and optimization theory. It is still studied intensively, due to many important applications such as transportation problem, chip design, evolution theory, molecular biology, etc. In this review we point out several significant directions of the Theory. We also state some open problems which solution seems to be crucial for the further development of the Theory. Minimal Networks can be considered as one-dimensional minimal surfaces. The simplest example of such a network is a shortest curve or, more generally, a geodesic. The first ones are global minima of the length functional considered on the curves connecting fixed boundary points. The second ones are the curves such that each sufficiently small part of them is a shortest curve. A natural generalization of the problem appears, if the boundary consists of three and more points, and additional branching points are permitted. Steiner minimal trees are analogues of the shortest curves, and locally minimal networks are generalizations of geodesics. We also include some results concerning so-called minimal fillings and minimal networks in the spaces of compacts.

4.1 Steiner Problem and Its Generalizations

We start with several historical remarks concerning the Steiner problem that generates Minimal Networks Theory.

A.O. Ivanov · A.A. Tuzhilin (✉)
Mechanical and Mathematical Faculty, Lomonosov Moscow State University, GSP-1, Leninskie Gory, Moscow 119991, Russian Federation
e-mail: tuz@mech.math.msu.su

A.O. Ivanov
Bauman Moscow Technical University, ul. Baumanskay 2-ya, 5, Moscow 105005, Russia
e-mail: aoiva@mech.math.msu.su

V.A. Sadovnichiy and M.Z. Zgurovsky (eds.), *Advances in Dynamical Systems and Control*, Studies in Systems, Decision and Control 69,
DOI 10.1007/978-3-319-40673-2_4

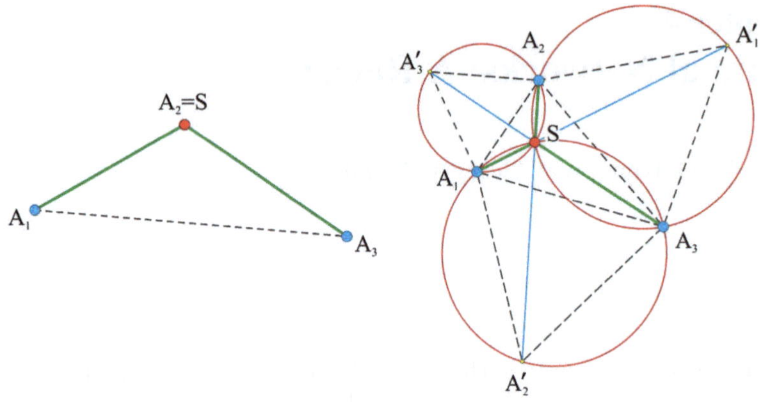

Fig. 4.1 Fermat problem's solution

4.1.1 Fermat Problem

One of the first versions of the Steiner problem had appeared long before Jacob Steiner. P. Fermat stated the following simplest case of the problem, see [1].

Problem 4.1 *(P. Fermat)* Qui hanc methodum non probaverit, ei proponitur: Datis tribus punctis, quartum reperire, a quo si ducantur tres rectae ad data puncta, summa trium harum rectarum sit minima quantitas.[1]

The solution to the Fermat problem is as follows. By A_1, A_2, A_3 we denote the given three points in the plane, and let S be the point we are looking for, which is referred as a *Fermat point*. If one of the angles of the triangle $A_1A_2A_3$ is more than or equal to 120°, then the point S coincides with the vertex of that angle. If all the angles of the triangle $A_1A_2A_3$ are less than 120°, then the location of the point S is uniquely defined by the following construction. By A'_k, $\{i, j, k\} = \{1, 2, 3\}$, we denote the point in the plane such that the triangle $A_iA_jA'_k$ is regular and intersects the triangle $A_1A_2A_3$ by the segment A_iA_j only. Then the circles circumscribed around the triangles $A_iA_jA'_k$, and the segments $[A'_iA_i]$ which are referred as the *Simpson lines*, all together intersect inside the triangle $A_1A_2A_3$ at the required point S, see Fig. 4.1. The lengths of the Simpson lines are equal to each other and are equal to the value $\sum_{i=1}^{3} |SA_i|$.

Remark 4.1 The intersection point of the circles does exist for an arbitrary triangle without any restrictions to its angles, and is usually referred as the *Torricelli point*. If one of the angles of the triangle is more than 120°, then the Torricelli point lies outside such a triangle. Also, for any triangle the straight lines passing through the

[1] Let the one that did not appreciate this method, solve the following problem: for given three points find the fourth one such that the total lengths of the three segments connecting it with the given three points takes a minimal value.

Simpson lines intersect a single point coinciding with the Torricelli point. If the angle of the triangle $A_1A_2A_3$ at the vertex A_i A_i is equal to 120°, then the Torricelli point coincides with A_i. Thus, for the triangles, whose angles does not exceed 120°, the Fermat point coincides with the Torricelli point. And if one of the angles of the triangle is more than 120°, then it is not so.

A natural generalization of the Fermat problem may be obtained by considering an arbitrary finite set of n points in the plane, instead of the three ones. Just this problem was studied by Steiner. Notice that for $n > 4$ the solution of such a generalized Fermat problem differs essentially from the initial case $n = 3$: in the case of $n = 3$ the above solution to Fermat problem gives an algorithm to construct the required point S by ruler and compass, but for $n \geq 5$ such algorithm does not exist (see, for example, [2]).

An alternative view to the Fermat problem leads to another natural generalization.

4.1.2 Graphs and Continuous Networks

We suppose that the reader is familiar with the main concepts of the Graph Theory, see for example [3]. But for convenience we include several general definitions and fix some basic notations.

For an arbitrary set V, by $V^{(k)}$ we denote the collection of all k-element subsets of V. Notice that the set $V^{(2)}$ consists of all non-ordered pairs of distinct elements of V, and $V^{(1)}$ is the family of single-element subsets of V, and hence, can be naturally identified with V.

Definition 4.1 A *graph* G is a triplet $G = (V, E, \partial)$ consisting of the sets V, E, and a mapping $\partial : E \to V^{(1)} \cup V^{(2)}$. Elements from V and E are called *vertices* and *edges* of the graph G, respectively, and ∂ is referred as *incidence* or *incidence mapping*. An edge e is called a *loop*, if $\partial(e) \in V^{(1)}$, and one says that *e connects the vertex v with itself*. If e is not a loop and $\partial(e) = \{u, v\}$, then one says that *e connects the vertices u and v*. An edge e is said to be *multiple*, if there exists another edge e' with $\partial(e) = \partial(e')$. A graph without loops and multiple edges is called *simple*.

Remark 4.2 If a graph (V, E, ∂) is simple, then the incidence mapping ∂ is an injection of the set E into $V^{(2)}$, therefore in this case E is often identified with $\partial(E) \subset V^{(2)}$, and an *edge e of a simple graph* is considered as the corresponding pair of vertices $\{u, v\} = \partial(e)$ which is denoted by uv for simplicity (notice that $uv = vu$ by definition). Under such identification, the mapping ∂ becomes unnecessary, and a *simple graph* is usually defined just as a pair (V, E), where $E \subset V^{(2)}$.

Remark 4.3 Usually we will consider *finite graphs*, i.e., the graphs with finite sets of edges and vertices. But many of the classical problems can be naturally generalized to the infinite graphs, see for example [4, 5].

Definition 4.2 A *path* in a graph G is a sequence $\gamma = v_{i_1}, e_{i_1}, v_{i_2}, \dots, e_{i_k} v_{i_{k+1}}$ of its vertices v_i and pairwise distinct edges e_j, such that each e_{i_m} connects the vertices v_{i_m} and $v_{i_{m+1}}$, $m = 1, \dots, k$. The path γ is said to be *connecting* the vertices v_{i_1} and $v_{i_{k+1}}$. If $v_{i_1} = v_{i_{k+1}}$, then the path γ is called a *cycle*. A graph is said to be *connected*, if any two its vertices are connected by a path. A connected graph without cycles is called a *tree*.

Definition 4.3 A graph $G = (V, E, \partial)$ is called *topological*, if E consists of topological segments, i.e., elements of E are topological spaces homeomorphic to a straight segment with the topology induced from $\mathbb{R}$.

Each topological graph G generates a topological space $T(G)$ in the following way: $T(G)$ is obtained from the disjoint union of the edges–segments of G by gluing their endpoints "as in the combinatorial graph" G.

Definition 4.4 Let $G = (V, E, \partial)$ be a topological graph, then a continuous mapping $\Gamma : V \sqcup (\sqcup_{e \in E} e) \to X$ such that for any $e \in E$ the continuous curve $\Gamma|_e$ connects the points from $\Gamma(\partial(e))$ is called a *(continuous) network Γ of the type G*, or a network *parameterized by the graph G*, in the topological space X. For each $v \in V$ the mapping $\Gamma|_v$ is called the *vertex of the network Γ corresponding to v*, and for each $e \in E$ the continuous curve $\Gamma|_e$ is called the *edge of the network Γ corresponding to e*. An edge $\Gamma|_e$ that maps the segment e onto a single point is called *degenerate*.

Remark 4.4 Each (continuous) network Γ of a type G in a topological space X generates a continuous mapping $\Gamma : T(G) \to X$. Conversely, each continuous mapping $\Gamma : T(G) \to X$ generates uniquely defined network Γ. Thus, continuous networks parameterized by G can be considered as continuous mappings from the corresponding topological space $T(G)$.

Remark 4.5 The concepts and properties that are defined for the parameterizing graph of a topological network usually attributed to the network itself. Thus the paths, cycles, incidence, degrees of vertices, connectivity, etc., are defined for networks.

We will consider boundary value problems, namely, we fix some subsets of the ambient space and study the networks connecting those subsets and being optimal in some reasonable sense. To give formal definitions, we extend the concept of graph assuming that for each graph G some subset ∂G of the vertex set is chosen. This subset is called the *boundary* of the graph. The vertices of a graph G belonging to its boundary are called *boundary* ones, and the remaining vertices are referred as *interior* ones.

Definition 4.5 The *boundary of a network Γ of a type G* is defined as the restriction of the mapping Γ on to the boundary ∂G of its parameterizing graph G.

Definition 4.6 Let M be an arbitrary subset of a topological space X. We say that a network Γ of a type G in X *connects the set M*, if $\Gamma(\partial G) = M$.

If a network is a mapping to a metric space, then one can define the length of each its edge as the length of the corresponding continuous curve, see, for example [6].

Definition 4.7 The *length* $|\Gamma|$ *of a network* Γ *in a metric space* is the sum of the lengths of all its edges.

Remark 4.6 The length of a network can be infinite as due to infinite length of some its edge, so as because of infinite number of its edges.

Now, the Fermat problem can be restated as the problem of finding a network of the least possible length connecting three given points in the Euclidean plane. Of course, there are much more networks connecting the points than the locations of the point S. However, if a shortest network Γ is found, then (1) all its nondegenerate edges are straight segments; (2) Γ contains neither nondegenerate loops, nor multiple edges; (3) Γ does not contain non-trivial cycles (non pointwise), and thus, Γ can be supposed to be a tree; (4) the tree Γ does not contains nondegenerate edges incident to interior vertices of degree 1, therefore, one can assume that all the vertices of Γ of degree 1 belong to the boundary; (5) the tree Γ can have interior vertices of degree 2, but the edges incident to such a vertex have to form an angle of $180°$, so each such pair of edges can be united into a single one, and the corresponding vertices of degree 2 can be removed from consideration.

Thus, a solution to the Fermat problem can be represented as a tree Γ, all whose vertices of degrees 1 and 2 belong to the three-point boundary. It is easy to see that there are two possibilities: (1) the tree Γ has four vertices, namely, three given points A_1, A_2, A_3 and additional vertex S connected by three edges with the vertices A_i (there are no other edges in the tree Γ); (2) the tree Γ has exactly three vertices A_1, A_2, A_3, and one of them is connected by edges with the other two (in that case the tree Γ consists exactly of the two edges).

The answer can be simplified even more: namely, the solution of the second type can be represented as the solution of the first type, where the additional vertex S coincides with the corresponding boundary point A_i. Thus, it is always possible to find a shortest network Γ among the networks of the first class, and minimization of the length is equivalent to minimization of the total distance from S to the given points A_i. So, the Fermat problem and the problem of finding a shortest network are equivalent.

4.1.3 Steiner Problem for Continuous Networks

The previous discussion leads to another generalization of the Fermat problem. This generalization apparently appeared first in papers of French mathematician J.D. Gergonne [7], who considered several points in the plane and described some constructions similar to Torricelli–Simpson construction, see Fig. 4.1 and well-known Melzak algorithm, see [8]. The case of four points was also studies actively by C.F. Gauss, H.C. Schumacher and K. Bopp [9]. The latter one stated the problem for an arbitrary number of points in the plane, considered locally minimal networks and understood the 120-degrees Principle, see Theorem 4.1. Notice that Bopp had used the Viviani's Theorem (the sum of distances from a point inside a regular triangle to its sides does not depend on the choice of the point), that he attributed to Jacob Steiner by a mistake, see also Remark 4.7. The modern statement of the problem for an arbitrary finite subset of a Euclidean space belongs to Jarnik and Kössler [10]. They proved an existence theorem and 120-degrees Principle. We will state this problem in the general case of metric spaces (other details and references can be found in a remarkable historical review [11]).

Let X be a metric space and M be some its subset. Consider all the networks in X connecting M and define the value $\mathrm{smt}(M)$ to be equal to the infimum of the lengths of all the networks.

Definition 4.8 Under the above assumptions, if $\mathrm{smt}(M) < \infty$ and there exists a network Γ connecting M and such that $|\Gamma| = \mathrm{smt}(M)$, then Γ is called a *shortest network*. If it is necessary to underline that Γ connects M, then such a network Γ is referred as a *shortest network on M*.

Problem 4.2 (*Jarnik, Kössler*) Find a shortest network connecting a given subset M of points of a metric space X (providing such a network does exist).

Remark 4.7 At present the Jarnik–Kössler problem stated above is referred as *Steiner Problem*, though J. Steiner worked on a similar but different problem (see above). The confusion had appeared due to the outstanding and very popular book [12].

Remark 4.8 As it has been already mentioned above, each shortest network can be parameterized by a nondegenerate tree, that is referred as a *Steiner minimal tree* (that explains the notation smt).

4.1.4 Local Structure of Shortest Trees. Locally Minimal Trees

Let us describe the structure of the shortest trees in small neighborhoods of their points. In fact, to do that it is necessary to solve the following problem: describe all the shortest networks of the "star–type", i.e. the ones that have exactly one interior vertex which is connected by edges with all the boundary vertices of the tree (the interior vertex could coincide with one of the boundary ones).

Consider the case of the Euclidean plane. It turns out that it suffices to consider the case of the star with three boundary vertices. Why is it so? The solution of Fermat problem implies that the angle between any two adjacent edges–segments of a shortest network can not be less than 120°, because otherwise a pair of such edges can be changed by a shorter tree that contradicts to minimality of the initial network. Thus, at most three edges of a shortest tree can be incident to its vertex. So, in fact, the solution of Fermat problem describes all possible ways of the edges adjacency in a shortest tree. The complete answer is as follows.

Theorem 4.1 (On the local structure of a shortest tree) *All the edges of a shortest tree in the Euclidean plane are straight segments, and the angles between the edges–segments are at least* 120°, *and hence, the degrees of the vertices of such trees are at most* 3. *The degrees of all the interior vertices are always equal to* 3, *and the adjacent edges form the angles of* 120°; *the degrees of the boundary vertices can be equal to* 1, 2, *or* 3, *and at a boundary vertex of degree* 2 *the segments meet by an angle of at at least* 120°, *and at a boundary vertex of degree* 3 *the edges meet as at an interior one.*

Definition 4.9 A tree in Euclidean plane, whose local structure is as in Theorem 4.1, i.e., a plane tree whose edges–segments meet by the angles of at least 120° and with the boundary containing all the vertices of degree 1 and 2, is called *locally minimal.*

Remark 4.9 Theorem 4.1 and Definition 4.9 imply that any shortest tree in Euclidean plane is locally minimal, therefore under studying of shortest trees one can restrict consideration to so called *Steiner trees* that are defined as the trees, whose edges are straight segments, vertices are of degree at most 3, and all the vertices of degrees 1 and 2 are boundary ones. For a fixed boundary, each such a tree is uniquely defined by its combinatorial structure and the location of its interior vertices.

Moreover, if one "splits" each vertex of degree 2 of a shortest tree by changing it by a boundary vertex of degree 1 and an interior vertex of degree 3 connected by a degenerate edge (see Fig. 4.2, left), and also "splits" each boundary vertex of degree 3 into two interior vertices of degree 3 and one boundary vertex of degree 1 connected as in Fig. 4.2, right, then the resulting tree has only the vertices of degrees 1 and 3, and the boundary of the resulting tree consists of all its vertices of degree

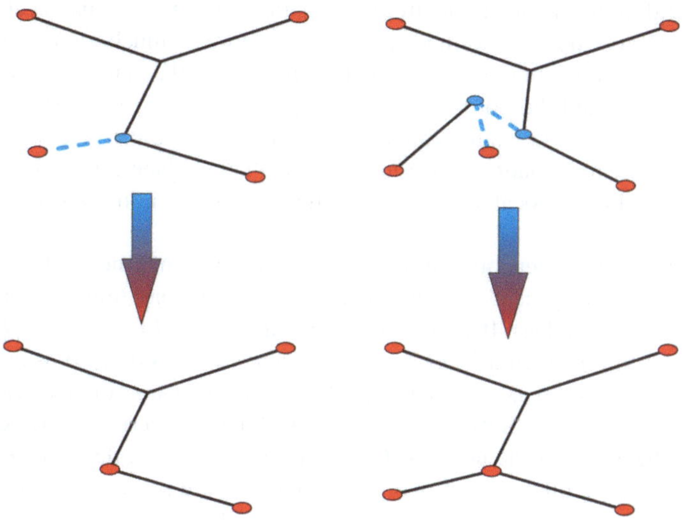

Fig. 4.2 Splitting of the boundary vertices of degree more than 1

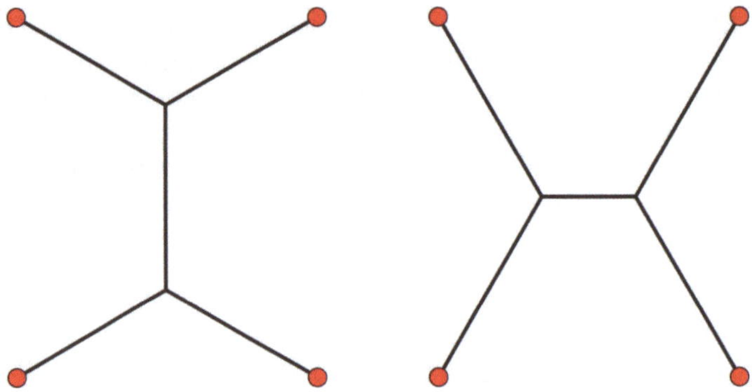

Fig. 4.3 Two locally minimal trees connecting the vertices of a rectangle

1. We call such trees *binary*. Thus, each shortest tree (and also each locally minimal tree) can be represented as a binary tree, some of whose edges might be degenerate.

Remark 4.10 Of course, a locally minimal tree need not be a shortest one. For example, the vertex set of a rectangle with distinct sides and such that the angle between its diagonals is greater than 60°, is connected by two locally minimal trees having distinct lengths. It is clear that the longest tree is not a shortest, see Fig. 4.3.

Now, let us pass to locally minimal and shortest trees in other ambient spaces. It turns out that their local structure in n-dimensional Euclidean space is the same (this fact can be easily proved using the solution of Fermat problem again, see, for example [13]). Similar result, see [14], is also valid for minimal trees in surfaces and, more general, in Riemannian manifolds (the straight segments must be changed by segments of geodesics). In normed spaces the situation is much more complicated, see [15]. For example, in so called Manhattan plane, i.e., the plane with the standard coordinates x, y and the norm $\|(x, y)\| = |x| + |y|$, shortest networks could have vertices of degree 4. For example, a finite part of the "coordinate cross" forms such a network. More information concerning the Manhattan plane can be found in [16]. In papers [15, 17–20] local structure of minimal networks in more general normed planes is studied.

Now let us give a general definition of a locally minimal network. Let Γ be a network in a metric space X, and $P \in \Gamma$ be its arbitrary point (either a vertex, or an interior point of an edge). If P lies inside an edge $\gamma : [a, b] \to X$, i.e., $P = \gamma(t_0)$ for some $t_0 \in (a, b)$, then a *local network centered at* P is defined as an arbitrary curve $\delta = \gamma|_{[\alpha,\beta]}$, where $t_0 \in (\alpha, \beta) \subset (a, b)$, and the *boundary* of this local network consists of the two end points of the curve δ. And if P is a vertex of the network Γ, and γ_i are the edges of the network Γ that are incident to P, then a *local network centered at* P is defined as a network, whose edges are some fragments δ_i of the edges γ_i containing the point P, and the *boundary* of this local network consists of all the ends of the curves δ_i, provided P is a boundary vertex of Γ, and of all the ends of this curves excluding P, provided P is non-boundary.

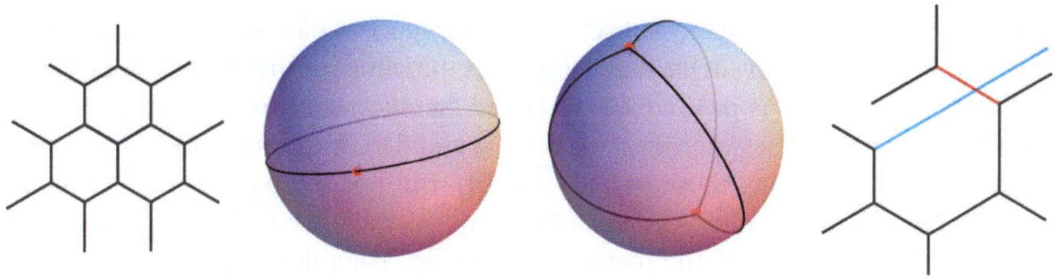

Fig. 4.4 Locally minimal networks with cycles, multiple edges and self-intersections

Definition 4.10 A (continuous) network in a metric space is called *locally minimal*, if for any its point there exists a local network centered at this point, which is shortest (with respect to its boundary).

Remark 4.11 It is easy to see that locally minimal networks consists of geodesics (locally shortest curves). Besides, each shortest network is locally minimal, and the converse statement is not valid.

Remark 4.12 A locally minimal network can have cycles, loops, multiple edges and self-intersections (see Fig. 4.4).

Remark 4.13 All vertices of degree 1 of a locally minimal network must belong to its boundary. All the interior vertices of degree 2 can be excluded except the only case, when the network is a closed curve. In the latter case we need a single interior vertex of degree 2 to get a topological graph with a singe loop, see Fig. 4.4.

4.1.5 Steiner Problem for Discrete Networks

Notice that not every metric space permits connection of its finite subsets by continuous curves. For example, such trouble appears if one consider a finite metric space X, or a space of words with so-called editorial distance, that is actively used in biology in problems related to genes and evolution, see [21]. But in those cases the Steiner problem can be also stated. To do that we re-define the concepts of a network and of the length of an edge. Let X be a metric space, and $M \subset X$.

Definition 4.11 By a *(discrete) network in* X we call an arbitrary connected combinatorial graph $\Gamma = (V, E, \partial)$ such that $V \subset X$. We say that a *network* Γ *connects* M, if $M \subset V$; and the set M is referred as the *boundary of the network* Γ and is denoted by $\partial\Gamma$. Put the *length of an edge* e *of the network* Γ to be equal to the distance in X between the vertices connected by the edge.

Now the *Steiner Problem* is exactly Problem 4.2, where networks are considered in the sense of Definition 4.11.

Remark 4.14 Assume that any subset of a metric space X can be connected by some network. Is it true that in such a case $\operatorname{smt}(M)$ is the same for the both Definitions 4.4 and 4.11? The answer is positive for so-called *length-metric spaces*, i.e., the spaces such that the distance between any two point is equal to the infimum of the lengths of all the curves connecting those points (see [6]). But not any path-connected metric space is a length metric one. As an example, consider a circle in the Euclidean plane with the metric induced by the metric of the plane. Then the distance between two points P and Q of the circle is equal to the Euclidean length $|PQ|$ of the straight segment $[P, Q]$, but the shortest curve is one of the arcs of the circle that is definitely longer than $|PQ|$.

4.2 Minimal Fillings

Up to now we have considered the Steiner problem for the network lying in an ambient metric space. Is it possible to state a similar problem without ambient space at all? Such version of the problem has been suggested in [22] by A. Ivanov and A. Tuzhilin, who extended M. Gromov's construction [23] of minimal fillings to the case of one-dimensional manifolds with singularities. Here we give the corresponding definitions.

Recall that a *weighted graph* $(G, \omega) = (V, E, \partial, \omega)$ is a graph $G = (V, E, \partial)$ endowed with a non-negative function ω on its edges that is referred as the *weight function*. For any subset E' of the edge set E of a weighted graph (V, E, ∂, ω) the *weight* $\omega(E')$ is defined as the sum of weights $\omega(e)$ of all the edges e from E'. In particular, the *weights of paths in* G are defined, so as the *weigh* $\omega(G) := \omega(E)$ of the graph as a whole. If a weighted graph (G, ω) is connected, then a *distance function* d_ω on the vertex set is defined as follows: for each pair u, v of vertices of the graph G the value $d_\omega(u, v)$ is defined as weight of a path connecting those vertices and having the least possible weight.

Let M be an arbitrary set, and $G = (V, E, \partial)$ be a connected graph. We say that the graph G *connects* M, if $M \subset V$.

Definition 4.12 A weighted connected graph (G, ω), connecting a metric space (M, ρ) is called a *filling of* M, if for any two points u and v from M the relation $\rho(u, v) \le d_\omega(u, v)$ holds.

By $\operatorname{mf}(M)$ we denote the greatest lower bound of the weights of all fillings of the space M.

Definition 4.13 A filling (G, ω) of a metric space M is called *minimal*, if $\omega(G) = \operatorname{mf}(M)$.

Problem 4.3 (*Ivanov and Tuzhilin*) Describe minimal fillings of finite metric spaces and find out the relations between minimal fillings and shortest trees.

4.3 Minimal Spanning Trees

This type of optimal networks often appear in different applications, because it is algorithmically simple.

Let M be a metric space. Consider all possible trees G with the vertex set M, for each of them calculate the length $|G|$ and put $\text{mst}(M)$ to be equal to the greatest lower bound of the numbers $|G|$ over all such trees.

Definition 4.14 If $\text{mst}(M) < \infty$, and G is a tree with the vertex set M such that $|G| = \text{mst}(M)$, then G is called a *minimal spanning tree* (that is a reason for the notation mst).

In fact, minimal spanning trees are often used as an approximation of the shortest trees, because there exists polynomial algorithm for there constructing, such as Kruskal's Algorithm, see for example [24].

Next statements demonstrate relations between the functions mst, smt (for discrete networks), and mf.

- Let M be a subset of a metric space X, and $\text{smt}(M) < \infty$. Then $\text{smt}(M)$ is equal to the infimum of the values $\text{mst}(W)$ over all W such that $M \subset W \subset X$.
- Let M be an arbitrary metric space such that $\text{mf}(M) < \infty$. Then $\text{mf}(M)$ is equal to the infimum of the values $\text{mst}(W)$ over all metric spaces W with finite $\text{mst}(W)$, such that M can be isometrically embedded into W.
- Let M be an arbitrary metric space such that $\text{mf}(M) < \infty$. Then $\text{mf}(M)$ is equal to the infimum of the numbers $\text{smt}(W)$ over all the pairs (W, X), where X is a metric space of cardinality at most continuum and $W \subset X$ is isometric to M.

In what follows all the types of optimal networks considered above, namely, shortest trees, locally minimal trees, minimal fillings, and minimal spanning trees are referred as *minimal networks*.

4.4 Properties of Minimal Networks

In this section we tell about some geometrical properties of minimal networks. Restrict ourselves to the case of the Euclidean plane.

4.4.1 Minimal Spanning Trees

In this section we collect several simple geometric properties of plane minimal spanning trees.

Proposition 4.1 *Let Γ be a minimal spanning tree connecting a finite subset M of the Euclidean plane. Then*

(1) *the tree Γ has no self-intersections;*
(2) *the angle between any two adjacent edges of the tree Γ is at least* $60°$*;*
(3) *if e is an arbitrary edge of the tree Γ, and Γ_1 and Γ_2 are the components of the forest $\Gamma \setminus e$ obtained from Γ by deleting the edge e, then the distance between the vertex sets of the trees Γ_i is equal to the length of the edge e.*

Remark 4.15 Property (3) form Proposition 4.1 remains valid in an arbitrary metric space.

4.4.2 Shortest Trees

Geometrical properties of shortest trees are studied much more better, see, for example [25]. Here we include just one well-known example, that have been recently generalized and developed.

Let $e = uv$ be an edge of a shortest tree Γ. By $L(e)$ we denote the intersection of the open circles or radius $|e|$ centered at u and v. The set $L(e)$ is referred as the *lune of the edge* e. The following classical result holds, see [25].

Proposition 4.2 *The lune of an edge e of a shortest tree Γ does not contain any points of Γ except the points from e. In other words,* $\Gamma \cap L(uv) = (u, v)$.

Remark 4.16 Ivanov and Tuzhilin stated the next problem: Describe possible structure of the intersection of a shortest tree Γ with a sufficiently small ε-neighborhood of the lune $L(e)$ of an edge e of Γ. This problem is completely solved by Ivanov, S'edina and Tuzhilin [26].

4.4.3 Locally Minimal Trees

In this section we demonstrate relations between the structure of locally minimal trees and geometry of their boundaries. This connection follows from a general fact concerning geometry of plane linear trees.

Let Γ be a plane graph, all whose edges are straight segments (such graphs are referred as *linear*). Define the *geometric boundary of linear graph* Γ as follows: a vertex v of Γ is a *boundary* one, if there exists a straight line ℓ passing through v and such that all the edges of Γ incident to v lie in a single open half-plane with respect to ℓ. All the remaining vertices of Γ are called *interior*. The set of all boundary vertices of Γ is referred as the *geometric boundary of* Γ and is denoted by $\partial\Gamma$, see Fig. 4.5.

Let Γ be a linear tree, and e and f be arbitrary edges of Γ. By γ we denote the unique path in Γ connecting e and f, and let $e_0 = e,\ e_1, \dots, e_m = f$ be consecutive edges of γ. Orient γ from e to f and consider each edge e_i as the corresponding vector in the plane. By $\alpha_i \in (-\pi, \pi)$ we denote the angle from e_{i-1} to e_i.

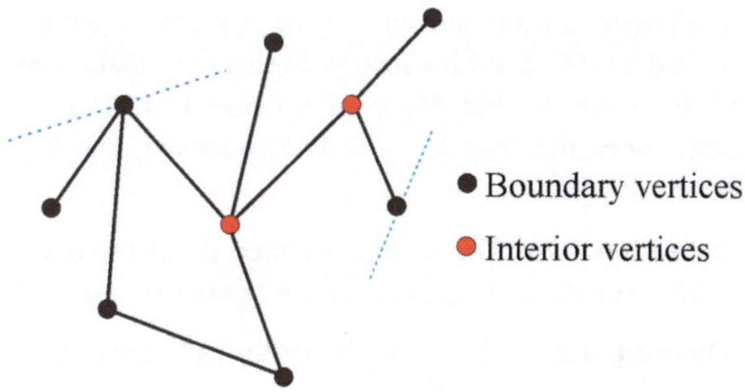

Fig. 4.5 Geometrical boundary of a linear graph

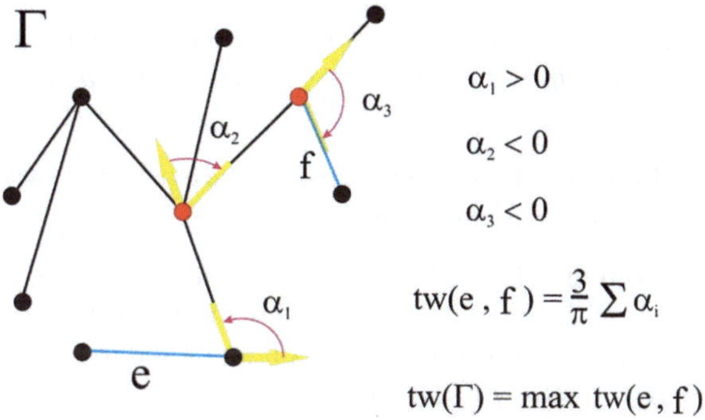

Fig. 4.6 The twisting number of a linear tree

Definition 4.15 The number $\alpha(e, f) = \sum_{i=1}^{m} \alpha_i$ is called the *twisting angle from* e to f, and the value $\alpha(\Gamma) = \max_{(e,f)} \alpha(e, f)$ is called the *twisting angle of the linear tree* Γ. Normalized twisting angles $\frac{3}{\pi}\alpha(e, f)$ and $\frac{3}{\pi}\alpha(\Gamma)$ are referred as the corresponding *twisting numbers* and are denoted by $\mathrm{tw}(e, f)$ and $\mathrm{tw}\Gamma$, see Fig. 4.6.

Remark 4.17 Since $\mathrm{tw}(e, f) = -\mathrm{tw}(f, e)$, then $\mathrm{tw}\Gamma$ is non-negative.

Remark 4.18 If Γ is a plane locally minimal binary tree, then $\mathrm{tw}(e, f)$ is equal to the difference between the number of "left" and "right" turns under the walk from e to f along Γ, see Fig. 4.7.

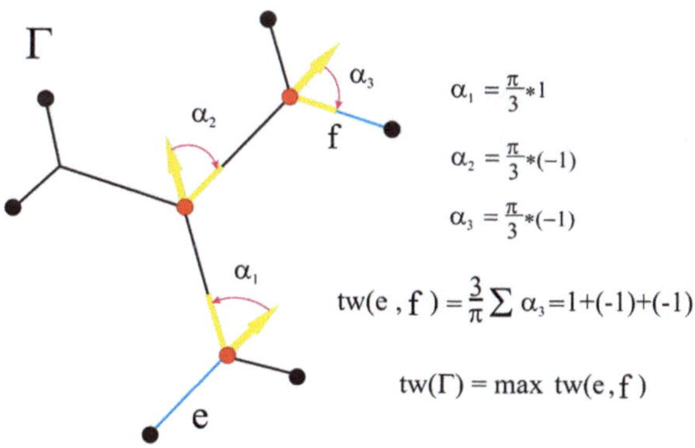

Fig. 4.7 The twisting number of a locally minimal binary tree

Further, for any subset $X \subset \mathbb{R}^2$ by conv$X$ we denote the convex hull of the set X, and let ∂X stand for its topological boundary. Let M be an arbitrary finite subset of the plane $\mathbb{R}^2$. Put $M_1 = M \cap \partial\mathrm{conv}M$ and $M_1' = M \setminus M_1$. If M_i' is defined and non-empty, then put $M_{i+1} = M_i' \cap \partial\mathrm{conv}M_i'$ and $M_{i+1}' = M_i' \setminus M_{i+1}$. It is clear that $M = \sqcup_{i=1}^k M_i$.

Definition 4.16 The set M_i is called the *convexity level of* M, and the number $k = k(M)$ is referred as the *number of convexity levels of the set* M.

Theorem 4.2 (Ivanov and Tuzhilin [27, 28]) *Let Γ be a plane linear tree with geometric boundary $\partial\Gamma$, and $k = k(\partial\Gamma)$ be the number of convexity levels of the set $\partial\Gamma$. Then* $\mathrm{tw}\Gamma \le 12(k-1)+6$. *If Γ is a locally minimal binary tree, then its geometric boundary coincides with the boundary of the binary tree defined above, i.e., with the set of vertices of degree* 1, *and a stronger estimate is valid, namely,* $\mathrm{tw}\Gamma \le 12(k-1)+5$.

Remark 4.18 demonstrates how the concept of twisting number can be transferred to the case of a plane binary tree.

Definition 4.17 If Γ is a plane binary tree, then for any ordered pair (e, f) of edges of Γ put $\mathrm{tw}(e, f)$ to be equal to the difference between the number of "left" and "right" turns under the walk along the tree Γ from e to f. As above, $\mathrm{tw}\Gamma = \max_{(e,f)} \mathrm{tw}(e, f)$, see Fig. 4.8.

Definition 4.18 A boundary of a locally minimal tree is called *convex*, if it has exactly one convexity level.

Definition 4.19 We say that a plane Steiner tree *has a convex minimal realization*, if there exists a planar equivalent plane locally minimal tree with a convex boundary.

Theorem 4.2 implies that the twisting number of plane binary tree having a convex minimal realization does not exceed 5. It turns out, that the converse statement also holds, and the proof is utter non-trivial (see [29–31]). Thus, the following result holds.

Theorem 4.3 (Ivanov and Tuzhilin [29–32]) *A plane binary tree has a convex minimal realization, if and only if its twisting number does not exceed* 5.

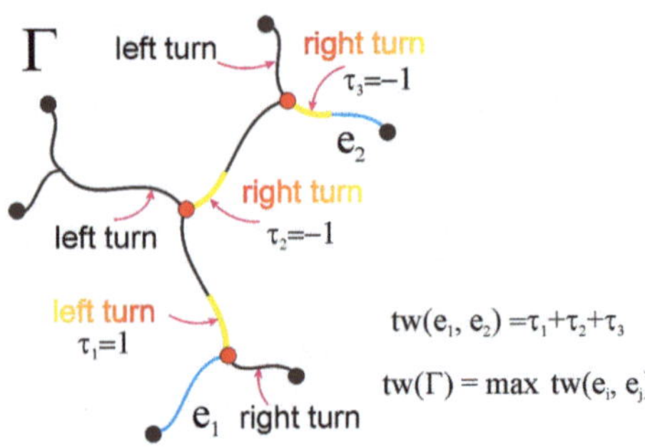

Fig. 4.8 The twisting number of plane binary tree

4.4.4 Minimal Fillings

In the present section we list several properties of minimal fillings.

4.4.4.1 Generic Spaces

To state the following result we need a concept of a generic finite metric space. To define this concept we fix a finite set M, enumerate its points, i.e., $M = \{p_1, \dots, p_n\}$, and consider an arbitrary metric ρ on M. Put $\rho_{ij} = \rho(p_i, p_j)$, write down all non-zero elements of the upper triangle of the distance matrix (ρ_{ij}) as a vector $v(\rho) = (\rho_{12}, \rho_{13}, \dots, \rho_{(n-1)n})$, and get a point in the space $\mathbb{R}^{n(n-1)/2}$. Notice that the set $\mathcal{M}_n = \{v(\rho)\}_\rho$ is an open cone in the positive orthant determined by all triangle inequalities of the form $\rho_{ij} + \rho_{jk} \ge \rho_{ik}$.

Definition 4.20 We say, that a property *holds for all generic finite metric spaces*, if for any n this property holds for all finite metric spaces from some everywhere dense subset of $\mathcal{M}_n$.

Theorem 4.4 (Eremin [33]) *Minimal filling of a generic finite metric space is a nondegenerate binary tree.*

4.4.4.2 Additive Spaces

Definition 4.21 A path connecting boundary vertices in a graph G is called *boundary*. A boundary path in a filling is said to be *exact*, if its weight is equal to the distance between its ends.

Definition 4.22 A metric space is called *additive*, if there exists its filling such that all its boundary paths are exact. Such a filling is called a *generating tree* of the corresponding additive space.

In other words the space is additive, if all the distances are generated by some weighted tree connecting it.

Remark 4.19 A finite metric space is additive, if and only if the following *four points rule* holds: for any four points x_1, x_2, x_3, x_4 of the space, considered as vertices of a tetrahedron, the three numbers $|x_ix_j| + |x_kx_l|$, $\{i, j, k, l\} = \{1, 2, 3, 4\}$, which are equal to the sum of the lengths of the opposite sides of the tetrahedron form the lengths of the sides of an isosceles triangle, whose base does not exceed the other sides. It is also well-known that a nondegenerate generating tree of an additive space is unique, see [34–37].

Proposition 4.3 (Ivanov and Tuzhilin [22]) *A nondegenerate generating tree of an additive space is its unique nondegenerate minimal filling.*

Let $M = \{p_1, \ldots, p_n\}$ be a finite metric space with a metric ρ, and π is a permutation on M. Put $p_{n+1} = p_1$,

$$p_\pi(M, \rho) = \frac{1}{2}\sum_{i=1}^{n} \rho\left(\pi(p_i), \pi(p_{i+1})\right),$$

and

$$p(M, \rho) = \min_\pi p_\pi(M, \rho).$$

Definition 4.23 The value $p_\pi(M, \rho)$ is called the *half-perimeter of the metric space* (M, ρ) *with respect to the permutation* π, and the value $p(M, \rho)$ is called the *half-perimeter of the metric space* (M, ρ).

Proposition 4.4 (Rubleva [38]) *The weight of a minimal filling of a metric space is equal to the half-perimeter of this space, if and only if the space is additive.*

Recall that an *Euler cycle* in a connected graph is a cycle containing all the edges and passing through each edge exactly once. Evenness of degrees of all the vertices is a necessary and sufficient condition for an Euler cycle existence in a connected graph.

Definition 4.24 The *doubling of a graph* $G = (V, E, d)$ is the graph $(V, E \sqcup E, d')$, where the restriction of d' onto each E coincides with d.

It is clear, that the degrees of all the vertices of the doubling are even, therefore the doubling of any connected graph contains an Euler cycle.

Let G be a binary tree and G' be its doubling. Consider an arbitrary Euler cycle C in G' and orient it. It is not difficult to show that C consists of a sequence of oriented boundary paths $\gamma_1, \ldots, \gamma_n$, where n is the number of boundary vertices (i.e., the vertices of degree 1) of G. Moreover, for any $v \in \partial G$ there exists unique path γ_i, such that v is its beginning vertex. Let $\pi_C : \partial G \to \partial G$ be the mapping that maps each vertex v onto the corresponding ending vertex of the unique path γ_i that goes out of v. The resulting permutation on ∂G is called the *walk around the tree* G. The following result is evident.

Proposition 4.5 *Let* G *be the binary generating tree for an additive space* (M, ρ) *and* π *be a walk around the tree* G. *Then the half-perimeter* $p_\pi(M, \rho)$ *is equal to the weight of the tree* G, *and hence, does not depend on the choice of the walk.*

Is the inverse result valid? The answer turns out to be negative. Ovsyannikov [39] described the set of all metric spaces satisfying that property.

4.4.4.3 Pseudo-additive Spaces

Let us permit the weights of edges of a filling be negative. For such a weighted tree (G, ω) the same function d_ω on the vertex set of G can be defined. But now, d_ω can take negative values.

Definition 4.25 A metric space (M, ρ) is called *pseudo-additive*, if there exists a weighted tree (G, ω), whose weights could be negative, connecting M and such that the relation $d_\omega(x, y) = \rho(x, y)$ holds for all $x, y \in M$.

Theorem 4.5 (Ovsyannikov [39]) *Let (M, ρ) be some metric space. Then the following properties are equivalent:*

- *there exists a binary tree G connecting M and such that the half-perimeters of all the walks around G are the same;*
- *the space M is pseudo-additive.*

Remark 4.20 Ovsyannikov found out that a pseudo-additivity criterion can be obtained from the four points rule, see Remark 4.19, as follows: we just need to omit the condition that the base of the isosceles triangle is longer than its other sides.

4.5 Classifications

In the present section we give a description of some classes of minimal networks for some classes of boundary sets.

4.5.1 Shortest Trees

The pioneer work of Jarnik and Kössler [10] contains not only a new problem statement, that becomes known as the Steiner problem, but also a solution to this problem for the case of regular n-gons in Euclidean plane for all n except $6 \le n \le 12$. This gap was closed in 1987 only by Du, Hwang and Weng [40].

Theorem 4.6 (Jarnik, Kössler [10], Du, Hwang and Weng [40]) *Let M be the vertex set of a regular n-gon in the Euclidean plane. Then SMT(M) consists from one (for $n = 3$), two (for $n = 4$), and n (for $n \ge 5$) trees, which are obtained one from another by some rotations around the center of the polygon M. For $n = 3, 4, 5$ those trees have the form shown in Fig. 4.9 and uniquely defined by its local structure*

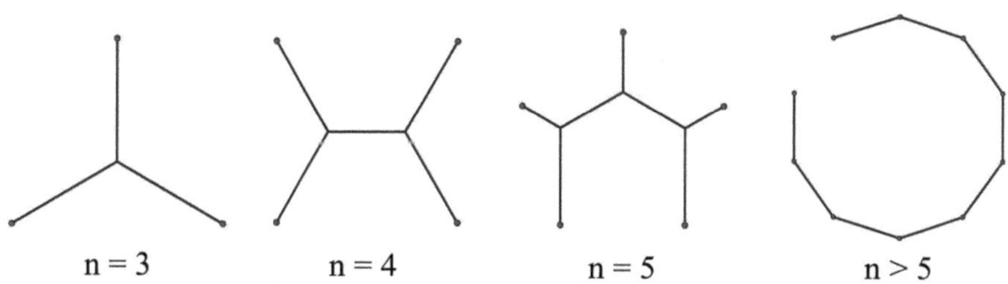

Fig. 4.9 Shortest trees connecting the vertices of regular n-gons

(Theorem 4.1). For $n \geq 6$ each of the trees consists of all the sides of the polygon M except one.

There exists a series of classifying results describing shortest trees connecting vertices of so-called “zig-zags” [41], “ladders” [42], polygons inscribed into a circle and having at most one “long” side [43, 44], “Chinese checkers board” [45]. Also, a conjecture concerning the structure of shortest trees connecting the vertices of a usual “chess board” is stated in [46].

Notice that in each of the above cases a proof is extremely non-trivial. Of course, for the boundaries consisting of a small number of points it is possible to proceed with some algorithm of Melzak type searching through all combinatorial possibilities. But for large sets even modern computers turns out to be powerless. A way out is either in finding some heuristics, such as a minimal spanning tree, or in elaboration a theory that gives an opportunity to improve the direct algorithm basing on some geometrical results. One example of such a theory is discussed above in Sect. 4.4.3, where a relation between geometry of a boundary set and possible structure of minimal networks connecting this set is established.

Another important problem is the one of constructing examples of shortest networks with some prescribed properties. Consider several results of that type for the case of Euclidean plane.

Due to definition, each shortest tree is locally minimal. As we have already mentioned above, the converse statement does not hold in general. The following question arises: Are there any structural obstacles for realization of a given locally minimal tree (a plane Steiner tree) as a shortest one? Here by the “realization” we understand a graph that is planar equivalent to the initial one. It is clear that shortest trees have no self intersections. So, the initial trees also must not have self-intersections. Are there some other restrictions? The following theorem gives the negative answer.

Theorem 4.7 ([47]) *Any plane (embedded, i.e. without self-intersection) Steiner tree is planar equivalent to some shortest tree.*

Notice that any locally minimal tree can be transformed to an embedded one by changing the lengths of edges only without changing their directions (this fact can be easily proved by induction). Thus, using Theorem 4.7, conclude that any locally minimal tree can be transformed to a shortest one by changing the lengths of edges.

It turns out that there is another way to transform a locally minimal tree without self-intersections into a shortest one.

Theorem 4.8 (Ivanov and Tuzhilin [48]) *Let Γ be a locally minimal tree without self-intersections in a Euclidean space. Then the edges of the tree Γ can be subdivided by boundary vertices of degree 2 in such a way that the resulting tree is a shortest one for the new boundary.*

Thus, each embedded locally minimal tree in a Euclidean space coincides as a subset with some shortest tree. In this sense the family of all shortest trees (considered as subsets of Euclidean space) coincides with the set of all embedded locally minimal trees.

4.5.2 Locally Minimal Trees

Usually, the set of locally minimal trees with a given boundary is essentially bigger than the set of the shortest trees with the same boundary. One of a still open problems (stated by Ivanov and Tuzhilin) is to describe all locally minimal networks connecting vertex sets of regular polygons in Euclidean plane $\mathbb{R}^2$ (recall, that the shortest trees are described by Theorem 4.6). Solving this problem Ivanov and Tuzhilin considered a more general problem on description of embedded locally minimal trees in $\mathbb{R}^2$, whose boundaries are the vertex sets of convex polygons. Such boundaries are referred as *convex* for shortness.

We give this description for an important particular case of binary trees.

4.5.2.1 Plane Locally Minimal Binary Trees with Convex Boundaries

For plane binary trees a twisting number has been defined above. In accordance with Theorem 4.3 the class we are interested in coincides with the class of plane binary trees with twisting number at most five. How can one describe all such trees? It turns out that it is essentially more convenient to study such trees in a "dual language" of so called "triangle tilings".

Consider a standard partition of the plane into regular triangles. Below that partition is referred as the *(triangle) tilling of the plane*, and the triangles forming the tiling are called *cells*. Two cells are called *adjacent*, if they have a common side. It is clear that for each cell there exist exactly three cells adjacent to it.

Definition 4.26 A finite set of cells is called a *tiling*. A cell's side located at the topological boundary of a tiling containing it is called *boundary*.

Let T be a tiling. By V we denote the set of all the centers of all its cells and all the centers of its boundary sides. By E we denote the set of all straight segments connecting the centers of adjacent cells and the center of each boundary side with the center of the unique cell containing this side.

Definition 4.27 The plane graph $G_T = (V, E)$ constructed above is called the *dual graph of the tiling* T, see Fig. 4.10.

Definition 4.28 A tiling is called a *tree tiling*, if its dual graph is a tree. The twisting number of the dual graph of a tree tiling is called the *twisting number of the tiling*.

Theorem 4.9 (On a tiling realization: Ivanov and Tuzhilin [29–32]) *Each plane binary tree, whose twisting number is at most five, is planar equivalent to the dual graph of a tree tiling.*

So, our problem is equivalent to description of all tree tilings, whose twisting number is at mots 5.

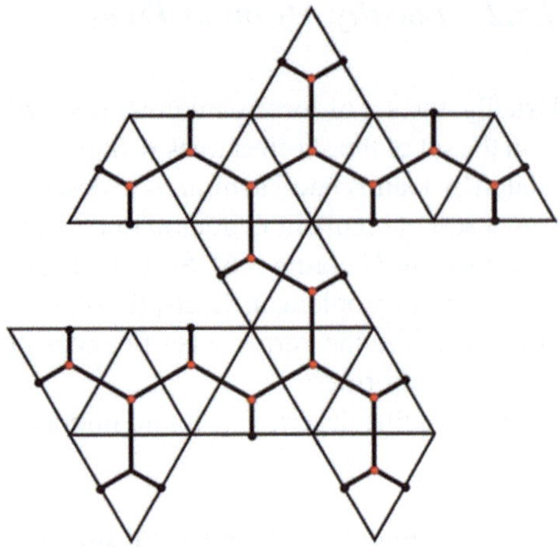

Fig. 4.10 A tiling T and its dual graph G_T

It turns out that each tiling can be represented as a union of some part having a "regular structure" (this part is referred as a skeleton of the tiling), and some cells glued to the skeleton in a "random way" (those cells are referred as growths).

Definition 4.29 A cell of a tiling is called *extreme*, if at least two its sides are boundary ones, see Fig. 4.11. If a cell has no boundary sides, than it is called *interior*. An extreme cell adjacent with an interior one is called a *growth*, see Fig. 4.12, and a tiling without growths is called a *skeleton*.

Let T be an arbitrary tiling. By R we denote the set of the cells of T obtained as follows: for each interior vertex of T chose one of adjacent extreme cells (if any) and put in R. Clearly, all cells from R are growths of the tiling T. By S we denote the tiling obtained from T by deleting all the cells from R. It is not difficult to verify that S is a skeleton, so, we have constructed a *decomposition of the tiling T into a skeleton S and growths R*. Notice that the set R, and hence, the skeleton S are not uniquely defined, because some interior vertices could be adjacent with several growths, see Fig. 4.12.

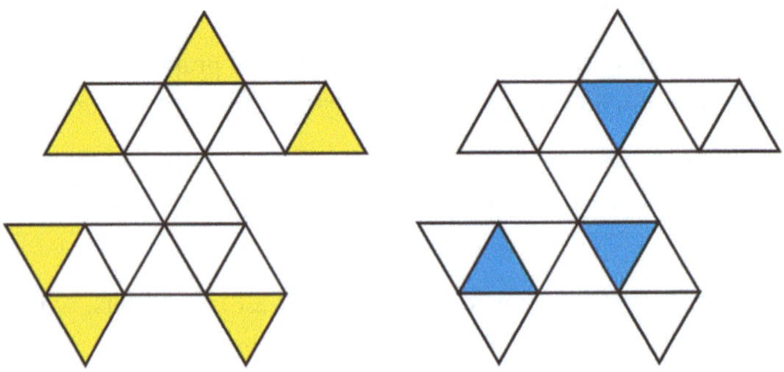

Fig. 4.11 Extreme (*left*) and interior (*right*) cells

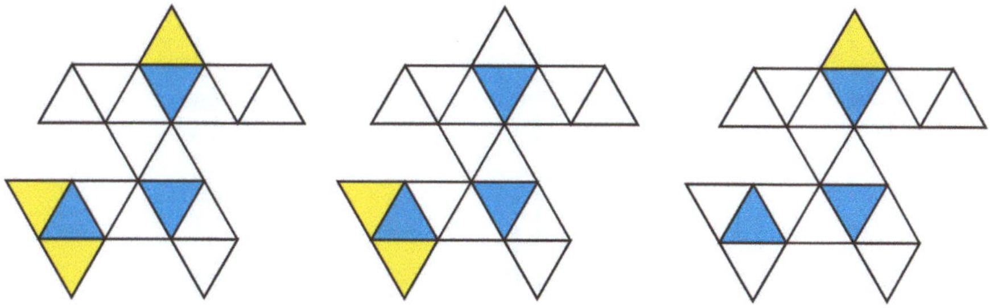

Fig. 4.12 Growths (*left*), paired (*middle*) and single (*right*)

For the tilings, whose twisting number does not exceed 5, Ivanov and Tuzhilin described geometrical structure of their skeletons and possible location of their growths. We do not include the complete answer (see details in [29–32]). Instead, we show how this theory works demonstrating several corollaries.

Definition 4.30 A connected components of the set of interior vertices of a skeleton are called *branching nodes*, and the connected components of a skeleton of a tree tiling, whose branching nodes are deleted, are called *linear parts*.

Let S be a skeleton of a tree tiling. Construct a plane graph, whose vertex set is formed by taking a single point inside each extreme cell and inside each branching node. Connect the chosen points by embedded curves each of which lies inside the union of the branching node (nodes) and the linear part adjacent to the nodes or the linear part adjacent to the node and containing the extreme cell. We demand in addition that the curves intersect each other by their ending points only. The resulting graph is called a *code of the skeleton S*, see Fig. 4.13.

Fig. 4.13 A code of a skeleton

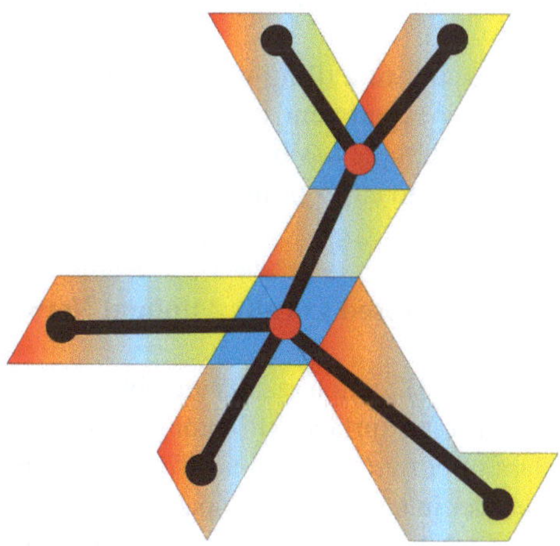

Fig. 4.14 All possible codes of skeletons, whose twisting number is at most 5

Theorem 4.10 (Ivanov and Tuzhilin [29–32]) *The codes of skeletons, whose twisting number is at most* 5 *are plane trees with at most six vertices of degree* 1 *and without vertices of degree* 2*. Thus, there are* 16 *such codes depicted in Fig. 4.14, up to a planar equivalence.*

Corollary 4.1 *Each skeleton, whose twisting number is at most* 5 *contains at most four branching nodes (and at most four interior cells) and at most* 9 *linear parts.*

This theory had been applied to problem of classification of locally minimal networks connecting the vertex sets of regular polygons. This problem turns out to be very hard, and a complete answer still is not obtained. Nevertheless, the case of binary trees that are dual graphs of skeletons had been completely studied. It turns out that there are two infinite and one finite series of such skeletons. See details in [49–52].

The case of a tiling with growths is even more complicated. By means of large computer experiment, two series of such tilings were found that seems to be infinite. Some technique has been elaborated in [53], but the problem remains open even in the predicted cases of the two series.

4.5.2.2 Closed Locally Minimal Networks on Surfaces

Some classification results are obtained on closed (compact without boundary) two-dimensional surfaces (Riemannian manifolds) of constant curvature. Notice that the class of *closed* minimal networks, i.e., the networks without boundary is natural to consider in such ambient spaces. The local structure description, see above, implies that such networks consist of geodesic segments meeting at common vertices by angles of 120°.

Here we consider the surfaces of non-negative curvature only, because in the case of negative curvature there are no classification results for today. Notice that the problem of closed locally minimal networks description in the constant curvature surfaces was stated by A.T. Fomenko. Let us pass to details.

Sphere and Projective Plane

The problem of closed locally minimal networks in the standard sphere appears naturally under studying of soap films singularities, see [54]. The complete answer had bee obtained by Heppes [55], who studied "regular" geodesic partitions of the sphere, see also [13].

Theorem 4.11 (Heppes [55]) *In the standard sphere there exist exactly* 10 *closed locally minimal networks, up to an isometry.*

Recall that the projective plane can be represented as the quotient of the standard sphere by identifying its antipodal points. Therefore, the list of closed locally minimal network in the projective plane can be obtained as the sublist of central-symmetrical closed locally minimal networks in the sphere. There are exactly three such networks, up to an isometry, see [13].

Flat Tori and Klein Bottles

Recall that each flat torus T^2 can be obtained by gluing of a parallelogram as is depicted in Fig. 4.15. It is clear that if such tori differ by a similarity, then the families

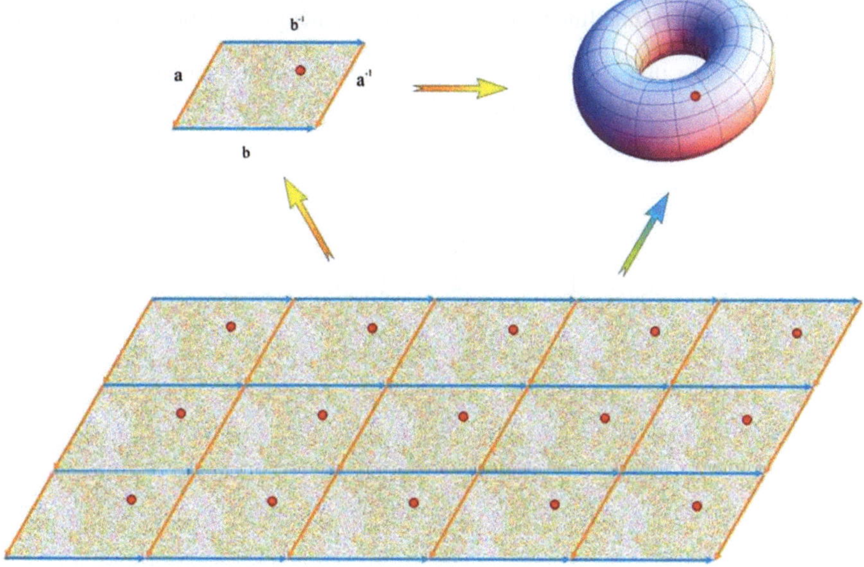

Fig. 4.15 Filling of the plane by parallelograms generates a locally isometric covering of a torus by the Euclidean plane

of closed locally minimal networks are also similar. Therefore, in what follows we assume that one side of the parallelograms is equal to 1. The standard filling of the Euclidean plane by copies of such a parallelogram generates a locally isometric covering $\nu : \mathbb{R}^2 \to T^2$.

Fix a Cartesian coordinates Oxy in the plane, which are matched with the filling in the following sense. Chose some parallelogram of the filling and let the origin O be one of its vertices and the abscissa axis Ox go along the unit side of the parallelogram (by e we denote the corresponding unit vector e). Also let the ordinate axis be directed into the half-plane generated by Ox that contains the parallelogram. So, the second side of the parallelogram oriented from the origin is given by the vector $f = (f_1, f_2)$, $f_2 > 0$, and the first side is the vector $e = (1, 0)$. By $T^2(f)$ we denote the corresponding flat torus underlying its dependence on the vector f. We also need the set $L(f) = \{m\,e + n\,f : m,\ n \in \mathbb{Z}\}$ that is referred as the *lattice of the torus* $T^2(f)$. It is clear that $L(f) = \nu^{-1}(\nu(O))$. Thus, our goal is to describe all closed locally minimal networks in the flat torus $T^2(f)$ for any f.

Let G be an arbitrary such network. By Γ we denote its lifting to the plane $\mathbb{R}^2$, see Fig. 4.16, i.e., $\Gamma = \nu^{-1}(G)$ (with evident partition into edges and vertices). It is clear that translations of $\mathbb{R}^2$ generates isometries of the torus $T^2(f)$, and hence, do not change geometry of G and Γ. Therefore, without loss of generality one can assume that one of the vertices of the network coincides with O.

Notice that each edge of the network Γ is parallel to one of the three straight lines, and so, there are exactly three classes of edges that are referred as *parallel classes*.

Definition 4.31 Infinite polygonal line emitted from a vertex of the network Γ and consisting of the edges of Γ belonging to at most two parallel classes is called a *net ray*. Each finite polygonal line that is contained in a net ray is called a *net geodesic*.

It is easy to see that six distinct net rays can be emitted from any vertex of Γ. Emit from O two net rays which are neighboring with respect to a walk around the

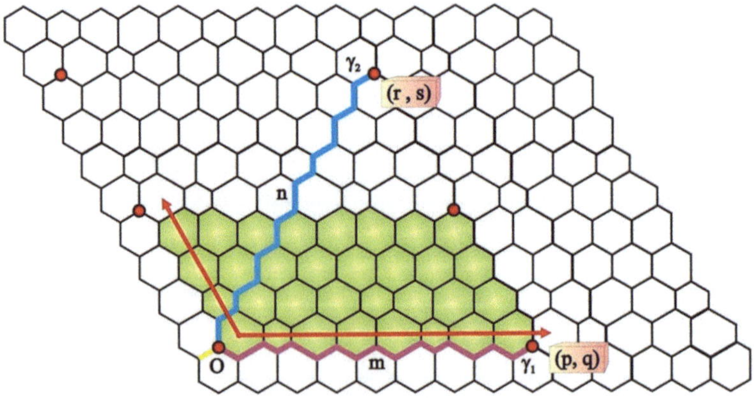

Fig. 4.16 Lifting of a minimal network to the covering plane

point O. By γ_1 and γ_2 we denote the parts of those net rays between O and the first point from $L(f)$ distinct from O. Let (p, q) and (r, s) be the integer coordinates of the end points of the net geodesics γ_1 and γ_2 (distinct from O) with respect to the basis (e, f), see Fig. 4.16. It turns out, see [13, 30, 56], that the numbers p and q are co-prime, so as the numbers r and s, and the vectors (p, q) and (r, s) are linearly independent.

It also can be shown that each net geodesic γ_i consists of an even number of edges of Γ. Let m and n be the numbers of pairs of edges forming γ_1 and γ_2, respectively. Put $M = \begin{pmatrix} p & r \\ q & s \end{pmatrix}$. Without loss of generality, assume that the net geodesics are ordered in such a way that $\det M > 0$. It turns out that the numbers m and n are divisible by $\det M$, see [13, 30, 56].

Put $m = u \det M$, $n = v \det M$, and form the integer matrix $g(M, m, n) = \begin{pmatrix} pv & ru \\ qv & su \end{pmatrix}$ with positive determinant. Notice that by each integer matrix $g = \begin{pmatrix} P & R \\ Q & S \end{pmatrix}$ with positive determinant a triplet (M, m, n) can be restored, where M is the matrix obtained from g by reducing the columns by there greatest common factors v and u, respectively, and $m = u \det M$ and $n = v \det M$.

Definition 4.32 The matrix $g(M, m, n)$ constructed by a network G is referred as the *type of the network* G.

The types of closed locally minimal networks in a torus $T^2(f)$ characterize the topological structure of the networks. It is easy to see that the type of a network remains the same under translations of the network described above. What other transformations preserve the type? It is easy to see that one can chose an arbitrary face of the network (such a face is a six-gone with angles of 120°) and move its vertices uniformly along the bisectors of the angles. It turns out that each network can be transformed into so-called *regular* one by such deformation, where a *regular network* is the one all whose faces are isometric to each other. Thus, the types of the networks defined above classify regular networks up to translations of the torus.

What other ambiguities do we have? It is clear that the type of the network depends also on the choice of the net geodesics γ_i. There are six possibilities, see Fig. 4.16. It is not difficult to calculate that under a change of net geodesics γ_i the type g changes by right multiplication by one of the matrices J^k, $J = \begin{pmatrix} 0 & -1 \\ 1 & 1 \end{pmatrix}$. Thus, we characterized each closed locally minimal network on $T^2(f)$ by an element of the quotient space $\mathcal{G}/\langle J \rangle$, where $\mathcal{G}$ stands for the set of all integer matrices with positive determinant, and $\langle J \rangle$ is the cyclic group of order 6 generated by the matrix J and acting on $\mathcal{G}$ by right multiplication.

Now, describe the types of closed locally minimal networks on a fixed flat torus $T^2(f)$. To do that we need to introduce the concept of a characteristic triangle.

Definition 4.33 For a type $g = \begin{pmatrix} P & R \\ Q & S \end{pmatrix}$ and a torus $T^2(f)$, the *characteristic triangle* is a triangle in the plane defined by its vertices O, $A = P\,e + Q\,f$, and $B = R\,e + S\,f$.

Remark 4.21 The characteristic triangles of the types $g\,J^k$ for a same torus $T^2(f)$ are equal to each other.

Theorem 4.12 (Ivanov, Ptitsyna and Tuzhilin [13, 30, 56]) *A closed locally minimal network of a type g does exist on a flat torus $T^2(f)$, if and only if all the angles of the corresponding characteristic triangle are less than* 120°.

Corollary 4.2 (Ivanov, Ptitsyna and Tuzhilin [13, 30, 56]) *For any matrix $g \in \mathcal{G}$ there exists a flat torus $T^2(f)$ and a closed locally minimal network of the type g on $T^2(f)$.*

Corollary 4.3 (Ivanov, Ptitsyna and Tuzhilin [13, 30, 56]) *On any flat torus there exist infinitely many closed locally minimal networks of different types.*

Corollary 4.4 (Ivanov, Ptitsyna and Tuzhilin [13, 30, 56]) *For any closed locally minimal network G on a flat torus $T^2(f)$ there exists a neighborhood U of the point f in the plane, such that for any $f' \in U$, a closed locally minimal network of the same type as G exists on the torus $T^2(f')$.*

To study the case of Klein bottles it suffices to use the well-known two-sheeted locally isomeric covering of a Klein bottle by a flat torus. It turns out that any closed locally minimal network on a Klein bottle can be lifted onto a flat torus glued form a rectangle, and such networks on tori are classified completely by the above results. The corresponding classification can be found in [13, 30, 57].

Disphenoids and Other Polyhedra

Similarly to the case of Klein bottles, the case of disphenoids, i.e., of the tetrahedra all whose faces are equal each other, can be reduced to the case of flat tori. Namely, there also exists a locally isometric double covering by a flat torus, but this covering have branch points at the vertices of the tetrahedron, see Fig. 4.17.

Fig. 4.17 Branched double covering of a disphenoid by a flat torus

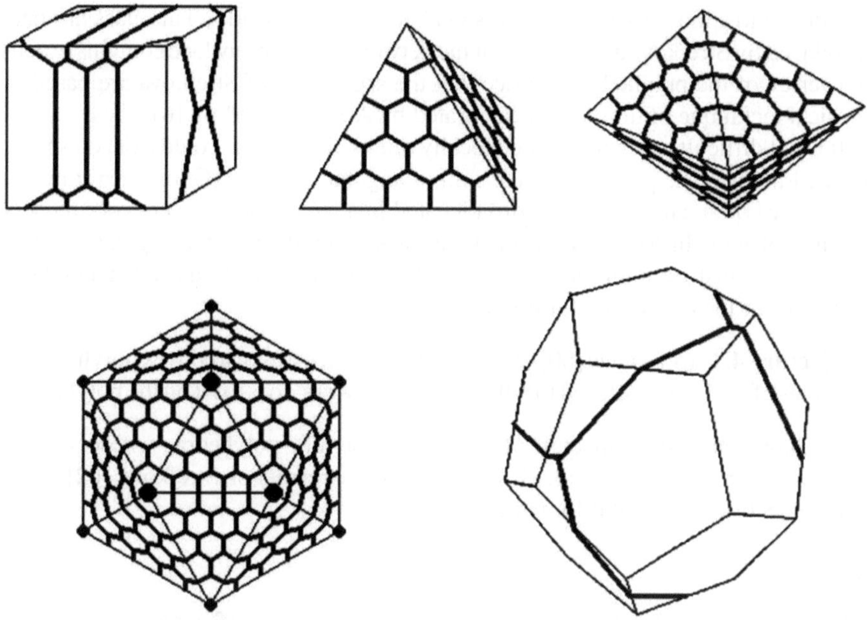

Fig. 4.18 Closed locally minimal networks on the surfaces of Platonic bodies

It is not difficult to show that each closed locally minimal network on the surface of a convex polyhedron can not pass through the vertices of the polyhedron, therefore, the branchings are inessential. The complete answer for the disphenoids can be found in [13, 30, 58].

Remark 4.22 Notice that the disphenoids are exactly the tetrahedra with the same curvature at the vertices, i.e., the tetrahedra of constant curvature.

Examples of closed locally minimal networks on the surfaces of all the Platonic bodies are shown in Fig. 4.18. Notice that the non-trivial example of such network on the surface of dodecahedron had been constructed by T. Anikeeva (Pavlyukevich).

But in this case a direct reduction to flat tori does not work, because either infinite-sheeted branched coverings arise, as in the case of octahedron, or it is possible to construct a two-sheeted covering but not by a torus, but by a torus with some holes. The corresponding covering by a plane with holes can be used to construct examples, and probably for classification, but no complete results have been obtained in this direction still even in the case of cube.

A huge contribution in description of closed locally minimal networks on the surfaces of convex polyhedra has been made by N.P. Strelkova. She has obtained a complete description of possible combinatorial structures of closed locally minimal networks on the surfaces of all convex polyhedra and also of the possible lengths of the edges of such networks, see [59, 60]. Besides, she has shown that the problem on description of all closed locally minimal networks on all convex polyhedra can be reduced to a more narrow class of so-called *simple* networks. This class has been introduced by Strelkova and consists of the networks, whose any face contains exactly one vertex of the polyhedron. Notice that the arguments of Strelkova are based on classical technique of unfoldings elaborated by A.D. Alexandrov [61].

It is not difficult to see that closed locally minimal networks could exist on special polyhedra only. The point is that the sum of angular excesses (Gauss curvatures) at the vertices of an ambient polyhedron located in a single face of a network must be multiple of $\pi/3$. In particular, if a network is simple, then all the angular excesses themselves must be multiples of $\pi/3$. A natural question: Is the latter condition sufficient for the existence of a simple network?

Conjecture 4.1 (*Strelkova* [60]) If all the angular excesses of a convex polyhedra are multiples of $\pi/3$, then a closed locally minimal network does exist on the polyhedron.

Strelkova proved Conjecture 4.1 for "almost all" polyhedra, whose angular excesses are multiples of $\pi/3$. Besides, an earlier results of Strelkova [62] imply that Conjecture 4.1 is valid for tetrahedra.

4.6 How to Calculate or Estimate the Length of a Minimal Network of a Given Topology Without Constructing the Network Itself?

In this section we collect several formulas permitting to calculate or estimate the weight or the length of a minimal network in terms of its boundary set.

4.6.1 The Length of a Minimal Spanning Tree

Let (M,ρ) be an arbitrary metric space. Generally speaking, we do not assume that M is a finite set. The questions are: When $\mathrm{mst}(M)<\infty$, and How to calculate $\mathrm{mst}(M)$ without constructing a minimal spanning tree. Here we list the results of Ivanov et al. [5].

Let N_1 and N_2 be arbitrary non-intersecting subsets of M. Put

$$\rho(N_1,N_2)=\inf\{\rho(x,y)\mid x\in N_1,\ y\in N_2\}$$

(if one of those sets is empty, then $\rho(N_1,N_2)=+\infty$).

Remark 4.23 This function ρ is not a metric on the subsets, see also Sect. 4.7.

For any $d\ge 0$ we put

$$\mathscr{P}(d)=\{N\subset M\mid \rho(N,M\setminus N)\ge d\}.$$

Notice that for any d the sets $\emptyset$ and M are contained in $\mathscr{P}(d)$, and also $\mathscr{P}(0)$ is the set of all subsets of M.

For each $x \in M$ put:

$$\mathscr{P}^x(d) = \{A \in \mathscr{P}(d) \mid x \in A\} \quad \text{and} \quad c_d(x) = \cap_{A\in\mathscr{P}^x(d)} A.$$

Notice that $c_0(x) = \{x\}$.

By $\mathscr{P}_d(M)$ we denote the set of all distinct $c_d(x)$. Notice that $\mathscr{P}_0(M)$ is the set of single-point subsets of M. It is not difficult to see, that the set $\mathscr{P}_d(M)$ is a partition of M, and if $d_1 \le d_2$, then the partition $\mathscr{P}_{d_1}(M)$ is a subpartition of $\mathscr{P}_{d_2}(M)$.

Put

$$\operatorname{diam}(M) = \sup\{\rho(x, y) \mid x \in M,\ y \in M\},$$
$$\operatorname{Diam}_d(M) = \sum_{c\in\mathscr{P}_d(M)} \operatorname{diam}(c).$$

For each set X by $\#X$ we denote the number of elements in X, provided X is finite, and put $\#X = +\infty$ otherwise.

For any $\lambda \ge 0$ put $\pi_\lambda(M) = \#\mathscr{P}_\lambda(M)$.

Theorem 4.13 (Ivanov et al. [5]) *Let (M, ρ) be an arbitrary metric space. Then* $\operatorname{mst}(M) < \infty$, *if and only if the following conditions hold*

- *the space (M, ρ) is bounded and at most countable;*
- $\int_0^{diamM} \pi_\lambda(M)\, d\lambda < \infty$;
- $\operatorname{Diam}_\lambda(M) \to 0$ *as* $\lambda \to 0$.

Moreover, for such spaces the following equality holds

$$\operatorname{mst}(M) = \int_0^{diamM} \pi_\lambda(M)\, d\lambda - \operatorname{diam}(M).$$

4.6.2 Maxwell Formula

The classical Maxwell Formula [25, 63, 64] calculates the length $\rho(\Gamma)$ of a given plane locally minimal binary tree Γ in terms of the coordinates A_i of its boundary vertices and the direction vectors n_i of the edges of Γ coming to these vertices:

$$\rho(\Gamma) = \sum_i \langle A_i, n_i\rangle,$$

where the angle brackets stand for the scalar product, see Fig. 4.19.

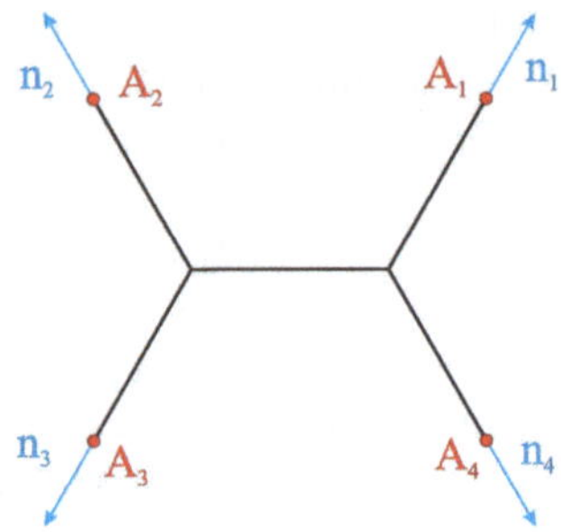

Fig. 4.19 Maxwell Formula

This formula remains valid in the case of an arbitrary locally minimal network Γ, if put n_i to be equal to the sum of direction vectors of all the edges of Γ coming to the vertex A_i. It is also valid in the Euclidean space of an arbitrary dimension.

Ivanov and Tuzhilin stated the following problem: *How to calculate the length of a plane locally minimal tree in terms of its boundary and topology only*? Here we give the solution of this problem for a bigger class of minimal parametric networks in $\mathbb{R}^n$, see [65].

Start with another version of discrete network definition that is more convenient in this context (compare with the definition given above).

Let $G = (V, E, \partial)$ be an arbitrary connected combinatorial graph with a boundary ∂G, and (X, ρ) be a metric space. Fix some embedding $\psi : \partial G \to X$ and call it a *boundary mapping*.

Definition 4.34 By a *network in X of the type G with the boundary ψ* we call a mapping $\Gamma : V \to X$ such that $\Gamma|_{\partial G} = \psi$. By $[G, \psi]$ we denote the set of all such networks.

For each edge $e \in E$ define its *length* $\rho_\Gamma(e)$ to be equal to 0 for a loop e, and to be equal to the number $\rho\,(\Gamma(u), \Gamma(v))$ for other edges $e = uv$. The *length* $\rho(\Gamma)$ *of a network* Γ is the value $\sum_{e\in E} \rho_\Gamma(e)$.

Thus, we defined a mapping $\rho : [G, \psi] \to \mathbb{R}$.

Let $I = V \setminus \partial G$ be the set of all interior vertices of the graph G, then each network $\Gamma \in [G, \psi]$ is uniquely defined by the location of the points $\Gamma(v)$, $v \in I$, and hence, if $m = \#I$, then $[G, \psi]$ can be naturally identified with the Cartesian product $\prod_{i=1}^m X$. It is easy to see that ρ is a continuous function on this space.

Put $\mathrm{mpn}\,(G, \psi) = \inf\{\rho\,(\Gamma) \mid \Gamma \in [G, \psi]\}$. Each network $\Gamma \in [G, \psi]$ such that $\rho\,(G) = \mathrm{mpn}\,(G, \psi)$ is called a *minimal parametric network of the type G*.

Let $G = (V, E)$ be a tree with a boundary $B = \partial G = \{v_1, \dots, v_n\}$. By this tree we construct a system of equations and inequalities on the variables from $\mathbb{R}^{nd}$ as follows. Let $\left(x_1^1, \dots, x_1^d, \dots, x_n^1, \dots, x_n^d\right)$ be the standard coordinates in $\mathbb{R}^{nd}$. Put $x_k = \left(x_k^1, \dots, x_k^d\right)$ and $x = (x_1, \dots, x_n)$.

For each $e \in E$ by $G \setminus e = G_1 \cup G_2$ we denote the forest obtained from the tree G by deleting the edge e. Let V_k be the vertex set of the tree G_k, and $B_k = B \cap V_k$, $k = 1, 2$. Chose any B_k, and let $B_k = \{v_{k_1}, \dots, v_{k_p}\}$. By σ_e we denote the inequality $\| \sum_q x_{k_q} \|^2 \le 1$. By σ we denote the equality $\sum_k x_k = 0$.

Consider the system S_G consisting of σ and all the inequalities σ_e, $e \in E$, and let $|S_G| \subset \mathbb{R}^{nd}$ stands for the set of all solutions to this system. It turns out that the set $|S_G|$ does not depend on the choice of B_k, and $|S_G|$ is a compact convex body in the hyperspace of $\mathbb{R}^{nd}$, defined by the equality σ.

Now consider an arbitrary boundary mapping $\psi : B \to \mathbb{R}^d$. Put $A_k = \psi(v_k)$ and $A = (A_1, \dots, A_n) \in \mathbb{R}^{nd}$. Define a function $\rho_\psi : \mathbb{R}^{nd} \to \mathbb{R}$ as follows: $\rho_\psi(x) = \langle x, A \rangle$.

Theorem 4.14 (Ivanov and Tuzhilin [65]) *Under the above assumptions,*

$$\mathrm{mpn}(G, \psi) = \max_{x \in |S_G|} \rho_\psi(x).$$

Remark 4.24 In accordance with its definition, the length of a minimal parametric network is a minimal value of the function defined on a linear space and having the form of a sum of square roots of sums of coordinates squares. This function is difficult for studying, but its domain is simple. Theorem 4.14 represents the same value as a maximal value of a linear function but not on the entire space, but on some its convex compact subset $|S_G|$. Thus, now the function is simple, but the domain is difficult.

Recently, A. Bannikova, D. Il'utko, and I. Nikonov generalized the classical Maxwell Formula and Theorem 4.14 to the case of extreme networks in normed spaces [66]. Notice that in contrast to Euclidean space locally minimal networks in a normed space need not be a local minimum of the length functional. The ones that are local minima are referred as *extreme networks*.

4.6.3 The Weight of a Minimal Filling

In [22] exact formulas for the weight of minimal fillings are obtained in some simple cases, such as three and four points metric spaces, regular simplices, etc. In this section we give a general formula calculating the weight of a minimal filling of a finite metric space obtained by A. Eremin [33].

Let $G = (V, E, d)$ be an arbitrary graph, and k be a positive integer. By G^{2k} we denote the graph $(V, \sqcup_{i=1}^{2k} E, d')$, where the restriction of d' onto each copy of E coincides with d. Notice that for $k = 1$ the graph G^2 is the doubling of the graph G, see Definition 4.24. It is clear, that the degrees of all the vertices of the graph G^{2k} are even, therefore if G is connected, then G^{2k} contains an Euler cycle.

Let (M, ρ) be a finite metric space, and G be a binary tree connecting $M = \partial G$. By $\mathcal{T}_2(M)$ we denote the set of all such trees G.

Let k be a positive integer. Consider an Euler cycle C in G^{2k}, that consists of consecutive boundary paths $\gamma_1, \dots, \gamma_N$ (recall that a path in a graph passes each edge at most once). This Euler cycle is called a *multitour of multiplicity k of the tree G*. By $O_\mu(G)$ we denote the set of all multitours of the tree G.

For each γ_i by v_i and w_i we denote its ending vertices. Put

$$p(C) = \frac{1}{2k}\sum_{i=1}^{N} \rho(v_i, w_i).$$

Theorem 4.15 (Eremin [33]) *Let (M, ρ) be a finite metric space. Then*

$$\mathrm{mf}(M) = \min_{G\in\mathcal{T}_2(M)} \max_{C\in O_\mu(G)} p(C).$$

Remark 4.25 The idea of such a formula belongs to Ivanov and Tuzhilin, who conjectured a similar formula for tours (i.e. multitours with $k = 1$). Eremin and Ovsyannikov constructed a counter example. Then Ivanov, Ovsyannikov, Strelkova, and Tuzhilin understand that the value mf(M) remains the same if negative weights on edges of fillings are permitted [67]. Such fillings are called *generalized.* At last, Eremin introduced multitours and proved the formula using the generalized fillings.

4.6.4 Ratios

Let M be an arbitrary finite metric space isometrically embedded into an ambient metric space X. Above the following three values are defined: mst(M) (the least possible length of spanning trees on M; to define it the embedding into X is not necessary), smt(M) (the least possible length of trees on X connecting M; here the embedding into X is necessary), mf(M) (the weight of a minimal filling; as in the first case the embedding into X is not necessary). Clearly, mst$(M) \geq$ smt$(M) \geq$ mf(M). Besides, in the important particular case when the ambient space X is a normed space all these three values are homogeneous with respect to dilatations. Therefore, it is natural to consider the ratios of the function to compare there values.

Those reasonings lead to the definition of the classical value: the *Steiner ratio* sr(M) *of the set M* is defined as smt(M)/mst(M), see [25]. The infimum sr(X) of the Steiner ratios over all finite subsets $M \subset X, \#M \geq 2$, is a non-trivial characteristic of the ambient space X and is referred as the *Steiner ratio of the space X*.

Remark 4.26 The Steiner ratio sr(M) had been defined to measure the relative error of the approximation of a shortest tree by a minimal spanning tree. The importance of such an approximation is explained by the fact that the algorithms constructing shortest trees work slowly, but the algorithms constructing minimal spanning trees are quite fast.

Two other ratios, namely $\mathrm{mf}(M)/\mathrm{smt}(M)$, and $\mathrm{mf}(M)/\mathrm{mst}(M)$ have been defined by Ivanov and Tuzhilin and referred as *Steiner subratio* $\mathrm{ssr}(M)$ and *Steiner–Gromov ratio* $\mathrm{sgr}(M)$. The both of them generate interesting characteristics of ambient space that are defied similarly to the Steiner ratio. Notice that $\mathrm{sgr}(M)$ does not use isometric embedding of M into an ambient space, but $\mathrm{sr}(M)$ and $\mathrm{ssr}(M)$ do use the embedding.

It is not difficult to verify that the three ratios take all the values from the segment $[1/2, 1]$, see [13] and Pakhomova [68]. But the exact calculation and even good estimates of each of the ratios are very non-trivial problems. The exact results are known to very few metric spaces such as Manhattan plane [69] and Lobachevski plane [70]. The Gilbert–Pollack conjecture [25] concerning the exact value of the Steiner ratio for the Euclidean plane remains open during about 30 years in spite of several attempts, see [71–74]. Several estimates concerning Riemannian manifolds and surface of tetrahedra obtained by mean of covering technique are received in [75, 76]. The detailed reviews can be found in [13, 77, 78].

4.7 Spaces of Compacts

The present Section is devoted to the Steiner problem in the space of compact metric spaces, endowed with Gromov–Hausdorff metric. Here we show that each boundary set consisting of finite metric spaces only, can be connected by a Steiner minimal tree. In the general case, the authors have solved the Steiner problem for 2-point boundaries [79], where the problem is equivalent to the fact that the ambient space is geodesic. General case of more than 2 boundary points has resisted to the authors attempts based on the Gromov pre-compactness criterion. Nevertheless, we hope that the technique we worked out will be useful for either proving the theorem, or for constructing a counterexample.

4.7.1 Main Definitions and Results

Let X be an arbitrary metric space. By $|xy|$ we denote the distance between points $x,\ y \in X$. Let $\mathscr{P}(X)$ be the family of all nonempty subsets of X. For $A,\ B \in \mathscr{P}(X)$ we put

$$d_H(A, B) = \max\left\{\sup_{a\in A}\inf_{b\in B}|ab|,\ \sup_{b\in B}\inf_{a\in A}|ab|\right\}.$$

The value $d_H(A, B)$ is called the *Hausdorff distance between A and B*.

Notice that $d_H(A, B)$ as may be equal to infinity (e.g., for $X = A = \mathbb{R}$ and $B = \{0\} \subset \mathbb{R}$), so as may vanish for non-equal A and B (e.g., for $X = \mathbb{R}$, $A = [a, b]$, and $B = [a, b)$).

Let $\mathscr{H}(X) \subset \mathscr{P}(X)$ denote the set of all nonempty closed bounded subsets of X. It is well-known, see, for example [6], that the restriction of d_H onto $\mathscr{H}(X)$ is a metric.

Let X and Y be metric spaces. The triple (X', Y', Z) consisting of a metric space Z and its subsets X' and Y' which are isometric to X and Y, respectively, is called a *realization of the pair* (X, Y). We put

$$d_{GH}(X, Y) = \inf\{r : \exists (X', Y', Z),\ d_H(X', Y') \le r\}.$$

The value $d_{GH}(X, Y)$ is called the *Gromov–Hausdorff distance between X and Y.*

By $\mathscr{M}$ we denote the set of all compact metric spaces considered up to an isometry. It is also well-known that the restriction of d_{GH} onto $\mathscr{M}$ is a metric, see [6].

The Gromov–Hausdorff distance can be effectively investigated in terms of correspondences.

Let X and Y be arbitrary nonempty sets. We put $\mathscr{P}(X, Y) = \mathscr{P}(X \times Y)$. The elements of $\mathscr{P}(X, Y)$ are called *relations* between X and Y. If $X' \subset X$ and $Y' \subset Y$ are nonempty subsets, and $\sigma \in \mathscr{P}(X, Y)$, then we put

$$\sigma|_{X' \times Y'} = \{(x, y) \in \sigma : x \in X',\ y \in Y'\}.$$

Notice that $\sigma|_{X' \times Y'}$ may be empty and, thus, may not belong to $\mathscr{P}(X', Y')$.

Let $\pi_X : (x, y) \mapsto x$ and $\pi_Y : (x, y) \mapsto y$ be the canonical projections. A relation $\sigma \in \mathscr{P}(X, Y)$ is called a *correspondence*, if the restrictions of π_X and π_Y onto σ are surjective. By $\mathscr{R}(X, Y)$ we denote the set of all correspondences between X and Y.

If X and Y are metric spaces, then for each relation $\sigma \in \mathscr{P}(X, Y)$ its *destortion* is defined as

$$\operatorname{dis}\sigma = \sup\{\big||xx'| - |yy'|\big| : (x, y),\ (x', y') \in \sigma\}.$$

Proposition 4.6 ([6]) *Let X and Y be metric spaces. Then*

$$d_{GH}(X, Y) = \frac{1}{2}\inf\{\operatorname{dis}R : R \in \mathscr{R}(X, Y)\}.$$

For a metric space X by $\operatorname{diam}X$ we denote its *diameter*: $\operatorname{diam}X = \sup\{|xy| : x, y \in X\}$.

Corollary 4.5 ([6]) *For any metric spaces X and Y such that the diameter of at least one of them is finite, we have*

$$d_{GH}(X, Y) \ge \frac{1}{2}|\operatorname{diam}X - \operatorname{diam}Y|.$$

A correspondence $R \in \mathscr{R}(X, Y)$ is called *optimal*, if $d_{GH}(X, Y) = \frac{1}{2}\operatorname{dis}R$. By $\mathscr{R}_{\text{opt}}(X, Y)$ we denote the set of all optimal correspondences between X and Y.

Proposition 4.7 ([80–82]) *For* $X, Y \in \mathcal{M}$ *we have* $\mathcal{R}_{opt}(X, Y) \neq \emptyset$.

Let $\mathcal{M}_n \subset \mathcal{M}$ consist of all metric spaces containing at most n points; let $\mathcal{M}(d) \subset \mathcal{M}$ consist of all spaces, whose diameters are at most d; at last, put $\mathcal{M}_n(d) = \mathcal{M}_n \cap \mathcal{M}(d)$.

Proposition 4.8 ([6]) *The space* $\mathcal{M}_n(d)$ *is compact.*

The technique developed in [30] for Riemannian manifolds can be obviously generalized to proper metric spaces.

Proposition 4.9 *Let* X *be a proper metric space. Then for each nonempty finite* $M \subset X$ *we have* $\mathrm{SMT}(M, X) \neq \emptyset$.

The next result follows from Propositions 4.8 and 4.9.

Corollary 4.6 *For any nonempty finite set* $M \subset \mathcal{M}_n(d)$ *we have*

$$\mathrm{SMT}\,(M, \mathcal{M}_n(d)) \neq \emptyset.$$

The above technique permits to prove the following Theorem.

Theorem 4.16 *For each* $M = \{m_1, \ldots, m_k\} \subset \mathcal{M}_n$ *we have*

$$\mathrm{SMT}(M, \mathcal{M}) \neq \emptyset.$$

References

1. Fermat de P., Tannery, H. (eds.): OeuPres, vol. 1, Paris 1891, Supplement: Paris 1922, p. 153 (1643)
2. Chanderjit B.: Limitations To Algorithm Solvability: Galois Methods and Models of Computation, Computer Science Technical Reports, Paper 486 (1986) http://docs.lib.purdue.edu/cstech/486
3. Harary, F.: Graph Theory. Addison-Wesley, MA (1969)
4. Ivanov, A.O., Tuzhilin, A.A.: Minimal Spanning Trees on Infinite Sets. Fund. i Prikl. Matem. **20**(2), 89–103 (2015). (in Russian, English translation to appear in J. of Math. Sci., 2016)
5. Ivanov, A.O., Nikonov, I.M., Tuzhilin, A.A.: Sets admitting connection by graphs of finite length. Matem. Sbornik **196**(6), 71–110 (2005). (Sbornik: Math., **196** (6), pp. 845–884)
6. Burago D., Burago Yu., and Ivanov S.: A Course in Metric Geometry, Graduate Studies in Math., **33**, A.M.S., Providence, RI (2001)
7. Gergonne, J.D.: Solutions purement gèomètriques des problèmes de minimis proposès aux pages 196, 232 et 292 de ce volume, et de divers autres problèmes analogues. Annales de Mathèmatiques pures et appliquèes **1**, 375–384 (1810)
8. Melzak, Z.A.: On the problem of Steiner. Canad. Math. Bull. **4**, 143–148 (1960)
9. Bopp, K.: Über das kürzeste Verbindungssystem zwischen vier Punkten. Universität Göttingen, PhDthesis (1879)
10. Jarnik, V., Kössler, M.: O minimalnich grafeth obeahujiicich n danijch bodu. Cas. Pest. Mat. a. Fys. **63**, 223–235 (1934)

11. Brazil, M., Graham, R.L., Thomas, D.A., Zachariasen, M.: On the history of the euclidean Steiner tree problem. Arch. Hist. Exact Sci. pp. 1–30 (2013)
12. Courant, R., Robbins, G.: What Is Mathematics?. Oxford University Press, London (1941)
13. Ivanov, A.O., Tuzhilin, A.A.: Extreme Networks Theory. In-t Komp. Issl, Moscow, Izhevsk (2003). [in Russian]
14. Ivanov, A.O., Tuzhilin, A.A.: Geometry of minimal networks and the one-dimensional plateau problem. Uspekhi Matem. Nauk **47**(2), 53–115 (1992). (Russian Math. Surv., **47** (2), pp. 59–131 (1992))
15. Ivanov, A.O., Tuzhilin, A.A.: Branching geodesics in normed spaces. Izv. RAN, Ser. Matem. **66**(5), 33–82 (2002). (Izvestiya: Math., **66** (5) pp. 905–948 (2002))
16. Ivanov, A.O., Van Hong, L., Tuzhilin, A.A.: Nontrivial critical networks. Singularities of lagrangians and a criterion for criticality. Matem. Zametki **69**(4), 566–580 (2001). (Math. Notes, **69** (4), pp. 514–526 (2001))
17. Swanepoel, K.: The local steiner problem in normed planes. Networks **36**(2), 104–113 (2000)
18. Il'yutko, D.P.: Locally minimal trees in n-normed spaces. Matem. Zametki **74**(5), 656–668 (2003). (Math. Notes, **74** (5), 619–629 (2003))
19. Il'yutko, D.P.: Branching extremals of the functional of λ-normed length. Matem. Sbornik **197**(5), 75–98 (2006). (Sbornik: Math., **197** (5), 705–726 (2006))
20. Il'yutko, D.P.: Geometry of extreme networks in λ-geometry Vestnik MGU. Math., Mech. **1**(4), 52–54 (2005). (Moscow Univ. Math. Bull., **60** (4) pp. 39–52 (2005))
21. Sankoff, D.: Minimal mutation trees of sequences. SIAM J. Appl. Math. **28**, 35–42 (1975)
22. Ivanov, A.O., Tuzhilin, A.A.: One-dimensional Gromov minimal filling problem. Matem. Sbornik **203**(5), 65–118 (2012). (Sbornik: Math., **203** (5), pp. 677–726 (2012))
23. Gromov, M.: Filling Riemannian manifolds. J. Diff. Geom. **18**(1), 1–147 (1983)
24. Cormen, Th.H., Leiserson, Ch.E., Rivest, R.L., Stein, C.: Introduction To Algorithms, 3rd edn. MIT Press, Cambridge (2009)
25. Gilbert, E.N., Pollak, H.O.: Steiner minimal trees. SIAM J. Appl. Math. **16**(1), 1–29 (1968)
26. Ivanov, A.O., S'edina, O.A., Tuzhilin, A.A.: The structure of minimal Steiner trees in the neighborhoods of the lunes of their edges. Matem. Zametki **91**(3), 352–370 (2012). (Math. Notes, **91** (3), pp. 339–353 (2012))
27. Ivanov, A.O., Tuzhilin, A.A.: The twist number of planar linear trees. Matem. Sbornik **187**(8), 41–92 (1996). (Sbornik: Math., **187** (8), pp. 1149–1195)
28. Ivanov, A.O.: The geometry of plane locally minimal binary trees. Matem. Sbornik **186**(9), 45–76 (1995). (Sbornik: Math., **186** (9), pp. 1271–1301 (1995))
29. Ivanov, A.O., Tuzhilin, A.A.: Minimal Surfaces. In: Fomenko, A. (ed.) The Steiner Problem for Convex Boundaries, General Case. Advances in Soviet Mathematics, pp. 15–92. American Mathematical Society, Providence (1993)
30. Ivanov, A.O., Tuzhilin, A.A.: Minimal Networks. Steiner Problem and Its Generalizations. CRC Press, Boca Raton (1994)
31. Ivanov, A.O., Tuzhilin, A.A.: Branching Geodesics. Geometry of Locally Minimal Networks. Russian Math. and Sci. Researches, vol. 5. Edwin–Mellen Press, Lewiston (1999). [in Russian]
32. Ivanov, A.O., Tuzhilin, A.A.: The Steiner problem in the plane or in plane minimal nets. Matem. Sbornik **182**(12), 1813–1844 (1991). (Math. of the USSR–Sbornik, **74** (2), pp. 555–582 (1993))
33. Eremin, AYu.: A formula for the weight of a minimal filling of a finite metric space. Matem. Sbornik **204**(9), 51–72 (2013). (Sbornik: Math., **204** (9), pp. 1285–1306 (2013))
34. Zaretskij, K.A.: Construction of a tree from the collection of distances between suspending vertices. Uspekhi Matem. Nauk **20**(6), 90–92 (1965). [in Russian]
35. Simões-Pereira, J.M.S.: A note on the tree realizability of a distance matrix. J. Comb. Theory **6**, 303–310 (1969)
36. Smolenskij, E.A.: About a linear denotation of graphs. Zh. Vychisl. Mat. Mat. Fiz. **2**(2), 371–372 (1962)
37. Hakimi, S.L., Yau, S.S.: Distance matrix of a graph and its realizability. Quart. Appl. Math. **12**, 305–317 (1975)

38. Rubleva, O.V.: The additivity criterion for finite metric spaces and minimal fillings, Vestnik MGU. Matem. Mech. **1**(2), 8–11 (2012). (Moscow Univ. Math. Bull., **67** (2), pp. 52–54 (2012))
39. Ovsyannikov, Z.N.: Pseudo-additive Metric Spaces and Minimal Fillings, Diploma Thesis. Mech. Math, MGU (2013)
40. Du, D.Z., Hwang, F.K., Weng, J.F.: Steiner minimal trees for regular polygons. Discrete Comput. geom. **2**, 65–84 (1987)
41. Du, D.Z., Hwang, F.K., Weng, J.F.: Steiner minimal trees on zig-zag lines. Trans. Am. Math. Soc. **278**(1), 149–156 (1983)
42. Chung, F.R.K., Graham, R.L.: Steiner trees for ladders. Ann. Discr. Math. (2), 173–200 (1978)
43. Du, D.Z., Hwang, F.K., Chao, S.C.: Steiner minimal tree for points on a circle. Proc. Am. Math. Soc. **95**(4), 613–618 (1985)
44. Rubinstein, J.H., Thomas, D.A.: Graham's problem on shortest networks for points on a circle. Algorithmica **7**, 193–218 (1992)
45. Du, D.Z., Hwang, F.K.: Steiner minimal trees on chinese checkerboards. Math. Mag. **64**(5), 332–339 (1991)
46. Chung, F.R.K., Gardner, M., Graham, R.L.: Steiner trees on a checkboard. Math. Mag. **62**(2), 83–96 (1989)
47. Ivanov, A.O., Tuzhilin, A.A.: Uniqueness of Steiner minimal trees on boundaries in general position. Matem. Sbornik **197**(9), 55–90 (2006). (Sbornik: Math., **197** (9), pp. 1309–1340 (2006))
48. Ivanov, A.O., Tuzhilin, A.A.: Stabilization of locally minimal trees. Matem. Zametki **86**(4), 512–524 (2009). (Math. Notes, **86** (4), pp. 483–492 (2009))
49. Ivanov, A.O., Tuzhilin, A.A.: Minimal Surfaces. In: Fomenko, A. (ed.) The Steiner Problem for Convex Boundaries, the Regular Case. Advances in Soviet Mathematics, vol. 15, pp. 93–131
50. Tuzhilin, A.A.: Minimal binary trees with regular boundary: the case of skeletons with four ends. Matem. Sbornik **187**(4), 117–159 (1996). (Sbornik: Math., **187** (4), pp. 581–622, (1996))
51. Tuzhilin, A.A.: Minimal binary trees with a regular boundary: the case of skeletons with five endpoints. Matem. Zametki **61**(6), 906–921 (1997). (Math. Notes, **61** (6), pp. 758–769 (1997))
52. Tuzhilin, A.A.: Complete classification of locally minimal binary trees with a regular boundary whose dual triangulations are skeletons. Fundam. Prikl. Mat. **2**(2), 511–562 (1996). [in Russian]
53. Ivanov A.O., Tuzhilin A.A.: Planar Local Minimal Binary Trees with Convex, Quasiregular, and Regular Boundaries, Sonderforschungsbereich 256 Preprint (1997)
54. Fomenko, A.T.: Topological Variational Problems. Izd-vo MGU, Moscow,1984. Gordon and Breach Science Publishers, New York (1990)
55. Heppes, A.: Isogonal Spherische Netze. Ann. Univ. Sci., Budapest, Sect. Math. **7**, 4–48 (1964)
56. Ivanov, A.O., Ptitsyna, I.V., Tuzhilin, A.A.: Classification of closed minimal networks on flat two-dimensional tori. Matem. Sbornik **183**(12), 3–44 (1992). (Sbornik. Math., **77** (2), pp. 391–425 (1994))
57. Ptitsyna, I.V.: Classification of closed minimal networks on flat klein bottles, Vestnik MGU. Ser. Matem. Mech. (2), 15–22 (1995). (Moscow Univ. Math. Bull., **50** (2), pp. 13–19 (1995))
58. Ptitsyna, I.V.: Classification of closed minimal networks on tetrahedra. Matem. Sbornik **185**(5), 119–138 (1994). (Sbornik. Math., **82** (1), pp. 101–116 (1995))
59. Strelkova, N.P.: Realization of plane graphs as closed locally minimal nets on convex polyhedra. Dokl. RAN **435**(1), 1–3 (2010). (Doklady Math., **82** (3), pp. 939–941 (2010))
60. Strelkova, N.P.: Closed locally minimal networks on surfaces of convex polyhedra. Model. Anal. Inf. Sist. **20**(5), 117–147 (2013). [in Russian]
61. Alexandrov, A.D.: Convex Polyhedra. Gos. Izd-vo Tekh.–Teor. Liter., Moscow–Leningrad, 1950. Springer, Berlin (2005)
62. Strelkova, N.P.: Closed locally minimal nets on tetrahedra. Matem. Sbornik **202**(1), 141–160 (2011). (Sbornik: Math., **202** (1), pp. 135–153 (2011))
63. Maxwell J.C.: Cambridge Philos. Mag. (1864)
64. Maxwell J.C.: Trans. Roy. Soc. vol. 26, Edinburgh (1869)
65. Ivanov, A.O., Tuzhilin, A.A.: Generalized Maxwell formula for the length of a minimal tree with a given topology, Vestnik MGU. Ser. Matem. Mech. **1**(3), 7–14 (2010). (Moscow Univ. Math. Bull., **65** (3), pp. 100–106 (2010))

66. Bannikova, A.G., Ilyutko, D.P., Nikonov, I.M.: The length of an extremal network in a normed space: Maxwell formula. Sovrem. Matem. Fundam. Napravl. **51**, 5–20 (2016). (J. of Math. Sci., **214** (5), pp. 593–608 (2016))
67. Ivanov, A.O., Ovsyannikov, Z.N., Strelkova, N.P., Tuzhilin, A.A.: One-dimensional minimal fillings with negative edge weights. Vestnik MGU, Ser. Matem. Mech. **1**(5), 3–8 (2012). (Moscow Univ. Math. Bull., **67** (5), pp. 189–194 (2012))
68. Pakhomova, A.S.: The estimates for Steiner subratio and Steiner–Gromov ratio. Vestnik MGU, Ser. Mat. Mech. (1), 17–25 (2014). (Moscow Univ. Math. Bull., **69** (1), pp. 16–23 (2014))
69. Hwang, F.K.: On Steiner minimal trees with rectilinear distance. SIAM J. Appl. Math. **30**, 104–114 (1976)
70. Innami, N., Kim, B.H.: Steiner ratio for hyperbolic surfaces. Proc. Jpn. Acad. **82**, 77–79 (2006)
71. Ivanov, A.O., Tuzhilin, A.A.: Steiner ratio. the state of the art. Math. Quest. Cybern. **11**, 27–48 (2002)
72. Du, D.-Z., Hwang, F.K.: A proof of the Gilbert–Pollak conjecture on the Steiner ratio. Algorithmica **7**, 121–135 (1992)
73. Innami, N., Kim, B.H., Mashiko, Y., Shiohama, K.: The Steiner ratio Gilbert–Pollak conjecture may still be open. Algorithmica **57**(4), 869–872 (2010)
74. Ivanov, A.O., Tuzhilin, A.A.: The Steiner ratio Gilbert–Pollak conjecture is still open. Clarification statement. Algorithmica **62**(1–2), 630–632 (2012)
75. Ivanov, A.O., Tuzhilin, A.A.: Branched coverings and steiner ratio. Int. Trans. Op. Res. **2**, 1–8 (2015)
76. Ivanov, A.O., Tuzhilin, A.A., Cieslik, D.: Steiner Ratio for Manifolds. Matem. Zametki **74**(3), 387–395 (2003). (Math. Notes, **74** (3), pp. 367–374 (2003))
77. Cieslik, D.: The Steiner Ratio of Metric Spaces (Report. http://www.math-inf.uni-greifswald.de/mathe/images/Boldt/cieslik-steiner-neu.pdf)
78. Ivanov, A.O., Tuzhilin, A.A.: Discrete Geometry and Algebraic Combinatorics. In: Barg, A., Musin, O. (eds.) Minimal Fillings of Finite Metric Spaces: The State of the Art. Contemporary Mathematics, vol. 625, pp. 9–35. AMS, Providence (2014)
79. Ivanov A.O., Nikolaeva N.K., Tuzhilin A.A.: The Gromov–Hausdorff Metric on the Space of Compact Metric Spaces is Strictly Intrinsic, arXiv e-prints, arXiv:1504.03830 (2015)
80. Ivanov A.O., Iliadis S., Tuzhilin A.A.: Realizations of Gromov–Hausdorff Distance, arXiv e-prints, arXiv:1603.08850, (2016)
81. Chowdhury S., Memoli F.: Constructing Geodesics on the Space of Compact Metric Spaces, arXiv e-prints, arXiv:1603.02385 (2016)
82. http://mathoverflow.net/questions/135184/for-which-metric-spaces-is-gromov-hausdorff-distance-actually-achieved?rq=1

Chapter 5
Generalized Pisot Numbers and Matrix Decomposition

Nikolai M. Dobrovol'skii, Nikolai N. Dobrovolsky, Irina N. Balaba, Irina Yu. Rebrova, Dmitrii K. Sobolev and Valentina N. Soboleva

Abstract We consider the linear fractional transformations of polynomials and the linear transformations of homogeneous binary forms and study their properties. A definition of generalized Pisot number is given. This definition differs from definition of Pisot numbers by the absence of a requirement to be integer. In the case of totally real algebraic fields reduced generalized numbers Pisot are reduced algebraic irrationalities. It is shown that for arbitrary real algebraic irrationality α of degree $n \geq 2$, a sequence of residual fractions α_m is a sequence of the reduced generalized numbers Pisot starting from some index $m_0 = m_0(\alpha)$. The asymptotic formula for conjugate numbers to residual fractions of generalized numbers Pisot is found. We study properties of the minimal polynomials of the residual fractions in the continued fraction expansion of the algebraic numbers. The recurrence formulas to find the minimum polynomials of the residual fractions using linear fractional transformations are given.

N.M. Dobrovol'skii (✉) · N.N. Dobrovolsky · I.N. Balaba · I.Y. Rebrova
Tula State Lev Tolstoy Pedagogical University, 125, Lenina pr., Tula 300026, Russia
e-mail: dobrovol@tspu.tula.ru

N.N. Dobrovolsky
e-mail: nikolai.dobrovolsky@gmail.com

I.N. Balaba
e-mail: ibalaba@mail.ru

I.Y. Rebrova
e-mail: i_rebrova@mail.ru

D.K. Sobolev · V.N. Soboleva
Moscow State Pedagogical University, 1/1 M. Pirogovskaya Str., Moscow 119991, Russian Federation
e-mail: co6ojib@gmail.com

V.N. Soboleva
e-mail: printsessa@gmail.com

V.A. Sadovnichiy and M.Z. Zgurovsky (eds.), *Advances in Dynamical Systems and Control*, Studies in Systems, Decision and Control 69,
DOI 10.1007/978-3-319-40673-2_5

5.1 Introduction

The history of the theory of continued fractions has more than three hundred years. To a significant degree, basic foundation of this theory was laid in the works of L. Euler and J. L. Lagrange. According to this theory, any real irrational number[1] α has unique infinite continued fraction expansion.

$$\alpha = \alpha_0 = q_0 + \cfrac{1}{q_1 + \cfrac{1}{\ddots + \cfrac{1}{q_k + \cfrac{1}{\ddots}}}} = q_0 + \cfrac{1}{q_1 + \cfrac{1}{\ddots + \cfrac{1}{q_k + \cfrac{1}{\alpha_{k+1}}}}}, \tag{5.1}$$

where incomplete quotients q_k and residual fractions α_k are uniquely determined by following conditions:

$$q_k = [\alpha_k], \quad k \geq 0; \quad \alpha_k = \frac{1}{\alpha_{k-1} - q_{k-1}}, \quad k \geq 1.$$

As usual, by P_k and Q_k, we denote numerator and denominator of kth-order convergent of continued fraction $\frac{P_k}{Q_k}$ for α. There are well-known recurrence equations

$$\begin{cases} P_k = q_k P_{k-1} + P_{k-2} \\ Q_k = q_k Q_{k-1} + Q_{k-2} \end{cases},$$

which continue to hold for $k \geq 0$, if we assume as usual that $P_{-1} = 1$, $P_{-2} = 0$ and $Q_{-1} = 0$, $Q_{-2} = 1$.

The analogous formulas hold for α and its residual fractions:

$$\begin{cases} \alpha = \frac{\alpha_{k+1}P_k + P_{k-1}}{\alpha_{k+1}Q_k + Q_{k-1}}, \\ \qquad\qquad\qquad\qquad k \geq -1. \\ \alpha_{k+1} = \frac{\alpha Q_{k-1} - P_{k-1}}{P_k - \alpha Q_k}, \end{cases} \tag{5.2}$$

Due to the well-known equality

$$P_k Q_{k-1} - P_{k-1} Q_k = (-1)^{k-1} \quad (k \geq -1),$$

one can easily prove by induction the relation between α and its residual fractions can be rewritten as

[1] Throughout this paper, by α we denote real irrational number.

$$\begin{cases} \alpha = \dfrac{P_k}{Q_k} + \dfrac{(-1)^k}{Q_k(\alpha_{k+1}Q_k + Q_{k-1})}, \\ \alpha_{k+1} = -\dfrac{Q_{k-1}}{Q_k} + \dfrac{(-1)^{k-1}}{Q_k(P_k - \alpha Q_k)} = -\dfrac{Q_{k-1}}{Q_k} + \dfrac{1}{Q_k|P_k - \alpha Q_k|}, \end{cases} \quad (k \geq 0).$$

Notice that the relation between convergents allows to describe the real irrationality of α in the form of alternating series

$$\alpha = q_0 + \sum_{\nu=1}^{\infty} \frac{(-1)^{\nu-1}}{Q_{\nu-1}Q_\nu}.$$

Very little is known about the continued fraction expansion of algebraic irrationalities of degree $n > 2$. It is one of the most difficult questions in the modern number theory. The various aspects of this theory can be seen in the papers [1–9, 15–17, 20].

The paper [22] describes the set of reduced algebraic irrationalities of n degree assigned that this set has the property of rational convexity. The paper [23] shows that generalized Pisot numbers have the analogous properties.

The minimal polynomials of residual fractions of continued fraction expansion of real algebraic irrationalities were investigated in [8]. The linear fractional transformations of the minimal polynomials of real algebraic irrationalities play a significant role in these researches. This is natural, since every number is equivalent to their residual fraction, and the equivalence is given by a unimodular linear fractional transformation.

The aim of this paper is the study the following questions: first, the properties of linear fractional transformations of polynomials in the wider context than in [8, 12], secondly, the properties of the minimal polynomial of residual fractions which arise during the work Lagrange algorithm for algebraic irrationalities of n-th degree. We are interested in both reduced algebraic irrationalities and generalized Pisot numbers in general.

Notice that the case of reduced algebraic irrationalities of degree n is closely connected with quadrature formulas with weights in K. K. Frolov's method (see [5–7, 13, 14]). The fact is that the reduced algebraic irrationalities generate totally real algebraic fields of degree n. If we consider a lattice similar to the lattice of integer conjugate algebraic numbers from a totally real algebraic field, then the points of polar lattice belonging to unit n-dimensional cube will form an algebraic net. These nets are used in Frolov's method solving the problem of constructing quadrature formulas that give the right order of decreasing norm of a linear fractional of error of approximate integration in the class E_s^α of periodic functions with rapidly decreasing Fourier coefficients.

Let α be a reduced cubic irrationality, that is, $\alpha^{(1)} = \alpha > 1$, and conjugate algebraic irrationalities satisfy the relationship $-1 < \alpha^{(3)} < \alpha^{(2)} < 0$. The concept of a reduced cubic irrationality is the natural extension of a reduced quadratic irrationality.

It is not difficult to see that a positive root α of the equation

$$x^3 - 4x^2 - 5x - 1 = 0$$

is a reduced cubic irrationality.

Indeed, for the polynomial $f(x) = x^3 - 4x^2 - 5x - 1$, we have:

$$f(-1) = f(0) = f(5) = -1, \quad f(6) = 41, \quad f\left(-\frac{1}{2}\right) = \frac{3}{8},$$

so $\alpha = \alpha^{(1)} > 5$, $-1 < \alpha^{(3)} < -\frac{1}{2}$, $-\frac{1}{2} < \alpha^{(2)} < 0$.

The matrix decompositions of algebraic irrationalities are considered in [10, 15, 17]. In particular, for cubic irrationality α satisfying the equation

$$f(t) = t^3 + at^2 + bt + c, \quad f(\alpha) = 0$$

the matrix decomposition is

$$\begin{pmatrix} \alpha \\ 1 \end{pmatrix} = \prod_{k=0}^{\infty} \left(\begin{pmatrix} t & -at^2 - 2bt - 3c \\ 1 & 3t^2 + 2at + b \end{pmatrix} \begin{pmatrix} 3k+2 & 0 \\ 0 & 3k+1 \end{pmatrix} \cdot \right.$$
$$\left. \cdot \begin{pmatrix} 3t^2 + 2at + b & -at^2 - 2bt - 3c \\ 1 & t \end{pmatrix} \begin{pmatrix} ab - 9c & 2b^2 - 6ac \\ 2a^2 - 6b & ab - 9c \end{pmatrix} \right) \tag{5.3}$$

It states that this matrix decomposition converges for t such that the difference $|t - \alpha|$ is small.

We will give a general definition of the convergence of matrix decomposition in Sect. 5.8.

Other aims of our paper are to get a new form of matrix decomposition of the reduced cubic irrationality α, to consider the realization of Lagrange algorithm of the expansion of this irrationality in the ordinary continued fraction, to construct conversion algorithm a matrix decomposition in ordinary continued fraction and to compare the results of these two algorithms.

Let us briefly consider the contents of this paper.

In Sect. 5.2, we introduce the necessary definitions and notations used throughout the paper.

In Sect. 5.3 is explicitly built some class of generalized Pisot numbers and reduced cubic irrationalities by specifying the minimum polynomial.

Section 5.4 is devoted to the consideration of linear fractional transformations of polynomials and linear transformations of homogeneous binary forms and detailed study of their properties.

In Sect. 5.5, we consider linear fractional transformations of the polynomials with integer coefficients and study their properties.

Section 5.6 describes the behavior of residual fractions and its conjugate numbers for continued fraction expansion of algebraic numbers. It is shown that for arbitrary real algebraic irrationality α of degree $n \geq 2$, a sequence of residual fractions α_m is a sequence of the reduced generalized Pisot numbers starting from some index $m_0 = m_0(\alpha)$. The asymptotic formula for conjugate numbers to residual fractions of generalized Pisot numbers is found. Using this formula, we get that conjugate numbers to residual fraction α_m focus around fraction $-\frac{Q_{m-2}}{Q_{m-1}}$ in the interval of radius $O\left(\frac{1}{Q_{m-1}^2}\right)$ in a case of totally real algebraic irrationality or in the interior circle of the same radius in the general case of real algebraic irrationality having complex conjugate numbers.

Section 5.7 is devoted to the study of the minimal polynomials of residual fractions. It is shown that a sequence of the minimal polynomials of the residual fractions is the sequence of the polynomials with equal discriminants.

In Sect. 5.8, we define a chain sequence of linear fractional transformations of the plane and give the interpretation of the received results in terms of these sequences.

Lagrange algorithm of infinite continued fraction expansion for arbitrary reduced irrationalities of degree n is considered in Sect. 5.9.

In Sect. 5.10, We suggest the modification of Lagrange algorithm which requires the calculation only two values of the minimal polynomial for determining the next incomplete quotient.

Section 5.11 describes the basic properties of the matrix decomposition.

Section 5.12 is devoted to the construction conversion algorithm a matrix decomposition in ordinary continued fraction.

In Sect. 5.13, we compare the results of the two algorithms for reduced cubic irrationality α.

In conclusion, perspective directions of research are formulated.

5.2 Notation and Preliminaries

We begin with the definition of a reduced algebraic irrationality of nth-degree and generalized Pisot number of nth degree. Here, we follow [9, 11, 22].

Definition 5.1 Let

$$f(x) = \sum_{k=0}^{n} a_k x^k \in \mathbb{Z}[x], \quad a_n > 0$$

be such irreducible polynomial with integer coefficients[2] that all its roots $\alpha^{(k)}$ $(k = 1, 2, \ldots, n)$ are different real numbers satisfying the following condition:

$$-1 < \alpha^{(n)} < \ldots < \alpha^{(2)} < 0, \quad \alpha^{(1)} > 1.$$

[2]By irreducible polynomial $f(x)$ with integer coefficients, we understand such polynomial that if $f(x) = g(x)h(x)$, where $\deg(g(x)) \leq \deg(h(x))$, and then $g(x) = \pm 1$, $h(x) = \mp f(x)$. In particular, irreducibility of a polynomial means $(a_0, \ldots, a_n) = 1$.

Then, an algebraic number $\alpha = \alpha^{(1)}$ is called reduced algebraic irrationality of nth degree.

Definition 5.2 Let

$$f(x) = \sum_{k=0}^{n} a_k x^k \in \mathbb{Z}[x], \quad a_n > 0$$

be such irreducible polynomial with integer coefficients that all its roots $\alpha^{(k)}$ ($k = 1, 2, \ldots, n$) satisfy the following condition:

$$|\alpha^{(j)}| < 1, \ (2 \leq j \leq n), \quad \alpha^{(1)} > 1,$$

Then, an algebraic number $\alpha = \alpha^{(1)}$ is called a generalized Pisot number of nth degree.

It is not hard to see that if $\alpha = \alpha^{(1)}$ is a reduced algebraic irrationality, then all n algebraic conjugate fields $\mathbb{Q}[\alpha^{(1)}], \ldots, \mathbb{Q}[\alpha^{(n)}]$ are real. It is clear that α is a generalized Pisot number, but generalized Pisot number need not to be a reduced algebraic irrationality. Indeed, $\beta = \beta^{(1)} = \left(\alpha^{(1)}\right)^2$ is a generalized Pisot number since $0 < \beta^{(j)} = \left(\alpha^{(j)}\right)^2 < 1$ $(2 \leq j \leq n)$, but it is not a reduced algebraic irrationality.

The definition of a generalized Pisot number differ from Pisot number by the absence of a requirement to be integer.

Note that for minimal polynomial $f(x)$ defining a reduced algebraic irrationality α of nth degree, we always have

$$a_0 < 0. \tag{5.4}$$

Indeed, the polynomial $f(x)$ has only one root α belonging to interval $[0; \infty)$; hence, for $x > \alpha$, we have $f(x) > 0$, so $f(0) < 0$. Besides, the following inequalities hold

$$a_n + a_{n-1} + \ldots + a_1 + a_0 = f(1) < 0, \tag{5.5}$$

$$a_n - a_{n-1} + \ldots + (-1)^{n-1} a_1 + (-1)^n a_0 = (-1)^n f(-1) > 0. \tag{5.6}$$

For generalized Pisot number, the inequalities (5.5) and (5.6) hold too. Indeed, the inequality (5.5) follows from the fact that minimal polynomial $f(x)$ with a leading coefficient $a_n > 0$ has exactly one root belonging to the interval $[1; +\infty)$. A lack of the roots of $f(x)$ on $(-\infty; -1]$ implies the inequality (5.6). In addition, for any generalized Pisot number α, there exists a natural number $q_0 = [\alpha]$ such that $f_0(q_0) < 0, f_0(q_0 + 1) > 0$.

Lemma 5.1 *For an arbitrary real algebraic irrationality α of degree n, its residual fraction α_1 is real algebraic irrationality of degree n satisfying the irreducible polynomial*

$$f_1(x) = \sum_{k=0}^{n} a_{k,1} x^k \in \mathbb{Z}[x], \quad a_{n,1} > 0,$$

where

$$a_{k,1} = \frac{b_k}{d_0},\ d_0 = (b_0, \ldots, b_n),\ b_k = -\sum_{m=n-k}^{n} a_m C_m^{m+k-n} q_0^{m+k-n},\ (0 \le k \le n),$$

and the following equality holds

$$f_1(x) = \frac{-f_0(q_0)}{d_0} \prod_{j=1}^{n} \left(x - \frac{1}{\alpha^{(j)} - q_0} \right).$$

If α is a reduced algebraic irrationality, then α_1 is a reduced algebraic irrationality too.

Proof Consider the polynomial

$$g(x) = -f\left(q_0 + \frac{1}{x}\right) \cdot x^n = \sum_{k=0}^{n} b_k x^k.$$

Since $\alpha = q_0 + \frac{1}{\alpha_1}$, it follows that $g(\alpha_1) = 0$.
We have

$$f\left(q_0 + \frac{1}{x}\right) \cdot x^n = \sum_{k=0}^{n} a_k x^{n-k} (q_0 x + 1)^k = \sum_{k=0}^{n} a_k x^{n-k} \sum_{m=0}^{k} C_k^m q_0^m x^m =$$

$$= \sum_{k=0}^{n} a_k \sum_{m=n-k}^{n} C_k^{m+k-n} q_0^{m+k-n} x^m = \sum_{k=0}^{n} x^k \sum_{m=n-k}^{n} a_m C_m^{m+k-n} q_0^{m+k-n},$$

so

$$b_k = -\sum_{m=n-k}^{n} a_m C_m^{m+k-n} q_0^{m+k-n}, \quad (0 \le k \le n)$$

and $b_n = -f(q_0)$. But $q_0 < \alpha, f(\alpha) = 0, a_n > 0$, and α is the unique positive root of the polynomial $f(x)$, so $f(q_0) < 0$, and therefore, $b_n > 0$.

Hence, by dividing the polynomial $g(x)$ by the greatest common divisor of its coefficients, we obtain an irreducible polynomial $f_1(x)$.

Further notice that the roots $\alpha^{(k)}$ $(k = 1, 2, \ldots, n)$ of the polynomial $f(x)$ correspond to the roots $\beta^{(k)}$ $(k = 1, 2, \ldots, n)$ of the polynomial $g(x)$ which are connected by the equalities

$$\alpha^{(k)} = q_0 + \frac{1}{\beta^{(k)}}, \quad \beta^{(k)} = \frac{1}{\alpha^{(k)} - q_0} \quad (k = 1, \ldots, n).$$

It follows that

$$-1 < \beta^{(k)} < 0 \quad (2 \le k \le n), \quad \beta^{(1)} > 1$$

and therefore, $\alpha_1 = \beta^{(1)}$ is a reduced algebraic irrationality of degree n. The lemma is proved. □

Using Lemma 5.1, we prove the following theorem by induction.

Theorem 5.1 *For an arbitrary real algebraic irrationality α of degree n, all its residual fractions α_m are also real algebraic irrationalities of degree n satisfying the irreducible polynomials*

$$f_m(x) = \sum_{k=0}^{n} a_{k,m} x^k \in \mathbb{Z}[x], \quad a_{n,m} > 0,$$

where

$$a_{k,m} = \frac{b_{k,m}}{d_m}, \quad d_m = (b_{0,m}, \ldots, b_{n,m}),$$

$$b_{k,m} = -\sum_{l=n-k}^{n} a_{l,m-1} C_l^{l+k-n} q_{m-1}^{l+k-n}, \quad (0 \le k \le n).$$

The polynomials $f_m(x)$ have the roots

$$\alpha_m^{(j)} = \frac{\alpha^{(j)} Q_{m-2} - P_{m-2}}{P_{m-1} - \alpha^{(j)} Q_{m-1}} \quad (1 \le j \le n) \tag{5.7}$$

and the following equalities hold

$$f_m(x) = \frac{-f_{m-1}(q_{m-1})}{d_{m-1}} \prod_{j=1}^{n} \left(x - \alpha_m^{(j)}\right).$$

If α is a reduced algebraic irrationality, then all its residual fractions α_m are the reduced algebraic irrationalities too.

It is easily shown that if $\alpha = \alpha_0 = \alpha^{(1)}$ is a generalized Pisot number, then a residual fraction α_1, where

$$\alpha_1 = \frac{1}{\alpha_0 - q_0}, \quad q_0 = [\alpha_0],$$

need not a generalized Pisot number.

Indeed, if $q_0 = 1$ and there is ν such that $|\alpha^{(\nu)} - q_0| < 1$, then for a conjugate number $\alpha_1^{(\nu)} = \frac{1}{\alpha^{(\nu)} - q_0}$ for residual fraction α_1, the inequality $|\alpha_1^{(\nu)}| < 1$ is false.

Let give the following definition.

Definition 5.3 A generalized Pisot number $\alpha = \alpha^{(1)}$ is called a reduced generalized Pisot number if supplementary conditions hold: For natural $q_0 = \left[\alpha^{(1)}\right]$, we have the following inequality:

$$|\alpha^{(j)} - q_0| > 1, \ (2 \le j \le n).$$

Lemma 5.2 *For an arbitrary reduced generalized Pisot number α of degree n, its residual fraction α_1 is also reduced generalized Pisot number of degree n, satisfying the irreducible polynomial*

$$f_1(x) = \sum_{k=0}^{n} a_{k,1} x^k \in \mathbb{Z}[x], \quad a_{n,1} > 0,$$

where

$$a_{k,1} = \frac{b_k}{d_0}, \ d_0 = (b_0, \dots, b_n), \ b_k = -\sum_{m=n-k}^{n} a_m C_m^{m+k-n} q_0^{m+k-n}, \ (0 \le k \le n)$$

and the following equality holds

$$f_1(x) = \frac{-f_0(q_0)}{d_0} \prod_{j=1}^{n} \left(x - \frac{1}{\alpha^{(j)} - q_0} \right).$$

Proof Indeed, the conjugate numbers $\alpha_1^{(\nu)} = \frac{1}{\alpha^{(\nu)} - q_0}$ to a residual fraction $\alpha_1 = \frac{1}{\alpha_0 - q_0}$ by Definition 5.3 satisfy the conditions $|\alpha_1^{(\nu)}| < 1$ $(2 \le \nu \le n)$, so a residual fraction α_1 is generalized Pisot number.

Now, we need to prove that α_1 is a reduced generalized Pisot number. Consider three possible cases.

I. Let $q_0 > 1$, then $\alpha^{(\nu)} - q_0 = -x_\nu + y_\nu i, x_\nu > q_0 - 1 \ge 1$,

$$\alpha_1^{(\nu)} = \frac{1}{\alpha^{(\nu)} - q_0} = \frac{-x_\nu - y_\nu i}{x_\nu^2 + y_\nu^2}.$$

So $\alpha_1^{(\nu)}$ lies in the left half-plane bounded by imaginary line. It follows that for $q_1 = \lfloor \alpha_1 \rfloor$, we have

$$\left| \alpha_1^{(\nu)} - q_1 \right| > 1 \quad (\nu = 2, \dots, n).$$

Therefore, in this case, α_1 is a reduced generalized Pisot number.

II. Let $q_0 = 1$ and $\alpha^{(\nu)} = -x_\nu + y_\nu i, x_\nu > 0, x_\nu^2 + y_\nu^2 < 1$, then

$$\alpha_1^{(\nu)} = \frac{1}{\alpha^{(\nu)} - q_0} = \frac{-x_\nu - 1 - y_\nu i}{(x_\nu + 1)^2 + y_\nu^2}, \quad \left|\alpha_1^{(\nu)}\right| < 1$$

$\alpha_1^{(\nu)}$ lies in the left half-plane bounded by imaginary line. So for $q_1 = [\alpha_1]$, we have

$$\left|\alpha_1^{(\nu)} - q_1\right| > 1 \quad (\nu = 2, \dots, n).$$

Hence, α_1 is a reduced generalized Pisot number.

III. Let $q_0 = 1$ and there exists ν such that $\alpha^{(\nu)} = x_\nu + y_\nu i$, $x_\nu > 0$, $x_\nu^2 + y_\nu^2 < 1$, then $(1 - x_\nu)^2 + y_\nu^2 > 1$. Thus, we have

$$\alpha_1^{(\nu)} = \frac{1}{\alpha^{(\nu)} - q_0} = \frac{x_\nu - 1 - y_\nu i}{(x_\nu - 1)^2 + y_\nu^2}, \quad \left|\alpha_1^{(\nu)}\right| < 1$$

and $\alpha_1^{(\nu)}$ lies in the left half-plane bounded by imaginary line. It follows that for $q_1 = [\alpha_1]$, we have

$$\left|\alpha_1^{(\nu)} - q_1\right| > 1 \quad (\nu = 2, \dots, n).$$

Therefore, in this case, α_1 is a reduced generalized Pisot number too.

Consider a polynomial $g(x) = -x^n f_0\left(q_0 + \frac{1}{x}\right)$. Since

$$g(x) = -x^n a_n \prod_{\nu=1}^{n}\left(q_0 + \frac{1}{x} - \alpha^{(\nu)}\right) = -a_n \prod_{\nu=1}^{n}(q_0 - \alpha^{(\nu)}) \prod_{\nu=1}^{n}\left(x - \frac{1}{\alpha^{(\nu)} - q_0}\right) =$$
$$= -f_0(q_0) \prod_{\nu=1}^{n}\left(x - \frac{1}{\alpha^{(\nu)} - q_0}\right),$$

the roots of $g(x)$ are a residual fraction α_1 and its conjugate algebraic numbers $\alpha_1^{(\nu)}$ $(2 \le \nu \le n)$.

By Taylor formula

$$f_0\left(q_0 + \frac{1}{x}\right) = f_0(q_0) + \sum_{\nu=1}^{n} \frac{f_0^{(\nu)}(q_0)}{\nu!} \frac{1}{x^\nu},$$

so that

$$g(x) = -f_0(q_0)x^n - \sum_{\nu=1}^{n} \frac{f_0^{(\nu)}(q_0)}{\nu!} x^{n-\nu} \in \mathbb{Z}[x].$$

This completes the proof. □

Remark 5.1 As will be shown below $d_0 = 1$, so

$$f_1(x) = -f_0(q_0)x^n - \sum_{\nu=1}^{n} \frac{f_0^{(\nu)}(q_0)}{\nu!} x^{n-\nu} \in \mathbb{Z}[x].$$

5.3 Some Class of Generalized Pisot Numbers and Reduced Cubic Irrationalities

Let describe some class of the reduced cubic irrationalities.

Consider for natural $p \geq 4$ the polynomials

$$f(p,x) = x(x+1)(x-p) - 1 = x^3 - (p-1)x^2 - px - 1.$$

A positive root $\alpha(p)$ of the equation $f(p,x) = 0$ is a reduced cubic irrationality. Indeed, for the polynomial $f(p,x) = x^3 - (p-1)x^2 - px - 1$, we have:

$$f(p,-1) = f(p,0) = f(p,p) = -1, \quad f(p+1) = p^2 + 3p + 1 > 0,$$
$$f\left(p, -\frac{1}{2}\right) = \frac{2p+1}{8} - 1 > 0,$$

so $p+1 > \alpha(p) = \alpha^{(1)} > p$, $-1 < \alpha^{(3)} < -\frac{1}{2}$, $-\frac{1}{2} < \alpha^{(2)} < 0$. Since the polynomial $f(p,x)$ has no rational roots, it is irreducible.

5.4 Linear Fractional Transformation of Polynomials and Linear Transformation of Forms

As usual, let $\mathbb{N}$ be the set of natural numbers, $\mathbb{Z}$ the ring of integers, $\mathbb{Q}$ the field of rational numbers, $\mathbb{R}$ the field of real numbers, and $\mathbb{C}$ the field of complex numbers.

Denote by $\mathbb{Z}[x]$, $\mathbb{Q}[x]$, $\mathbb{R}[x]$, $\mathbb{C}[x]$ the corresponding rings of polynomials and by $\mathbb{PZ}[X,Y]$, $\mathbb{PQ}[X,Y]$, $\mathbb{PR}[X,Y]$, $\mathbb{PC}[X,Y]$ the corresponding multiplication groups of homogeneous forms.

It is clear that

$$\mathbb{Z}[x] \subset \mathbb{Q}[x] \subset \mathbb{R}[x] \subset \mathbb{C}[x],$$
$$\mathbb{PZ}[X,Y] \subset \mathbb{PQ}[X,Y] \subset \mathbb{PR}[X,Y] \subset \mathbb{PC}[X,Y].$$

Let $\mathbb{K}$ be one of the sets $\mathbb{Z}$, $\mathbb{Q}$, $\mathbb{R}$, or $\mathbb{C}$. If

$$\vec{a} = (a_0, a_1, \ldots, a_n) \in \mathbb{K}^{n+1},$$

then

$$f_{\vec{a}}(x) = \sum_{\nu=0}^{n} a_\nu x^\nu \in \mathbb{K}[x], \quad F_{\vec{a}}(X, Y) = \sum_{\nu=0}^{n} a_\nu X^\nu Y^{n-\nu} \in \mathbb{PK}[X, Y].$$

It is clear that the following equality holds

$$F_{\vec{a}}(X, Y) = Y^n f_{\vec{a}}\left(\frac{X}{Y}\right). \tag{5.8}$$

The formula (5.8) specifies bijection φ between $\mathbb{K}$-module $\mathbb{K}_n[x]$ of all polynomials of degree less or equal to n and $\mathbb{K}$-module $\mathbb{PK}_n[X, Y]$ of all homogeneous form of n order.[3]

Denote by $\mathcal{M}_2(\mathbb{K})$ the ring of quadratic matrixes of second order whose elements belong to $\mathbb{K}$. Let $\mathcal{M}_2^*(\mathbb{K})$ be a multiplication group of $\mathcal{M}_2(\mathbb{K})$, i.e., a set of all nondegenerate matrixes and $\mathcal{U}_2(\mathbb{K})$ be a set of all unimodular matrixes. Thus, we have

$$M = \begin{pmatrix} A & B \\ C & D \end{pmatrix} \in \mathcal{M}_2(\mathbb{K}), \quad \text{if} \quad A, B, C, D \in \mathbb{K};$$
$$M \in \mathcal{M}_2^*(\mathbb{K}), \quad \text{if} \quad \det M = AD - BC \neq 0;$$
$$M \in \mathcal{U}_2(\mathbb{K}), \quad \text{if} \quad \det M = \pm 1.$$

Definition 5.4 For nondegenerate matrix $M = \begin{pmatrix} A & B \\ C & D \end{pmatrix} \in \mathcal{M}_2^*(\mathbb{K})$, linear fractional transformation M of a polynomial $f_{\vec{a}}(x) \in \mathbb{K}[x]$ is called the transformation given by the formula

$$M(f_{\vec{a}}(x)) = (Cx + D)^n f_{\vec{a}}\left(\frac{Ax + B}{Cx + D}\right).$$

Definition 5.5 For nondegenerate matrix $M = \begin{pmatrix} A & B \\ C & D \end{pmatrix} \in \mathcal{M}_2^*(\mathbb{K})$, linear transformation M of a form $F_{\vec{a}}(X, Y) \in \mathbb{PK}[X, Y]$ is called the transformation given by formula

$$M(F_{\vec{a}}(X, Y)) = F_{\vec{a}}(AX + BY, CX + DY).$$

Obviously, that unity matrix E specifies identity transformations:

$$E(f_{\vec{a}}(x)) = f_{\vec{a}}(x), \quad E(F_{\vec{a}}(x)) = F_{\vec{a}}(X, Y). \tag{5.9}$$

For any matrix $M = \begin{pmatrix} A & B \\ C & D \end{pmatrix} \in \mathcal{M}_2(\mathbb{K})$, let define matrix $M^{(n+1)} \in \mathcal{M}_{n+1}(\mathbb{K})$ by equality

[3]Here we suppose that only null form belongs to all $\mathbb{PK}_n[X, Y]$.

$$M^{(n+1)} = \begin{pmatrix} D^n & C_n^1 CD^{n-1} & \dots & C_n^{n-1}C^{n-1}D & C^n \\ BD^{n-1} & m(1,1) & \dots & m(1,n-1) & AC^{n-1} \\ \vdots & \vdots & \ddots & \vdots & \vdots \\ B^{n-1}D & m(n-1,1) & \dots & m(n-1,n-1) & A^{n-1}C \\ B^n & C_n^1 AB^{n-1} & \dots & C_n^{n-1}A^{n-1}B & A^n \end{pmatrix} = $$
$$= (\, m(\nu, j)\,)_{\nu=0,\dots,n}^{j=0,\dots,n}, \qquad (5.10)$$

where

$$m(\nu, j) = \sum_{\lambda=\max(0,j-\nu)}^{\min(n-\nu,j)} C_\nu^{j-\lambda} A^{j-\lambda} B^{\nu-j+\lambda} C_{n-\nu}^{\lambda} C^\lambda D^{n-\nu-\lambda}.$$

Lemma 5.3 *For any nondegenerate matrix $M \in \mathscr{M}_2^* (\mathbb{K})$, the following equality holds*

$$M(f_{\vec{a}}(x)) = f_{\vec{b}}(x), \qquad (5.11)$$

where

$$\vec{b} = \vec{a} \cdot M^{(n+1)}. \qquad (5.12)$$

Proof Indeed,

$$M(f_{\vec{a}}(x)) = (Cx + D)^n f_{\vec{a}} \left(\frac{Ax + B}{Cx + D} \right) = \sum_{\nu=0}^{n} a_\nu (Ax + B)^\nu (Cx + D)^{n-\nu} =$$

$$= \sum_{\nu=0}^{n} a_\nu \sum_{\mu=0}^{\nu} C_\nu^\mu A^\mu B^{\nu-\mu} x^\mu \sum_{\lambda=0}^{n-\nu} C_{n-\nu}^\lambda C^\lambda D^{n-\nu-\lambda} x^\lambda = \sum_{\nu=0}^{n} a_\nu \sum_{j=0}^{n} x^j m(\nu, j),$$

where

$$m(\nu, j) = \sum_{\lambda=\max(0,j-\nu)}^{\min(n-\nu,j)} C_\nu^{j-\lambda} A^{j-\lambda} B^{\nu-j+\lambda} C_{n-\nu}^{\lambda} C^\lambda D^{n-\nu-\lambda}.$$

So

$$M(f_{\vec{a}}(x)) = \sum_{j=0}^{n} b_j x^j$$

and

$$b_j = \sum_{\nu=0}^{n} a_\nu \sum_{\lambda=\max(0,j-\nu)}^{\min(n-\nu,j)} C_\nu^{j-\lambda} C_{n-\nu}^{\lambda} A^{j-\lambda} B^{\nu-j+\lambda} C^\lambda D^{n-\nu-\lambda} = \sum_{\nu=0}^{n} a_\nu m(\nu, j).$$

Hence, $\vec{b} = \vec{a} \cdot M^{(n+1)}$ as we wanted to show. □

By Lemma 5.3, it follows that any linear fractional transformation with matrix $M \in \mathcal{M}_2^*(\mathbb{K})$ maps $\mathbb{K}_n[x]$ into itself.

Lemma 5.4 *For any nondegenerate matrix $M \in \mathcal{M}_2^*(\mathbb{K})$, the following equality holds*

$$M(F_{\vec{a}}(X, Y)) = F_{\vec{b}}(X, Y), \tag{5.13}$$

where

$$\vec{b} = \vec{a} \cdot M^{(n+1)}. \tag{5.14}$$

Proof Indeed,

$$M(F_{\vec{a}}(X, Y)) = F_{\vec{a}}(AX + BY, CX + DY) = \sum_{\nu=0}^{n} a_\nu (AX + BY)^\nu (CX + DY)^{n-\nu} =$$

$$= \sum_{\nu=0}^{n} a_\nu \sum_{\mu=0}^{\nu} C_\nu^\mu A^\mu B^{\nu-\mu} X^\mu Y^{\nu-\mu} \sum_{\lambda=0}^{n-\nu} C_{n-\nu}^{\lambda} C^\lambda D^{n-\nu-\lambda} X^\lambda Y^{n-\nu-\lambda} =$$

$$= \sum_{\nu=0}^{n} a_\nu \sum_{j=0}^{n} X^j Y^{n-j} m(\nu, j),$$

where

$$m(\nu, j) = \sum_{\lambda=\max(0,j-\nu)}^{\min(n-\nu,j)} C_\nu^{j-\lambda} A^{j-\lambda} B^{\nu-j+\lambda} C_{n-\nu}^{\lambda} C^\lambda D^{n-\nu-\lambda}.$$

So

$$M(F_{\vec{a}}(X, Y)) = \sum_{j=0}^{n} b_j X^j Y^{n-j}$$

and

$$b_j = \sum_{\nu=0}^{n} a_\nu \sum_{\lambda=\max(0,j-\nu)}^{\min(n-\nu,j)} C_\nu^{j-\lambda} C_{n-\nu}^{\lambda} A^{j-\lambda} B^{\nu-j+\lambda} C^\lambda D^{n-\nu-\lambda} = \sum_{\nu=0}^{n} a_\nu m(\nu, j).$$

Hence, $\vec{b} = \vec{a} \cdot M^{(n+1)}$ as we wanted to show. □

By Lemma 5.4, it follows that any linear fractional transformation with matrix $M \in \mathscr{M}_2^*(\mathbb{K})$ maps $\mathbb{PK}_n[X, Y]$ into itself.

Using Lemmas 5.3 and 5.4, we obtain the following theorem.

Theorem 5.2 *For any nondegenerate matrix $M \in \mathscr{M}_2^*(\mathbb{K})$, bijection φ is defined by the formula (5.8) preserved, that is, if*

$$M(f_{\vec{a}}(x)) = f_{\vec{b}}(x), \tag{5.15}$$

then

$$M(F_{\vec{a}}(X, Y)) = F_{\vec{b}}(X, Y). \tag{5.16}$$

Proof Indeed, by Lemmas 5.3 and 5.4, the vector $\vec{b}$ in the formulas (5.15) and (5.16) is the same. This completes the proof. □

Denote by $\mathbb{K}_n^*[x]$ a set of all polynomials of degree n with $a_0 \neq 0$ and denote by $\mathbb{PK}_n^*[X, Y]$ a set of all nondegenerate homogeneous forms of order n, that is, such forms $F_{\vec{a}}(X, Y)$ that $a_n \neq 0$ and $a_0 \neq 0$.

By fundamental theorem of algebra, any polynomial $f_{\vec{a}}(x) \in \mathbb{K}_n^*[x]$ has n roots $\alpha^{(1)}, \ldots, \alpha^{(n)} \in \mathbb{C}$ and we have the following decomposition in $\mathbb{C}_n^*[x]$:

$$f_{\vec{a}}(x) = a_n \left(x - \alpha^{(1)}\right) \ldots \left(x - \alpha^{(n)}\right).$$

Turning to the forms, we get two decompositions

$$\begin{aligned} F_{\vec{a}}(X, Y) &= a_n \left(X - \alpha^{(1)} Y\right) \ldots \left(X - \alpha^{(n)} Y\right) = \\ &= a_0 \left(\beta^{(1)} X + Y\right) \ldots \left(\beta^{(n)} X + Y\right); \\ \beta^{(\nu)} &= \frac{-1}{\alpha^{(\nu)}} \quad \nu = 1, \ldots, n. \end{aligned}$$

Thus, a binary form $F_{\vec{a}}(X, Y)$ has n root lines

$$X - \alpha^{(\nu)} Y = 0 \quad (\nu = 1, \ldots, n),$$

on which the form becomes zero.

Lemma 5.5 *For any polynomials $f_{\vec{a}}(x)$, $g_{\vec{b}}(x)$ and any linear fractional transformation with matrix $M \in \mathscr{M}_2^*(\mathbb{K})$, we have the equation*

$$M(f_{\vec{a}}(x) g_{\vec{b}}(x)) = M(f_{\vec{a}}(x)) M(g_{\vec{b}}(x)).$$

Proof Indeed, if $\deg(f(x)) = k$, $\deg(g(x)) = l$ and $n = k + l$, then

$$M(f_{\vec{a}}(x)g_{\vec{b}}(x)) = (Cx+D)^n f_{\vec{a}}\left(\frac{Ax+B}{Cx+D}\right) g_{\vec{b}}\left(\frac{Ax+B}{Cx+D}\right) =$$
$$= \left((Cx+D)^k f_{\vec{a}}\left(\frac{Ax+B}{Cx+D}\right)\right)\left((Cx+D)^l g_{\vec{b}}\left(\frac{Ax+B}{Cx+D}\right)\right) = M(f_{\vec{a}}(x))M(g_{\vec{b}}(x)).$$

This completes the proof. □

Lemma 5.6 *For any forms $F_{\vec{a}}(X,Y)$, $G_{\vec{b}}(X,Y)$ and any linear transformation with matrix $M \in \mathscr{M}_2^*(\mathbb{K})$, we have the equation*

$$M(F_{\vec{a}}(X,Y)G_{\vec{b}}(X,Y)) = M(F_{\vec{a}}(X,Y))M(G_{\vec{b}}(X,Y)).$$

Proof Indeed,

$$M(F_{\vec{a}}(X,Y)G_{\vec{b}}(X,Y)) = F_{\vec{a}}\,(Ax+B, Cx+D)\, G_{\vec{b}}\,(Ax+B, Cx+D) =$$
$$= M(F_{\vec{a}}(X,Y))M(G_{\vec{b}}(X,Y)).$$

This completes the proof. □

Lemma 5.7 *For any linear fractional transformation with matrix $M = \begin{pmatrix} A & B \\ C & D \end{pmatrix} \in \mathscr{M}_2^*(\mathbb{K})$ and any polynomial $f(x)$, having the roots $\alpha^{(\nu)}$ ($A \neq C\alpha^{(\nu)}$ for all $\nu = 1, \dots, n$), the polynomial*

$$M(f(x)) = \sum_{\nu=0}^{n} b_\nu x^\nu$$

has the following roots

$$\beta^{(\nu)} = \frac{D\alpha^{(\nu)} - B}{A - C\alpha^{(\nu)}} \quad (1 \le \nu \le n), \quad b_n = \begin{cases} C^n f\left(\frac{A}{C}\right), & \text{if } C \neq 0, \\ a_n A^n, & \text{if } C = 0, \end{cases}$$
$$b_0 = \begin{cases} D^n f\left(\frac{B}{D}\right), & \text{if } D \neq 0, \\ a_n B^n, & \text{if } D = 0. \end{cases}$$

Proof Indeed, if

$$f(x) = \sum_{\nu=0}^{n} a_\nu x^\nu = a_n \prod_{\nu=1}^{n} \left(x - \alpha^{(\nu)}\right),$$

then for $C \neq 0$

$$M(f(x)) = (Cx+D)^n a_n \prod_{\nu=1}^{n} \left(\frac{Ax+B}{Cx+D} - \alpha^{(\nu)}\right) =$$
$$= a_n \prod_{\nu=1}^{n} \left(Ax + B - C\alpha^{(\nu)}x - D\alpha^{(\nu)}\right) =$$

$$= a_n \prod_{\nu=1}^{n} \left(A - C\alpha^{(\nu)}\right) \prod_{\nu=1}^{n} \left(x - \frac{D\alpha^{(\nu)} - B}{A - C\alpha^{(\nu)}}\right) =$$

$$= a_n C^n \prod_{\nu=1}^{n} \left(\frac{A}{C} - \alpha^{(\nu)}\right) \prod_{\nu=1}^{n} (x - \beta^{(\nu)}) = C^n f\left(\frac{A}{C}\right) \prod_{\nu=1}^{n} (x - \beta^{(\nu)}).$$

For $C = 0$, we have:

$$M(f(x)) = D^n a_n \prod_{\nu=1}^{n} \left(\frac{Ax + B}{D} - \alpha^{(\nu)}\right) =$$

$$= a_n \prod_{\nu=1}^{n} \left(Ax + B - D\alpha^{(\nu)}\right) = a_n A^n \prod_{\nu=1}^{n} \left(x - \frac{D\alpha^{(\nu)} - B}{A - C\alpha^{(\nu)}}\right) =$$

$$= a_n A^n \prod_{\nu=1}^{n} (x - \beta^{(\nu)})$$

and first, the statement of lemma is proved.

For $D \neq 0$, we have $b_0 = M(f(0)) = D^n f\left(\frac{B}{D}\right)$.

If $D = 0$, then $b_0 = M(f(0)) = a_n B^n$.

This completes the proof. □

Thus, the roots of a polynomial $f(x)$ are conversed in the roots of a polynomial $M(f(x))$ by the linear fractional transformation of the complex plane

$$M^*(z) = \frac{Dz - B}{-Cz + A}$$

with matrix

$$M^* = \begin{pmatrix} D & -B \\ -C & A \end{pmatrix}.$$

Lemma 5.8 *For any linear transformation with matrix* $M = \begin{pmatrix} A & B \\ C & D \end{pmatrix} \in \mathscr{M}_2^*(\mathbb{K})$ *and any form* $F(X, Y)$ *with the root lines* $X - \alpha^{(\nu)} Y = 0$ $(\nu = 1, \dots, n)$*, the form*

$$M(F(X, Y)) = \sum_{\nu=0}^{n} b_\nu X^\nu Y^{n-\nu}$$

has the root lines

$$(A - C\alpha^{(\nu)})X - (D\alpha^{(\nu)} - B)Y = 0 \quad (1 \leq \nu \leq n), \quad b_n = F(A, C),$$
$$b_0 = F(B, D).$$

Proof Indeed, if

$$F(X,Y)=\sum_{\nu=0}^{n}a_\nu X^\nu Y^{n-\nu}=a_n\prod_{\nu=1}^{n}(X-\alpha^{(\nu)}Y),$$

Then,

$$M(F(X,Y))=a_n\prod_{\nu=1}^{n}\left(AX+BY-\alpha^{(\nu)}(CX+DY)\right)=$$
$$=a_n\prod_{\nu=1}^{n}\left((A-C\alpha^{(\nu)})X-(D\alpha^{(\nu)}-B)Y\right).$$

It follows that the root lines have the forms which are listed in the lemma.

Since

$$b_n=M(F(1,0))=F(A,C)$$
$$b_0=M(F(0,1))=F(B,D)$$

the lemma is completely proved. □

From the lemma proved above, it immediately follows that the root lines of a form $F(X,Y)$ are conversed in the root lines of a form $M(F(X,Y))$ by the linear transformation of the two-dimensional complex space.

$$M^*(X,Y)=(DX-BY,-CX+AY)$$

with matrix

$$M^*=\begin{pmatrix}D & -B\\ -C & A\end{pmatrix}.$$

Lemma 5.9 *For composition* $\circ$ *of linear fractional transformations, the following equality holds*

$$M_1\circ M=M\cdot M_1,$$

where $\cdot$ *is matrix multiplication, and at the same time, the roots of polynomials are transformed by the rule*

$$(M_1\circ M)^*=M_1^*\cdot M^*.$$

Proof Indeed, let

$$M=\begin{pmatrix}A & B\\ C & D\end{pmatrix},\quad M_1=\begin{pmatrix}A_1 & B_1\\ C_1 & D_1\end{pmatrix}$$

and $g(x) = M(f(x))$, then

$$M_1 \circ M(f(x)) = (C_1x + D_1)^n g\left(\frac{A_1x + B_1}{C_1x + D_1}\right) =$$

$$= (C_1x + D_1)^n \left(C\frac{A_1x + B_1}{C_1x + D_1} + D\right)^n f\left(\frac{A\frac{A_1x+B_1}{C_1x+D_1} + B}{C\frac{A_1x+B_1}{C_1x+D_1} + D}\right) =$$

$$= ((CA_1 + DC_1)x + (CB_1 + DD_1))^n \cdot$$

$$\cdot f\left(\frac{(AA_1 + BC_1)x + (AB_1 + BD_1)}{(CA_1 + DC_1)x + (CB_1 + DD_1)}\right) = M_2(f(x)),$$

where

$$M_2 = \begin{pmatrix} AA_1 + BC_1 & AB_1 + BD_1 \\ CA_1 + DC_1 & CB_1 + DD_1 \end{pmatrix} = M \cdot M_1$$

First, statement of the lemma is found.

Let $\alpha^{(1)}, \dots, \alpha^{(n)}$ be the roots of a polynomial $f(x)$, $\beta^{(1)}, \dots, \beta^{(n)}$ the roots of $M(f(x))$, and $\gamma^{(1)}, \dots, \gamma^{(n)}$ the roots of $(M_1 \circ M)(f(x))$, then

$$\beta^{(\nu)} = M^*\left(\alpha^{(\nu)}\right) = \frac{D\alpha^{(\nu)} - B}{-C\alpha^{(\nu)} + A},$$

$$\gamma^{(\nu)} = M_1^*\left(\beta^{(\nu)}\right) = \frac{D_1\beta^{(\nu)} - B_1}{-C_1\beta^{(\nu)} + A_1} = \frac{D_1\frac{D\alpha^{(\nu)}-B}{-C\alpha^{(\nu)}+A} - B_1}{-C_1\frac{D\alpha^{(\nu)}-B}{-C\alpha^{(\nu)}+A} + A_1} =$$

$$= \frac{D_1(D\alpha^{(\nu)} - B) - B_1(-C\alpha^{(\nu)} + A)}{-C_1(D\alpha^{(\nu)} - B) + A_1(-C\alpha^{(\nu)} + A)} =$$

$$= \frac{(D_1D + B_1C)\alpha^{(\nu)} - (D_1B + B_1A)}{-(C_1D + A_1C)\alpha^{(\nu)} + (C_1B + A_1A)} = M_2^*\left(\alpha^{(\nu)}\right),$$

where

$$M_2^* = \begin{pmatrix} CB_1 + DD_1 & -(AB_1 + BD_1) \\ -(CA_1 + DC_1) & AA_1 + BC_1 \end{pmatrix} = M_1^* \cdot M^*.$$

The lemma is completely proved. □

Lemma 5.10 *For composition* $\circ$ *of linear transformations of the forms, the following equality holds*

$$M_1 \circ M = M \cdot M_1,$$

where $\cdot$ *is the matrix multiplication, and at the same time, the root lines are transformed by the rule*

$$(M_1 \circ M)^* = M_1^* \cdot M^*.$$

Proof Indeed, let

$$M = \begin{pmatrix} A & B \\ C & D \end{pmatrix}, \quad M_1 = \begin{pmatrix} A_1 & B_1 \\ C_1 & D_1 \end{pmatrix}$$

and $G(X, Y) = M(F(X, Y))$, then

$$\begin{aligned} M_1 \circ M(F(X, Y)) &= G(A_1X + B_1Y, C_1X + D_1Y) = \\ = F(A(A_1X + B_1Y) &+ B(C_1X + D_1Y), C(A_1X + B_1Y) + D(C_1X + D_1Y)) = \\ = F((AA_1 + BC_1)X &+ (AB_1 + BD_1)Y, (CA_1 + DC_1)X + (CB_1 + DD_1)Y) = \\ &= M_2(F(X, Y)), \end{aligned}$$

where

$$M_2 = \begin{pmatrix} AA_1 + BC_1 & AB_1 + BD_1 \\ CA_1 + DC_1 & CB_1 + DD_1 \end{pmatrix} = M \cdot M_1$$

and first, statement of the lemma is found.

Let

$X - \alpha^{(\nu)}Y = 0$ $(\nu = 1, \ldots, n)$ be the root lines for a form $F(X, Y)$,

$X - \beta^{(\nu)}Y = 0$ $(\nu = 1, \ldots, n)$ be the root lines for a form $M(F(X, Y))$, and

$X - \gamma^{(\nu)}Y = 0$ $(\nu = 1, \ldots, n)$ be the root lines for a form $(M_1 \circ M)(F(X, Y))$.

Then, first, collection of the root lines is conversed to second by the linear transformation $M^*(X, Y) = (DX - BY, -CX + AY)$, and the second collection is conversed to third by the linear transformation $M_1^*(X, Y) = (D_1X - B_1Y, -C_1X + A_1Y)$. Therefore, first, collection is conversed to third by the composition

$$\begin{aligned} &(M_1 \circ M)(F(X, Y)) = \\ &= (D_1(DX - BY) - B_1(-CX + AY), -C_1(DX - BY) + A_1(-CX + AY)) = \\ &= ((D_1D + B_1C)X - (D_1B + B_1A)Y, -(C_1D + A_1C)X + (C_1B + A_1A)Y) = \\ &= M_2(F(X, Y)), \end{aligned}$$

where

$$M_2^* = \begin{pmatrix} CB_1 + DD_1 & -(AB_1 + BD_1) \\ -(CA_1 + DC_1) & AA_1 + BC_1 \end{pmatrix} = M_1^* \cdot M^*,$$

and the lemma is completely proved. □

Recall the definition of the discriminant $D(f)$ of a polynomial

$$f(x) = a_nx^n + a_{n-1}x^{n-1} + \ldots + a_1x + a_0, \quad a_n \neq 0,$$

having the roots $\alpha^{(1)}, \ldots, \alpha^{(n)}$. By definition

$$D(f) = a_n^{2n-2} \prod_{\nu<\mu} \left(\alpha^{(\nu)} - \alpha^{(\mu)}\right)^2.$$

Similarly, a discriminant $D(F)$ of a form

$$F(X, Y) = a_n X^n + a_{n-1} X^{n-1} Y + \ldots + a_1 X Y^{n-1} + a_0 Y^n, \quad a_n \neq 0, a_0 \neq 0,$$

having the root lines $X - \alpha^{(\nu)} Y = 0$ $(\nu = 1, \ldots, n)$ is given by

$$D(F) = a_n^{2n-2} \prod_{\nu<\mu} \left(\alpha^{(\nu)} - \alpha^{(\mu)}\right)^2 = a_0^{2n-2} \prod_{\nu<\mu} \left(\beta^{(\nu)} - \beta^{(\mu)}\right)^2.$$

The discriminant of the form is well defined, since by Vieta theorem $\alpha^{(1)} \ldots \alpha^{(n)} = (-1)^n \frac{a_0}{a_n}$ and for $\beta^{(\nu)} = \frac{-1}{\alpha^{(\nu)}}$ $(\nu = 1, \ldots, n)$, we have:

$$a_0^{2n-2} \prod_{\nu<\mu} \left(\beta^{(\nu)} - \beta^{(\mu)}\right)^2 = a_0^{2n-2} \prod_{\nu<\mu} \frac{\left(\alpha^{(\nu)} - \alpha^{(\mu)}\right)^2}{\left(\alpha^{(\nu)} \alpha^{(\mu)}\right)^2} =$$

$$= a_0^{2n-2} \frac{\prod_{\nu<\mu} \left(\alpha^{(\nu)} - \alpha^{(\mu)}\right)^2}{\left(\prod_{\nu=1}^{n} \alpha^{(\nu)}\right)^{2(n-1)}} = a_n^{2n-2} \prod_{\nu<\mu} \left(\alpha^{(\nu)} - \alpha^{(\mu)}\right)^2.$$

Theorem 5.3 *For any linear fractional transformation with matrix* $M = \begin{pmatrix} A & B \\ C & D \end{pmatrix} \in \mathcal{M}_2^*(\mathbb{K})$ *and any polynomial* $f(x)$ *with roots* $\alpha^{(\nu)}$ $(A \neq C\alpha^{(\nu)}$, $(\nu = 1, \ldots, n)$ *and the polynomial* $M(f(x))$, *the following holds*

$$(\det M)^{n(n-1)} D(f) = D(M(f)).$$

Proof Indeed, by Lemma 5.7 for $C \neq 0$ so that

$$D(M(f)) = \left(C^n f\left(\frac{A}{C}\right)\right)^{2n-2} \prod_{\nu<\mu} \left(\beta^{(\nu)} - \beta^{(\mu)}\right)^2 =$$

$$= a_n^{2n-2} \left(\prod_{\nu=1}^{n} (A - C\alpha^{(\nu)})\right)^{2n-2} \prod_{\nu<\mu} \left(\frac{D\alpha^{(\nu)} - B}{A - C\alpha^{(\nu)}} - \frac{D\alpha^{(\mu)} - B}{A - C\alpha^{(\mu)}}\right)^2 =$$

$$= a_n^{2n-2} \frac{\left(\prod_{\nu=1}^{n} (A - C\alpha^{(\nu)})\right)^{2n-2}}{\left(\prod_{\nu=1}^{n} (A - C\alpha^{(\nu)})\right)^{2n-2}} \prod_{\nu<\mu} \left((DA - BC)(\alpha^{(\nu)} - \alpha^{(\mu)})\right)^2 =$$

$$= (\det M)^{n(n-1)} D(f)$$

and in this case, the theorem is true.

If $C = 0$, then $\det M = DA$ and

$$D(M(f)) = (A^n a_n)^{2n-2} \prod_{\nu<\mu} \left(\beta^{(\nu)} - \beta^{(\mu)}\right)^2 =$$

$$= a_n^{2n-2} A^{n(2n-2)} \prod_{\nu<\mu} \left(\frac{D\alpha^{(\nu)} - B}{A} - \frac{D\alpha^{(\mu)} - B}{A}\right)^2 =$$

$$= a_n^{2n-2} \prod_{\nu<\mu} \left(DA(\alpha^{(\nu)} - \alpha^{(\mu)})\right)^2 = (\det M)^{n(n-1)} D(f)$$

and the theorem is completely proved. □

5.5 Linear Fractional Transformation of Integer Polynomials

Denote by $\mathbb{P}_n[x]$ a set of all irreducible integer polynomials $f(x) \in \mathbb{Z}[x]$ of degree n. In other words, if $f(x) \in \mathbb{P}_n[x]$, then

$$f(x) = a_n x^n + a_{n-1} x^{n-1} + \ldots + a_1 x + a_0, \quad a_n \neq 0 \neq a_0, \quad a_j \in \mathbb{Z} \quad (0 \leq j \leq n)$$

and the equations $f(x) = g(x)h(x)$, $\deg(g(x)) \leq \deg(h(x))$ imply that $g(x) \equiv \pm 1$, $h(x) \equiv \mp f(x)$. In particular, any irreducible polynomial is primitive, that is, $(a_0, \ldots, a_n) = 1$.

By $\mathbb{PP}_n[X, Y]$, we denote a set of all irreducible binary integer form $F(X, Y) \in \mathbb{PZ}[X, Y]$ of order n. This means that if $F(X, Y) \in \in \mathbb{PZ}_n[X, Y]$, then

$$F(X, Y) = a_n X^n + a_{n-1} X^{n-1} Y + \ldots + a_1 X Y^{n-1} + a_0 Y^n,$$
$$a_n \neq 0 \neq a_0, \quad a_j \in \mathbb{Z} \quad (0 \leq j \leq n)$$

and the equations $F(X, Y) = G(X, Y)H(X, Y), \deg(G(X, Y)) \leq \deg(H(X, Y))$ imply that $G(X, Y) \equiv \pm 1$, $H(X, Y) \equiv \mp F(X, Y)$. In particular, any irreducible form is primitive, that is, $(a_0, \ldots, a_n) = 1$.

According to H. Weyl [21], denote by $Ct(f)$ the content of polynomial $f(x)$ and by $Ct(F)$ the content of form F. Thus, $Ct(f) = Ct(F) == (a_0, \ldots, a_n)$.

Lemma 5.11 *For any linear fractional transformation with unimodular matrix $M \in \mathcal{U}_2(\mathbb{Z})$, the following holds*

$$Ct(f) = Ct(M(f)).$$

Proof Indeed, let

$$M = \begin{pmatrix} A & B \\ C & D \end{pmatrix}, \qquad M^{-1} = M_1 = \begin{pmatrix} A_1 & B_1 \\ C_1 & D_1 \end{pmatrix}$$

and

$$f(x)=\sum_{\nu=0}^{n}a_\nu x^\nu,\quad M(f(x))=\sum_{\nu=0}^{n}b_\nu x^\nu,\quad a_\nu,b_\nu\in\mathbb{Z}\quad(0\le\nu\le n).$$

Then, we get the following relation between the coefficients a_ν and b_ν

$$M(f(x))=\sum_{\nu=0}^{n}b_\nu x^\nu=\sum_{\nu=0}^{n}a_\nu(Ax+B)^\nu(Cx+D)^{n-\nu}=$$

$$=\sum_{\nu=0}^{n}a_\nu\sum_{\mu=0}^{\nu}C_\nu^\mu A^\mu B^{\nu-\mu}x^\mu\sum_{\lambda=0}^{n-\nu}C_{n-\nu}^\lambda C^\lambda D^{n-\nu-\lambda}x^\lambda=$$

$$=\sum_{\nu=0}^{n}a_\nu\sum_{\lambda=0}^{n}x^\lambda\sum_{\mu=\max(0,\lambda+\nu-n)}^{\min(\nu,\lambda)}C_\nu^\mu C_{n-\nu}^{\lambda-\mu}A^\mu B^{\nu-\mu}C^{\lambda-\mu}D^{n+\mu-\nu-\lambda};$$

$$b_\lambda=\sum_{\nu=0}^{n}a_\nu\sum_{\mu=\max(0,\lambda+\nu-n)}^{\min(\nu,\lambda)}C_\nu^\mu C_{n-\nu}^{\lambda-\mu}A^\mu B^{\nu-\mu}C^{\lambda-\mu}D^{n+\mu-\nu-\lambda};$$

$$a_\lambda=\sum_{\nu=0}^{n}b_\nu\sum_{\mu=\max(0,\lambda+\nu-n)}^{\min(\nu,\lambda)}C_\nu^\mu C_{n-\nu}^{\lambda-\mu}A_1^\mu B_1^{\nu-\mu}C_1^{\lambda-\mu}D_1^{n+\mu-\nu-\lambda}.$$

This means that $Ct(f)|Ct(M(f))$ and $Ct(M(f))|Ct(f)$, and therefore,

$$Ct(f)=Ct(M(f)).$$

This completes the proof. □

In view of the bijection φ given by the Eq. (5.8) for any linear unimodular transformation of integral forms, we obtain $Ct(F)=Ct(M(F))$.

Lemma 5.12 *The image of any irreducible polynomial $f(x)$ under the linear fractional transformation with unimodular matrix $M\in\mathcal{U}_2^*$ is a irreducible polynomial.*

Proof Indeed, by Lemma 5.5, a linear fractional transformation conserves the product, so the linear fractional transformation with unimodular matrix having an inverse transformation converts a primitive polynomial into primitive and irreducible polynomial into irreducible. □

In view of the bijection φ, the similar statement is true for irreducible form.

Theorem 5.4 *For any linear fractional transformation with unimodular matrix $M=\begin{pmatrix}A&B\\C&D\end{pmatrix}\in\mathcal{U}_2^*(\mathbb{Z})$ and any polynomial $f(x)$ with the roots $\alpha^{(\nu)}$ ($A\ne C\alpha^{(\nu)}$, $\nu=1,\ldots,n$), the following holds*

$$D(f)=D(M(f)).$$

Proof Indeed, since $\det M = \pm 1$ by Theorem 5.3, we have

$$D(M(f)) = (\det M)^{n(n-1)} D(f) = D(f).$$

This completes the proof. □

In view of the bijection φ for any linear transformation of the form $F \in \mathbb{PZ}_n[X, Y]$ with unimodular matrix $M \in \mathcal{U}_2^*(\mathbb{Z})$, we obtain $D(M(F)) = D(F)$.

5.6 Behavior of Residual Fractions and Its Conjugate Numbers

Let $\alpha = \alpha^{(1)}, \alpha^{(2)}, \dots, \alpha^{(n)}$ be the roots of some irreducible integer polynomial.

Denote by

$$\delta(\alpha) = \min_{2 \le j \le n} \left|\alpha^{(1)} - \alpha^{(j)}\right| > 0.$$

It is clear that $\delta(\alpha)$ is well defined as all the roots are distinct.

For $m \ge 1$, we define θ_{m-1} $(0 < \theta_{m-1} < 1)$ using the equation

$$\alpha = \alpha^{(1)} = \frac{P_{m-1}}{Q_{m-1}} + \frac{(-1)^{m-1}\theta_{m-1}}{Q_{m-1}Q_m}.$$

It is easy to calculate that

$$\theta_{m-1} = \frac{Q_m}{\alpha_m Q_{m-1} + Q_{m-2}}.$$

A residual fraction $\alpha_m = \alpha_m^{(1)}$ has an expansion

$$\alpha_m = \alpha_m^{(1)} = q_m + \cfrac{1}{q_{m+1} + \cfrac{1}{\ddots + \cfrac{1}{q_k + \cfrac{1}{\ddots}}}} > 1 \quad (m \ge 1).$$

Theorem 5.5 *Let $\alpha = \alpha_0$ be a real root of irreducible integer polynomial*

$$f_0(x) = a_n x^n + a_{n-1}x^{n-1} + \ldots + a_1 x + a_0 \in \mathbb{Z}[x], \quad a_n > 0,$$

$\alpha = \alpha^{(1)}, \alpha^{(2)}, \dots, \alpha^{(n)}$ *its root and let*

$$\alpha = \alpha_0 = q_0 + \cfrac{1}{q_1 + \cfrac{1}{\ddots + \cfrac{1}{q_k + \cfrac{1}{\ddots}}}}$$

be continued fraction expansion of α.

Suppose that a sequence of the polynomials $f_m(x)$ $(m \geq 1)$ *is defined by the recurrence relations*

$$f_m(x) = \varepsilon_m x^n f_{m-1}\left(q_{m-1} + \frac{1}{x}\right), \quad \textit{where} \quad \varepsilon_m = \operatorname{sign}(f_{m-1}(q_{m-1})).$$

Then,

(1) $f_m(x) = \sum_{k=0}^{n} a_{k,m}x^k \in \mathbb{Z}[x], \quad a_{n,m} > 0$ *and*

$$a_{k,m} = \varepsilon_m \sum_{\nu=n-k}^{n} a_{\nu,m-1} C_\nu^{\nu+k-n} q_{m-1}^{\nu+k-n} = \varepsilon_m \frac{f_{m-1}^{(n-k)}(q_{m-1})}{(n-k)!} \quad (0 \leq k \leq n); \tag{5.17}$$

(2) the polynomials $f_m(x)$ *have the roots*

$$\alpha_m^{(j)} = \frac{\alpha^{(j)} Q_{m-2} - P_{m-2}}{P_{m-1} - \alpha^{(j)} Q_{m-1}} \quad (1 \leq j \leq n); \tag{5.18}$$

(3)

$$f_m(x) = \varepsilon_m f_{m-1}(q_{m-1}) \prod_{j=1}^{n} \left(x - \alpha_m^{(j)}\right); \tag{5.19}$$

(4) there exits index $m_0 = m_0(\alpha)$ *such that for any* $m \geq m_0$*, a residual fraction* $\alpha_m = \alpha_m^{(1)}$ *is a reduced generalized Pisot number and*

$$\frac{Q_{m-2}}{Q_{m-1}} = \cfrac{1}{q_{m-1} + \cfrac{1}{\ddots + \cfrac{1}{q_2 + \cfrac{1}{q_1}}}}, \tag{5.20}$$

$$\alpha_m^{(j)} = -\frac{Q_{m-2}}{Q_{m-1}} + \frac{(-1)^m}{Q_{m-1}^2\left(\frac{(-1)^m \theta_{m-1}}{Q_{m-1} Q_m} + \alpha^{(1)} - \alpha^{(j)}\right)} \quad (2 \leq j \leq n). \tag{5.21}$$

Proof Consider a sequence of the polynomials

$$f_m(x) = \varepsilon_m x^n f_{m-1}\left(q_{m-1} + \frac{1}{x}\right) \quad (m \geq 1).$$

Using Taylor formula, we get

$$f_{m-1}(x) = \sum_{\nu=0}^{n} \frac{f_{m-1}^{(\nu)}(q_{m-1})}{\nu!}(x - q_{m-1})^{\nu},$$

so

$$f_m(x) = \varepsilon_m \sum_{\nu=0}^{n} \frac{f_{m-1}^{(\nu)}(q_{m-1})}{\nu!} x^{n-\nu}.$$

It is easy to see that for the coefficients of the polynomials

$$f_m(x) = \sum_{\nu=0}^{n} a_{\nu,m} x^{\nu}$$

we obtain

$$a_{\nu,m} = \varepsilon_m \frac{f_{m-1}^{(n-\nu)}(q_{m-1})}{(n-\nu)!} = \varepsilon_m \sum_{k=n-\nu}^{n} a_{k,m-1} C_k^{n-\nu} q_{m-1}^{k+\nu-n}.$$

And the statement (5.17) is true.

If

$$\alpha_{m-1}^{(j)} = \frac{\alpha^{(j)} Q_{m-3} - P_{m-3}}{P_{m-2} - \alpha^{(j)} Q_{m-2}} \quad (1 \leq j \leq n)$$

are the roots of a polynomial $f_{m-1}(x)$, then

$$\alpha_m^{(j)} = \frac{1}{\alpha_{m-1}^{(j)} - q_{m-1}} = \frac{\alpha^{(j)} Q_{m-2} - P_{m-2}}{P_{m-1} - \alpha^{(j)} Q_{m-1}} \quad (1 \leq j \leq n),$$

and we get (5.18) and (5.19).

The Eq. (5.20) is well known.

To prove the last statement, we write (5.18) in the form

$$\alpha_m^{(j)} = \frac{Q_{m-2}}{Q_{m-1}} \cdot \frac{\alpha^{(j)} - \frac{P_{m-2}}{Q_{m-2}}}{\frac{P_{m-1}}{Q_{m-1}} - \alpha^{(j)}} \quad (1 \leq j \leq n). \tag{5.22}$$

For $j = 1$, we have the inequality $\alpha_m^{(1)} > 1$, which follows from the definition of a residual fraction

Let now $2 \le j \le n$, then

$$\alpha_m^{(j)} = \frac{Q_{m-2}}{Q_{m-1}}\left(-1+\frac{\frac{P_{m-1}}{Q_{m-1}}-\frac{P_{m-2}}{Q_{m-2}}}{\frac{P_{m-1}}{Q_{m-1}}-\alpha^{(j)}}\right) = \frac{Q_{m-2}}{Q_{m-1}}\left(-1+\frac{\frac{(-1)^m}{Q_{m-1}Q_{m-2}}}{\frac{P_{m-1}}{Q_{m-1}}-\alpha^{(j)}}\right) =$$

$$= \frac{Q_{m-2}}{Q_{m-1}}\left(-1+\frac{(-1)^m}{Q_{m-1}Q_{m-2}\left(\frac{P_{m-1}}{Q_{m-1}}-\alpha^{(j)}\right)}\right) =$$

$$= -\frac{Q_{m-2}}{Q_{m-1}} + \frac{(-1)^m}{Q_{m-1}^2\left(\frac{(-1)^m\theta_{m-1}}{Q_{m-1}Q_m}+\alpha^{(1)}-\alpha^{(j)}\right)}. \tag{5.23}$$

There exists m_0 such that

$$\left|\frac{(-1)^m\theta_{m-1}}{Q_{m-1}Q_m}\right| \le \frac{\delta(\alpha)}{2}, \quad \frac{2}{Q_{m-1}\delta(\alpha)} < 1,$$

for all $m \ge m_0$.

So for all $m \ge m_0$, we have

$$|\alpha_m^{(j)}| \le \frac{Q_{m-2}}{Q_{m-1}}\left(1+\frac{2}{Q_{m-1}Q_{m-2}\delta(\alpha)}\right) = \frac{Q_{m-2}}{Q_{m-1}} + \frac{2}{Q_{m-1}^2\delta(\alpha)} < 1, \tag{5.24}$$

it follows that $\alpha_m^{(1)}$ is a generalized Pisot number.

We now show that the inequalities $|q_m - \alpha_m^{(j)}| > 1$ hold for all $2 \le j \le n$. Consider two possible cases.

I. Let $\alpha^{(j)}$ be a real algebraic number. Then,

$$-1 < -\frac{Q_{m-2}}{Q_{m-1}} - \frac{2}{Q_{m-1}^2\delta(\alpha)} \le \alpha_m^{(j)} = -\frac{Q_{m-2}}{Q_{m-1}} +$$

$$+\frac{(-1)^m}{Q_{m-1}^2\left(\frac{(-1)^m\theta_{m-1}}{Q_{m-1}Q_m}+\alpha^{(1)}-\alpha^{(j)}\right)} \le -\frac{Q_{m-2}}{Q_{m-1}} + \frac{2}{Q_{m-1}^2\delta(\alpha)} < 0,$$

so

$$q_m - \alpha_m^{(j)} > 1$$

and an inequality holds for this algebraic conjugate to a residual fraction α_m.

II. Let now $\alpha^{(j)}$ be a complex algebraic number. Then, complex algebraic conjugate $\alpha_m^{(j)}$ for a residual fraction α_m lies within the circle of radius less than $\frac{1}{Q_{m-1}}$ and center $-\frac{Q_{m-2}}{Q_{m-1}}$. It follows that $|q_m - \alpha_m^{(j)}| > 1$, and in this case, the necessary inequality holds too.

This concludes the proof. □

5.7 Minimal Polynomials of Residual Fractions

Theorem 5.6 *Let $\alpha = \alpha_0$ be a real root of irreducible integer polynomial*

$$f_0(x) = a_n x^n + a_{n-1}x^{n-1} + \ldots + a_1 x + a_0 \in \mathbb{Z}[x], \quad a_n > 0,$$

$\alpha = \alpha^{(1)}, \alpha^{(2)}, \ldots, \alpha^{(n)}$ its root and let

$$\alpha = \alpha_0 = q_0 + \cfrac{1}{q_1 + \cfrac{1}{\ddots + \cfrac{1}{q_k + \cfrac{1}{\ddots}}}}$$

be continued fraction expansion of α.

Then, the sequence of the discriminants $D(f_m)$ of the minimal polynomials $f_m(x)$ of residual fractions $\alpha_m = \alpha_m^{(1)}$ is integer and stationary.

Proof Indeed, since all polynomials $f_m(x) \in \mathbb{Z}[x]$ and according to the property of discriminant (see [18], p. 34), it follows that $D(f_m) \in \mathbb{Z}$. By Theorem 5.3, so that $D(f_{m-1}) = D(f_m)$.

This completes the proof. □

Theorem 5.7 *Let $\alpha = \alpha_0$ be a real root of irreducible integer polynomial*

$$f_0(x) = a_n x^n + a_{n-1}x^{n-1} + \ldots + a_1 x + a_0 \in \mathbb{Z}[x], \quad a_n > 0,$$

$\alpha = \alpha^{(1)}, \alpha^{(2)}, \ldots, \alpha^{(n)}$ its root and let

$$\alpha = \alpha_0 = q_0 + \cfrac{1}{q_1 + \cfrac{1}{\ddots + \cfrac{1}{q_k + \cfrac{1}{\ddots}}}}$$

be continued fraction expansion of α.

If α is reduced generalized Pisot number, then the minimal polynomial $f_m(x)$ of a residual fraction α_m is as follows:

$$f_m(x) = (-1)^m (Q_{m-1}x + Q_{m-2})^n f_0\left(\frac{P_{m-1}x + P_{m-2}}{Q_{m-1}x + Q_{m-2}}\right) = \sum_{\nu=0}^{n} a_{\nu,m} x^{\nu}, \qquad (5.25)$$

where

$$a_{n,m} = Q_{m-1}^n \left| f_0\left(\frac{P_{m-1}}{Q_{m-1}}\right)\right|, \quad a_{0,m} = -Q_{m-2}^n \left| f_0\left(\frac{P_{m-2}}{Q_{m-2}}\right)\right|, \tag{5.26}$$

$$a_{\nu,m} = Q_{m-1}^{\nu} Q_{m-2}^{n-\nu} \sum_{\mu=0}^{n-\nu} \frac{f_0^{(\mu)}\left(\frac{P_{m-1}}{Q_{m-1}}\right)}{\mu!} \frac{(-1)^{m+(m-1)\mu}}{(Q_{m-2}Q_{m-1})^{\mu}} C_{n-\mu}^{\nu} \quad (0 \leq \nu \leq n), \tag{5.27}$$

$$a_{n-1,m} = Q_{m-1}^{n-1} Q_{m-2} \left(n \left| f_0\left(\frac{P_{m-1}}{Q_{m-1}}\right)\right| - \frac{1}{Q_{m-2}Q_{m-1}} f_0'\left(\frac{P_{m-1}}{Q_{m-1}}\right)\right). \tag{5.28}$$

Proof The proof is by induction on m.

For $m = 0$, we have

$$P_{-1} = 1,\ P_{-2} = 0,\ Q_{-1} = 0,\ Q_{-2} = 1,$$
$$(Q_{-1}x + Q_{-2})^n f_0\left(\frac{P_{-1}x + P_{-2}}{Q_{-1}x + Q_{-2}}\right) = f_0(x)$$

and the equality (5.25) is true.

Assume the statement is true for $m \geq 0$, then

$$f_m(x) = (-1)^m M_m\left(f_0(x)\right), \quad M_m = \begin{pmatrix} P_{m-1} & P_{m-2} \\ Q_{m-1} & Q_{m-2} \end{pmatrix}.$$

Since $a_{n,m} > 0$ and α_m is a reduced generalized Pisot number, it follows that $f_m(q_m) < 0$ and

$$f_{m+1}(x) = -x^n f_m\left(q_m + \frac{1}{x}\right) = -M_m'\left(f_m(x)\right), \quad M_m' = \begin{pmatrix} q_m & 1 \\ 1 & 0 \end{pmatrix}.$$

Note that

$$M_m \cdot M_m' = \begin{pmatrix} P_{m-1} & P_{m-2} \\ Q_{m-1} & Q_{m-2} \end{pmatrix} \begin{pmatrix} q_m & 1 \\ 1 & 0 \end{pmatrix} = \begin{pmatrix} q_m P_{m-1} + P_{m-2} & P_{m-1} \\ q_m Q_{m-1} + Q_{m-2} & Q_{m-1} \end{pmatrix} = M_{m+1}.$$

By the inductive assumption and Lemma 5.9,

$$f_{m+1}(x) = -M_m'\left((-1)^m M_m\left(f_0(x)\right)\right) = (-1)^{m+1}\left(M_m \cdot M_m'\right)\left(f_0(x)\right) =$$
$$= (-1)^{m+1} M_{m+1}\left(f_0(x)\right),$$

and the equality (5.25) is proved.

We now start on the proof of (5.26).

By Lemma 5.7, so that

$$a_{n,m} = (-1)^m Q_{m-1}^n f_0\left(\frac{P_{m-1}}{Q_{m-1}}\right).$$

If m is even, then $\frac{P_{m-1}}{Q_{m-1}} > \alpha$ and $f_0\left(\frac{P_{m-1}}{Q_{m-1}}\right) > 0$. If m is odd, then $\frac{P_{m-1}}{Q_{m-1}} < \alpha$ and $f_0\left(\frac{P_{m-1}}{Q_{m-1}}\right) < 0$. Therefore,

$$(-1)^m Q_{m-1}^n f_0\left(\frac{P_{m-1}}{Q_{m-1}}\right) = Q_{m-1}^n \left|f_0\left(\frac{P_{m-1}}{Q_{m-1}}\right)\right|$$

and the equality for $a_{n,m}$ is proved.

Similarly,

$$a_{0,m} = (-1)^m Q_{m-2}^n f_0\left(\frac{P_{m-2}}{Q_{m-2}}\right) = -Q_{m-2}^n \left|f_0\left(\frac{P_{m-2}}{Q_{m-2}}\right)\right|$$

and the equalities (5.26) are proved.

To prove (5.27), we notice that

$$\frac{P_{m-1}x + P_{m-2}}{Q_{m-1}x + Q_{m-2}} = \frac{P_{m-1}}{Q_{m-1}} + \frac{(-1)^{m-1}}{Q_{m-1}(Q_{m-1}x + Q_{m-2})}.$$

Using Taylor formula, we get

$$(Q_{m-1}x + Q_{m-2})^n f_0\left(\frac{P_{m-1}x + P_{m-2}}{Q_{m-1}x + Q_{m-2}}\right) = (Q_{m-1}x + Q_{m-2})^n f_0\left(\frac{P_{m-1}}{Q_{m-1}}\right) +$$

$$+\sum_{\nu=1}^{n} \frac{f_0^{(\nu)}\left(\frac{P_{m-1}}{Q_{m-1}}\right)}{\nu!} \frac{(-1)^{(m-1)\nu}(Q_{m-1}x + Q_{m-2})^{n-\nu}}{Q_{m-1}^{\nu}} = Q_{m-1}^n f_0\left(\frac{P_{m-1}}{Q_{m-1}}\right) x^n +$$

$$+f_0\left(\frac{P_{m-1}}{Q_{m-1}}\right) \sum_{\nu=0}^{n-1} C_n^\nu Q_{m-1}^\nu Q_{m-2}^{n-\nu} x^\nu + \sum_{\mu=1}^{n} \frac{f_0^{(\mu)}\left(\frac{P_{m-1}}{Q_{m-1}}\right)}{\mu!} \frac{(-1)^{(m-1)\mu}}{Q_{m-1}^{\mu}} \cdot$$

$$\cdot \sum_{\nu=0}^{n-\mu} C_{n-\mu}^\nu Q_{m-1}^\nu Q_{m-2}^{n-\mu-\nu} x^\nu = Q_{m-1}^n f_0\left(\frac{P_{m-1}}{Q_{m-1}}\right) x^n + f_0\left(\frac{P_{m-1}}{Q_{m-1}}\right) \cdot$$

$$\cdot \sum_{\nu=0}^{n-1} C_n^\nu Q_{m-1}^\nu Q_{m-2}^{n-\nu} x^\nu + \sum_{\nu=0}^{n-1} x^\nu Q_{m-1}^\nu Q_{m-2}^{n-\nu} \sum_{\mu=1}^{n-\nu} C_{n-\mu}^\nu \frac{(-1)^{(m-1)\mu}}{(Q_{m-2}Q_{m-1})^\mu} \frac{f_0^{(\mu)}\left(\frac{P_{m-1}}{Q_{m-1}}\right)}{\mu!} =$$

$$= Q_{m-1}^n f_0\left(\frac{P_{m-1}}{Q_{m-1}}\right) x^n + \sum_{\nu=0}^{n-1} x^\nu Q_{m-1}^\nu Q_{m-2}^{n-\nu} \sum_{\mu=0}^{n-\nu} C_{n-\mu}^\nu \frac{(-1)^{(m-1)\mu}}{(Q_{m-2}Q_{m-1})^\mu} \frac{f_0^{(\mu)}\left(\frac{P_{m-1}}{Q_{m-1}}\right)}{\mu!},$$

and the equality (5.27) is proved.

For $\nu = n-1$, we obtain

$$a_{n-1,m} = (-1)^m Q_{m-1}^{n-1} Q_{m-2} \sum_{\mu=0}^{1} C_{n-\mu}^{n-1} \frac{(-1)^{(m-1)\mu}}{(Q_{m-2}Q_{m-1})^{\mu}} \frac{f_0^{(\mu)}\left(\frac{P_{m-1}}{Q_{m-1}}\right)}{\mu!} =$$

$$= (-1)^m Q_{m-1}^{n-1} Q_{m-2} \left(n f_0\left(\frac{P_{m-1}}{Q_{m-1}}\right) + \frac{(-1)^{m-1}}{Q_{m-2}Q_{m-1}} f_0'\left(\frac{P_{m-1}}{Q_{m-1}}\right)\right) =$$

$$= Q_{m-1}^{n-1} Q_{m-2} \left(n \left| f_0\left(\frac{P_{m-1}}{Q_{m-1}}\right)\right| - \frac{1}{Q_{m-2}Q_{m-1}} f_0'\left(\frac{P_{m-1}}{Q_{m-1}}\right)\right)$$

and the equality (5.28) is proved.

Finally, we shall verify that (5.27) implies (5.26).

Indeed, for $\nu = n$, we get

$$Q_{m-1}^{\nu} Q_{m-2}^{n-\nu} \sum_{\mu=0}^{n-\nu} \frac{f_0^{(\mu)}\left(\frac{P_{m-1}}{Q_{m-1}}\right)}{\mu!} \frac{(-1)^{m+(m-1)\mu}}{(Q_{m-2}Q_{m-1})^{\mu}} C_{n-\mu}^{\nu} =$$

$$= Q_{m-1}^{n} f_0\left(\frac{P_{m-1}}{Q_{m-1}}\right)(-1)^m = a_{n,m}$$

and the first of the equalities (5.26) is proved.

Similarly, for $\nu = 0$, we get

$$Q_{m-1}^{\nu} Q_{m-2}^{n-\nu} \sum_{\mu=0}^{n-\nu} \frac{f_0^{(\mu)}\left(\frac{P_{m-1}}{Q_{m-1}}\right)}{\mu!} \frac{(-1)^{m+(m-1)\mu}}{(Q_{m-2}Q_{m-1})^{\mu}} C_{n-\mu}^{\nu} =$$

$$= (-1)^m Q_{m-2}^{n} \sum_{\mu=0}^{n} \frac{f_0^{(\mu)}\left(\frac{P_{m-1}}{Q_{m-1}}\right)}{\mu!} \left(\frac{P_{m-2}}{Q_{m-2}} - \frac{P_{m-1}}{Q_{m-1}}\right)^{\mu} =$$

$$= (-1)^m Q_{m-2}^{n} f_0\left(\frac{P_{m-2}}{Q_{m-2}}\right) = a_{0,m}$$

and the second of the equalities (5.26) is proved. □

Lemma 5.13 *Let α be a root of minimal polynomial*

$$f_0(x) = a_n x^n + \ldots + a_1 x + a_0 \in \mathbb{P}_n[x].$$

Then,

$$f_0^{(\nu)}(\alpha) \neq 0 \quad (\nu = 1, \ldots, n).$$

Proof Indeed, since $f_0(x) \in \mathbb{P}_n[x]$ and $a_n \neq 0$, it follows that $f_0^{(n)}(x) = n! a_n \neq 0$,

Let $1 \le \nu \le n-1$ and $g(x) = f_0^{(\nu)}(x)$, $g(\alpha) = 0$. Since $g(x) \in \mathbb{Z}[x]$, $f_0(x)$, and $g(x)$ have the same root, we get $(f_0(x), g(x)) \ne 1$. This contradicts irreducible minimal polynomial. This completes the proof. □

Denote by $c(\alpha, \varepsilon) > 0$ the constant in Roth's theorem [19]. Thus, for any integer p and natural q, the following inequality holds

$$\left|\alpha - \frac{p}{q}\right| \ge \frac{c(\alpha, \varepsilon)}{q^{2+\varepsilon}}. \tag{5.29}$$

Let

$$\Delta(\alpha) = \max_{2 \le j \le n} |\alpha^{(1)} - \alpha^{(j)}|.$$

Lemma 5.14 *Let α be a real irrationality of degree $n > 2$ and let*

$$f_0(x) = a_n x^n + \ldots + a_1 x + a_0 \in \mathbb{P}_n[x]$$

be a minimal polynomial.

Then, for any convergent $\frac{P_m}{Q_m}$ to α, the following inequalities hold

$$a_n \frac{c(\alpha, \varepsilon)\left(\frac{\delta(\alpha)}{2}\right)^{n-1}}{Q_m^{2+\varepsilon}} < \left| f_0\left(\frac{P_m}{Q_m}\right)\right| < a_n \frac{(1+\Delta(\alpha))^{n-1}}{Q_m^2}. \tag{5.30}$$

Proof Indeed,

$$\left| f_0\left(\frac{P_m}{Q_m}\right)\right| = a_n \prod_{j=1}^{n} \left|\frac{P_m}{Q_m} - \alpha^{(j)}\right| = a_n \left|\frac{P_m}{Q_m} - \alpha\right| \prod_{j=2}^{n} \left|\frac{P_m}{Q_m} - \alpha + \alpha^{(1)} - \alpha^{(j)}\right|.$$

Next, we note that

$$\frac{c(\alpha, \varepsilon)}{Q_m^{2+\varepsilon}} < \left|\frac{P_m}{Q_m} - \alpha\right| < \frac{1}{Q_m^2},$$

$$\frac{\delta(\alpha)}{2} < \left|\frac{P_m}{Q_m} - \alpha + \alpha^{(1)} - \alpha^{(j)}\right| < 1 + \Delta(\alpha) \quad (2 \le j \le n).$$

This completes the proof. □

By Lemma 5.14 and Theorem 5.7, it follows that for $n > 2$, the highest coefficient $a_{n,m}$ of the minimal polynomial $f_m(x)$ for reduced generalized Pisot number α increases as quantity of order $O\left(Q_{m-1}^{n-2-\varepsilon}\right)$.

Indeed, for $m > m_0$, we have

$$a_n \frac{c(\alpha, \varepsilon)\left(\frac{\delta(\alpha)}{2}\right)^{n-1}}{Q_m^{2+\varepsilon}} < \left| f_0\left(\frac{P_m}{Q_m}\right)\right| < a_n \frac{(1+\Delta(\alpha))^{n-1}}{Q_m^2},$$

$$a_n c(\alpha, \varepsilon)\left(\frac{\delta(\alpha)}{2}\right)^{n-1} Q_{m-1}^{n-2-\varepsilon} < a_{n.m} = Q_{m-1}^n \left| f_0\left(\frac{P_{m-1}}{Q_{m-1}}\right)\right| <$$
$$< a_n(1+\Delta(\alpha))^{n-1} Q_{m-1}^{n-2}.$$

Denote by

$$A_\nu(\alpha) = \sum_{j=2}^{n} \frac{1}{\left(\alpha^{(1)} - \alpha^{(j)}\right)^\nu}, \quad \nu = 1, 2, \ldots.$$

Theorem 5.8 *Let α be a real irrationality of degree $n > 2$ and let*

$$f_0(x) = a_n x^n + \ldots + a_1 x + a_0 \in \mathbb{P}_n[x]$$

be a minimal polynomial.

Then, for $m > m_0$ for any convergent $\frac{P_m}{Q_m}$ to reduced generalized Pisot number α and residual fraction α_m, the following relations hold

$$\alpha_m = -\frac{Q_{m-2}}{Q_{m-1}} + \frac{f_0'\left(\frac{P_{m-1}}{Q_{m-1}}\right)}{Q_{m-1}^2 \left| f_0\left(\frac{P_{m-1}}{Q_{m-1}}\right)\right|} + (-1)^{m-1}\frac{\lambda_m}{Q_{m-1}^2}, \tag{5.31}$$

where

$$\lambda_m = A_1(\alpha) + \frac{(-1)^{m-1}\theta_{m-1}}{Q_{m-1}Q_m} A_2(\alpha)\varepsilon_m, \quad |\varepsilon_m| < 2. \tag{5.32}$$

Proof Indeed, by Vieta theorem, we have

$$\alpha_m^{(1)} + \ldots + \alpha_m^{(n)} = -\frac{a_{n-1,m}}{a_{n,m}}.$$

The formulas (5.26) and (5.28) imply

$$\alpha_m^{(1)} + \ldots + \alpha_m^{(n)} = -n\frac{Q_{m-2}}{Q_{m-1}} + \frac{f_0'\left(\frac{P_{m-1}}{Q_{m-1}}\right)}{Q_{m-1}^2 \left| f_0\left(\frac{P_{m-1}}{Q_{m-1}}\right)\right|}.$$

Using (5.23), we get

$$\alpha_m^{(1)} + \ldots + \alpha_m^{(n)} = \alpha_m - (n-1)\frac{Q_{m-2}}{Q_{m-1}} + \sum_{j=2}^{n} \frac{(-1)^m}{Q_{m-1}^2 \left(\frac{(-1)^m \theta_{m-1}}{Q_{m-1}Q_m} + \alpha^{(1)} - \alpha^{(j)}\right)}.$$

It now follows that

$$\alpha_m = -\frac{Q_{m-2}}{Q_{m-1}} + \frac{f_0'\left(\frac{P_{m-1}}{Q_{m-1}}\right)}{Q_{m-1}^2 \left|f_0\left(\frac{P_{m-1}}{Q_{m-1}}\right)\right|} + (-1)^{m-1}\frac{\lambda_m}{Q_{m-1}^2},$$

where

$$\lambda_m = \sum_{j=2}^{n} \frac{1}{\frac{(-1)^m \theta_{m-1}}{Q_{m-1}Q_m} + \alpha^{(1)} - \alpha^{(j)}}.$$

Next, we note that for $m > m_0$

$$\frac{1}{\frac{(-1)^m \theta_{m-1}}{Q_{m-1}Q_m} + \alpha^{(1)} - \alpha^{(j)}} = \frac{1}{\alpha^{(1)} - \alpha^{(j)}} - (-1)^m \frac{\theta_{m-1}}{Q_{m-1}Q_m} \cdot$$

$$\cdot \frac{1}{\left(\frac{(-1)^m \theta_{m-1}}{Q_{m-1}Q_m} + \alpha^{(1)} - \alpha^{(j)}\right)\left(\alpha^{(1)} - \alpha^{(j)}\right)} = \frac{1}{\alpha^{(1)} - \alpha^{(j)}} +$$

$$+(-1)^{m-1}\frac{\theta_{m-1}}{Q_{m-1}Q_m}\frac{\varepsilon}{\left(\alpha^{(1)} - \alpha^{(j)}\right)^2},$$

where $|\varepsilon| < 2$. Finally, we obtain

$$\lambda_m = A_1(\alpha) + \frac{(-1)^{m-1}\theta_{m-1}}{Q_{m-1}Q_m} A_2(\alpha)\varepsilon_m, \quad |\varepsilon_m| < 2.$$

This completes the proof. □

5.8 Chain Sequence of Linear Fractional Transformations of Plane

Recall the definition of a convergence of sequence of the integral matrixes to a number given in [15].

Definition 5.6 We say that a matrix decomposition

$$\prod_{k=0}^{\infty} \begin{pmatrix} a_k & b_k \\ c_k & d_k \end{pmatrix}$$

convergence to a number α if for matrixes

$$M_n = \prod_{k=0}^{n} \begin{pmatrix} a_k & b_k \\ c_k & d_k \end{pmatrix} = \begin{pmatrix} A_n & B_n \\ C_n & D_n \end{pmatrix}$$

the following relation holds

$$\lim_{n\to\infty} \frac{A_n}{C_n} = \lim_{n\to\infty} \frac{B_n}{D_n} = \alpha.$$

In this case, we write

$$\begin{pmatrix} \alpha \\ 1 \end{pmatrix} = \prod_{k=0}^{\infty} \begin{pmatrix} a_k & b_k \\ c_k & d_k \end{pmatrix}.$$

The theory of matrix representations of real numbers is systematically stated in [11]. We are now interested in the case of ordinary continued fractions. If α is expanded in a continued fraction (5.1), then we obtain the following matrix decomposition

$$\begin{pmatrix} \alpha \\ 1 \end{pmatrix} = \prod_{\nu=0}^{\infty} \begin{pmatrix} q_\nu & 1 \\ 1 & 0 \end{pmatrix}, \tag{5.33}$$

because

$$M_m = \prod_{\nu=0}^{m} \begin{pmatrix} q_\nu & 1 \\ 1 & 0 \end{pmatrix} = \begin{pmatrix} P_m & P_{m-1} \\ Q_m & Q_{m-1} \end{pmatrix} \quad (m \geq 0)$$

and the sequence of the matrixes M_m converges to α by the properties of the convergents.

Consider now an arbitrary linear fractional transformation of complex plane with matrix M:

$$M = \begin{pmatrix} A & B \\ C & D \end{pmatrix}, \quad w = M(z) = \frac{Az+B}{Cz+D}.$$

By (5.2), it follows that irrational number α and a residual fraction α_{k+1} relate to inverse linear fractional transformation:

$$M_k = \begin{pmatrix} P_k & P_{k-1} \\ Q_k & Q_{k-1} \end{pmatrix}, \quad M_k^* = \begin{pmatrix} Q_{k-1} & -P_{k-1} \\ -Q_k & P_k \end{pmatrix}$$

$$\begin{cases} \alpha = M_k(\alpha_{k+1}) \\ \alpha_{k+1} = M_k^*(\alpha) \end{cases}. \tag{5.34}$$

Analyzing the formula (5.7) for the root of a minimal polynomial $f_m(x)$, we conclude that they are obtained from the roots of a given minimal polynomial under the linear fractional transformation M^*_{k-1}.

Define the following concept.

Definition 5.7 Let α be a real irrationality. A chain sequence of the first kind of linear fractional transformations for the polynomials is called a sequence

$$\left\{ M_\nu(\alpha) = \begin{pmatrix} P_\nu(\alpha) & P_{\nu-1}(\alpha) \\ Q_\nu(\alpha) & Q_{\nu-1}(\alpha) \end{pmatrix} \middle| \, \nu = 0, 1, \dots \right\},$$

where $P_\nu(\alpha)$ is a numerator and $Q_\nu(\alpha)$ is a denominator of νth-order convergent of α.

A chain sequence of the first kind of linear fractional transformations of complex plane is called a sequence

$$\left\{ M^*_\nu(\alpha) = \begin{pmatrix} Q_{\nu-1}(\alpha) & -P_{\nu-1}(\alpha) \\ -Q_\nu(\alpha) & P_\nu(\alpha) \end{pmatrix} \middle| \, \nu = 0, 1, \dots \right\}.$$

To understand the effect of focusing the algebraic conjugate numbers for a residual fraction α_m around fraction $-\frac{Q_{m-2}}{Q_{m-1}}$, we prove the following lemmas.

Lemma 5.15 *Let $M^*(z)$ be an arbitrary linear fractional transformation of complex plane with unimodular matrix M^*:*

$$M^* = \begin{pmatrix} D & -B \\ -C & A \end{pmatrix} \quad A, B, C, D \in \mathbb{Z},\ |AD - BC| = 1,\ C \neq 0.$$

Then,

(1) *the image of the exterior of the circle $K\left(\frac{A}{C}, 1\right) = \left\{z \mid \left|z - \frac{A}{C}\right| \geq 1\right\}$ is the inside of the circle $K\left(-\frac{D}{C}, \frac{1}{C^2}\right)$ with center deleted;*
(2) *the image of the circumference $C\left(\frac{A}{C}, 1\right) = \left\{z \mid \left|z - \frac{A}{C}\right| = 1\right\}$ is the circumference $C\left(-\frac{D}{C}, \frac{1}{C^2}\right)$;*
(3) *the image of the inside of the circle $K\left(\frac{A}{C}, 1\right)$ with center deleted is the exterior of the circle $K\left(-\frac{D}{C}, \frac{1}{C^2}\right)$;*
(4) *the image of the ring*

$$R\left(\frac{A}{C}, 1, r\right) = \left\{ z \,\middle|\, r < \left| z - \frac{A}{C} \right| < 1 \right\} \quad (0 < r < 1)$$

is the ring $R\left(-\frac{D}{C}, \frac{1}{rC^2}, \frac{1}{C^2}\right)$;
(5) *the point $z = \frac{A}{C}$ is a pole of the linear fractional transformation $M^*(z)$ with residue $\frac{1}{C^2}$.*

Proof Indeed,

$$M^*(z) = \frac{Dz - B}{-Cz + A} = -\frac{D}{C} + \frac{AD - BC}{C(A - Cz)}, \quad \left| M^*(z) + \frac{D}{C} \right| = \frac{1}{C^2 \left| \frac{A}{C} - z \right|},$$

hence, all the statements of the lemma hold. □

Consider the linear fractional transformation $N^*(z)$ with matrix

$$N^* = \begin{pmatrix} C & D \\ 0 & C \end{pmatrix} \quad C, D \in \mathbb{Z}, C \neq 0, \quad N^*(z) = \frac{Cz + D}{C} = z + \frac{D}{C}.$$

It is easy to see that

$$M_1^* = N^* \cdot M^* = \begin{pmatrix} C & D \\ 0 & C \end{pmatrix} \begin{pmatrix} D & -B \\ -C & A \end{pmatrix} = \begin{pmatrix} 0 & AD - BC \\ -C^2 & AC \end{pmatrix}$$

Lemma 5.16 *Let*

$$M = \begin{pmatrix} A & B \\ C & D \end{pmatrix}, \quad N = \begin{pmatrix} C & -D \\ 0 & C \end{pmatrix}, \quad A, B, C, D \in \mathbb{Z}, \quad C \neq 0,$$

$M_1 = N \circ M$ be the linear fractional transformation of polynomials with matrix

$$M_1 = M \cdot N = \begin{pmatrix} A & B \\ C & D \end{pmatrix} \begin{pmatrix} C & -D \\ 0 & C \end{pmatrix} = \begin{pmatrix} AC & BC - AD \\ C^2 & 0 \end{pmatrix}$$

and let $\beta^{(1)}, \dots, \beta^{(n)}$ be the roots of the polynomial $g(x) = M_1(f(x))$.

Then,

$$g(x) = M_1(f(x)) = C^{2n} x^n f\left(\frac{A}{C} + \frac{BC - AD}{C^2 x}\right) =$$
$$= C^{2n} f\left(\frac{A}{C}\right) x^n + \sum_{\nu=1}^{n} \frac{f^{(\nu)}\left(\frac{A}{C}\right)}{\nu!} C^{2(n-\nu)} x^{n-\nu} (BC - AD)^{\nu} \tag{5.35}$$

and

$$\beta^{(\nu)} = M_1^*\left(\alpha^{(\nu)}\right) = \frac{AD - BC}{C^2\left(\frac{A}{C} - \alpha^{(\nu)}\right)} \quad (1 \leq \nu \leq n). \tag{5.36}$$

Proof Indeed, if $h(x) = M(f(x))$, then

$$h(x) = (Cx + D)^n f\left(\frac{Ax + B}{Cx + D}\right), \quad g(x) = M_1(f(x)) = C^n h\left(x - \frac{D}{C}\right) =$$
$$= C^n \left(C\left(x - \frac{D}{C}\right) + D\right)^n f\left(\frac{A\left(x - \frac{D}{C}\right) + B}{C\left(x - \frac{D}{C}\right) + D}\right) =$$

$$= \left(C^2x\right)^n f\left(\frac{ACx + (BC - AD)}{C^2x}\right) = M_1(f(x)).$$

Using Taylor formula, we get

$$\left(C^2x\right)^n f\left(\frac{ACx + (BC - AD)}{C^2x}\right) = \left(C^2x\right)^n f\left(\frac{A}{C} + \frac{BC - AD}{C^2x}\right) =$$

$$= C^{2n}f\left(\frac{A}{C}\right)x^n + \sum_{\nu=1}^{n} \frac{f^{(\nu)}\left(\frac{A}{C}\right)}{\nu!} C^{2(n-\nu)}x^{n-\nu}(BC - AD)^\nu$$

and the equality (5.35) is proved.

Then, we have

$$g(x) = \left(C^2x\right)^n a_n \prod_{\nu=1}^{n} \left(\frac{ACx + (BC - AD)}{C^2x} - \alpha^{(\nu)}\right) =$$

$$= a_n \prod_{\nu=1}^{n} \left((AC - C^2\alpha^{(\nu)})x - (AD - BC)\right) =$$

$$= C^{2n}f\left(\frac{A}{C}\right) \prod_{\nu=1}^{n} \left(x - \frac{AD - BC}{C^2\left(\frac{A}{C} - \alpha^{(\nu)}\right)}\right) = C^{2n}f\left(\frac{A}{C}\right) \prod_{\nu=1}^{n} \left(x - \beta^{(\nu)}\right)$$

and this proves the equality (5.36). □

5.9 Lagrange Algorithm for Reduced Algebraic Irrationality of Degree *n*

Recall the definition of a reduced algebraic irrationality of degree n given in Sect. 5.2 (Definition 5.1). For the infinite continued fraction expansion (5.1) of a reduced algebraic irrationality α, the following theorem holds.

Theorem 5.9 *The incomplete quotient q_k is uniquely determined as a natural number satisfying the following condition:*

$$f_k(q_k) < 0, \quad f_k(q_k + 1) > 0.$$

Proof Indeed, since $f_k(\alpha_k) = 0$, $q_k < \alpha_k < q_k + 1$, $a_{n,k} > 0$, and α_k is the unique positive root of the polynomial $f_k(x)$, it follows that $f_k(q_k) < 0$ and $f_k(q_{k+1}) > 0$. □

It is easily shown that we need to calculate $O(\ln q_k)$ values of $f_k(x)$ for computation q_k. Indeed, consider a sequence $f_k(1), f_k(2), \ldots, f_k(2^m), f_k(2^{m+1})$, where $m = \left[\log_2(q_k)\right]$. It is clear that $f_k(2^j) < 0$ for $0 \le j \le m$ and $f_k(2^{m+1}) > 0$. Then, using the method of interval bisection contract segment $[2^m; 2^{m+1}]$ to segment $[q_k; q_k + 1]$, that will require to compute yet m values of $f_k(x)$.

Thus, the description of the version Lagrange algorithm for calculating the incomplete quotients in continued fraction expansion of a reduced algebraic irrationality α of degree n is completed.

Theorem 5.1 generalizes to continued fraction of arbitrary totally real algebraic irrationality α of degree n. First, we prove the following lemma.

Lemma 5.17 *Let*

$$f(x) = \sum_{k=0}^{n} a_k x^k \in \mathbb{Z}[x], \quad a_n > 0$$

be arbitrary irreducible integer polynomial, all of whose roots $\alpha^{(k)}$ $(k = 1, 2, \ldots, n)$ *are different real numbers satisfying the following condition:*

$$\alpha^{(n)} < \ldots < \alpha^{(2)} < \alpha^{(1)}.$$

Suppose that for integer number q, the following inequalities hold:

$$\begin{cases} \alpha^{(k)} < q & \text{for } k \geq k_0, \\ q < \alpha^{(k)} < q + 1 & \text{for } k_0 > k \geq k_1, \\ \alpha^{(k)} > q + 1 & \text{for } k_1 > k \geq 1, \end{cases}$$

Then, the polynomial

$$g(x) = -f\left(q + \frac{1}{x}\right) \cdot x^n = \sum_{k=0}^{n} b_k x^k.$$

has roots $\beta^{(k)} = \frac{1}{\alpha^{(k)} - q}$ $(k = 1, 2, \ldots, n)$ *satisfying the following inequalities:*

$$\begin{cases} \beta^{(k)} < 0 & \text{for } k \geq k_0, \\ 1 < \beta^{(k)} & \text{for } k_0 > k \geq k_1, \\ 0 < \beta^{(k)} < 1 & \text{for } k_1 > k \geq 1. \end{cases}$$

Proof The proof is similar to that of Lemma 5.1. □

Theorem 5.10 *For arbitrary totally real algebraic irrationality* α *of degree n, all of its residual fractions* α_m *are reduced algebraic irrationalities of nth degree starting with some index* $m_0 + 1$.

Proof Let $\alpha = \alpha^{(j)}$ and

$$\alpha^{(n)} < \ldots < \alpha^{(2)} < \alpha^{(1)}$$

be the real roots of irreducible integer polynomial

$$f(x) = \sum_{k=0}^{n} a_k x^k \in \mathbb{Z}[x], \quad a_n > 0.$$

Let $q_0 = [\alpha]$, $k_{0,0} = k_0$, and $k_{1,0} = k_1$ are defined as in Lemma 5.17 for $q = q_0$. Then, $k_{0,0} > j \geq k_{1,0}$ and the polynomial

$$f_1(x) = -f\left(q_0 + \frac{1}{x}\right) \cdot x^n = \sum_{k=0}^{n} a_{k,1} x^k$$

has roots

$$\alpha_1^{(n)} < \ldots < \alpha_1^{(2)} < \alpha_1^{(1)},$$

among which there are $n + 1 - k_{0,0}$ negative roots, $k_{1,0} - 1$ positive roots less than 1, and $k_{0,0} - k_{1,0}$ positive roots more than 1.

Notice that $\alpha_1 = \alpha_1^{(j_1)}$ and $k_{0,0} - k_{1,0} \geq j_1 \geq 1$.

Let integer polynomial $f_m(x)$ for residual fraction $\alpha_m = \alpha_m^{(j_m)}$ be determined. Then, defining $q = q_m = [\alpha_m]$, $k_{0,m} = k_0$, and $k_{1,m} = k_1$ as in Lemma 5.17, we get $k_{0,m} > j_m \geq k_{1,m}$ and a polynomial

$$f_{m+1}(x) = -f_m\left(q_m + \frac{1}{x}\right) \cdot x^n = \sum_{k=0}^{n} a_{k,m+1} x^k$$

has roots

$$\alpha_{m+1}^{(n)} < \ldots < \alpha_{m+1}^{(2)} < \alpha_{m+1}^{(1)},$$

among which there are $n + 1 - k_{0,m}$ negative roots, $k_{1,m} - 1$ positive roots less than 1, and $k_{0,m} - k_{1,m}$ positive roots more than 1.

It is clear that $\alpha_{m+1} = \alpha_{m+1}^{(j_{m+1})}$ and $k_{0,m} - k_{1,m} \geq j_{m+1} \geq 1$.

By the proof of Lemma 5.17, it follows that $j_1 \geq j_2 \geq \ldots \geq j_m \geq \ldots$; $k_{0,1} \geq \geq k_{0,2} = k_{0,1} - k_{1,0} + 1 \geq \ldots \geq k_{0,m} = k_{0,m-1} - k_{1,m-1} + 1 \geq \ldots$.

The numbers $k_{0,m}, k_{1,m}$ have a simple meaning: For $k_{0,m} > \nu \geq k_{1,m}$, $\alpha_m^{(\nu)}$ is mth residual fraction for $\alpha^{(\nu+j-j_m)}$. By unique continued fraction expansion, it follows that there exists m_0 such that for $0 \leq k < m_0 - 1$, the incomplete quotients q_k are the same for $\alpha^{(\nu)}$ if $k_2 \geq \nu \geq k_3$, $k_2 \geq j \geq k_3$, and the incomplete quotients q_{m_0-1} for $\alpha = \alpha^{(j)}$ differ from corresponding incomplete quotients for $\alpha^{(\nu)}$ if $k_2 \geq \nu \geq k_3$. This implies that $k_{0,m_0-1} = k_{1,m_0-1} + 1$, $k_{0,m_0} = 2$, $k_{1,m_0} = 1$. Thus, $\alpha_{m_0+1} = \alpha_{m_0+1}^{(1)}$ is a reduced algebraic irrationality.

This completes the proof (Fig. 5.1). □

```
cfki1(p,n) := | a ← (−1  −p  −p + 1  1)^T, q_{floor((n−1)/40),39} ← 0, q_{0,0} ← p, D^⟨0⟩ ← a, D^⟨n⟩ ← a
              | l0 ← 0, l1 ← 1, b ← p
              | for k ∈ 1..n − 1
              |   | r ← b, a ← (−a3  −a3·3·r − a2  −a3·3·r^2 − a2·2·r − a1  −a3·r^3 − a2·r^2 − a1·r − a0)^T
              |   | d ← −a0
              |   | for j ∈ 1..3
              |   |   | m ← d, l ← aj
              |   |   | r ← m, m ← l, l ← r  if m > l
              |   |   | while m > 0
              |   |   |   r ← floor(l/m), r1 ← l − r·m, l ← m, m ← r1
              |   |   | d ← l
              |   | b ← 1, c ← 2, D^⟨k⟩ ← a
              |   | while [(a3·c + a2)·c + a1]·c + a0 < 0
              |   |   b ← c, c ← 2·c
              |   | m ← b
              |   | while m ≥ 1
              |   |   | r ← b + m, f ← [(a3·r + a2)·r + a1]·r + a0
              |   |   | b ← r  if f < 0
              |   |   | c ← r  otherwise
              |   |   | m ← m/2
              |   | q_{l0,l1} ← b, l1 ← l1 + 1
              |   | l0 ← l0 + 1, l1 ← 0  if l1 = 40
              | q
```

Fig. 5.1 Shows the program text for computing the incomplete quotients for a reduced cubic irrationalities $\alpha(p)$. For given natural $p \geq 4$, this program computes n incomplete quotients of continued fraction expansion for $\alpha(p)$ in the form of a table of 40 values in one line

5.10 Modification Lagrange Algorithm for Continued Fraction Expansion of Algebraic Number

The importance of generalized Pisot number for the Lagrange algorithm for continued fraction expansion is explained by the following lemma.

Lemma 5.18 *If*

$$f_0(x) = a_n x^n + a_{n-1}x^{n-1} + \ldots + a_1 x + a_0 \in \mathbb{Z}[x], \quad a_n \geq 1$$

is a minimal polynomial for generalized Pisot number $\alpha^{(1)} = \alpha_0$, *and then for continued fraction expansion*

$$\alpha^{(1)} = \alpha_0 = q_0 + \cfrac{1}{q_1 + \cfrac{1}{\ddots + \cfrac{1}{q_n + \cfrac{1}{\ddots}}}}$$

the following inequality holds

$$\left[-\frac{a_{n-1}}{a_n}\right] + 1 - n \leq q_0 < -\frac{a_{n-1}}{a_n} + n - 1. \tag{5.37}$$

Proof Indeed, by Vieta formula, we have:

$$-\frac{a_{n-1}}{a_n} = \alpha^{(1)} + \alpha^{(2)} + \ldots + \alpha^{(n)}.$$

Since the minimal polynomial $f_0(x)$ is irreducible, it follows that

$$\alpha^{(2)} + \alpha^{(3)} + \ldots + \alpha^{(n)} \neq 0,$$

otherwise, $\alpha^{(1)} = -\frac{a_{n-1}}{a_n} \in \mathbb{Q}$, which contradicts irreducibility of $f_0(x)$.

Since $\alpha^{(1)}$ is Pisot number

$$|\alpha^{(j)}| < 1, \quad (2 \leq j \leq n).$$

So

$$0 < |\alpha^{(2)} + \ldots + \alpha^{(n)}| < n - 1$$

and

$$-\frac{a_{n-1}}{a_n} + 1 - n < \alpha^{(1)} < -\frac{a_{n-1}}{a_n} + n - 1.$$

But $q_0 < \alpha^{(1)} < q_0 + 1$, so the statements of the lemma hold. □

Thus, Theorem 5.5 and Lemma 5.18 imply that starting from some m_0, all partial quotients q_m $(m \geq m_0)$ require for their computations no more than $O(\ln n)$ calculation values of the $f_m(x)$. This result can be significantly intensified using the asymptotic formula (5.21) for conjugate numbers to the residual fractions.

Theorem 5.11 *Let $\alpha = \alpha_0$ be a real root of a irreducible integer polynomial*

$$f_0(x) = a_n x^n + a_{n-1}x^{n-1} + \ldots + a_1 x + a_0 \in \mathbb{Z}[x],\, a_n > 0,$$

$\alpha = \alpha^{(1)}, \alpha^{(2)}, \ldots, \alpha^{(n)}$ *its root and*

$$\alpha = \alpha_0 = q_0 + \cfrac{1}{q_1 + \cfrac{1}{\ddots + \cfrac{1}{q_k + \cfrac{1}{\ddots}}}} \text{andlet}$$

be continued fraction expansion of α.

Suppose that a sequence of the polynomials $f_m(x)$ for the residual fractions α_m is defined by (5.17) and $m_0 = m_0(\alpha, \varepsilon)$ is defined from the inequality

$$\frac{2(n-1)}{Q_{m_0-1}\delta(\alpha)} < \varepsilon. \tag{5.38}$$

Then, for any $m > m_0$, the following equalities hold

$$q_m = \begin{cases} q_m^* & \text{for } f_m\left(q_m^* + 1\right) > 0 \text{ and } f_m\left(q_m^*\right) < 0, \\ q_m^* + 1 & \text{for } f_m\left(q_m^* + 1\right) < 0, \\ q_m^* - 1 & \text{for } f_m\left(q_m^*\right) > 0, \end{cases} \tag{5.39}$$

where

$$q_m^* = \left[-\frac{f'_{m-1}(q_{m-1})}{f_{m-1}(q_{m-1})} + \frac{(n-1)Q_{m-2}}{Q_{m-1}} \right].$$

Proof By Theorem 5.5, so that

$$f_m(x) = -f_{m-1}(q_{m-1})x^n - \frac{f'_{m-1}(q_{m-1})}{1!}x^{n-1} - \sum_{\nu=2}^{n} \frac{f^{(\nu)}_{m-1}(q_{m-1})}{\nu!}x^{n-\nu}.$$

Using Vieta formula, we get

$$-\frac{f'_{m-1}(q_{m-1})}{f_{m-1}(q_{m-1})} = \alpha_m^{(1)} + \sum_{j=2}^{n} \left(-\frac{Q_{m-2}}{Q_{m-1}} + \frac{(-1)^m}{Q_{m-1}^2 \left(\frac{(-1)^m \theta_{m-1}}{Q_{m-1}Q_m} + \alpha^{(1)} - \alpha^{(j)} \right)} \right).$$

Therefore,

$$\alpha_m^{(1)} = -\frac{f'_{m-1}(q_{m-1})}{f_{m-1}(q_{m-1})} + \frac{(n-1)Q_{m-2}}{Q_{m-1}} + \Delta,$$

where

$$\Delta = \sum_{j=2}^{n} \left(\frac{(-1)^{m-1}}{Q_{m-1}^2 \left(\frac{(-1)^m \theta_{m-1}}{Q_{m-1}Q_m} + \alpha^{(1)} - \alpha^{(j)} \right)} \right)$$

and

$$|\Delta| < \frac{2(n-1)}{Q_{m-1}^2 \delta(\alpha)} < \frac{\varepsilon}{Q_{m-1}}.$$

Since $f_m(x) > 0$ for $x > \alpha_m^{(1)}$, $f_m(x) < 0$ for $1 \le x < \alpha_m^{(1)}$ and $q_m^* - 1 < \alpha_m^{(1)} < q_m^* + 2$, there exist three possible cases:

(1) if $f_m(q_m^* + 1) < 0$, then $q_m = q_m^* + 1$;
(2) if $f_m(q_m^*) < 0$ and $f_m(q_m^* + 1) > 0$, then $q_m = q_m^*$;
(3) if $f_m(q_m^*) > 0$, then $q_m = q_m^* - 1$.

This completes the proof. □

5.11 Properties of Matrix Decomposition

We will consider further only nonnegative integer nondegenerate matrix.

Notice some simple properties of the matrix decomposition.

Lemma 5.19 *Let*

$$\begin{pmatrix} \alpha \\ 1 \end{pmatrix} = \prod_{k=0}^{\infty} \begin{pmatrix} a_k & b_k \\ c_k & d_k \end{pmatrix}$$

be a convergent matrix decomposition, $i_1 < \ldots i_n < \ldots$ *be arbitrary monotonic sequence of natural numbers and* $i_0 = 0$.

If the matrixes m_k *are defined by the equalities*

$$m_k = \prod_{j=i_k}^{i_{k+1}-1} \begin{pmatrix} a_j & b_j \\ c_j & d_j \end{pmatrix} \quad (k = 0, 1, \ldots),$$

then matrix product

$$\prod_{k=0}^{\infty} m_k$$

converges to α.

Proof Indeed, if

$$\prod_{k=0}^{n} \begin{pmatrix} a_k & b_k \\ c_k & d_k \end{pmatrix} = \begin{pmatrix} A_n & B_n \\ C_n & D_n \end{pmatrix} = M_n \quad \text{and} \quad \begin{pmatrix} \alpha \\ 1 \end{pmatrix} = \lim_{n\to\infty} M_n,$$

then

$$\lim_{n\to\infty} \frac{A_n}{C_n} = \lim_{n\to\infty} \frac{B_n}{D_n} = \alpha,$$

hence

$$\lim_{k\to\infty} \frac{A_{i_k-1}}{C_{i_k-1}} = \lim_{k\to\infty} \frac{B_{i_k-1}}{D_{i_k-1}} = \alpha.$$

Applying the associative law of a matrix product, we get

$$\prod_{k=0}^{n} m_k = \prod_{k=0}^{n} \left(\prod_{j=i_k}^{i_{k+1}-1} \begin{pmatrix} a_j & b_j \\ c_j & d_j \end{pmatrix} \right) = \prod_{k=0}^{i_{n+1}-1} \begin{pmatrix} a_k & b_k \\ c_k & d_k \end{pmatrix} =$$
$$= \begin{pmatrix} A_{i_{n+1}-1} & B_{i_{n+1}-1} \\ C_{i_{n+1}-1} & D_{i_{n+1}-1} \end{pmatrix} = M_{i_{n+1}-1}.$$

Thus, the matrix product

$$\prod_{k=0}^{\infty} m_k$$

converges to α. □

Lemma 5.20 *Let*

$$\begin{pmatrix} \alpha \\ 1 \end{pmatrix} = \prod_{k=0}^{\infty} \begin{pmatrix} a_k & b_k \\ c_k & d_k \end{pmatrix}$$

be a convergent matrix decomposition and

$$\begin{pmatrix} a & b \\ c & d \end{pmatrix}, \quad a, b, c, d \geq 0, \quad \det \begin{pmatrix} a & b \\ c & d \end{pmatrix} \neq 0.$$

Then, the matrix product

$$\begin{pmatrix} a & b \\ c & d \end{pmatrix} \prod_{k=0}^{\infty} \begin{pmatrix} a_k & b_k \\ c_k & d_k \end{pmatrix}$$

converges to $\frac{a\alpha+b}{c\alpha+d}$.

Proof Indeed, if

$$\prod_{k=0}^{n} \begin{pmatrix} a_k & b_k \\ c_k & d_k \end{pmatrix} = \begin{pmatrix} A_n & B_n \\ C_n & D_n \end{pmatrix},$$

then

$$\lim_{n\to\infty} \frac{A_n}{C_n} = \lim_{n\to\infty} \frac{B_n}{D_n} = \alpha$$

Hence,

$$\begin{pmatrix} a & b \\ c & d \end{pmatrix} \prod_{k=0}^{n} \begin{pmatrix} a_k & b_k \\ c_k & d_k \end{pmatrix} = \begin{pmatrix} aA_n + bC_n & aB_n + bD_n \\ cA_n + dC_n & cB_n + dD_n \end{pmatrix}$$

and

$$\lim_{n\to\infty} \frac{aA_n + bC_n}{cA_n + dC_n} = \lim_{n\to\infty} \frac{a\frac{A_n}{C_n} + b}{c\frac{A_n}{C_n} + d} = \frac{a\alpha + b}{c\alpha + d} = \lim_{n\to\infty} \frac{a\frac{B_n}{D_n} + b}{c\frac{B_n}{D_n} + d} = \lim_{n\to\infty} \frac{aB_n + bD_n}{cB_n + dD_n}.$$

Since all matrixes are nonnegative and $\alpha > 0$, lemma is proved □

Lemma 5.21 *Let*

$$\begin{pmatrix} \alpha \\ 1 \end{pmatrix} = \prod_{k=0}^{\infty} \begin{pmatrix} a_k & b_k \\ c_k & d_k \end{pmatrix}$$

be a convergent matrix decomposition.

Then, for any $n > 0$, the matrix product

$$\prod_{k=n}^{\infty} \begin{pmatrix} a_k & b_k \\ c_k & d_k \end{pmatrix}$$

converges to β_n and $\alpha = \frac{A_{n-1}\beta_n + B_{n-1}}{C_{n-1}\beta_n + D_{n-1}}$.

Proof The statement of the lemma follows from the preceding Lemma 5.20 for $a = A_{n-1}$, $b = B_{n-1}$, $c = C_{n-1}$, and $d = D_{n-1}$. □

Lemma 5.22 *Let*

$$\begin{pmatrix} \alpha \\ 1 \end{pmatrix} = \prod_{k=0}^{\infty} \begin{pmatrix} a_k & b_k \\ c_k & d_k \end{pmatrix}$$

be a convergent matrix decomposition, $i_1 < \dots i_n < \dots$ be arbitrary monotonic sequence integer nonnegative numbers and

$$\begin{pmatrix} a_{i_j} & b_{i_j} \\ c_{i_j} & d_{i_j} \end{pmatrix} = \begin{pmatrix} f_j & 0 \\ 0 & f_j \end{pmatrix} \quad (j = 1, 2, \dots).$$

Then, the matrix product

$$\prod_{j=1}^{\infty} \prod_{k=i_{j-1}+1}^{i_j-1} \begin{pmatrix} a_k & b_k \\ c_k & d_k \end{pmatrix},$$

converges to α (here $i_0 = -1$)

Proof Indeed, let

$$M_n = \prod_{k=0}^{n} \begin{pmatrix} a_k & b_k \\ c_k & d_k \end{pmatrix} = \begin{pmatrix} A_n & B_n \\ C_n & D_n \end{pmatrix},$$

$$M'_m = \prod_{j=1}^{m} \prod_{k=i_{j-1}+1}^{i_j-1} \begin{pmatrix} a_k & b_k \\ c_k & d_k \end{pmatrix} = \begin{pmatrix} A'_m & B'_m \\ C'_m & D'_m \end{pmatrix},$$

$$F_m = \prod_{j=1}^{m} f_j.$$

Then,

$$M_{i_m} = \begin{pmatrix} A_{i_m} & B_{i_m} \\ C_{i_m} & D_{i_m} \end{pmatrix} = F_m M'_m = \begin{pmatrix} F_m A'_m & F_m B'_m \\ F_m C'_m & F_m D'_m \end{pmatrix}.$$

Hence,

$$\alpha = \lim_{m\to\infty} \frac{A_{i_m}}{C_{i_m}} = \lim_{m\to\infty} \frac{A'_m}{C'_m} = \lim_{m\to\infty} \frac{B'_m}{D'_m} = \lim_{m\to\infty} \frac{B_{i_m}}{D_{i_m}}$$

and the lemma is proved. □

Lemma 5.23 *Let*

$$\begin{pmatrix} \alpha \\ 1 \end{pmatrix} = \prod_{k=0}^{\infty} \begin{pmatrix} a_k & b_k \\ c_k & d_k \end{pmatrix}$$

be a matrix decomposition that converges to an irrational number α.

Then, all matrixes contained in the decomposition are nondegenerate.

Proof Assuming the converse, let

$$\begin{pmatrix} a_n & b_n \\ c_n & d_n \end{pmatrix}$$

be a degenerate matrix. Then, the matrix

$$M_n = \prod_{k=0}^{n} \begin{pmatrix} a_k & b_k \\ c_k & d_k \end{pmatrix} = \begin{pmatrix} A_n & B_n \\ C_n & D_n \end{pmatrix}$$

is degenerate too, that is, $\frac{B_n}{D_n} = \frac{A_n}{C_n}$ or $C_n = mA_n$, $D_n = mB_n$. Calculating

$$\begin{pmatrix} A_{n+1} & B_{n+1} \\ C_{n+1} & D_{n+1} \end{pmatrix} = \begin{pmatrix} A_n & B_n \\ C_n & D_n \end{pmatrix} \cdot \begin{pmatrix} a_{n+1} & b_{n+1} \\ c_{n+1} & d_{n+1} \end{pmatrix},$$

we get $\frac{A_{n+1}}{C_{n+1}} = \frac{B_{n+1}}{D_{n+1}} = \frac{A_n}{C_n}$.

Thus, $\frac{A_k}{C_k} = \frac{B_k}{D_k} = \frac{A_n}{C_n}$ for $k \geq n$. This contradiction proves the lemma. □

Denote by $\Delta_n = \det M_n = A_n D_n - B_n C_n$, $\delta_n = a_n d_n - b_n c_n$.

Lemma 5.24 *Let*

$$\prod_{k=0}^{\infty} \begin{pmatrix} a_k & b_k \\ c_k & d_k \end{pmatrix}$$

be an infinite matrix decomposition, and all matrixes included in the decomposition are nondegenerate integer positive with a condition

$$\delta_k < 0, \quad \min\left(\frac{|\delta_n|}{a_n d_n}, \frac{|\delta_n|}{c_n b_n}\right) \leq \delta < 1.$$

Then, matrix product converges.

Proof First, we notice that

$$\begin{aligned} \Delta_n = \det M_n = A_n D_n - B_n C_n = \\ = (A_{n-1} a_n + B_{n-1} c_n)(C_{n-1} b_n + D_{n-1} d_n) - \\ -(A_{n-1} b_n + B_{n-1} d_n)(C_{n-1} a_n + D_{n-1} c_n) = \\ = (A_{n-1} D_{n-1} - B_{n-1} C_{n-1})(a_n d_n - b_n c_n) = \Delta_{n-1} \delta_n. \end{aligned}$$

Hence, we get $\Delta_n = (-1)^{n+1} |\Delta_n| \quad (n = 0, 1, \ldots)$.

Consider the differences $\frac{A_n}{C_n} - \frac{B_n}{D_n}$ for $n = 0, 1, \ldots$. We obtain:

$$\frac{A_n}{C_n} - \frac{B_n}{D_n} = \frac{\Delta_n}{C_n D_n} = \frac{\Delta_{n-1} \delta_n}{(C_{n-1} a_n + D_{n-1} c_n)(C_{n-1} b_n + D_{n-1} d_n)},$$

$$\left|\frac{A_n}{C_n}-\frac{B_n}{D_n}\right| \le \frac{|\Delta_{n-1}|}{C_{n-1}D_{n-1}} \min\left(\frac{|\delta_n|}{a_n d_n}, \frac{|\delta_n|}{c_n b_n}\right) < \frac{|\Delta_{n-1}|\delta}{C_{n-1}D_{n-1}};$$

$$\frac{A_n}{C_n}-\frac{A_{n-1}}{C_{n-1}} = \frac{(A_{n-1}a_n + B_{n-1}c_n)C_{n-1} - A_{n-1}(C_{n-1}a_n + D_{n-1}c_n)}{C_n C_{n-1}} = \\ = \frac{c_n(B_{n-1}C_{n-1} - A_{n-1}D_{n-1})}{C_n C_{n-1}} = \frac{-c_n\Delta_{n-1}}{C_n C_{n-1}};$$

$$\frac{B_n}{D_n}-\frac{B_{n-1}}{D_{n-1}} = \frac{(A_{n-1}b_n + B_{n-1}d_n)D_{n-1} - B_{n-1}(C_{n-1}b_n + D_{n-1}d_n)}{D_n D_{n-1}} = \\ = \frac{b_n(A_{n-1}D_{n-1} - B_{n-1}C_{n-1})}{D_n D_{n-1}} = \frac{b_n\Delta_{n-1}}{D_n D_{n-1}};$$

$$\frac{A_n}{C_n}-\frac{B_{n-1}}{D_{n-1}} = \frac{(A_{n-1}a_n + B_{n-1}c_n)D_{n-1} - (C_{n-1}a_n + D_{n-1}c_n)B_{n-1}}{C_n D_{n-1}} = \\ = \frac{a_n(A_{n-1}D_{n-1} - B_{n-1}C_{n-1})}{C_n D_{n-1}} = \frac{a_n\Delta_{n-1}}{C_n D_{n-1}};$$

$$\frac{B_n}{D_n}-\frac{A_{n-1}}{C_{n-1}} = \frac{(A_{n-1}b_n + B_{n-1}d_n)C_{n-1} - (C_{n-1}b_n + D_{n-1}d_n)A_{n-1}}{D_n C_{n-1}} = \\ = \frac{-d_n(A_{n-1}D_{n-1} - B_{n-1}C_{n-1})}{D_n C_{n-1}} = \frac{-d_n\Delta_{n-1}}{D_n C_{n-1}}.$$

It follows that

$$\frac{A_0}{C_0} < \frac{B_0}{D_0}, \quad \frac{A_{2k}}{C_{2k}} < \frac{B_{2k}}{D_{2k}}, \quad \frac{B_{2k+1}}{D_{2k+1}} < \frac{A_{2k+1}}{C_{2k+1}},$$

$$\left[\frac{A_0}{C_0}; \frac{B_0}{D_0}\right] \supset \left[\frac{B_1}{D_1}; \frac{A_1}{C_1}\right] \supset \left[\frac{A_2}{C_2}; \frac{B_2}{D_2}\right] \supset \ldots \supset \\ \supset \left[\frac{A_{2k}}{C_{2k}}; \frac{B_{2k}}{D_{2k}}\right] \supset \left[\frac{B_{2k+1}}{D_{2k+1}}; \frac{A_{2k+1}}{C_{2k+1}}\right] \supset \left[\frac{A_{2k+2}}{C_{2k+2}}; \frac{B_{2k+2}}{D_{2k+2}}\right] \supset \ldots$$

Thus, we have the contracting sequence of the embedded segment. This implies that the sequences of its ends converge to the same limit. This completes the proof □

We now notice that by the proof of this lemma, there exist two monotonic sequences of the fractions converging to α:

$$\frac{A_0}{C_0} < \frac{B_1}{D_1} < \frac{A_2}{C_2} < \ldots < \frac{A_{2k}}{C_{2k}} < \frac{B_{2k+1}}{D_{2k+1}} < \frac{A_{2k+2}}{C_{2k+2}} < \ldots, \tag{5.40}$$

$$\frac{B_0}{D_0} > \frac{A_1}{C_1} > \frac{B_2}{D_2} > \ldots > \frac{B_{2k}}{D_{2k}} > \frac{A_{2k+1}}{C_{2k+1}} > \frac{B_{2k+2}}{D_{2k+2}} > \ldots. \tag{5.41}$$

Consider the following sequence of matrixes

$$M_n = \begin{pmatrix} 2\cdot 2^{n+1} + (-1)^{n+1} & 2\cdot 2^n + (-1)^n \\ 2^{n+1} & 2^n \end{pmatrix} \quad (n \geq 0). \tag{5.42}$$

It is easy to see that

$$M_0 = \begin{pmatrix} 3\,3 \\ 2\,1 \end{pmatrix}, \quad M_n = M_{n-1} \cdot \begin{pmatrix} 1\,1 \\ 2\,0 \end{pmatrix} \quad (n \geq 1).$$

Indeed,

$$\begin{pmatrix} 2\cdot 2^n + (-1)^n & 2\cdot 2^{n-1} + (-1)^{n-1} \\ 2^n & 2^{n-1} \end{pmatrix} \cdot \begin{pmatrix} 1\,1 \\ 2\,0 \end{pmatrix} = \\ = \begin{pmatrix} 2\cdot 2^{n+1} + (-1)^{n+1} & 2\cdot 2^n + (-1)^n \\ 2^{n+1} & 2^n \end{pmatrix}.$$

This implies that

$$M_n = \begin{pmatrix} 3\,3 \\ 2\,1 \end{pmatrix} \cdot \begin{pmatrix} 1\,1 \\ 2\,0 \end{pmatrix}^n$$

and the matrix product

$$\begin{pmatrix} 3\,3 \\ 2\,1 \end{pmatrix} \cdot \prod_{k=1}^{\infty} \begin{pmatrix} 1\,1 \\ 2\,0 \end{pmatrix}$$

converges to 2, that is,

$$\begin{pmatrix} 2 \\ 1 \end{pmatrix} = \begin{pmatrix} 3\,3 \\ 2\,1 \end{pmatrix} \cdot \prod_{k=1}^{\infty} \begin{pmatrix} 1\,1 \\ 2\,0 \end{pmatrix}. \tag{5.43}$$

The sequence of the matrixes (5.42) and matrix product (5.43) show that not every matrix product can be converted into ordinary continued fraction.

Assign the class of the matrixes $\mathfrak{M}^+$ and the subclasses $\mathfrak{M}^+(q)$, $\mathfrak{M}^\pm$, $\mathfrak{M}^*$ $\mathfrak{M}^*(q)$ $(q \in \mathbb{N})$.

Definition 5.8 We say that an integer nonnegative matrix M belongs to a class $\mathfrak{M}^+$ if

$$M = \begin{pmatrix} a\,b \\ c\,d \end{pmatrix}, \; a \geq c \geq 0, \; b \geq d \geq 0, \quad \det M = ad - bc \neq 0, \quad a, b, c, d \in \mathbb{Z}. \tag{5.44}$$

Put

$$\mathfrak{M}^\pm = \left\{ M \in \mathfrak{M}^+ \,|\, \det M < 0 \right\}, \tag{5.45}$$

and

$$\mathfrak{M}^+(q) = \left\{ M \in \mathfrak{M}^+ \,\middle|\, \left[\frac{a}{c}\right] = \left[\frac{b}{d}\right] = q \right\}, \tag{5.46}$$

Lemma 5.25 *(1) $\mathfrak{M}^+$ is a multiplicative semigroup.*
(2) For any matrixes $M, K, L \in \mathfrak{M}^\pm$, we have the following

$$M \cdot K \cdot L \in \mathfrak{M}^\pm.$$

Proof Indeed, if

$$M = \begin{pmatrix} a & b \\ c & d \end{pmatrix}, \quad K = \begin{pmatrix} e & f \\ g & h \end{pmatrix}, \quad M, K \in \mathfrak{M}^+,$$

then

$$ae + bg \geq ce + dg \geq 0, \quad af + bh \geq cf + dh \geq 0, \quad \det(MK) = \det M \det K \neq 0.$$

Hence, $M \cdot K \in \mathfrak{M}^+$ and the statement (1) is proved.

If

$$\begin{pmatrix} a & b \\ c & d \end{pmatrix} = M \cdot K \cdot L,$$

then $a \geq c \geq 0,\ b \geq d \geq 0$, and $\det(M \cdot K \cdot L) < 0$. This completes the proof. □

Definition 5.9 We say that an integer nonnegative matrix M belongs to a class $\mathfrak{M}^*$, if

$$M = \begin{pmatrix} a & b \\ c & d \end{pmatrix} \in \mathfrak{M}^\pm \quad \text{and} \quad \left[\frac{a}{c}\right] = \left[\frac{b}{d}\right] \in \mathbb{N}. \tag{5.47}$$

Put

$$\mathfrak{M}^*(q) = \left\{ M \in \mathfrak{M}^* \,\middle|\, \left[\frac{a}{c}\right] = \left[\frac{b}{d}\right] = q \right\}. \tag{5.48}$$

It is clear that

$$\mathfrak{M}^* = \bigcup_{q=1}^{\infty} \mathfrak{M}^*(q).$$

Lemma 5.26 *Let*

$$M = \begin{pmatrix} a & b \\ c & d \end{pmatrix} \in \mathfrak{M}^*(q)$$

and

$$K = \begin{pmatrix} a_1 & b_1 \\ c_1 & d_1 \end{pmatrix}$$

be an arbitrary nondegenerate integer matrix satisfying the condition $\det K > 0$, a_1, b_1, c_1, $d_1 \geq 0$.

Then, $M \cdot K \in \mathfrak{M}^*(q)$.

Proof Indeed, by the condition, we have $a = qc + r$, $0 \leq r < c$, $b = qd + s$, $0 \leq s < d$. Then,

$$M \cdot K = \begin{pmatrix} aa_1 + bc_1 & ab_1 + bd_1 \\ ca_1 + dc_1 & cb_1 + dd_1 \end{pmatrix} =$$

$$= \begin{pmatrix} q(ca_1 + dc_1) + ra_1 + sc_1 & q(cb_1 + dd_1) + rb_1 + sd_1 \\ ca_1 + dc_1 & cb_1 + dd_1 \end{pmatrix}.$$

Since $\det M \cdot K < 0, 0 \leq ra_1 + sc_1 < ca_1 + dc_1, 0 \leq rb_1 + sd_1 < cb_1 + dd_1$, it follows

$$\left[\frac{aa_1 + bc_1}{ca_1 + dc_1}\right] = \left[\frac{ab_1 + bd_1}{cb_1 + dd_1}\right] = q$$

This completes the proof. □

Theorem 5.12 *Let*

$$\prod_{k=0}^{\infty} \begin{pmatrix} a_k & b_k \\ c_k & d_k \end{pmatrix} = \prod_{k=0}^{\infty} m_k, \quad m_k = \begin{pmatrix} a_k & b_k \\ c_k & d_k \end{pmatrix} \tag{5.49}$$

be an infinite matrix decomposition and $m_k \in \mathfrak{M}^*$ *for all k.*

Then, the matrix product converges to $\alpha > 1$.

If in addition α *is an irrational number, then for any matrix* $m \in \mathfrak{M}^+ \setminus \mathfrak{M}^*$ *and a natural* $n \in \mathbb{N}$ *there exists* $t \geq n$ *such that*

$$m \prod_{k=n}^{t} \begin{pmatrix} a_k & b_k \\ c_k & d_k \end{pmatrix} \in \mathfrak{M}^*.$$

Proof Put $q_k = \left[\frac{a_k}{c_k}\right] = \left[\frac{b_k}{d_k}\right]$ and $\alpha_k = \left\{\frac{a_k}{c_k}\right\}$, $\beta_k = \left\{\frac{b_k}{d_k}\right\}$. Then, $a_k = (q_k + \alpha_k) \cdot c_k$, $b_k = (q_k + \beta_k)d_k$ $\delta_k = a_k d_k - b_k c_k = c_k d_k(\alpha_k - \beta_k) < 0$. So

$$\min\left(\frac{|\delta_k|}{a_k d_k}, \frac{|\delta_k|}{c_k b_k}\right) = \min\left(\frac{\beta_k - \alpha_k}{q_k + \alpha_k}, \frac{\beta_k - \alpha_k}{q_k + \beta_k}\right) < \frac{\beta_k}{1 + \beta_k} < \frac{1}{2}.$$

By Lemma 5.24, the matrix product (5.49) converges to $\alpha > q_0 \geq 1$.

Let

$$m = \begin{pmatrix} a & b \\ c & d \end{pmatrix}, \quad M_{n,t} = \prod_{k=n}^{t} \begin{pmatrix} a_k & b_k \\ c_k & d_k \end{pmatrix} = M_{n-1}^{-1} M_t = \begin{pmatrix} A_{n,t} & B_{n,t} \\ C_{n,t} & D_{n,t} \end{pmatrix}.$$

By Lemma 5.21, we get

$$\lim_{t\to\infty}\frac{A_{n,t}}{C_{n,t}} = \lim_{t\to\infty}\frac{B_{n,t}}{D_{n,t}} = \beta_n.$$

Since α is an irrational number, β_n is an irrational number too for any natural number n.

Note that

$$m\cdot M_{n,t} = \begin{pmatrix} aA_{n,t}+bC_{n,t} & aB_{n,t}+bD_{n,t} \\ cA_{n,t}+dC_{n,t} & cB_{n,t}+dD_{n,t} \end{pmatrix}$$

and

$$\lim_{t\to\infty}\frac{aA_{n,t}+bC_{n,t}}{cA_{n,t}+dC_{n,t}} = \lim_{t\to\infty}\frac{aB_{n,t}+bD_{n,t}}{cB_{n,t}+dD_{n,t}} = \frac{a\beta_n+b}{c\beta_n+d} \notin \mathbb{Q}.$$

Therefore, there exists natural number t_0 such that for any $t \geq t_0$, the following equality holds

$$\left[\frac{aA_{n,t}+bC_{n,t}}{cA_{n,t}+dC_{n,t}}\right] = \left[\frac{a\beta_n+b}{c\beta_n+d}\right] = \left[\frac{aB_{n,t}+bD_{n,t}}{cB_{n,t}+dD_{n,t}}\right].$$

This proves the theorem if we put

$$t = \begin{cases} t_0 & \text{for} \quad \det m\cdot(-1)^{t_0-n} > 0, \\ t_0+1 & \text{for} \quad \det m\cdot(-1)^{t_0-n} < 0. \end{cases}$$

□

Lemma 5.27 *Let*

$$M = \begin{pmatrix} a & b \\ c & d \end{pmatrix} \in \bigcup_{q=1}^{\infty}\mathfrak{M}^+(q).$$

Then, M can be represented in the form

$$M = \left(\prod_{k=0}^{n}\begin{pmatrix} q_k & 1 \\ 1 & 0 \end{pmatrix}\right)\cdot K, \qquad (5.50)$$

where

$$K = \begin{pmatrix} e & f \\ g & h \end{pmatrix} \in \mathfrak{M}^+ \setminus \bigcup_{q=1}^{\infty}\mathfrak{M}^+(q).$$

Proof First, observe that if $M \in \mathfrak{M}^+(q)$, then

$$M = \begin{pmatrix} q & 1 \\ 1 & 0 \end{pmatrix} \cdot \begin{pmatrix} c & d \\ a - qc & b - qd \end{pmatrix}$$

and

$$\begin{pmatrix} c & d \\ a - qc & b - qd \end{pmatrix} \in \mathfrak{M}^+.$$

This representation is unique. Since $\max(c, d) > \max(a - qc, b - qd), \max(a, b) \geq \max(c, d)$, it follows that if

$$\begin{pmatrix} c & d \\ a - qc & b - qd \end{pmatrix} \in \bigcup_{q=1}^{\infty} \mathfrak{M}^+(q).$$

We continue factorization separating the factors of the form

$$\begin{pmatrix} q & 1 \\ 1 & 0 \end{pmatrix}.$$

This procedure breaks off after finite number of steps. The remaining matrix K will belong to the set

$$\mathfrak{M}^+ \setminus \bigcup_{q=1}^{\infty} \mathfrak{M}^+(q).$$

□

5.12 Conversion Algorithm of Matrix Decomposition in Ordinary Continued Fraction

Consider the matrix decomposition (5.3) for $\alpha(p)$. If $t = p, a = -p + 1, b = -p$ and $c = -1$, then we get

$$\begin{aligned} \begin{pmatrix} \alpha(p) \\ 1 \end{pmatrix} = \prod_{k=0}^{\infty} \Bigg(& \begin{pmatrix} p & p^3 + p^2 + 3 \\ 1 & p^2 + p \end{pmatrix} \begin{pmatrix} 3k + 2 & 0 \\ 0 & 3k + 1 \end{pmatrix} \cdot \\ & \cdot \begin{pmatrix} p^2 + p & p^3 + p^2 + 3 \\ 1 & p \end{pmatrix} \begin{pmatrix} p^2 - p + 9 & 2p^2 - 6p + 6 \\ 2p^2 + 2p + 2 & p^2 - p + 9 \end{pmatrix} \Bigg) = \\ = \prod_{k=0}^{\infty} & M(p, k), \end{aligned} \tag{5.51}$$

where

$$M(p,k)=\begin{pmatrix} p & p^3+p^2+3 \\ 1 & p^2+p \end{pmatrix}\begin{pmatrix} 3k+2 & 0 \\ 0 & 3k+1 \end{pmatrix}\begin{pmatrix} p^2+p & p^3+p^2+3 \\ 1 & p \end{pmatrix}\cdot$$
$$\cdot\begin{pmatrix} p^2-p+9 & 2p^2-6p+6 \\ 2p^2+2p+2 & p^2-p+9 \end{pmatrix}\begin{pmatrix} \frac{1}{3} & 0 \\ 0 & \frac{1}{3} \end{pmatrix}=\begin{pmatrix} A_k(p) & B_k(p) \\ C_k(p) & D_k(p) \end{pmatrix},$$
$$A_k(p)=(27+9p+33p^2+32p^3+8p^4+10p^5+4p^6)k+$$
$$+9+5p+16p^2+16p^3+4p^4+5p^5+2p^6,$$
$$B_k(p)=(18+36p+12p^2+24p^3+8p^4+4p^5+2p^6)k+$$
$$+6+21p+5p^2+12p^3+4p^4+2p^5+p^6,$$
$$C_k(p)=(6+24p+26p^2+8p^3+10p^4+4p^5)k+$$
$$+4+13p+14p^2+4p^3+5p^4+2p^5,$$
$$D_k(p)=(27+9p+21p^2+8p^3+4p^4+2p^5)k+$$
$$+18+4p+11p^2+4p^3+2p^4+p^5.$$

The program on Fig. 5.2 implements an algorithm of transition from matrix decomposition $\alpha(5)$ to a conventional continuous fraction.

Using the symbolic computation, we obtain

$$M(4,k)=\begin{pmatrix} 31311k+15645 & 16226k+8106 \\ 7686k+3864 & 3983k+2002 \end{pmatrix},$$
$$M(5,k)=\begin{pmatrix} 103647k+51809 & 52248k+26111 \\ 20526k+10294 & 10347k+5188 \end{pmatrix}$$

The value of $M(5,k)$ is used in the given program.

Lemma 5.28 *The program in Fig. 5.2 realizes the conversion algorithm of a matrix decomposition in continued fraction.*

Proof Indeed, first observe that

$$\left[\frac{103647k+51809}{20526k+10294}\right]=5+\left[\frac{1017k+319}{20526k+10294}\right]=5,$$
$$\left[\frac{52248k+26111}{10347k+5188}\right]=5+\left[\frac{513k+171}{10347k+5188}\right]=5$$

and

$$\begin{pmatrix} 103647k+51809 & 52248k+26111 \\ 20526k+10294 & 10347k+5188 \end{pmatrix}\in\mathfrak{M}^*.$$

```
cfki(n) := | M ← (1 0; 0 1), j ← 0, l ← 0
           | for k ∈ 0..n
           |   | M ← M·(103647·k + 51809   52248·k + 26111;
           |   |         20526·k + 10294   10347·k + 5188)
           |   | A ← M0,0, B ← M0,1, C ← M1,0, D ← M1,1
           |   | r ← floor(A/C)
           |   | while r = floor(B/D)
           |   |   | r1 ← A − C·r, A ← C, C ← r1
           |   |   | r1 ← B − D·r, B ← D, D ← r1
           |   |   | ql,j ← r, j ← j + 1
           |   |   | l ← l + 1, j ← 0  if j = 40
           |   |   | r ← floor(A/C), M ← (A B; C D)
           |   | a ← (A B C D)^T
           |   | d ← a0
           |   | for kk ∈ 1..3
           |   |   | p ← d, nn ← akk
           |   |   | r ← p, p ← nn, nn ← r  if p > nn
           |   |   | while p > 0
           |   |   |   r ← floor(nn/p), r1 ← nn − r·p, nn ← p, p ← r1
           |   |   | d ← nn
           |   | M ← M·(1/d)
           | q
```

Fig. 5.2 Describes conversion algorithm of a matrix decomposition $\alpha(5)$ in ordinary continued fraction

By Theorem 5.12, the matrix decomposition

$$\prod_{k=0}^{\infty}\begin{pmatrix} 103647k+51809 & 52248k+26111 \\ 20526k+10294 & 10347k+5188 \end{pmatrix}$$

converges.

We observe further that the outside loop *for* $k \in 0..n$ realizes the calculation of the product

$$\prod_{k=0}^{n}\begin{pmatrix} 103647k+51809 & 52248k+26111 \\ 20526k+10294 & 10347k+5188 \end{pmatrix}$$

and separate the product

$$\prod_{j=0}^{J}\begin{pmatrix} q_j & 1 \\ 1 & 0 \end{pmatrix}$$

using inner loop *while* $r = floor\left(\frac{B}{D}\right)$.

The auxiliary loop *for* $kk \in 1..3$ allows to reduce numbers in the matrix M, if it is possible. By Lemma 5.22, the division of all elements of a matrix by common divisor does not change the value of the matrix decomposition. Therefore, based on Theorem 5.12 and Lemma 5.27, the given program computes the partial quotients. □

5.13 Results of Symbolic Computation

The symbolic computations based on programs in Figs. 5.1 and 5.2 show that these programs provide the same partial quotients. The calculations using the program based on matrix decomposition are faster.

The calculations *cfki*(100) give the values of 592 partial quotients, and *cfki*(200) give the values of 1194 partial quotients. Since the results are presented in the matrix form containing 40 elements in each row, the last elements of the last line may be zero (Figs. 5.3 and 5.4).

$$\text{submatrix}(St,0,1,0,19) \rightarrow \begin{pmatrix} 0 & 1 & 2 & 3 & 4 & 5 & 6 & 7 & 8 & 9 & 10 & 11 & 12 & 13 & 14 & 15 & 16 & 17 & 18 & 19 \\ 6 & 484 & 196 & 114 & 75 & 51 & 44 & 30 & 17 & 20 & 11 & 11 & 14 & 9 & 6 & 10 & 5 & 10 & 5 & 2 \end{pmatrix}$$

$$\text{submatrix}(St,0,1,20,39) \rightarrow \begin{pmatrix} 20 & 21 & 22 & 23 & 24 & 25 & 27 & 28 & 29 & 30 & 33 & 34 & 35 & 36 & 37 & 40 & 42 & 43 & 44 & 47 \\ 5 & 3 & 2 & 2 & 1 & 1 & 1 & 4 & 1 & 4 & 1 & 2 & 1 & 2 & 1 & 1 & 1 & 1 & 2 & 2 \end{pmatrix}$$

$$\text{submatrix}(St,0,1,40,59) \rightarrow \begin{pmatrix} 54 & 55 & 60 & 63 & 68 & 78 & 79 & 82 & 84 & 87 & 93 & 95 & 97 & 111 & 120 & 123 & 128 & 129 & 134 & 140 \\ 1 & 1 & 2 & 1 & 2 & 2 & 2 & 2 & 1 & 1 & 1 & 1 & 1 & 1 & 1 & 1 & 1 & 1 & 1 & 1 \end{pmatrix}$$

$$\text{submatrix}(St,0,1,60,75) \rightarrow \begin{pmatrix} 141 & 154 & 164 & 176 & 180 & 201 & 228 & 234 & 244 & 255 & 288 & 425 & 467 & 1333 & 1813 & 3139 \\ 1 & 2 & 1 & 1 & 1 & 1 & 1 & 1 & 1 & 1 & 1 & 1 & 1 & 1 & 1 & 1 \end{pmatrix}$$

Fig. 5.3 Gives the distribution of values of partial quotients taking into account the value zero that are not partial quotients

Fig. 5.4 The program of the calculations of the distribution of values of partial quotients

```
R := cfki(200)
St := | t4000 ← 0
      | for k ∈ 0..29
      |   for j ∈ 0..39
      |     r ← R_k,j , t_r ← t_r + 1
      | j ← 0
      | for k ∈ 0..4000
      |   N_0,j ← k, N_1,j ← t_k, j ← j + 1  if  t_k > 0
      | N
```

5.14 Conclusion

The results of this paper show that reduced algebraic irrationalities in case of totally real algebraic fields and generalized Pisot numbers in general case play a fundamental role for the continued fraction expansion of algebraic irrationalities. Starting with some index, all residual fractions are the reduced algebraic numbers in the first case and generalized Pisot numbers in the second case.

Theorem 5.12 implies that starting with the number m_0, to calculate the next partial quotient is sufficient to calculate the two values of the minimal polynomial $f_m(x)$. There is a recurrence formula for calculating the next partial quotient.

Apparently, it is of interest to further study the focus conjugate to the residual fraction α_m around the fraction $-\frac{Q_{m-2}}{Q_{m-1}}$.

Consider the conjugation spectrum of irrational number α, that is, the set of all conjugate to residual fractions. The conjugation spectrum is infinite if $n > 2$, and it is a finite set if $n = 2$.

If we will call the rational conjugate spectrum of real algebraic numbers the set of all fractions of the form $-\frac{Q_{m-2}}{Q_{m-1}}$, then the natural question arises about its structure.

In the quadratic case, there is a finite set of limit points for the rational conjugate spectrum that is conjugate spectrum. What is in general case?

From the results of this paper, we see that the theory of linear fractional transformations of polynomials is closely related to the theory of linear transformations of homogeneous binary form. The second theory is simpler in many respects, and the proof of many statements is shorter.

Such relation is not casual. Apparently, the theory of linear fractional transformations of polynomials is connected with Diophantine approximations of the first kind, and the linear transformations of homogeneous forms are connected with Diophantine approximations of the second kind.

Acknowledgments This research was supported by the Russian Foundation for Basic Research (Grant Nos 15-01-01540, 15-41-03262, 15-41-03263).

References

1. Aleksandrov, A. G.: Computer investigation of continued fractions. In: Algorithmic studies in combinatorics, pp. 142–161. Nauka, Moscow (1978)
2. Berestovskii, V.N., Nikonorov, YuG: Continued fractions, the Group GL(2, Z), and Pisot numbers. Siberian Adv. Math. **17**(4), 268–290 (2007)
3. Bruno, A.D.: Continued fraction expansion of algebraic numbers. Zh. Vychisl. Mat. Mat. Fiz. **4**, 211–221 (1964)
4. Bruno, A.D.: Universal generalization of the continued fraction algorithm. Cheb. Sb. **16**(2), 35–65 (2015)
5. Dobrovol'skii, N.M.: Hyperbolic Zeta function lattices. Available from VINITI, Moscow, No 6090–84 (1984)
6. Dobrovol'skii, N.M.: Quadrature formulas for classes $E_s^\alpha(c)$ and $H_s^\alpha(c)$. Available from VINITI, Moscow, No 6091–84 (1984)
7. Dobrovol'skii, N.M.: About the modern problems of the theory of hyperbolic zeta-functions of lattices. Cheb. Sb. **16**(1), 176–190 (2015)
8. Dobrovol'skii, N.M., Dobrovol'skii, N.N.: About minimal polynomial residual fractions for algebraic irrationalities. Cheb. Sb. **16**(3), 147–182 (2015)
9. Dobrovol'skii, N.M., Yushina, E.I.: On the reduction of algebraic irrationalities. In: Algebra and Applications: Proceedings of the International Conference on Algebra, Dedicated to the 100th Anniversary of L. A. Kaloujnine, Nalchik, KBSU, pp. 44–46 (2014)
10. Dobrovol'skii, N.M., Dobrovol'skii, N.N., Yushina, E.I.: On a matrix form of a theorem of Galois on purely periodic continued fractions. Cheb. Sb. **13**(3), 47–52 (2012)
11. Dobrovol'skii, N.M., Sobolev, D.K., Soboleva, V.N.: On the matrix decomposition of a reduced cubic irrational. Cheb. Sb. **14**(1), 34–55 (2013)
12. Dobrovol'skii, N.M., Dobrovol'skii, N.N., Balaba, I.N., Rebrova, I.Yu., Polyakova, N.S.: Linear-fractional transformation of polynomials. In: Algebra, Number Theory and Discrete Geometry: Contemporary Issues and Applications: Proceedings of XIII International conference. A supplementary volume, Tula, pp. 134–149 (2015)
13. Frolov, K.K.: Upper bounds for the errors of quadrature formulae on classes of functions. Dokl. Akad. Nauk SSSR. **231**(4), 818–821 (1976)
14. Frolov, K.K.: Quadrature formula on the classes of the functions. Ph.D. thesis. Computer Centre of the Academy of Sciences of USSR (1979)
15. Podsypanin, V.D.: On the expansion of irrationalities of the fourth degree in the continued fraction. Cheb. Sb. **8**(3), 43–46 (2007)
16. Podsypanin, E.V.: A generalization of the algorithm for continued fractions related to the algorithm of Viggo Brunn. J. Sov. Math. **16**, 885–893 (1981)

17. Podsypanin, E.V.: On the expansion of irrationalities of higher degrees in the generalized continued fraction (Materials V. D. Podsypanin) the manuscript of 1970. Cheb. Sb. **8**(3), 47—49 (2007)
18. Prasolov, V.V.: Polynomials. Algorithms and Computation in Mathematics Algorithms and Computation in Mathematics, vol. 11. Springer, Berlin (2004)
19. Roth, K.F.: Rational approximations to algebraic numbers. Mathematika (1955). doi:10.1112/S0025579300000644
20. Trikolich, E.V., Yushina, E.I.: Continued fractions for quadratic irrationalities from the field $\mathbb{Q}(\sqrt{5})$. Cheb. Sb. **10**(1), 77–94 (2009)
21. Weyl, H.: Algebraic Theory of Numbers. Princeton University Press, Princeton (1940)
22. Yushina, E.I.: About some the reduction of algebraic irrationalities. In: Modern Problems of Mathematics, Mechanics, Computer Science: Proceedings of the Regional Scientific Student Conference, Tula, pp. 66–72 (2015)
23. Yushina, E.I.: On some generalized Pisot number. In: University of the XXI Century: Research within Academic Schools: Proceedings of the Russian conference, Tula, pp. 161–170 (2015)

Chapter 6
On the Periodicity of Continued Fractions in Hyperelliptic Fields

Gleb V. Fedorov

Abstract Let L be a function field of a hyperelliptic curve defined over an arbitrary field characteristic different from 2. We construct an arithmetic of continued fractions of an arbitrary quadratic irrationality in field of formal power series with respect to linear finite valuation. The set of infinite valuation and finite linear valuation of L is denoted by S. As an application, we have found a relationship between the issue of the existence of nontrivial S-units in L and periodicity of continued fractions of some key elements of L.

6.1 Introduction

Let K be a field of characteristic different from 2, and let $f(x) \in K[x]$ be a square free polynomial of odd degree $2s+1, s \geq 1$. Given an irreducible polynomial $h(x) \in K[x]$, we use v_h to denote the corresponding valuation of the field $K(x)$. Suppose that v_h has two extensions to the field $L = K(x)(\sqrt{f})$, namely v_h^- and v_h^+. We set $S = \{\mathrm{v}_h^-, \mathrm{v}_\infty\}$, where v_∞ is the infinite valuation of the field L.

An elementary introduction to some of the theory of hyperelliptic curves over finite fields of arbitrary characteristic may be found, for example, in [7].

The multiplicative group $\mathcal{O}_S^*$ of the ring $\mathcal{O}_S$ of S-integer elements of L is called *the group of S-units*.

The article [6] given a positive answer for two questions:

- Is there a relationship between the existence of nontrivial S-units and the periodicity of the expansion of an appropriate element of in a continued fraction?

G.V. Fedorov (✉)
Mechanics and Mathematics Faculty, Moscow State University,
Moscow 119991, Russia
e-mail: glebonyat@mail.ru

G.V. Fedorov
Research Institute of System Development, Russian Academy of Sciences,
Nakhimovskii Pr. 36, Korp. 1, Moscow 117218, Russia

V.A. Sadovnichiy and M.Z. Zgurovsky (eds.), *Advances in Dynamical Systems and Control*, Studies in Systems, Decision and Control 69,
DOI 10.1007/978-3-319-40673-2_6

- Is it possible to calculate a fundamental S-unit by using convergents for a continued fraction expansion of an element of L (as in the case of a finite field of constants studied in [2] and [3])?

In this paper, we construct the arithmetic of the general continued fractions, necessary for the proof of the main results of [6].

If $f(x)$ is a polynomial of even degree, an arithmetic of continued fractions with respect to infinite valuation described in articles [1] and [10]. Moreover, the relationship between fundamental units of the hyperelliptic field $L = K(x)(\sqrt{f})$ and continued fractions is considered there. Another approach to this case uses only the Riemann–Roch theorem for curves and manipulations of divisors related with continued fractions (see [4]).

6.2 Continued Fractions

Let us define $\mathcal{O}_h = \{\omega \in K(x) : \mathrm{v}_h(\omega) \geq 0\}$ is the valuation ring of the valuation v_h of the field $K(x)$, and $\rho_h = \{\omega \in K(x) : \mathrm{v}_h(\omega) > 0\}$ is an ideal of the valuation v_h. We fix the set Σ of the representative of related classes $\mathcal{O}_h$ by ρ_h, so that $\Sigma = \{\omega \in K[x] : \deg \omega < \deg h\}$. Then, we can consider the set

$$\Sigma((h)) = \Sigma_K((h)) = \left\{ \sum_{j=m}^{\infty} b_j h^j \; : \; b_j \in \Sigma, \; m \in \mathbb{Z} \right\}.$$

The set $\Sigma((h))$ is called *the set of a formal power series.*

Let $\alpha \in \Sigma((h))$, then α has the form

$$\alpha = \sum_{j=m}^{\infty} b_j h^j.$$

We introduce the notation

$$[\alpha] = [\alpha]_h = \begin{cases} \sum_{j=m}^{0} b_j h^j, & if \; m \leq 0, \\ 0, & if \; m > 0. \end{cases}$$

We set $\alpha_0 = \alpha$ and $a_0 = [\alpha_0]$. For $j \in \mathbb{N}$, we define elements α_j and a_j by induction as follows: If $\alpha_j - a_j \neq 0$, then

$$\alpha_j = \frac{1}{\alpha_{j-1} - a_{j-1}} \in \Sigma((h)), \quad a_j = [\alpha_j].$$

As a result, we obtain a *continued fraction*, for which we use the standard brief notation $[a_0, a_1, a_2, \ldots]$. Note that a_j can be considered as an element of the field $K(x)$.

We set $p_{-2} = 0,\ p_{-1} = 1,\ q_{-2} = 1, and\ q_{-1} = 0$ and define elements $p_j, q_j \in K(x)$ by induction as

$$p_j = a_j p_{j-1} + p_{j-2}, \quad q_j = a_j q_{j-1} + q_{j-2}, \quad j \geq 0, \tag{6.1}$$

then $p_j/q_j = [a_0, a_1, a_2, \ldots, a_j]$ is *the* j*th convergent* of α. The standard way we can show (see [8]) that for $j \geq -1$, the following relations hold

$$q_j p_{j-1} - p_j q_{j-1} = (-1)^j, \tag{6.2}$$

$$q_j \alpha - p_j = \frac{(-1)^j}{q_j \alpha_{j+1} + q_{j-1}}, \tag{6.3}$$

$$\alpha = \frac{p_j \alpha_{j+1} + p_{j-1}}{q_j \alpha_{j+1} + q_{j-1}}. \tag{6.4}$$

By the construction, for $j \geq 1$ we have $\nu_h\left(a_j\right) = \nu_h\left(\alpha_j\right) < 0$. From (6.1) by induction, we easily obtain relations

$$\nu_h\left(q_j\right) = \nu_h\left(a_j\right) + \nu_h\left(q_{j-1}\right) = \sum_{i=1}^{j} \nu_h\left(a_i\right),$$

$$\nu_h\left(p_j\right) = \nu_h\left(a_j\right) + \nu_h\left(p_{j-1}\right) = \sum_{i=0}^{j} \nu_h\left(a_i\right).$$

From (6.3), we have

$$\nu_h\left(q_j\alpha - p_j\right) = -\nu_h\left(q_{j+1}\right) = -\nu_h\left(a_{j+1}\right) - \nu_h\left(q_j\right) > -\nu_h\left(q_j\right), \tag{6.5}$$

or equivalently,

$$\nu_h\left(\alpha - \frac{p_j}{q_j}\right) = -\nu_h\left(q_{j+1}\right) - \nu_h\left(q_j\right) > -2\nu_h\left(q_j\right).$$

Thus, $\lim\limits_{j\to\infty} p_j/q_j = \alpha$, i. e., the convergents converge to α.

In [3], it was shown that an effective connection between the nontrivial S-units in $\mathcal{O}_S$ and the expansion of $\sqrt{f}$ or of elements related to in a continued fraction is possible only if $\deg h = 1$.

Below, we will assume that $\deg h = 1$.

Suppose that $\overline{K(x)_h}$ *is the completion of the field* $K(x)$ *with respect to the valuation* ν_h. *In the case of* $\deg h = 1$, *it is easily to proof that* $\overline{K(x)_h} = K((h)) = \Sigma((h))$. *Since by assumption* ν_h *has two extensions to the field* $L = K(x)(\sqrt{f})$, *then the field L has two embeddings into* $\overline{K(x)}_h$. *We fix one of this embedding, so that every element of the field L has the unique formal power series in* $K((h))$.

The continued fraction $[a_0, a_1, a_2, \ldots]$ of an element $\alpha \in K((h))$ is finite if and only if $\alpha \in K(x)$ (see [3], Proposition 5.1). In a standard way, we can show that if the continued fraction $[a_0, a_1, \ldots]$ for some $\alpha \in K((h))$ is periodic, then α is a quadratic irrationality. In the case of an infinite field K and valuation v_∞, the converse is not always true (see [1]). However, in the case of a field $K = \mathbb{F}_q$ and $\deg h = 1$ an assertion holds: If $\alpha \in K((h))$ is a quadratic irrationality, the continued fraction for the α is periodic (see [3]).

6.3 Some Relations with Continued Fractions

Let α *is a root of the polynomial*

$$H(X) = \lambda_2 X^2 + 2\lambda_1 X + \lambda_0, \qquad \text{where } \lambda_0, \lambda_1, \lambda_2 \in K[x]. \tag{6.6}$$

We define $\overline{\alpha}$ *is* conjugate *of the element* α, and $d = \lambda_1^2 - \lambda_2\lambda_0$ *is the shortened discriminant of the polynomial* (6.6). *We assume that* d/f *is a perfect square in the field* $K(x)$, i. e., $\alpha \in L$. *Let* $\alpha = [a_0, a_1, \ldots]$ *is a decomposition of* α *into a continued fraction, with respect to the valuation* v_h^-.

For all $j \geq 0$, *we denote* $s_j = -\mathrm{v}_h(a_j)$ *and* $t_j = -\mathrm{v}_h(q_j)$. *Since* $\mathrm{v}_h(f) = 0$, *we can define* $t = \frac{1}{2}\mathrm{v}_h(d) \geq 0$, *so that* $t \in \mathbb{Z}$.

By the construction of a continued fraction, for $j \geq 0$ we have

$$s_j \geq 1, \quad t_j = \sum_{i=1}^{j} s_i, \quad \mathrm{v}_h(p_j) = -t_j - s_0.$$

The element $\beta \in L$ is called *reduced* with respect to the valuation v_h^-, if $\mathrm{v}_h^-(\beta) < 0$ and $\mathrm{v}_h^-(\overline{\beta}) > 0$.

Proposition 6.1 *The element* $\alpha + a_0$ *is reduced if and only if* $\mathrm{v}_h(\lambda_0) < \mathrm{v}_h(\lambda_2) < \mathrm{v}_h(\lambda_1)$.

Proof By the construction of a continued fraction, we have $\mathrm{v}_h^-(\alpha - a_0) > 0$.

Assume $\alpha + a_0$ is reduced, then $\mathrm{v}_h^-(\alpha + a_0) < 0$ and $\mathrm{v}_h^-(\overline{\alpha} + a_0) > 0$. By virtue of Vieta's formulas, we have

$$\mathrm{v}_h\left(\frac{\lambda_1}{\lambda_2}\right) = \mathrm{v}_h((\alpha - a_0) + (\overline{\alpha} + a_0)) \geq \min\left(\mathrm{v}_h^-(\alpha - a_0),\ \mathrm{v}_h^-(\overline{\alpha} + a_0)\right) > 0, \tag{6.7}$$

from which $\mathrm{v}_h(\lambda_2) < \mathrm{v}_h(\lambda_1)$. Without loss of generality, we can assume that $\alpha = \frac{-\lambda_1 + \sqrt{d}}{\lambda_2}$. Since $\mathrm{v}_h^-(\alpha + a_0) < 0$, it follows that

$$\mathrm{v}_h(a_0) = \mathrm{v}_h^-(\alpha) = \mathrm{v}_h^-\left(\frac{-\lambda_1 + \sqrt{d}}{\lambda_2}\right) < 0, \tag{6.8}$$

but from the inequality (6.7), we have

$$\mathrm{v}_h(a_0) = \mathrm{v}_h^-\left(\frac{\sqrt{d}}{\lambda_2}\right) < 0, \tag{6.9}$$

therefore

$$\mathrm{v}_h\left(\lambda_1^2 - \lambda_2\lambda_0\right) = \mathrm{v}_h(d) < 2\mathrm{v}_h(\lambda_2), \tag{6.10}$$

it means that $\mathrm{v}_h(\lambda_0) < \mathrm{v}_h(\lambda_2)$.

Conversely, if $\mathrm{v}_h(\lambda_0) < \mathrm{v}_h(\lambda_2) < \mathrm{v}_h(\lambda_1)$, then the inequalities (6.8), (6.9), and (6.10) hold, and by the construction of a continued fraction $\mathrm{v}_h^-(\alpha - a_0) > 0$, hence, $\mathrm{v}_h^-(\alpha + a_0) < 0$. Then, we write

$$\mathrm{v}_h^-(\overline{\alpha} + a_0) = \mathrm{v}_h^-\left((a_0 - \alpha) - \frac{2\lambda_1}{\lambda_2}\right) \geq \min\left(\mathrm{v}_h^-(\alpha - a_0),\ \mathrm{v}_h\left(\frac{2\lambda_1}{\lambda_2}\right)\right) > 0,$$

and it was to be proved.

Let $H(X, Y) = \lambda_2 X^2 + 2\lambda_1 XY + \lambda_0 Y^2$. *For* $j \geq -1$, *we denote*

$$A_j = (-1)^{j+1} H(p_j, q_j), \quad B_j = (-1)^j (\lambda_2 p_{j-1} p_j + 2\lambda_1 p_{j-1} q_j + \lambda_0 q_{j-1} q_j). \tag{6.11}$$

The explicit form of A_j and B_j for $j = -1$ and $j = 0$ is

$$A_{-1} = \lambda_2, \quad B_{-1} = 0, \qquad A_0 = -(\lambda_2 a_0^2 + 2\lambda_1 a_0 + \lambda_0), \quad B_0 = \lambda_2 a_0 + 2\lambda_1. \tag{6.12}$$

Proposition 6.2 *For* $j \geq -1$ *the following identity holds*

$$\alpha_{j+1} = \frac{B_j + \lambda_2 \alpha}{A_j}, \tag{6.13}$$

Proof From (6.1), we can write

$$\alpha_{j+1} = -\frac{p_{j-1} - \alpha q_{j-1}}{p_j - \alpha q_j} = -\frac{(p_{j-1} \quad \alpha q_{j-1})(p_j - \overline{\alpha} q_j)}{(p_j - \alpha q_j)(p_j - \overline{\alpha} q_j)},$$

then with the notation (6.11), it follows that

$$\alpha_{j+1} = \frac{(-1)^j \lambda_2 \big(p_j p_{j-1} - (\alpha + \overline{\alpha}) q_j p_{j-1} + \alpha\overline{\alpha} q_j q_{j-1} + \alpha(q_j p_{j-1} - p_j q_{j-1})\big)}{A_j},$$

and by virtue of (6.2), we have (6.13).

In the article [3], it is proved that starting from a certain number j, quantities A_j and B_j are polynomials. We need a more rigorous statement.

Proposition 6.3 *For $j \geq -1$ quantities A_j and B_j are polynomials, i. e., $A_j, B_j \in K[x]$.*

Proof Without loss of generality, we can assume that

$$\alpha = \frac{-\lambda_1 + \sqrt{d}}{\lambda_2}.$$

In the case $j = -1$, we have $A_{-1} = \lambda_2$ and $B_{-1} = 0$; consequently, the statement is obvious. In the case $j = 0$ by the construction, we have

$$0 < \mathrm{v}_h^- (\alpha - a_0) = \mathrm{v}_h^- \left(\frac{\sqrt{d} - \lambda_1 - a_0\lambda_2}{\lambda_2} \right),$$

and taking into account (6.12), we obtain

$$0 < \mathrm{v}_h \left(\frac{d - (\lambda_1 + a_0\lambda_2)^2}{\lambda_2} \right) = \mathrm{v}_h (A_0).$$

From the equation $\alpha(\lambda_2\alpha + 2\lambda_1) = -\lambda_0$ it follows that $\mathrm{v}_h (\lambda_2\alpha) \geq 0$ hence $\mathrm{v}_h (a_0) + \mathrm{v}_h (\lambda_2) \geq 0$, so we have $\mathrm{v}_h (B_0) \geq 0$.

Now, we assume that $j \geq 1$. By the construction of (6.1) and (6.11), we conclude that $A_j, B_j \in K(x)$ are rational functions, and their denominators can be just the kind of h^n for some $n \in \mathbb{Z}$. Thus, for $A_j, B_j \in K[x]$, it suffices to show that $\mathrm{v}_h \left(A_j\right) \geq 0$ and $\mathrm{v}_h \left(B_j\right) \geq 0$.

Since $H(X, Y) = \lambda_2(X - \alpha Y)(X - \overline{\alpha} Y)$, and taking into account (6.5), we have

$$\begin{aligned} \mathrm{v}_h \left(A_j\right) &= \mathrm{v}_h^- \left(\lambda_2(p_j - \alpha q_j)(p_j - \overline{\alpha} q_j)\right) \\ &= \mathrm{v}_h (\lambda_2) - \mathrm{v}_h \left(a_{j+1}\right) + \mathrm{v}_h^- \left(\frac{p_j}{q_j} - \overline{\alpha} \right). \end{aligned} \tag{6.14}$$

If $\mathrm{v}_h^- \left(\sqrt{d}/\lambda_2\right) \leq 0$, then

$$\mathrm{v}_h^- \left(\frac{p_j}{q_j} - \alpha \right) > 0 \geq \mathrm{v}_h^- (\alpha - \overline{\alpha}) = \mathrm{v}_h^- \left(\frac{2\sqrt{d}}{\lambda_2} \right) = \frac{1}{2} \mathrm{v}_h (d) - \mathrm{v}_h (\lambda_2).$$

Hence

$$\mathrm{v}_h^-\left(\frac{p_j}{q_j}-\overline{\alpha}\right)=\mathrm{v}_h^-\left(\frac{p_j}{q_j}-\alpha+\alpha-\overline{\alpha}\right)=\mathrm{v}_h^-\left(\alpha-\overline{\alpha}\right).$$

Thus, from (6.14) we have $\mathrm{v}_h\left(A_j\right)=\frac{1}{2}\mathrm{v}_h\left(d\right)-\mathrm{v}_h\left(a_{j+1}\right)>0$.

If $\mathrm{v}_h^-\left(\sqrt{d}/\lambda_2\right)>0$, then

$$0<\mathrm{v}_h^-\left(\frac{p_j}{q_j}-\alpha\right)=\mathrm{v}_h^-\left(\frac{p_j}{q_j}+\frac{\lambda_1}{\lambda_2}+\frac{\sqrt{d}}{\lambda_2}\right),$$

it follows that $\mathrm{v}_h^-\left(\frac{p_j}{q_j}+\frac{\lambda_1}{\lambda_2}\right)>0$. Therefore

$$\mathrm{v}_h^-\left(\frac{p_j}{q_j}-\overline{\alpha}\right)\geq\min\left(\mathrm{v}_h\left(\frac{p_j}{q_j}+\frac{\lambda_1}{\lambda_2}\right),\ \mathrm{v}_h^-\left(\frac{\sqrt{d}}{\lambda_2}\right)\right)>0.$$

Again, from (6.14), we have $\mathrm{v}_h\left(A_j\right)>\mathrm{v}_h^-\left(\lambda_2\right)-\mathrm{v}_h\left(a_{j+1}\right)>0$.

Let us find a lower bound for $\mathrm{v}_h\left(B_j\right)$. From (6.13), it follows that $B_j=A_j\alpha_{j+1}-\lambda_2\alpha$. We have already seen that $\mathrm{v}_h\left(\lambda_2\alpha\right)\geq 0$. From the bound of $\mathrm{v}_h^-\left(A_j\right)$, we have $\mathrm{v}_h^-\left(A_j\alpha_{j+1}\right)=\mathrm{v}_h\left(A_ja_{j+1}\right)\geq 0$. Hence

$$\mathrm{v}_h\left(B_j\right)\geq\min\left(\mathrm{v}_h^-\left(A_j\alpha_{j+1}\right),\ \mathrm{v}_h^-\left(\lambda_2\alpha\right)\right)\geq 0.$$

Note that the condition $\mathrm{v}_h\left(\lambda_0\right)<\mathrm{v}_h\left(\lambda_2\right)<\mathrm{v}_h\left(\lambda_1\right)$ in the Proposition 6.1 implies that $\mathrm{v}_h^-\left(\sqrt{d}/\lambda_2\right)<0$; therefore, in this case, we have

$$\mathrm{v}_h\left(A_j\right)=\frac{1}{2}\mathrm{v}_h\left(d\right)-\mathrm{v}_h\left(a_{j+1}\right). \tag{6.15}$$

Proposition 6.4 *Let* $\gamma=\deg h-1$, *then we have*

$$\deg A_j\leq\max\left((2j+2)\gamma+\deg\lambda_2,\ (2j+1)\gamma+\deg\lambda_1,\ 2j\gamma+\deg\lambda_0\right),$$

$$\deg B_j\leq\max\left((2j+1)\gamma+\deg\lambda_2,\ 2j\gamma+\deg\lambda_1,\ (2j-1)\gamma+\deg\lambda_0\right).$$

Proof To assess the degree of polynomials $A_j,\ B_j$, we need to take into account that

$$\mathrm{v}_\infty\left(p_j\right)\geq-(j+1)\gamma,\quad \mathrm{v}_\infty\left(q_j\right)\geq-j\gamma,$$

from which we obtain required inequalities.

Thus, in the case $\deg h=1$ degree of polynomials $A_j,\ B_j$ do not exceed $\Lambda=\max\{\deg\lambda_0,\ \deg\lambda_1,\ \deg\lambda_2\}$.

Without loss of generality, we will make all calculations concerning the "variable" h.

Proposition 6.5 *When $j \geq 0$ the following identities hold*

$$(B_j - \lambda_1) + (B_{j-1} - \lambda_1) = a_j A_{j-1}, \tag{6.16}$$
$$(B_j - \lambda_1)^2 + A_j A_{j-1} = d. \tag{6.17}$$

Proof By the construction of a continued fraction $[a_0, a_1, \ldots]$, the following identity holds

$$\alpha_j = a_j + \frac{1}{\alpha_{j+1}}. \tag{6.18}$$

Let us substitute in place α_j and α_{j+1} in (6.18) the expression (6.13) and present it a common denominator

$$A_j A_{j-1} = (B_j + \lambda_2\alpha)(B_{j-1} - a_j A_{j-1} + \lambda_2\alpha). \tag{6.19}$$

Opening the parenthesis, we have

$$A_j A_{j-1} = B_j B_{j-1} - a_j A_{j-1} B_j + \lambda_2\alpha(B_j + B_{j-1} - a_j A_{j-1}) + \lambda_2^2\alpha^2. \tag{6.20}$$

Let us substitute the expression for the roots $H(X)$

$$\alpha,\ \overline{\alpha} = \frac{-\lambda_1 \pm \sqrt{d}}{\lambda_2}, \qquad d = \lambda_1^2 - \lambda_0\lambda_2, \tag{6.21}$$

in (6.20) and equate the coefficients of $\sqrt{d}$, and then, we get the recursive relation (6.16) for B_j. Let us substitute the expression (6.16) in the second bracket (6.19) and use the identity $\lambda_2\alpha^2 + 2\lambda_1\alpha + \lambda_0 = 0$, then

$$B_j^2 - 2\lambda_1 B_j + A_j A_{j-1} + \lambda_0\lambda_2 = 0, \tag{6.22}$$

from which it follows (6.17).

Let us define the *involution* $\iota : L \to L$ as an automorphism of the field L, acts as follows:

$$\iota(\omega_1 + \omega_2\sqrt{f}) = \omega_1 - \omega_2\sqrt{f}, \quad \omega_1,\ \omega_2 \in K(x).$$

Proposition 6.6 *For $j \geq 0$ elements α_{j+1} and $\iota\alpha_{j+1} = \overline{\alpha_{j+1}}$ are roots of the equation*

$$A_j X^2 - 2(B_j - \lambda_1)X - A_{j-1} = 0 \tag{6.23}$$

which has the discriminant (6.17). *It is true identity*

$$\alpha_{j+1} = -\frac{A_{j-1}}{B_j + \lambda_2 \overline{\alpha}}.$$

Proof By virtue of (6.13) and Vieta's formulas, we have

$$\alpha_{j+1} = \frac{B_j + \lambda_2 \alpha}{A_j}, \qquad \overline{\alpha_{j+1}} = \frac{B_j + \lambda_2 \overline{\alpha}}{A_j},$$

$$\alpha_{j+1} + \overline{\alpha_{j+1}} = \frac{2B_j - 2\lambda_1}{A_j}, \qquad \alpha_{j+1} \cdot \overline{\alpha_{j+1}} = \frac{B_j^2 - 2\lambda_1 B_j + \lambda_0 \lambda_2}{A_j^2} = -\frac{A_{j-1}}{A_j}.$$

Again, using Vieta's theorem, we obtain that α_{j+1} and $\overline{\alpha_{j+1}}$ are roots of the equation (6.23). Since (6.22) and (6.17), it follows the recursive function for A_j

$$A_j = \frac{2\lambda_1 B_j - B_j^2 - \lambda_0 \lambda_2}{A_{j-1}} = \frac{d - (B_j - \lambda_1)^2}{A_{j-1}}. \tag{6.24}$$

Let us write (6.24) into (6.13), then we have one more expression for α_{j+1}:

$$\alpha_{j+1} = \frac{A_{j-1}(B_j + \lambda_2 \alpha)}{d - (B_j - \lambda_1)^2} = -\frac{A_{j-1}}{B_j + \lambda_2 \overline{\alpha}}. \tag{6.25}$$

Proposition 6.7 *For $j \geq 0$ it is true identities*

$$\nu_h \left(A_j\right) = t + s_{j+1} \geq 1, \qquad \nu_h \left(B_j - \lambda_1\right) = t. \tag{6.26}$$

If $B_{j,i} \in K$ are coefficients of the polynomial $B_j = B_j(h)$, then

$$B_{j,i} - \frac{\lambda_{1,i}}{2} = \pm d_{i-t}, \quad i = 0, 1, \ldots, t + s_j + s_{j+1} - 1, \tag{6.27}$$

where $\lambda_{1,i} \in K$ are coefficients of the polynomial $\lambda_1 = \lambda_1(h)$ and

$$\sqrt{d} = \sum_{i=0}^{\infty} d_{i-t} h^i,$$

where $d_i = 0$ when $-t \leq i < 0$.

Proof Relations (6.26) follow from (6.15) and (6.17).

By virtue of (6.25), we have

$$B_j - \lambda_1 \mp \sqrt{d} = B_j + \lambda_2 \overline{\alpha} = -\frac{A_{j-1}}{\alpha_{j+1}}, \tag{6.28}$$

where the sign of $\sqrt{d}$ we choose depending on the sign of α and $\overline{\alpha}$ in (6.21) (everywhere positive sign or everywhere negative). If we compare in (6.28), the coefficients of the first powers of h up to degree $t + s_j + s_{j+1} - 1$, then we obtain relations (6.27). Also, we can get relations (6.27) by the formula (6.17).

Proposition 6.8 *Suppose we are given a polynomial $\omega \in K[x]$,*

$$\omega = c_0 + c_1 h + \ldots + c_n h^n, \quad c_i \in \Sigma, \; c_0 \neq 0, \; c_n \neq 0,$$

and its expansion in formal power series

$$\sqrt{\omega} = \sum_{i=0}^{\infty} \omega_i h^i, \quad \omega_i \in \Sigma,$$

where $\omega_{s_0+1} = \ldots = \omega_{s_0+\delta} = 0$ and $\omega_{s_0+\delta+1} \neq 0$ for some $s_0, \delta \in \mathbb{N}$. Then, we have $\delta < \max(s_0, \; n - s_0)$.

Proof Let $c_i = 0$ when $i > n$. Then, the following relations hold

$$c_i = \sum_{r=0}^{i} \omega_r \omega_{i-r}.$$

We assume that $\delta \geq \max(s_0, \; n - s_0)$, i. e. $\delta + s_0 + 1 > \max(2s_0, \; n)$. The coefficient $c_{\delta+s_0+1}$ satisfies the relation:

$$0 = c_{\delta+s_0+1} = \sum_{r=0}^{\delta+s_0+1} \omega_r \omega_{i-r} = 2\omega_0 \omega_{\delta+s_0+1}.$$

Since the conditions imply $\omega_{\delta+s_0+1} \neq 0$ and $\omega_0 = \sqrt{c_0} \neq 0$, we have a contradiction.

The following example shows that the upper bound for δ in the Proposition 6.8 is attained.

Example 6.1 Let us consider the polynomial $\omega \in K[x]$ of the following form

$$\omega = c^2 + ex^n, \quad c \neq 0, \; b \neq 0$$

and its expansion in $K((x))$:

$$\sqrt{\omega} = \sum_{i=0}^{\infty} \omega_i x^i, \quad \omega_i \in K.$$

It follows that $\omega_0 = c$ and $\omega_r = 0$ for any $r \in \mathbb{N}$, $r \not\equiv 0 \pmod{n}$ and $\omega_n = b/2c$. When $r = nl$ and $l \geq 2$, we have recursive relations

$$\omega_{nl} = -\frac{1}{2c}\sum_{i=1}^{l-1}\omega_{ni}\omega_{n(l-i)} = -\frac{n_l\omega_n^l}{(2c)^{l-1}} = -\frac{n_l b^l}{(2c)^{2l-1}}, \quad n_l \in \mathbb{N},$$

which are easy to set by induction. Thus, if $s_0 = 0$ or $s_0 = n$, then it follows that $\delta = n - 1 = \max(s_0,\ n - s_0) - 1$.

Proposition 6.9 *The sequence of polynomials A_j satisfies the recursive relation*

$$A_j = A_{j-2} + a_j(B_{j-1} - B_j), \quad j \geq 1. \tag{6.29}$$

Proof Let us write (6.17) for two consecutive numbers $j-1$ and j, and then subtract their:

$$0 = (B_j - \lambda_1)^2 - (B_{j-1} - \lambda_1)^2 + A_jA_{j-1} - A_{j-1}A_{j-2},$$

$$(B_j + B_{j-1} - 2\lambda_1)(B_j - B_{j-1}) = A_{j-1}(A_{j-2} - A_j).$$

If we substitute the expression (6.16) in the first bracket, then we obtain (6.29).

Proposition 6.10 *The incomplete partial a_j, $j \geq 1$, of a continued fraction $[a_0, a_1, a_2, \ldots]$ of the quadratic irrationality α satisfies the quadratic equation*

$$A_{j-1}X^2 - 2(B_{j-1} - \lambda_1)X + A_j - A_{j-2} = 0.$$

Moreover, roots of this equation have the form

$$a_j = \frac{(B_{j-1} - \lambda_1) + (B_j - \lambda_1)}{A_{j-1}}, \qquad a_j' = \frac{B_{j-1} - B_j}{A_{j-1}}. \tag{6.30}$$

Proof Let us substitute the expression (6.16) in (6.29), then we obtain the relation an analogue of (6.23), namely

$$A_{j-1}a_j^2 - 2(B_{j-1} - \lambda_1)a_j + A_j - A_{j-2} = 0,$$

which have the shortened discriminant

$$(B_{j-1} - \lambda_1)^2 - A_{j-1}(A_j - A_{j-2}) = d - A_{j-1}A_j = (B_j - \lambda_1)^2,$$

and roots have the form (6.30).

Proposition 6.11 *Following relations hold*

$$\lambda_2 = (B_j - 2\lambda_1)\frac{q_j}{p_j} + A_j\frac{q_{j-1}}{p_j}, \tag{6.31}$$

$$\frac{d}{\lambda_2} = B_j\frac{p_j}{q_j} + A_j\frac{p_{j-1}}{q_j} + \frac{\lambda_1^2}{\lambda_2}, \tag{6.32}$$

when $j \geq 0$.

Proof It follows from (6.4) that

$$\alpha_{j+1} = -\frac{\alpha q_{j-1} - p_{j-1}}{\alpha q_j - p_j} = -\frac{(\pm\sqrt{d} - \lambda_1)q_{j-1} - \lambda_2 p_{j-1}}{(\pm\sqrt{d} - \lambda_1)q_j - \lambda_2 p_j}, \tag{6.33}$$

where we are using (6.21). Let us put this expression into (6.13) and present a common denominator

$$(B_j - \lambda_1 \pm \sqrt{d})((\pm\sqrt{d} - \lambda_1)q_j - \lambda_2 p_j) = -A_j((\pm\sqrt{d} - \lambda_1)q_{j-1} - \lambda_2 p_{j-1}).$$

If we open the brackets and equate the coefficients of $\sqrt{d}$ and all the rest, we obtain desired identities (6.31) and (6.32).

Proposition 6.12 *The relation holds*

$$\alpha_1\alpha_2 \ldots \alpha_{j+1} = \frac{(-1)^j}{\alpha q_j - p_j}, \qquad j \geq 0.$$

Proof It follows by induction with using (6.33).

6.4 Best Approximations

For $p, q \in K[x]$, we denote

$$\varphi_h\left(\frac{p}{q}\right) = r - \mathrm{v}_h(q), \quad \text{where} \quad r = \max\left(\left[\frac{\deg p}{\deg h}\right], \left[\frac{\deg q}{\deg h}\right]\right).$$

Recall that an irreducible fraction $p/q \in K(h)$ is called *a best approximation* for α if for any other irreducible fraction $a/b \neq p/q$ such that $\varphi_h(u/w) \geq \varphi_h(p/q)$, we have

$$\mathrm{v}_h^-\left(\alpha - \frac{p}{q}\right) > \mathrm{v}_h^-\left(\alpha - \frac{a}{b}\right).$$

In the case $\deg h = 1$, the following assertions hold (see [3], Theorems 5.4 and 5.6):

(1) The fraction p/q is the best approximation to α if and only if

$$\mathrm{v}_h^-\left(\alpha - \frac{p}{q}\right) > -2\mathrm{v}_h(q);$$

(2) nth convergent p_n/q_n to α is the best approximation to α;
(3) If the fraction a/b is the best approximation to α, then there is a convergent p_j/q_j to α and a constant $c \in K^*$, that $a = cp_j$ and $b = cq_j$.

Proposition 6.13 *If the equation*

$$\omega_1^2 - \frac{f}{h^{2s_0}}\omega_2^2 = bh^m$$

has a solution for $1 \le s_0 \le \deg f - 1$ *such that* $\omega_1, \omega_2 \in K[x]$, $\nu_h(\omega_1) = 0$ *for some* $m \in \mathbb{N}$ *and* $b \in K$, *then* ω_1/ω_2 *is a best approximation for* α *and, therefore,* $\omega_1/\omega_2 = p_{n-1}/q_{n-1}$ *for some convergent* p_{n-1}/q_{n-1} *of* $\alpha = f/h^{2s_0}$.

The proof of Proposition 6.13 is similar to that given in [3], Sect. 5.2.

6.5 Properties of Periodic and Quasiperiodic Continued Fractions

We say that the continued fraction of an element $\beta \in L$ is *quasiperiodic* if there is $l \in \mathbb{N}_0$ and $\tau \in \mathbb{N}$ such that $\beta_l = c\beta_{l+\tau}$, where $c \in K^*$ and β_j are quotients of the continued fraction β. The least τ is called *the quasiperiod length.*

Proposition 6.14 *Let* $\lambda_1 = 0$ *and the continued fraction of* $\alpha + a_0$ *is pure quasiperiodic with the quasiperiod length n, i.e. the number* $n \in \mathbb{N}$ *is minimal such that for some constant* $c \in K^*$ *we have* $\alpha_n = c(\alpha_0 + a_0)$. *Then*

- *in the case* $n = 2k$ *we have only* $c = 1$, *i.e. the continued fraction of* $\alpha + a_0$ *is pure periodic with the quasiperiod length n, and besides*

$$\alpha + a_0 = [\overline{2a_0; a_1, \ldots, a_k,\ a_k, a_{k-1}, \ldots, a_1}]; \tag{6.34}$$

- *in the case* $n = 2k + 1$ *and* $c = 1$ *the continued fraction of* α *is periodic with the quasiperiod length n, and besides*

$$\alpha + a_0 = [\overline{2a_0; a_1, \ldots, a_k,\ a_{k+1},\ a_k, \ldots, a_1}]; \tag{6.35}$$

- *in the case* $n = 2k + 1$ *and* $c \ne 1$ *the continued fraction of* α *is periodic with the quasiperiod length* $2n$, *and besides*

$$\alpha + a_0 = \Big[\overline{2a_0; a_1, \ldots, a_k,\ c^{(-1)^k}a_k, c^{(-1)^{k-1}}a_{k-1}, \ldots, c^{-1}a_1,}$$
$$\overline{2ca_0, c^{-1}a_1, \ldots, c^{(-1)^k}a_k,\ a_k \ldots, a_1}\Big]. \tag{6.36}$$

The proof of the Proposition 6.14 is similar to that of Lemma 4.1 in [1].

6.6 Preliminary Details

Now, let $\lambda_2 = h^{2s_0}$, $\lambda_1 = 0$, and $\lambda_0 = f$, where $1 \le s_0 < \deg d$. Thus, $\alpha \in L$ is a root of the polynomial

$$H(X) = h^{2s_0} X^2 - f,$$

with the shortened discriminant $d = h^{2s_0} f$. Let $\alpha = [a_0, a_1, \dots]$ be the continued fraction expansion of α.

The Proposition 6.3 implies that A_j and B_j, defined in (6.11), are polynomials for all $j \ge 0$. In the current case, we have

$$A_j = (-1)^{j+1} (h^{2s_0} p_j^2 - f q_j^2), \tag{6.37}$$

$$B_j = (-1)^j (h^{2s_0} p_j p_{j-1} - f q_j q_{j-1}).$$

By virtue of (6.26), it follows that

$$\nu_h\left(A_j\right) = s_0 + s_{j+1}, \qquad \nu_h\left(B_j\right) = s_0, \tag{6.38}$$

$$s_0 + s_{j+1} \le \deg A_j,$$

$$\deg A_j,\ \deg B_j \le \max(2s_0,\ \deg f), \tag{6.39}$$

where s_j satisfy relations

$$\nu_h\left(a_j\right) = -s_j < 0, \quad \nu_h\left(p_j\right) = s_0 + t_j, \quad \nu_h\left(q_j\right) = t_j, \quad t_j = \sum_{r=1}^{j} s_r.$$

The result of the Proposition 6.7 implies that first few coefficients of the polynomial B_j is consistent with coefficients of the formal power series of $\sqrt{f}$.

We will use formulas (6.13), (6.16), and (6.17), which in this case take the form

$$\alpha_{j+1} = \frac{B_j + h^{s_0}\sqrt{f}}{A_j}, \tag{6.40}$$

$$B_j + B_{j-1} = a_j A_{j-1}, \tag{6.41}$$

$$A_j A_{j-1} + B_j^2 = h^{2s_0} d. \tag{6.42}$$

In particular, the last relation implies that

$$\deg A_j + \deg A_{j-1} = \max(2 \deg B_j,\ 2s_0 + \deg f). \tag{6.43}$$

If $0 < s_0 < \deg f$, then the Proposition 6.13 implies that a solution $\omega_1, \omega_2 \in K[x]$ of the norm equation

$$N_{L/K(x)}(\omega_1 - \alpha\omega_2) = \omega_1^2 - \omega_2^2 \frac{f}{h^{2s_0}} = bh^m, \quad m \in \mathbb{N}, \quad b \in K^*, \tag{6.44}$$

gives the best approximation of α and $\omega_1/\omega_2 = p_n/q_n$ for some convergent p_n/q_n of α. Note that the presence of solutions of the Eq. (6.44) is equivalent to the presence of solutions of the canonical normal equation

$$\omega_1^2 - \omega_2^2 f = bh^m, \tag{6.45}$$

where $\omega_1, \omega_2 \in K[x]$ are polynomials and $m \in \mathbb{N}$, $b \in K^*$.

Proposition 6.15 *There is the least number $n \geq 2$ that for which $\mathrm{v}_h(A_{n-1}) = \deg A_{n-1}$ if and only if there is the least number m for which the norm equation (6.45) has a solution $\omega_1, \omega_2 \in K[x]$ such that $\mathrm{v}_h(\omega_1) = 0$ and $b \in K^*$.*

Proof Let us assume that $\mathrm{v}_h(A_{n-1}) = \deg A_{n-1}$ and $n \geq 2$ is the least such number. Multiplying the expression (6.37) by $h^{2t_{n-1}}$, we obtain the same equation as (6.45), where $m = s_0 + s_n + 2t_{n-1}$. Note that if Eq. (6.45) is valid for a smaller value of m, then, by Proposition 6.13, there exists a convergent $p_{r-1}/q_{r-1} = \omega_1 h^{-s_0}/\omega_2$ of α; therefore, dividing (6.45) by $h^{2t_{r-1}}$, we would obtain $\mathrm{v}_h(A_{r-1}) = \deg A_{r-1}$ with $r < n$, which contradicts the condition that n is minimal.

Conversely, we write the norm equation (6.45) in the form

$$\omega_1^2 - \frac{f}{h^{2s_0}}(\omega_2 h^{s_0})^2 = b_0 h^m.$$

By Proposition 6.13, there exists a convergent $p_{n-1}/q_{n-1} = \omega_1 h^{-s_0}/\omega_2$ of $\alpha = \sqrt{f}/h^{s_0}$, where $\omega_1 = p_{n-1}h^{t_{n-1}+s_0}$ and $\omega_2 = q_{n-1}h^{t_{n-1}}$. According to (6.37), we obtain

$$A_{n-1} = (-1)^n b_0 h^{m-2t_{n-1}} = bh^{s_0+s_n}.$$

Obviously, the minimality of m implies that $n \geq 2$ is minimal.

It follows from [3] Sect. 2 that for a nontrivial S-unit $u = \omega_1 + \omega_2 f$, the norm equation (6.45) holds. We refer to m as the *degree* of the S-unit u. The converse is also true: If there is a solution $\omega_1, \omega_2 \in K[x]$ of the norm equation (6.45), then $u = \omega - \sqrt{f}\omega_2$ or $u = \omega + \sqrt{f}\omega_2$ is an S-unit of the field L. Thus, we can speak about S-units as solutions of norm equation (6.45).

6.7 The Periodic Continued Fraction

Suppose that $s_0 = s$ or $s_0 = s + 1$.

Theorem 6.1 *The following conditions are equivalent:*

- *$n \geq 2$ is the least number for which $v_h(A_{n-1}) = \deg A_{n-1}$;*
- *the continued fraction of $\alpha + a_0 \in K((h))$ is purely periodic with period length n or $2n$.*

Proof Now, we show that the first condition implies the second condition.

We introduce the notation

$$\beta(s_0) = \frac{\sqrt{f}}{h^{s_0}} + \left[\frac{\sqrt{f}}{h^{s_0}}\right] = \alpha + a_0.$$

Let n be the least number for which $\deg A_{n-1} = v_h(A_{n-1}) = s_0 + s_n$. Taking into account (6.38), for all $j \geq 0$ we denote $\hat{A}_j, \hat{B}_j \in K[h]$ as follows:

$$\hat{A}_j = h^{-s_0} A_j, \quad \hat{B}_j = h^{-s_0} B_j.$$

From (6.43), we have

$$s_n + \deg \hat{A}_n = \max\Big(2 \deg \hat{B}_n,\ 2s + 1\Big).$$

The relation (6.39) implies that

$$s_n,\ \deg \hat{A}_n,\ \deg \hat{B}_n \leq \max\big(s_0,\ 2s + 1 - s_0\big) = s + 1.$$

Thus, $s_n = s + 1$ if and only if $\deg \hat{A}_n = \deg \hat{B}_n = s$ or $\deg \hat{A}_n = \deg \hat{B}_n = s + 1$; otherwise, $s_n = s$. Each of this cases implies $\deg B_n \leq s_0 + s_n$. Therefore, by virtue of the Proposition 6.7, all coefficients in the polynomial B_n equal corresponding coefficients in the power series expansion of $\sqrt{f}$ in $K((h))$. Thus, by Proposition 6.6, we have $\alpha_n = \beta(s)/b$ or $\alpha_n = \beta(s+1)/b$.

The Proposition 6.14 implies the pure periodicity of the continued fraction $\alpha + a_0$ with period length n or $2n$.

Conversely, let us prove that the second condition implies the first condition.

It follows from the pure periodicity of $\alpha + a_0$ that there exists a number $r \geq 2$ such that $A_{r-1} = bA_0 = bh^{2s_0}$ for some $b \in K^*$. Hence, there exists a minimal n, $2 \leq n \leq r$, for which $\deg A_{n-1} = v_h(A_{n-1})$.

This completes the proof of the theorem.

Appendix

Note, that the existence of S-units in the hyperelliptic field L is equivalent to the following assertion: The class of the divisor $\mathscr{D} = v_h^+ - v_\infty$ has finite order m in the group $\Delta^0(L)$ of zeroth-degree divisor classes of the field L, where m is the degree of the fundamental S-unit (see [9]).

It follows from the Proposition 6.13 on a best approximation for $\frac{\sqrt{d}}{h^{s_0}}$ that, to find a fundamental S-unit, it suffices to examine the continued fraction expansion of $\frac{\sqrt{d}}{h^{s_0}}$ only one element at $s_0 = s$ or $s_0 = s + 1$.

The results of the paper make it possible to construct a fast algorithm for constructing a fundamental S-unit in the field L. Take A_{-1}, A_0, B_0 defined by (6.12). Note that, we do not have to calculate p_j and q_j. For $j = 1, 2, \ldots$, we cyclically perform following steps:

- Verify the equality deg $A_{j-1} = \nu_h\left(A_{j-1}\right)$; if it holds, then the cycle is terminated;
- Calculate $a_j = [\alpha_j]$ from A_{j-1} and B_{j-1} by using (6.40);
- Calculate B_j and A_j by formulas in Propositions 6.5 (or formula (6.41)) and 6.9, respectively, and proceed to the first step.

The well-known algorithm for finding torsion points in Jacobian of the hyperelliptic curve, using addition of divisors, is given in [5].

References

1. Adams, W.W., Razar, M.J.: Multiples of points on elliptic curves and continued fractions. Proc. London Math. Soc. **41**(3), 481–498 (1980)
2. Benyash-Krivets, V.V., Platonov, V.P.: Continued fractions and S-units in hyperelliptic fields. Russ. Math. Surv. **63**(2), 357–359 (2008)
3. Benyash-Krivets, V.V., Platonov, V.P.: Groups of S-units in hyperelliptic fields and continued fractions. Sb. Math. **200**(11), 1587–1615 (2009)
4. Berry, T.G.: On periodicity of continued fractions in hyperelliptic function fields. Arch. Math. **55**, 259–266 (1990)
5. Cantor, D.G.: Computing in the Jacobian of a Hyperelliptic Curve. Math. comput. **48**(177), 95–101 (1987)
6. Fedorov, G.V., Platonov, V.P.: S-Units and Periodicity of Continued Fractions in Hyperelliptic Fields. Dokl. Math. **92**(3), 1–4 (2015)
7. Koblitz, N.: "Algebraic aspect of cryptography", with an appendix on Hyperelliptic curves. In: Menezes, A.J., Yi-Hong Wu., Zuccherato, R.J. (eds.) Algorithms and Computation in Mathematics, vol. 3, Springer, Heidelberg (1999)
8. Lang, S.: Introduction to Diophantine Approximations. Columbia University, New York (1966)
9. Platonov, V.P.: Arithmetic of quadratic fields and torsion in Jacobians. Dokl. Math. **81**(1), 55–57 (2010)
10. Platonov, V.P.: Number-theoretic properties of hyperelliptic fields and the torsion problem in Jacobians of hyperelliptic curves over the rational number field. Russ. Math. Surv. **69**(1), 1–34 (2014)

Chapter 7
Method of Resolving Functions for the Differential-Difference Pursuit Game for Different-Inertia Objects

Lesia V. Baranovska

Abstract The paper is devoted to the differential-difference pursuit game for different-inertia objects. An approach to the solution of this problem based on the method of resolving functions is proposed. The guaranteed time of the game termination is found, and corresponding control law is constructed. The results are illustrated by a model example.

7.1 Differential-Difference Games of Pursuit. Problem Statement

We consider the pursuit game, whose dynamics is described by the system of differential-difference equations of retarded type (see [1, 2]):

$$\dot{z}(t) = Az(t) + Bz(t-\tau) + \phi(u, v), \quad z \in \mathbb{R}^n, \quad u \in U, \quad v \in V, \tag{7.1}$$

where A and B are square constant matrices of order n; U and V are nonempty compacts sets; the function $\phi(u, v)$, $\phi : U \times V \to \mathbb{R}^n$, is jointly continuous in its variables; $\tau = \text{const} > 0$.

Let $z(t)$ be a solution of Eq. (7.1) under the initial condition

$$z(t) = z^0(t), \quad -\tau \leq t \leq 0, \tag{7.2}$$

where function $z^0(t)$ is absolutely continuous on $[-\tau, 0]$.

The piece of the trajectory $z^t(\cdot)$, where

$$z^t(\cdot) = \{z(t+s), \quad -\tau \leq s \leq 0\}$$

L.V. Baranovska (✉)
Institute for Applied System Analysis, National Technical University of Ukraine "Kyiv Polytechnic Institute", 37 Peremogy Ave., Building 35, Kyiv 03056, Ukraine
e-mail: lesia@baranovsky.org

V.A. Sadovnichiy and M.Z. Zgurovsky (eds.), *Advances in Dynamical Systems and Control*, Studies in Systems, Decision and Control 69,
DOI 10.1007/978-3-319-40673-2_7

will be referred to as the state of system (7.1) at the moment t. The game is evolving on the closed time interval $[0, T]$.

The terminal set has cylindrical form, i.e.,

$$M^* = M_0 + M, \tag{7.3}$$

where M_0 is a linear subspace in $\mathbb{R}^n$; and M is a compact set from $L = M_0^{\perp}$ (the orthogonal complement of M_0 in $\mathbb{R}^n$).

The players choose their controls in the form of certain functions. In such a way, the pursuer and evader affect the process (7.1), pursuing their own goals. The goal of the pursuer (u) is in the shortest time to bring a trajectory of the process to a certain closed set M^*; the goal of the evader (v) is to avoid a trajectory of the process from meeting with the terminal set (7.3) on a whole semi-infinite interval of time or if is impossible to maximally postpone the moment of meeting.

Now we describe what kind of information is available to the pursuer in the course of the game.

Denote by Ω_U, Ω_V the sets of Lebesgue measurable functions $u(t)$, $v(t)$, $u(t) \in U$, $v(t) \in V$, $t \geq 0$, respectively. A mapping that puts into correspondence to a state $z^0(\cdot)$ some element in Ω_V is called an open-loop strategy of the evader, specific realization of this strategy for a given initial state $z^0(\cdot)$ of process (7.1) is called an open-loop control. In the process of the game (7.1), (7.3), the evader applies open-loop controls $v(\cdot) \in \Omega_V$.

Function

$$u(t) = u\left(z^0(\cdot), t, v(t)\right),$$

such that $v(\cdot) \in \Omega_V$ implies $u(\cdot) \in \Omega_U$ is called counter-control of pursuer corresponding to initial state $z^0(\cdot)$. We assume that the pursuer chooses his control in the form

$$u(t) = u\left(z^0(\cdot), t, v_t(\cdot)\right), \quad t \geq 0,$$

where $v_t(\cdot) = \{v(s) : s \in [0, t], v(\cdot) \in \Omega_V\}$, and $u(\cdot) \in \Omega_U$.

Under these hypotheses, we will play the role of the pursuer and find sufficient conditions on the parameters of the problem (7.1), (7.3), insuring the game termination for certain guaranteed time.

Let π be the orthogonal projector from $\mathbb{R}^n$ onto the subspace L. Consider the multi-valued mapping

$$W(t, v) = \pi K(t) \phi(U, v), \quad W(t) = \bigcap_{v \in V} W(t, v),$$

where $K(t)$ is a matrix-valued function which satisfies conditions of the following lemma.

Lemma 7.1 *Suppose that $z^0(\cdot)$ is absolutely continuous on $[-\tau, 0]$, $\varphi(u, v)$ is jointly continuous in its variables. Let $z(t)$ be the continuous solution of the system (7.1) under the initial condition (7.2) and given admissible controls $u(t)$, $v(t)$. Then*

$$z(t) = z^0(0)K(t) + B\int_{-\tau}^{0} z^0(s)K(t-s-\tau)\,ds + \int_0^t \phi(u(s), v(s))K(t-s)\,ds,$$

where $K(t)$ is the unique matrix function with the properties (see [3]):

(1) $K(t) = 0, \quad t < 0$;
(2) $K(0) = E$, *E is the identity matrix;*
(3) The function $K(t)$ is of class C^0 on $[0, +\infty)$;
(4) $K(t)$ *satisfies* $\dot{K}(t) = AK(t) + BK(t-\tau)$, $t > 0$.

Condition 7.1 (Pontryagin's condition) *The mapping $W(t) \neq \emptyset$ for all $t \geq 0$.*

Remark 7.1 For the linear process ($\phi(u, v) = u - v$)

$$W(t) = \pi K(t)U \overset{*}{-} \pi K(t)V,$$

where $\overset{*}{-}$ is a geometric subtraction of the sets (Minkowski' difference) (see [4]).

For the game, described above, satisfying Pontryagin's condition, the notion of resolving function was introduced in [1, 2, 5], through which the time of game termination was defined. The resolving function outlines the course of the game and at the instant of time at which the integral of this function turns into unity the game trajectory hits the terminal set. Sufficient conditions for solvability of the pursuit problem were derived (see [1, 2, 5]). The process of pursuit is divided into two parts. The method of the resolving function (see [6]) as such is working only on the first interval of time $[0, t_*]$, t_* being the instant of switching, on which the pursuer constructs his control on the basis of information on the prehistory of the evader's control. As soon as at some instant of time t_* the integral of the resolving function turns into unity, the process of pursuit switches to Pontryagin's First Direct Method realized within the class of counter-controls. That is why from the instant of switching to the rating time of the game termination the resolving function is set equal to zero.

7.2 Case of Different-Inertia Objects

We will show that for different-inertia objects Pontryagin's condition fails on some interval of time. The problem "Boy and Crocodile" is an example (see [6]). First consider ordinary differential game:

$$\ddot{x}(t) = u(t), \qquad x \in \mathbb{R}^n, \quad \|u\| \leq 1,$$
$$\dot{y}(t) = v(t), \qquad y \in \mathbb{R}^n, \quad \|v\| \leq 1.$$

Setting $z_1 = x - y, \quad z_2 = \dot{x}$, we come to the system

$$\begin{aligned} \dot{z}_1(t) &= z_2(t) - v\,, \qquad && z_1 \in \mathbb{R}^n\,, \quad n \geq 2, \\ \dot{z}_2(t) &= u\,, && z_2 \in \mathbb{R}^n. \end{aligned}$$

The pursuer ("crocodile," u) is clumsy because of boundness of his trajectory curvature radius though he may gather a high speed. The evader ("boy", v) is inertia-less though his speed is limited (see [7]).

We consider the analog of this game in the case when dynamics of the game is described by the following system of differential-difference equations:

$$\begin{aligned} \dot{z}_1(t) &= z_2(t - \tau) - v\,, \qquad && z_1 \in \mathbb{R}^n\,, \quad n \geq 2, \\ \dot{z}_2(t) &= u\,, && z_2 \in \mathbb{R}^n, \end{aligned} \tag{7.4}$$

where $\|u\| \leq \rho\,, \quad \rho > 0\,, \quad \|v\| \leq \sigma\,, \quad \sigma > 0.$

The terminal set is given by the equality $\|z_1\| \leq l$.

In the model example under study (7.1), the matrices A, B, and the control domains U, V take the following forms, respectively,

$$A = 0,\ B = \begin{pmatrix} 0 & E_n \\ 0 & 0 \end{pmatrix};$$

$$U = \left\{ \begin{pmatrix} 0 \\ u \end{pmatrix}, \quad u \in \mathbb{R}^n : \quad \|u\| \leq \rho \right\}, \quad V = \left\{ \begin{pmatrix} v \\ 0 \end{pmatrix}, \quad v \in \mathbb{R}^n : \quad \|v\| \leq \sigma \right\}.$$

The terminal set is

$$M^* = \left\{ z = (z_1\,, z_2) \in \mathbb{R}^{2n} : \quad \|z_1\| \leq l \right\}.$$

Here

$$M_0 = \left\{ z = (z_1\,, z_2) \in \mathbb{R}^{2n} : \quad z_1 = 0 \right\}, \quad L = \left\{ z = (z_1\,, z_2) \in \mathbb{R}^{2n} : \quad z_2 = 0 \right\},$$

$M = \left\{ z = (z_1\,, z_2) \in \mathbb{R}^{2n} : \quad \|z_1\| \leq l\,, \ z_2 = 0 \right\}.$

The operator of orthogonal projection $\pi : \mathbb{R}^{2n} \to L$ is defined by matrix $\begin{pmatrix} E_n & 0 \\ 0 & 0 \end{pmatrix}$, therefore $\pi z = \begin{pmatrix} E_n & 0 \\ 0 & 0 \end{pmatrix} \begin{pmatrix} z_1 \\ z_2 \end{pmatrix} = \begin{pmatrix} z_1 \\ 0 \end{pmatrix}$.

Matrix function $K(t)$ is unique and enjoys the properties:

(1) $K(t) = 0\,, \ t < 0$;
(2) $K(0) = E_{2n}$;
(3) $K(t)$ is continuous on $[0\,, \ +\infty)$;
(4) when $t > 0$ $K(t)$ satisfies the equation $\left[\dot{K}(t)\right] = [B] \cdot [K(t - \tau)]$.

This equation can be rewritten as follows:

$$\begin{pmatrix} \dot{K}_{11}(t) & \dot{K}_{12}(t) \\ \dot{K}_{21}(t) & \dot{K}_{22}(t) \end{pmatrix} \otimes E_n = \begin{pmatrix} 0 & E_n \\ 0 & 0 \end{pmatrix} \cdot \begin{pmatrix} K_{11}(t-\tau) & K_{12}(t-\tau) \\ K_{21}(t-\tau) & K_{22}(t-\tau) \end{pmatrix} \otimes E_n$$
$$= \begin{pmatrix} K_{21}(t-\tau) & K_{22}(t-\tau) \\ 0 & 0 \end{pmatrix} \otimes E_n,$$

where $K_{ij}(\cdot)$ are numerical functions.

By the derivative of a matrix function is meant the matrix obtained by replacing all its elements by their derivatives (see [8]).

With account of these conditions, we deduce:

$$\begin{cases} \dot{K}_{11}(t) = K_{21}(t-\tau)\,, \\ \dot{K}_{12}(t) = K_{22}(t-\tau)\,, \\ \dot{K}_{21}(t) = 0\,, \\ \dot{K}_{22}(t) = 0\,, \\ K_{12}(0) = K_{21}(0) = 0\,, \\ K_{11}(0) = K_{22}(0) = 1\,. \end{cases}$$

Below given is the solution to this system:

$$K_{11}(t) = 1\,, \quad K_{12}(t) = t, \quad K_{21}(t) = 0\,, \quad K_{22}(t) = 1.$$

Taking into account condition (a) we obtain an explicit form of the matrix function:

$$[K(t)] = \begin{pmatrix} K_{11}(t) & K_{12}(t) \\ 0 & K_{22}(t) \end{pmatrix} \otimes E_n,$$

where $K_{11}(t) = K_{22}(t) = \left[\begin{matrix} 1\,, & t \geq 0\,, \\ 0\,, & t < 0\,, \end{matrix}\right.$ $K_{12}(t) = \left[\begin{matrix} t\,, & t \geq 0\,, \\ 0\,, & t < 0. \end{matrix}\right.$

Thus, for $t \geq 0$ the matrix function is of the form:

$$[K(t)] = \begin{pmatrix} 1 & t \\ 0 & 1 \end{pmatrix} \otimes E_n.$$

Set $\gamma(t) \equiv 0$. For the linear process (7.4), we have

$$W(t) = \pi\,[K(t)]\,V \overset{*}{-} \pi\,[K(t)]\,U\,.$$

$$\pi\,[K(t)]\,V = \begin{pmatrix} E_n & 0 \\ 0 & 0 \end{pmatrix} \cdot \begin{pmatrix} E_n & t \\ 0 & E_n \end{pmatrix} \cdot \begin{pmatrix} v \\ 0 \end{pmatrix} = \begin{pmatrix} v \\ 0 \end{pmatrix} \otimes E_n,$$

$$\pi\,[K(t)]\,U = \begin{pmatrix} E_n & 0 \\ 0 & 0 \end{pmatrix} \cdot \begin{pmatrix} E_n & t \\ 0 & E_n \end{pmatrix} \cdot \begin{pmatrix} 0 \\ u \end{pmatrix} = \begin{pmatrix} tu \\ 0 \end{pmatrix} \otimes E_n.$$

Then

$$W(t) = (\rho t - \sigma) S = \emptyset \quad \text{for} \quad t \in \left[0, \frac{\sigma}{\rho}\right).$$

Thus, for this game Pontryagin's condition fails on some interval of time.

7.3 Modification of Pontryagin's Condition

Let for the differential-difference game (7.1), (7.3) Pontryagin's condition does not hold and let us suppose that the matrix function $K(t)$ satisfies assumptions of Lemma 7.1. We introduce multi-valued mappings

$$\bar{W}(t, v) = \pi K(t) \phi(U, D(t) v), \quad \bar{W}(t) = \bigcap_{v \in V} \bar{W}(t, v), \quad t \geq 0, \tag{7.5}$$

where $D(t)$, $t \in [0, +\infty)$ is some matrix function.

Condition 7.2 (modification of Pontryagin's condition) *There exists a continuous matrix function* $D(t)$, $t \in [0, +\infty)$, *such that the multi-valued mapping* $\bar{W}(t) \neq \emptyset$ *for all* $t \geq 0$.

Denote

$$\bar{\phi}(t, u, v) = \phi(u, v) - \phi(u, D(t) v), \quad t \geq 0, \ u \in U, \ v \in V; \tag{7.6}$$

$$M(t) = M \overset{*}{-} \int_0^t \pi K(s) \bar{\phi}(s, U, V) ds, \quad t \geq 0.$$

We see that the mapping $M(t)$ is upper semi-continuous as a geometric difference of two continuous multi-valued mappings (see [4]).

Condition 7.3 *For the above-mentioned matrix function* $D(t)$, *the multi-valued mapping* $M(t)$ *is nonempty for all* $t \geq 0$.

By virtue of the assumptions on the process parameters, the multi-valued mapping $\bar{W}(t, v)$ is continuous on the set $[0, +\infty) \times V$. Consequently, as follows from Condition 7.2, the mapping $\bar{W}(t)$ is upper semi-continuous (see [9]) and therefore Borelian (see [9]). Hence, there exists at least one Borelian selection $\gamma(t)$, $\gamma(t) \in \bar{W}(t)$, $t \geq 0$ (see [9–12]).

Let us denote by $\bar{G}$ the set of all Borelian selections of the multi-valued mapping $\bar{W}(t)$. For fixed $\gamma(\cdot) \in \bar{G}$, we put

$$\xi\left(t, z^0(\cdot), \gamma(\cdot)\right)$$

$$= \pi K(t) z^0(0) + \int_{-\tau}^{0} \pi K(t-s-\tau) B z^0(s) ds + \int_0^t \gamma(s) ds \tag{7.7}$$

and consider the resolving function

$$\begin{gathered} \bar{\alpha}\left(t, s, z^0(\cdot), v, \gamma(\cdot)\right) = \sup\{\alpha \geq 0 : \\ \left[\bar{W}(t-s, v) - \gamma(t-s)\right] \cap \alpha\left[M(t) - \xi\left(t, z^0(\cdot), \gamma(\cdot)\right)\right] \neq \emptyset\}. \end{gathered} \tag{7.8}$$

It is easy to see that since $0 \in \bar{W}(t-s, v) - \gamma(t-s)$, $v \in V$, $t \geq s \geq 0$, then $\bar{\alpha}\left(t, s, z^0(\cdot), v, \gamma(\cdot)\right) = +\infty$ for all $s \in [0, t]$, $v \in V$, if $\xi\left(t, z^0(\cdot), \gamma(\cdot)\right) \in M(t)$. If for some $t \geq 0$ $\xi\left(t, z^0(\cdot), \gamma(\cdot)\right) \notin M(t)$, then function (7.8) assumes finite values and, what is more, it is bounded for $s \in [0, t]$ and $v \in V$ (see [9]). From the analysis above there follows that function $\inf_{v\in V} \bar{\alpha}\left(t, s, z^0(\cdot), v, \gamma(\cdot)\right)$ is summable for $s \in [0, t]$ (see [9]).

Let us define

$$\bar{T}\left(z^0(\cdot), \gamma(\cdot)\right)$$

$$= \inf\left\{t \geq 0 : \int_0^t \inf_{v\in V} \bar{\alpha}\left(t, s, z^0(\cdot), v, \gamma(\cdot)\right) ds \geq 1\right\}, \quad \gamma(\cdot) \in \bar{G}.$$

If the inequality in the curly brackets is not satisfied for all $t \geq 0$, we set $\bar{T}\left(z^0(\cdot), \gamma(\cdot)\right) = +\infty$.

If $\xi\left(t, z^0(\cdot), \gamma(\cdot)\right) \in M$, then $\inf_{v\in V} \bar{\alpha}\left(t, s, z^0(\cdot), v, \gamma(\cdot)\right) \equiv +\infty$, $s \in [0, t]$, and it seems natural to set the integral in the formula above equal to $+\infty$. Then, the inequality in definition of the function $\bar{T}\left(z^0(\cdot), \gamma(\cdot)\right)$ is readily satisfied.

Theorem 7.1 *Let the conflict controlled process (7.1), (7.3) satisfies Conditions 7.2, 7.3, the set M be convex,* $\bar{T} = \bar{T}\left(z^0(\cdot), \gamma^0(\cdot)\right) < +\infty$ *for the given initial state* $z^0(\cdot)$ *and some selection* $\gamma^0(\cdot) \in \bar{G}$. *Then a trajectory of the process (7.1), (7.3) can be brought by the pursuer from* $z^0(\cdot)$ *to the terminal set* M^* *at the moment* $\bar{T}$ *under arbitrary admissible controls of the evader.*

Proof Let $v(s)$, $v(s) \in V$, $s \in \left[0, \bar{T}\right]$ be an arbitrary measurable function. First consider the case when $\xi\left(\bar{T}, z^0(\cdot), \gamma^0(\cdot)\right) \notin M(t)$. We introduce the controlling function (see [6, 13, 14])

$$\bar{h}(t) = 1 - \int_0^t \bar{\alpha}\left(\bar{T}, s, z^0(\cdot), v(s), \gamma^0(\cdot)\right) ds, \quad t \geq 0.$$

From the definition of time $\bar{T}$, there follows that there exists a switching time $t_* = t_*(v(\cdot))$, $0 < t_* \leq \bar{T}$, such that $\bar{h}(t_*) = 0$.

Let us describe the rules by which the pursuer constructs his control on the so-called active and the passive parts, $[0, t_*)$ and $[t_*, \bar{T}]$, respectively.

Consider the multi-valued mapping

$$U_1(s,v) = \{u \in U : \pi K(\bar{T}-s)\,\phi\,(u, D(\bar{T}-s)\,v) - \gamma^0(\bar{T}-s) \\ \in \bar{\alpha}(\bar{T}, s, z^0(\cdot), v, \gamma^0(\cdot))\,[M(\bar{T}) - \xi(\bar{T}, z^0(\cdot), \gamma^0(\cdot))]\}. \tag{7.9}$$

From assumptions concerning the process (7.1), (7.3) parameters, with account of properties of the resolving function, it follows that the mapping $U_1(s, v)$ is Borelian on the set $[0, T] \times V$. Then selection

$$u_1(s, v) = \text{lex min}\, U_1(s, v)$$

appears as a jointly Borelian function in its variables (see [9]). The pursuer's control on the interval $[0, t_*)$ is constructed in the following form:

$$u(s) = u_1(s, v(s)).$$

Being superposition of Borelian and measurable functions, it is also measurable (see [4, 9]).

Set

$$\bar{\alpha}(t, s, z^0(\cdot), v, \gamma^0(\cdot)) = 0, \quad s \in [t_*, T].$$

Then the mapping

$$U_2(s, v) \\ = \{u \in U : \pi K(\bar{T}-s)\,\phi(u, v) - \gamma^0(\bar{T}-s) = 0\}, \quad s \in [t_*, \bar{T}], \; v \in V$$

is Borelian in its variables, and its selection

$$u_2(s, v) = \text{lex min}\, U_2(s, v)$$

is also Borelian.

On the interval $[t_*, T]$, we set the pursuer's control equal to

$$u(s) = u_2(s, v(s)). \tag{7.10}$$

It is measurable function too (see [4, 9]).

Let $\xi(\bar{T}, z^0(\cdot), \gamma^0(\cdot)) \in M(\bar{T})$. In this case, we choose the pursuer's control on the interval $[0, \bar{T}]$ in the form (7.10).

Thus, the rules are defined, to which the pursuer should follow in constructing his control. We will now show that if the pursuer follows these rules in the course of the game, a trajectory of process (7.1) hits the terminal set at the time $\bar{T}$ under arbitrary admissible controls of the evader.

The Cauchy formula (see Lemma 7.1) for the system (7.1) implies the representation

$$\pi z\left(\bar{T}\right) = \pi K\left(\bar{T}\right) z^0(0) + \int_{-\tau}^{0} \pi K\left(\bar{T} - s - \tau\right) B z^0(s)\, ds$$
$$+ \int_0^{\bar{T}} \pi K\left(\bar{T} - s\right) \phi\left(u(s),\ v(s)\right) ds. \tag{7.11}$$

First we examine the case when $\xi\left(\bar{T},\ z^0(\cdot),\ \gamma^0(\cdot)\right) \notin M\left(\bar{T}\right)$. By adding and subtracting from the right-hand side of Eq. (7.11), the values

$$\int_0^{\bar{T}} \pi K\left(\bar{T} - s\right) \phi\left(u(s),\ D\left(\bar{T} - s\right) v(s)\right) \mathrm{d}s, \quad \int_0^{\bar{T}} \gamma^0\left(\bar{T} - s\right) \mathrm{d}s$$

one can deduce

$$\pi z\left(\bar{T}\right)$$
$$= \left[\pi K\left(\bar{T}\right) z^0(0) + \int_{-\tau}^{0} \pi K\left(\bar{T} - s - \tau\right) B z^0(s)\, \mathrm{d}s + \int_0^{\bar{T}} \gamma^0\left(\bar{T} - s\right) \mathrm{d}s\right]$$
$$+ \int_0^{\bar{T}} \left[\pi K\left(\bar{T} - s\right) \phi\left(u(s),\ D\left(\bar{T} - s\right) v(s)\right) - \gamma^0\left(\bar{T} - s\right)\right] \mathrm{d}s$$
$$+ \int_0^{\bar{T}} \left[\pi K\left(\bar{T} - s\right) \phi\left(u(s),\ v(s)\right) - \pi K\left(\bar{T} - s\right) \phi\left(u(s),\ D\left(\bar{T} - s\right) v(s)\right)\right] \mathrm{d}s.$$

Taking into account formulas (7.6), (7.7), (7.9), we come to the inclusion

$$\pi z\left(\bar{T}\right) \in \xi\left(\bar{T},\ z^0(\cdot),\ \gamma^0(\cdot)\right)$$
$$\times \left(1 - \int_0^{\bar{T}} \bar{\alpha}\left(\bar{T},\ s,\ z^0(\cdot),\ v(s),\ \gamma^0(\cdot)\right) \mathrm{d}s\right)$$
$$+ \int_0^{\bar{T}} \bar{\alpha}\left(\bar{T},\ s,\ z^0(\cdot),\ v(s),\ \gamma^0(\cdot)\right) M\left(\bar{T}\right) \mathrm{d}s$$
$$+ \int_0^{\bar{T}} \pi K\left(\bar{T} - s\right) \bar{\phi}\left(\bar{T} - s,\ u(s),\ v(s)\right) \mathrm{d}s, \tag{7.12}$$

where the following equation is taken into account

$$\bar{\alpha}\left(\bar{T},\ s,\ z^0(\cdot),\ v(s),\ \gamma^0(\cdot)\right) = 0, \quad s \in \left[t_*,\ \bar{T}\right].$$

If $\int_0^{\bar{T}} \bar{\alpha}\left(\bar{T}, s, z^0(\cdot), v(s), \gamma^0(\cdot)\right) ds = 1$ and that the set M is convex then

$$\int_0^{\bar{T}} \bar{\alpha}\left(\bar{T}, s, z^0(\cdot), v(s), \gamma^0(\cdot)\right) M\left(\bar{T}\right) ds \subset M\left(\bar{T}\right)$$

and inclusion (7.12) implies inclusion $\pi z\left(\bar{T}\right) \in M$.

Let $\xi\left(\bar{T}, z^0(\cdot), \gamma^0(\cdot)\right) \in M\left(\bar{T}\right)$. By adding and subtracting from the right-hand side of the Cauchy formula (7.11) the integral $\int_0^{\bar{T}} \gamma^0(T-s)\, ds$, we have the following:

$$\pi z\left(\bar{T}\right)$$

$$= \left[\pi K\left(\bar{T}\right) z^0(0) + \int_{-\tau}^{0} \pi K\left(\bar{T}-s-\tau\right) B z^0(s)\, ds + \int_0^{\bar{T}} \gamma^0\left(\bar{T}-s\right) ds\right]$$

$$+ \int_0^{\bar{T}} \left[\pi K\left(\bar{T}-s\right) \phi(u(s), v(s)) - \gamma^0\left(\bar{T}-s\right)\right] ds.$$

Then, using the rule of the pursuer control for the case when $\xi(\bar{T}, z^0(\cdot), \gamma^0(\cdot)) \in M(\bar{T})$, we derive the pursuer control in the form of (7.10). Taking into account expression (7.7) and definition of $M\left(\bar{T}\right)$, we come to the inclusion $\pi z\left(\bar{T}\right) \in M$.

Corollary 7.1 *Assume that the pursuit differential-difference game (7.1), (7.3) is linear* $(\phi(u, v) = u - v)$, *Conditions 7.2, 7.3 hold,* $\pi K(t) U = r(t) S$, $M(t) = q(t) S$, *where* $r(t)$, $r : \mathbb{R} \to \mathbb{R}$, $q(t)$, $q : \mathbb{R} \to \mathbb{R}$, *are continuous nonnegative numerical functions, and S is the unit ball from the subspace L, centered at zero. Then when* $\xi\left(t, z^0(\cdot), \gamma(\cdot)\right) \notin q(t) S$, *the resolving function (7.8) is the largest root of the quadratic equation for* α

$$\begin{aligned} \left\| \pi K(t-s) D(t-s) v + \gamma(t-s) - \alpha \xi\left(t, z^0(\cdot), \gamma(\cdot)\right) \right\| \\ = r(t-s) + \alpha q(t). \end{aligned} \tag{7.13}$$

Proof We will use the matrix function $D(t)$, $t \in [0, +\infty)$. In linear case $\phi(U, D(t) v) = U - D(t) v$. Then the multi-valued mapping in (7.5) reduces to the form

$$\bar{W}(t-s, v) = \pi K(t-s) U - \pi K(t-s) D(t-s) v.$$

Taking into account the assumptions of Corollary 7.1, we deduce from expression (7.8) that the resolving function $\bar{\alpha}\left(t, s, z^0(\cdot), v, \gamma(\cdot)\right)$ for fixed valued of its arguments is the maximal number α such that

$$\begin{aligned} &\left[r(t-s) S - \pi K(t-s) D(t-s) v - \gamma(t-s)\right] \cap \\ &\alpha\left[q(t) S - \xi\left(t, z^0(\cdot), \gamma(\cdot)\right)\right] \neq \emptyset. \end{aligned}$$

The last expression is equivalent to the inclusion

$$\pi K(t-s) D(t-s) v + \gamma(t-s) - \alpha\xi\left(t, z^0(\cdot), \gamma(\cdot)\right) \in \left[r(t-s) + \alpha q(t)\right] S.$$

Due to the linearity of the left-hand side of this inclusion in α, the vector $\pi K(t-s) D(t-s) v + \gamma(t-s) - \alpha\xi\left(t, z^0(\cdot), \gamma(\cdot)\right)$ at the maximal value of α lies on the boundary of the ball $\left[r(t-s) + \alpha q(t)\right] S$. In other words, the length of this vector is equal to the radius of this ball that is demonstrated by (7.13).

7.4 Example

Let us examine some analog of the game "Boy and Crocodile" with dynamics described by the system of differential-difference equations

$$\begin{aligned} \dot{z}_1(t) &= z_2(t-\tau) - v, \qquad && z_1 \in \mathbb{R}^n, \quad n \geq 2, \\ \dot{z}_2(t) &= u, && z_2 \in \mathbb{R}^n, \end{aligned}$$

$$\|u\| \leq \rho, \quad \rho > 0, \quad \|v\| \leq \sigma, \quad \sigma > 0.$$

The initial state is

$$z(t) = z^0(t) = \left(z_1^0(t), z_2^0(t)\right), \quad -\tau \leq t \leq 0.$$

The pursuit is completed when $\| z_1 \| \leq l$. In accordance with the Eq. (7.1) $A = 0$, $B = \begin{pmatrix} 0 & E_n \\ 0 & 0 \end{pmatrix}$; $z(t) = Bz(t-\tau) + u - v$.

The control domains are

$$U = \left\{ \begin{pmatrix} 0 \\ u \end{pmatrix}, \; u \in \mathbb{R}^n : \; \|u\| \leq \rho \right\}, \quad V = \left\{ \begin{pmatrix} v \\ 0 \end{pmatrix}, \; v \in \mathbb{R}^n : \; \|v\| \leq \sigma \right\}.$$

Here the terminal set M^* has the form

$$M^* = \left\{ z = (z_1, z_2) \in \mathbb{R}^{2n} : \; \|z_1\| \leq l \right\},$$

$$M_0 = \left\{ z = (z_1, z_2) \in \mathbb{R}^{2n} : \; z_1 = 0 \right\}, \quad L = \left\{ z = (z_1, z_2) \in \mathbb{R}^{2n} : \; z_2 = 0 \right\}$$

and $M = \left\{ z = (z_1, z_2) \in \mathbb{R}^{2n} : \; \|z_1\| \leq l, \; z_2 = 0 \right\}$.

Thus, the operator of orthogonal projection is defined by matrix $\pi = \begin{pmatrix} E_n & 0 \\ 0 & 0 \end{pmatrix}$.

The fundamental matrix has the form $[K(t)] = \begin{pmatrix} 1 & t \\ 0 & 1 \end{pmatrix} \otimes E_n$. Then, we have

$W(t) = (\rho t - \sigma) S = \emptyset$ for $t \in \left[0,\ \frac{\sigma}{\rho}\right)$. This means that Pontryagin's condition fails on this interval of time.

Set

$$D(t) = \begin{cases} \frac{\rho}{\sigma} t E_n, & 0 \le t < \frac{\sigma}{\rho}, \\ E_n, & t \ge \frac{\sigma}{\rho}. \end{cases}$$

Consider the multi-valued mapping $\bar{W}(t)$ of the form (7.5). In this example

$$\bar{W}(t) = \pi K(t) U \overset{*}{-} \pi K(t) D(t) V, \quad \pi K(t) U = \rho t S,$$

$$\pi K(t) D(t) V$$
$$= \begin{pmatrix} E_n & 0 \\ 0 & 0 \end{pmatrix} \cdot \begin{pmatrix} E_n & tE_n \\ 0 & E_n \end{pmatrix} \cdot \begin{pmatrix} \frac{\rho}{\sigma} tE_n & 0 \\ 0 & \frac{\rho}{\sigma} tE_n \end{pmatrix} \cdot \begin{pmatrix} v \\ 0 \end{pmatrix}$$
$$= \begin{pmatrix} \frac{\rho}{\sigma} tE_n & \frac{\rho}{\sigma} tE_n \\ 0 & 0 \end{pmatrix} \cdot \begin{pmatrix} v \\ 0 \end{pmatrix} = \begin{pmatrix} \frac{\rho}{\sigma} tE_n v \\ 0 \end{pmatrix} = \rho t S, \quad 0 \le t < \frac{\sigma}{\rho};$$

$$\pi K(t) D(t) V = \sigma S, \quad t \ge \frac{\sigma}{\rho}.$$

Finally, $\bar{W}(t) = \begin{cases} \{0\}, & t \in \left[0,\ \frac{\sigma}{\rho}\right], \\ (\rho t - \sigma) S, & t \in \left(\frac{\sigma}{\rho},\ +\infty\right). \end{cases}$ Therefore, Condition 7.2 is satisfied for all $t \ge 0$.

Denote

$$\bar{\phi}(t, U, V) = \phi(U, V) - \phi(U, D(t) V) = \begin{pmatrix} E_n - D(t) \\ 0 \end{pmatrix} \otimes V.$$

In the case, when $t \in \left[0,\ \frac{\sigma}{\rho}\right]$, we have

$$\pi K(t) \bar{\phi}(t, U, V) = \begin{pmatrix} E_n & tE_n \\ 0 & 0 \end{pmatrix} \cdot \begin{pmatrix} E_n \left(1 - t\frac{\rho}{\sigma}\right) \\ 0 \end{pmatrix} \cdot \begin{pmatrix} v \\ 0 \end{pmatrix}$$
$$= \begin{pmatrix} E_n \left(1 - t\frac{\rho}{\sigma}\right) \\ 0 \end{pmatrix} \cdot \begin{pmatrix} v \\ 0 \end{pmatrix} = \left(1 - t\frac{\rho}{\sigma}\right) \cdot \sigma S = (\sigma - t\rho) S.$$

Otherwise, when $t > \frac{\sigma}{\rho}$ we have $\pi K(t) \bar{\phi}(t, U, V) = 0$.

We now proceed to constructing the multi-valued mapping

$$M(t) = M \overset{*}{-} \int_0^t \pi K(t_1) \bar{\phi}(t_1, U, V)\, dt_1.$$

If

$$\int_0^t (\sigma - \rho t_1)\, S dt_1 = S\sigma t_1|_{t_1=0}^t - \rho \left.\frac{t_1^2}{2} S\right|_{t_1=0}^t = \left(-\frac{\rho t^2}{2} + \sigma t\right) S,$$

then

$$M(t) = \begin{cases} \left(\frac{\rho t^2}{2} - \sigma t + l\right) S, & t \in \left[0, \frac{\sigma}{\rho}\right]; \\ \left(l - \frac{\sigma^2}{2\rho}\right) \cdot S, & t > \frac{\sigma}{\rho}. \end{cases}$$

The mapping $M(t)$ is nonempty, provided $\frac{\rho t^2}{2} - \sigma t + l \geq 0$ for $t \in \left[0, \frac{\sigma}{\rho}\right]$ and $l - \frac{\sigma^2}{2\rho} \geq 0$ for $t > \frac{\sigma}{\rho}$.

Denote that the discriminant of the quadratic polynomial of the former inequality is equal to $D = \sigma^2 - 2\rho l$ and the account of the latter one yields $D \leq 0$. In other words, the disparities in the performance of the second branch of the parabola quadratic polynomial of the first inequality lie above (or intersect at one point) of the $t-$axis. Thus, the inequality $l - \frac{\sigma^2}{2\rho} \geq 0$ provides the inequality $\frac{\rho t^2}{2} - \sigma t + l \geq 0,\ t \geq 0$. Hence, the inequality $l - \frac{\sigma^2}{2\rho} \geq 0$ is sufficient the Condition 7.3 to hold.

Thus, if $l - \frac{\sigma^2}{2\rho} \geq 0$ then $M(t) \neq \emptyset$ for $t \geq 0$.

Let us analyze the case when $l - \frac{\sigma^2}{2\rho} < 0$. The multi-valued mapping $M(t) \neq \emptyset$, if

$$\frac{\rho t^2}{2} - \sigma t + l \geq 0 \quad \Leftrightarrow \quad t \in \left[0, \frac{\sigma - \sqrt{\sigma^2 - 2\rho l}}{\rho}\right] \cup \left[\frac{\sigma + \sqrt{\sigma^2 - 2\rho l}}{\rho}, +\infty\right)$$

and $M(t) = \emptyset$ for $t \in \left(\frac{\sigma - \sqrt{\sigma^2 - 2\rho l}}{\rho}, \frac{\sigma + \sqrt{\sigma^2 - 2\rho l}}{\rho}\right)$.

Hence, in the case $l - \frac{\sigma^2}{2\rho} < 0$ we seek the time of the game termination on the interval $t \in \left[0, \frac{\sigma - \sqrt{\sigma^2 - 2\rho l}}{\rho}\right]$.

Let us choose selection $\gamma^0(t) \equiv 0$ in $\bar{W}(t)$. Then

$$\xi\left(t, z^0(\cdot), 0\right) = \pi K(t) z^0(0) + \int_{-\tau}^0 \pi K(t - s - 1) B z^0(s) ds.$$

$$\xi\left(t, z^0(\cdot), 0\right) = \begin{pmatrix} E_n & 0 \\ 0 & 0 \end{pmatrix} \cdot \begin{pmatrix} E_n & tE_n \\ 0 & E_n \end{pmatrix} \cdot z^0(0)$$
$$+ \int_{-\tau}^0 \begin{pmatrix} E_n & (t - s - 1) E_n \\ 0 & 0 \end{pmatrix} \cdot \begin{pmatrix} 0 & E_n \\ 0 & 0 \end{pmatrix} \cdot z^0(s) ds$$
$$= \begin{pmatrix} E_n & tE_n \\ 0 & 0 \end{pmatrix} \cdot \begin{pmatrix} z_1^0(0) \\ z_2^0(0) \end{pmatrix} + \int_{-\tau}^0 \begin{pmatrix} 0 & E_n \\ 0 & 0 \end{pmatrix} \cdot \begin{pmatrix} z_1^0(s) \\ z_2^0(s) \end{pmatrix} ds.$$

The integral of the matrix is defined as the result of the "element-wise" integration (see [15]). We have

$$\xi\left(t\,,\,z^0\left(\,\cdot\,\right)\,,\,0\right)=\begin{bmatrix} z_1^0\,(0)+tz_2^0\,(0)+\int_{-\tau}^0 z_2^0\,(s)\,\mathrm{d}s, \\ 0 \end{bmatrix}, \tag{7.14}$$

$$z_i^0\left(\,\cdot\,\right)=\begin{bmatrix} z_{i1}^0\left(\,\cdot\,\right) \\ \vdots \\ z_{in}^0\left(\,\cdot\,\right) \end{bmatrix}_{i=1\,,\,2},$$

$[\xi\,(\cdots)]$ being the block matrix of dimension $2n\times 1$.

1. Let $0\leq t\leq\frac{\sigma}{\rho}$.

$$\begin{aligned}\bar{W}\,(t-s\,,\,v)&=\pi K\,(t-s)\,\phi\,(U\,,\,D\,(t-s)\,v)\\ &=\pi K\,(t-s)\,U-\pi K\,(t-s)\,D\,(t-s)\,v\\ &=\rho\,(t-s)\,S-\tfrac{\rho}{\sigma}\,(t-s)\,v=\rho\,(t-s)\left(S-\tfrac{v}{\sigma}\right).\end{aligned}$$

Then, the resolving function has the form

$$\begin{aligned}&\bar{\alpha}\left(t\,,\,s\,,\,z^0\left(\,\cdot\,\right)\,,\,v\,,\,0\right)=\sup\{\alpha\geq 0\,:\,\rho\,(t-s)\left(S-\tfrac{v}{\sigma}\right)\cap\\ &\alpha\cdot\left[\left(\tfrac{\rho t^2}{2}-\sigma t+l\right)S-\xi\left(t\,,\,z^0\left(\,\cdot\,\right)\,,\,0\right)\right]\neq\emptyset\},\quad v\in V.\end{aligned}$$

To find the resolving function, we make use of the Corollary 7.1. We seek continuous nonnegative functions $r\,(t)$ and $q\,(t)$, such that the equalities

$$\pi K\,(t)\,U=r\,(t)\,S,\quad M\,(t)=q\,(t)\,S$$

are satisfied.

We see that

$$\pi K\,(t)\,U=\begin{pmatrix} E_n & tE_n \\ 0 & 0 \end{pmatrix}\cdot\begin{pmatrix} 0 \\ u \end{pmatrix}=\begin{pmatrix} tE_nu \\ 0 \end{pmatrix}=t\cdot\rho S.$$

Let us set $r\,(t)=\rho t\,,\ q\,(t)=\frac{\rho t^2}{2}-\sigma t+l$. Then, in view of the Corollary 7.1, the resolving function $\bar{\alpha}\left(t\,,\,s\,,\,z^0\left(\,\cdot\,\right)\,,\,v\,,\,0\right)$ is the largest root of the quadratic equation for α

$$\left\|\,\pi K\,(t-s)\,D\,(t-s)\,v-\alpha\xi\left(t\,,\,z^0\left(\,\cdot\,\right)\,,\,\gamma\left(\,\cdot\,\right)\right)\right\|=r\,(t-s)+\alpha q\,(t)\,,$$

provided that $\xi\left(t\,,\,z^0\left(\,\cdot\,\right)\,,\,0\right)\neq q\,(t)\,S$.

This equation can be rewritten as follows:

$$\|\pi K(t-s)D(t-s)v\|^2 - 2\alpha\left(\pi K(t-s)D(t-s)v,\ \xi\left(t,\ z^0(\cdot),\ \gamma(\cdot)\right)\right) + \alpha^2\cdot\left\|\xi\left(t,\ z^0(\cdot),\ \gamma(\cdot)\right)\right\|^2 = r^2(t-s) + 2\alpha\cdot r(t-s)q(t) + \alpha^2 q^2(t),$$

$$\alpha^2\cdot\left(\left\|\xi\left(t,\ z^0(\cdot),\ \gamma(\cdot)\right)\right\|^2 - q^2(t)\right) - 2\alpha\left[\left(\pi K(t-s)D(t-s)v,\ \xi\left(t,\ z^0(\cdot),\ \gamma(\cdot)\right)\right) - r(t-s)q(t)\right] + \|\pi K(t-s)D(t-s)v\|^2 - r^2(t-s) = 0.$$

It is clear that $\min\limits_{v\in V}\bar{\alpha}\left(t,\ s,\ z^0(\cdot),\ v,\ 0\right) = 0$, vector

$$v = -\frac{\sigma}{\left\|\xi\left(t,\ z^0(\cdot),\ 0\right)\right\|}\cdot\xi\left(t,\ z^0(\cdot),\ 0\right)$$

furnishes the minimum, and the inequality in the definition of time $\bar{T}\left(z^0(\cdot),\ 0\right)$ fails for $t\in\left[0,\ \frac{\sigma}{\rho}\right]$.

If $\xi\left(t,\ z^0(\cdot),\ 0\right)\in M(t)$ then $\bar{\alpha}\left(t,\ s,\ z^0(\cdot),\ v,\ 0\right) = +\infty$. The least instant, at which the above inclusion holds true, satisfies the equation

$$\left\|\xi\left(t,\ z^0(\cdot),\ 0\right)\right\| = \frac{\rho t^2}{2} - \sigma t + l,\quad 0\le t\le\frac{\sigma}{\rho}. \tag{7.15}$$

Thus, on the interval $\left[0,\ \frac{\sigma}{\rho}\right]$ the time of game termination $\bar{T}\left(z^0(\cdot),\ 0\right)$ appears as the least positive root of Eq. (7.15).

2. Let $t > \frac{\sigma}{\rho}$.

(a) Let us analyze the case when $t-s\le\frac{\sigma}{\rho}$. Then $D(t-s) = \frac{\rho}{\sigma}(t-s)$. If $r(t) = \rho t$ then $q(t) = l - \frac{\sigma^2}{2\rho}$. In view of the Corollary 7.1, when $\xi\left(t,\ z^0(\cdot),\ 0\right)\notin\left(l-\frac{\sigma^2}{2\rho}\right)S$ the resolving function

$$\bar{\alpha}\left(t,\ s,\ z^0(\cdot),\ v,\ 0\right) = \sup\{\alpha\ge 0:\ \rho(t-s)\left(S-\tfrac{v}{\sigma}\right)\cap \alpha\cdot\left[\left(l-\tfrac{\sigma^2}{2\rho}\right)S - \xi\left(t,\ z^0(\cdot),\ 0\right)\right]\ne\emptyset\}$$

appears as the largest root of the following quadratic equation for α

$$\left\|\frac{\rho}{\sigma}(t-s)v - \alpha\cdot\xi\left(t,\ z^0(\cdot),\ 0\right)\right\| = \alpha\cdot\left(l-\frac{\sigma^2}{2\rho}\right) + \rho(t-s).$$

It is clear that

$$\min_{v\in V}\bar{\alpha}\left(t,\ s,\ z^0(\cdot),\ v,\ 0\right) = 0,\quad t-s\le\frac{\sigma}{\rho}. \tag{7.16}$$

The minimum is attained on the vector

$$v = -\frac{\sigma}{\left\| \xi\left(t\,,\,z^0\left(\,\cdot\,\right)\,,\,0\right) \right\|}\cdot \xi\left(t\,,\,z^0\left(\,\cdot\,\right)\,,\,0\right).$$

In this case the inequality in the definition of time $\bar{T}\left(z^0\left(\,\cdot\,\right)\,,\,0\right)$ is not satisfied.

Let $\xi\left(t\,,\,z^0\left(\,\cdot\,\right)\,,\,0\right) \in M\left(t\right)$. Then the time of game $\bar{T}\left(z^0\left(\,\cdot\,\right)\,,\,0\right)$ termination appears as the least positive root of the equation

$$\left\| \xi\left(t\,,\,z^0\left(\,\cdot\,\right)\,,\,0\right) \right\| = l - \frac{\sigma^2}{2\rho}, \quad t > \frac{\sigma}{\rho}.$$

(b) Let us analyze the case when $t - s > \frac{\sigma}{\rho}$. Then $D\left(t - s\right) = E_n$ and $\bar{W}\left(t - s\,,\,v\right) = \left(t - s\right)\rho S - v$.

The resolving function is presented by the formula

$$\begin{aligned}&\bar{\alpha}\left(t\,,\,s\,,\,z^0\left(\,\cdot\,\right)\,,\,v\,,\,0\right)\\&= \sup\left\{\alpha \geq 0\,:\,\left[\rho\left(t - s\right)S - v\right] \cap \alpha\cdot\left[\left(l - \tfrac{\sigma^2}{2\rho}\right)S - \xi\left(t\,,\,z^0\left(\,\cdot\,\right)\,,\,0\right)\right] \neq \emptyset\right\},\\&v \in V.\end{aligned}$$

When $\xi\left(t\,,\,z^0\left(\,\cdot\,\right)\,,\,0\right) \notin \left(l - \frac{\sigma^2}{2\rho}\right)S$ it is the largest root of the quadratic equation for α

$$\left\| v - \alpha\cdot\xi\left(t\,,\,z^0\left(\,\cdot\,\right)\,,\,0\right) \right\| = \rho\left(t - s\right) + \alpha\cdot\left(l - \frac{\sigma^2}{2\rho}\right).$$

The minimum is attained on the vector $v = -\frac{\sigma}{\left\| \xi\left(t\,,\,z^0\left(\cdot\right)\,,\,0\right) \right\|}\cdot\xi\left(t\,,\,z^0\left(\,\cdot\,\right)\,,\,0\right)$ and

$$\begin{aligned}&\min_{v\in V}\bar{\alpha}\left(t\,,\,s\,,\,z^0\left(\,\cdot\,\right)\,,\,v\,,\,0\right)\\&= \frac{\rho(t-s)-\sigma}{\left\| \xi\left(t\,,\,z^0(\cdot)\,,\,\gamma(\cdot)\right) \right\| - \left(l - \frac{\sigma^2}{2\rho}\right)}, \quad t - s > \frac{\sigma}{\rho}.\end{aligned} \tag{7.17}$$

Let us evaluate the instant of the game termination in the case $t > \frac{\sigma}{\rho}$. To the end, we perform calculations, taking account of formulas (7.16), (7.17).

$$\begin{aligned}&\int_0^t \min_{\|v\|\leq\sigma}\bar{\alpha}\left(t\,,\,s\,,\,z^0\left(\cdot\right)\,,\,v\,,\,0\right)ds = \int_0^{t-\frac{\sigma}{\rho}} \min_{\|v\|\leq\sigma}\bar{\alpha}\left(t\,,\,s\,,\,z^0\left(\cdot\right)\,,\,v\,,\,0\right)ds\\&+ \int_{t-\frac{\sigma}{\rho}}^t \min_{\|v\|\leq\sigma}\bar{\alpha}\left(t\,,\,s\,,\,z^0\left(\cdot\right)\,,\,v\,,\,0\right)ds = 1.\end{aligned}$$

We observe that if $0 \leq s \leq t - \frac{\sigma}{\rho}$, then $t - s > \frac{\sigma}{\rho}$. Therefore, in the first integral (integrating in s from 0 to $t - \frac{\sigma}{\rho}$), the integrand is expressed by the relation (7.17).

If $t - \frac{\sigma}{\rho} \leq s \leq t$ then $t - s \leq \frac{\sigma}{\rho}$. In this case, the integrand in the integral in s from $t - \frac{\sigma}{\rho}$ to t can be expressed by relation (7.16).

Therefore,

$$\int_0^t \min_{\|v\|\le\sigma} \bar{\alpha}\left(t\,,\,s\,,\,z^0(\cdot)\,,\,v\,,\,0\right) ds = \int_0^{t-\frac{\sigma}{\rho}} \min_{\|v\|\le\sigma} \bar{\alpha}\left(t\,,\,s\,,\,z^0(\cdot)\,,\,v\,,\,0\right) ds + 0$$
$$= \int_0^{t-\frac{\sigma}{\rho}} \frac{\rho(t-s)-\sigma}{\|\xi(t,z^0(\cdot),0)\|-l+\frac{\sigma^2}{2\rho}} ds = 1.$$

Upon integration of the above expression, we come to the formulas

$$\int_0^{t-\frac{\sigma}{\rho}} [\rho\,(t-s)-\sigma]\,ds = \left\|\xi\left(t\,,\,z^0(\,\cdot\,)\,,\,0\right)\right\| - l + \frac{\sigma^2}{2\rho},$$

$$\rho t\left(t-\frac{\sigma}{\rho}\right) - \frac{\rho}{2}\cdot\left(t-\frac{\sigma}{\rho}\right)^2 - \sigma\left(t-\frac{\sigma}{\rho}\right) = \left\|\xi\left(t\,,\,z^0(\,\cdot\,)\,,\,0\right)\right\| - l + \frac{\sigma^2}{2\rho},$$

$$\frac{\rho t^2}{2} - \sigma t + l = \left\|\xi\left(t\,,\,z^0(\,\cdot\,)\,,\,0\right)\right\|.$$

Thus, this pursuit game for different-inertia objects, under given initial condition

$$z^0(t) = \left\{\left(z_1^0(t)\,,\,z_2^0(t)\right) \,:\, z_1^0(t) \in \mathbb{R}^n\,,\, z_2^0(t) \in \mathbb{R}^n\,,\quad -\tau \le t \le 0\right\},$$

may be completed by the pursuer at the time $\bar{T} = \bar{T}\left(z^0(\,\cdot\,)\right)$, which is the least root of the quadratic equation for t

$$\frac{\rho t^2}{2} - \sigma t + l = \left\| z_1^0(0) + t z_2^0(0) + \int_{-\tau}^0 z_2^0(s)\,ds \right\|,\quad 0 \le t \le \frac{\sigma}{\rho},$$

when $l - \frac{\sigma^2}{2\rho} \ge 0$, or, when $t > \frac{\sigma}{\rho}$ at the time $\bar{T}$, the least root of following quadratic equations for t

$$\frac{\rho t^2}{2} - \sigma t + l = \left\| z_1^0(0) + t z_2^0(0) + \int_{-\tau}^0 z_2^0(s)\,ds \right\|,$$

$$\left\| z_1^0(0) + t z_2^0(0) + \int_{-\tau}^0 z_2^0(s)\,ds \right\| = l - \frac{\sigma^2}{2\rho}.$$

Otherwise, when $l - \frac{\sigma^2}{2\rho} < 0$, in the case $t \in \left[0\,,\ \frac{\sigma-\sqrt{\sigma^2-2\rho l}}{\rho}\right]$, $\bar{T}$ is the least root of the quadratic equation

$$\frac{\rho t^2}{2} - \sigma t + l = \left\| z_1^0(0) + t z_2^0(0) + \int_{-\tau}^0 z_2^0(s)\,ds \right\|.$$

We now dwell upon the issue of existence of roots of these equations.

1. If $l - \frac{\sigma^2}{2\rho} \geq 0$, then, under condition $z_2^0(0) = 0$, in the case $\left\| z_1^0(0) + \int_{-\tau}^{0} z_2^0(s)\,ds \right\| \geq l - \frac{\sigma^2}{2\rho}$ the time of the game termination is finite.
   Clearly, if $z_2^0(0) \neq 0$, then the time of the game termination is finite for all initial states.
2. Let $l - \frac{\sigma^2}{2\rho} < 0$. Then the game can be completed at the instant $T \leq \frac{\sigma - \sqrt{\sigma^2 - 2\rho l}}{\rho}$ in condition that $\left\| z_1^0(0) + \int_{-\tau}^{0} z_2^0(s)\,ds \right\| \leq l$.

Acknowledgments The author is grateful to Academician Zgurovsky M.Z. for the possibility of the publication and to Professor Kasyanov P.O. for assistance in publishing this article.

References

1. Chikrii, A.A., Baranovskaya, L.V.: A type of controlled system with delay. Cybern. Comput. Technol. **107**, 1–8 (1998)
2. Baranovskaya, L.V., Chikrii, Al.A.: On one class of difference-differential group pursuit games. In: Multiple Criteria and Game Problems Under Uncertainty, Proceedings of the Fourth International Workshop, vol. 11, pp. 8–14 September 1996. Moscow (1996)
3. Bellman, R., Cooke, K.L.: Differential-Difference Equations, Academic Press, New York (1963)
4. Pshenitchnyi, B.N.: Convex Analysis and Extremal Problems. Nauka, Moscow (1980)
5. Baranovskaya, G.G., Baranovskaya, L.V.: Group pursuit in quasilinear differential-difference games. J. Autom. Inf. Sci. **29**(1), 55–62 (1997). doi:10.1615/JAutomatInfScien.v29.i1.70
6. Chikrii, A.A.: Conflict-Controlled Processes. Springer Science & Business Media, Heidelberg (2013)
7. Isaacs, R.: Differential Games: A Mathematical Theory with Applications to Warfare and Pursuit, Control and Optimization. Wiley, New York (1965)
8. Gantmacher, F.R.: The Theory of Matrices. Chelsea Publishing Company, New York (1960)
9. Joffe, A.D., Tikhomirov, V.P.: Theory of Extremal Problems. North Holland, Amsterdam (1979)
10. Clarke, F.H.: Optimization and Nonsmooth Analysis. Wiley, New York (1983)
11. Aubin, J-P., Frankowska, He.: Set-Valued Analysis. Birkhause, Bostom (1990)
12. Chikrii, A.A.: Multi-valued mappings and their selections in game control problems. J. Autom. Inform. Sci., Scripta Technica Inc., New York. **27**(1), 3–14 (1994)
13. Chikrii, A.A.: Quasilinear controlled processes under conflict, dynamical systems. J. Math. Sci. **80**(1), 1489–1518 (1996)
14. Baranovskaya, L.V.: A method of resolving functions for one class of pursuit problems. East.-Eur. J. Enterp. Technol. **74**(4), 4–8 (2015). doi:10.15587/1729-4061.2015.39355
15. Lancaster, P.: Theory of Matrices. Academic Press, New York (1969)

Part II
Discrete and Continuous Dynamical Systems

Chapter 8
Characterization of Pullback Attractors for Multivalued Nonautonomous Dynamical Systems

Jacson Simsen and José Valero

Abstract In this paper we provide a review of the general results on pullback attractors for multivalued nonautonomous dynamical systems, completing at the same time some gaps in the theory. Also, when the attraction of a class of families of sets rather than just bounded sets is considered, we obtain the characterization of the pullback attractor as the union of all complete trajectories belonging to this class. Finally, an application to a reaction-diffusion equation without uniqueness of solutions is given.

8.1 Introduction

The theory of pullback attractors for nonautonomous dynamical systems in both single-valued and multivalued cases has been developed by several authors in the last years (see [3–11, 13–15, 17, 18, 20, 21, 23, 24] and the references therein).

In this paper we focus on the multivalued situation, when more than one solution can correspond to a given initial data. Such dynamical systems naturally appear in important models of the Mathematical Physics such as the three-dimensional Navier–Stokes system or reaction-diffusion equations. It is worth pointing out that, unlike the autonomous case, several approaches are possible in order to provide sufficient and necessary conditions for the existence of pullback attractors. Such results are scattered in the literature, so it would be nice to have in hand a whole picture of

J. Simsen
Instituto de Matemática e Computação, Universidade Federal de Itajubá,
Av. BPS n. 1303, Bairro Pinheirinho, Itajubá, MG 37500-903, Brazil
e-mail: jacson@unifei.edu.br

J. Simsen
Fakultät für Mathematik, Universität Duisburg-Essen,
Thea-Leymann-Str. 9, 45127 Essen, Germany

J. Valero (✉)
Centro de Investigación Operativa, Universidad Miguel Hernández de Elche,
03202 Elche (Alicante), Spain
e-mail: jvalero@umh.es

V.A. Sadovnichiy and M.Z. Zgurovsky (eds.), *Advances in Dynamical Systems and Control*, Studies in Systems, Decision and Control 69,
DOI 10.1007/978-3-319-40673-2_8

all these theories. On the other hand, an important feature of global attractors is its characterization as the union of all complete trajectories of the system satisfying a certain property. Such description was developed in [9] for pullback attractors which attract bounded subsets of the phase space and are backwards bounded, but nothing is known so far with respect to the attraction of families of sets.

Our goal is threefold.

First, we intend to make a review of the theory of pullback attractors for multivalued processes by putting together several theorems which have been published in different papers to date. In this way, we offer a common perspective of such results. On top of that, we complete some small gaps in the theory that were not covered by the above mentioned papers.

Second, when the attraction of a class of families of sets rather than just bounded sets is considered, we study the characterization of the pullback attractor as the union of all complete trajectories belonging to this class. Such description of the attractor gives us an insight of its internal structure, which is a first step leading to the understanding of the dynamics inside the pullback attractor. We highlight that such result is new even in the single-valued case. Also, we observe that in order to obtain this type of results we ought to use the framework of generalized process, that is, multivalued processes generated by a set of functions satisfying suitable properties, which in applications are given by the solutions of a differential equation.

Third, we apply the theorem about the characterization of the pullback attractor to a reaction-diffusion equation lacking uniqueness of solutions of the Cauchy problem.

8.2 Pullback Attraction of Bounded Sets

In this section we will give a review of some results from [4, 9, 13, 17, 24] concerning the existence and structure of pullback attractors for multivalued processes in the case when we consider the attraction of bounded sets of the phase space, completing at the same time some small gaps of the theory. In this way, we put together the results published so far in different papers and offer a common perspective. We will consider the particular case of multivalued processes generated by generalized processes.

Let (X, ρ) be a complete metric space with the metric ρ and let $\mathscr{P}(X)$, $\mathscr{B}(X)$ and $\mathscr{K}(X)$ denote, respectively, the set of non-empty, non-empty and bounded and non-empty and compact subsets of X. For $x \in X$, $A, B \in \mathscr{P}(X)$ and $\varepsilon > 0$ we set $dist(x, A) := \inf_{a\in A} \{\rho(x, a)\}$, $dist(A, B) := \sup_{a\in A} \{\rho(a, B)\}$, $\mathscr{O}_\varepsilon(A) := \{z \in X; \rho(z, A) < \varepsilon\}$. Also, let $\mathbb{R}_d = \{(t, \tau) \in \mathbb{R}^2 : \tau \leq t\}$.

We recall first the definition of generalized processes, which were introduced at first in [2].

Let us denote $W_\tau = C([\tau, \infty); X)$ and let $\mathscr{G} = \{\mathscr{G}(\tau)\}_{\tau\in\mathbb{R}}$ consists of maps $\varphi \in W_\tau$. Let us consider the following conditions:

(C1) For any $\tau \in \mathbb{R}$ and $x \in X$ there exists $\varphi \in \mathscr{G}(\tau)$ such that $\varphi(\tau) = x$.

(C2) $\varphi_s = \varphi|_{[\tau+s,\infty)} \in \mathscr{G}(\tau + s)$ for any $s \geq 0$, $\varphi \in \mathscr{G}(\tau)$ (translation property).

$(C3)$ Let $\varphi, \psi \in \mathscr{G}$ be such that $\varphi \in \mathscr{G}(\tau)$, $\psi \in \mathscr{G}(r)$ and $\varphi(s) = \psi(s)$ for some $s \geq r \geq \tau$. Then the function θ defined by

$$\theta(t) := \begin{cases} \varphi(t), & t \in [\tau, s], \\ \psi(t), & t \in [s, \infty), \end{cases}$$

belongs to $\mathscr{G}(\tau)$ (concatenation property).

$(C4)$ For any sequence $\varphi^n \in \mathscr{G}(\tau)$ such that $\varphi^n(\tau) \to \varphi_0$ in X, there exists a subsequence φ^{n_k} and $\varphi \in \mathscr{G}(\tau)$ such that

$$\varphi^{n_k}(t) \to \varphi(t), \forall t \geq \tau.$$

If $(C1)$–$(C2)$, $(C4)$ hold, then we say that $\mathscr{G}$ is a generalized process. If, moreover, $(C1)$–$(C4)$ hold, then $\mathscr{G}$ is an exact (or strict) generalized process.

We define the multivalued map $U_{\mathscr{G}} : \mathbb{R}_d \times X \to P(X)$ associated with the family $\mathscr{G}$ in the following way:

$$U_{\mathscr{G}}(t, \tau, x) = \{\varphi(t) : \varphi \in \mathscr{G}(\tau), \ \varphi(\tau) = x\}. \tag{8.1}$$

If $(C1)$–$(C2)$ hold, then the map $U_{\mathscr{G}}$ is a multivalued process, that is:

(1) $U_{\mathscr{G}}(t, t, x) = x$ for all $t \in \mathbb{R}$, $x \in X$;
(2) $U_{\mathscr{G}}(t, \tau, x) \subset U_{\mathscr{G}}(t, s, U_{\mathscr{G}}(s, \tau, x))$ for all $-\infty < \tau \leq s \leq t < \infty$, $x \in X$, where $U_{\mathscr{G}}(s, \tau, C) = \cup_{x \in C} U_{\mathscr{G}}(s, \tau, x)$ for any $C \subset X$.

If, moreover, $(C3)$ also holds, then the map $U_{\mathscr{G}}$ is a strict multivalued process, which means that $U_{\mathscr{G}}$ is a multivalued process and, additionally, in the second property a strict equality holds: $U_{\mathscr{G}}(t, \tau, x) = U_{\mathscr{G}}(t, s, U_{\mathscr{G}}(s, \tau, x))$ (see [24, Theorem 12.1]).

The orbit $\gamma^{\xi}(t, E)$ for $\xi \leq t$ and the ω-limit set $\omega(t, E)$ at time t for $E \subset X$ are defined by:

$$\gamma^{\xi}(t, E) = \bigcup_{s \leq \xi} U_{\mathscr{G}}(t, s, E),$$

$$\omega(t, E) = \bigcap_{\xi \leq t} \overline{\gamma^{\xi}(t, E)}.$$

Arguing as in [4, Lemma 5] one can show that the ω-limit set is characterized as follows:

$$\omega(t, E) = \left\{ z \in X : \exists \{\xi_n\}_{n \in \mathbb{N}}, s_n \to -\infty \text{ such that } \xi_n \in U_{\mathscr{G}}(t, s_n, E) \text{ and } \xi_n \to z \right\}.$$

In order to study the structure of the pullback attractor the concept of complete trajectory plays an important role.

Definition 8.1 The map $\psi : \mathbb{R} \to X$ is called a complete trajectory through $x \in X$ if $\psi(\tau) = x$ for some $\tau \in \mathbb{R}$ and $\psi_s = \psi\mid_{[\tau+s,\infty)} \in \mathcal{G}(\tau + s)$ for all $s \in \mathbb{R}$.

It is obvious that every complete trajectory ψ satisfies

$$\psi\,(t) \in U_{\mathcal{G}}\,(t, s, \psi\,(s))\,, \textit{for any } s \leq t.$$

Let us recall several concepts related to invariance of a family of sets $A = \{A(t)\}_{t\in\mathbb{R}}$. We say that:

- A is positively invariant if $U_{\mathcal{G}}(t, \tau, A(\tau)) \subset A(t)$ for all $-\infty < \tau \leq t < \infty$;
- A is negatively invariant if $A(t) \subset U_{\mathcal{G}}(t, \tau, A(\tau))$ for all $-\infty < \tau \leq t < \infty$;
- A is invariant if $U_{\mathcal{G}}(t, \tau, A(\tau)) = A(t)$ for all $-\infty < \tau \leq t < \infty$;
- A is quasi-invariant (or weakly invariant) if for each $\tau \in \mathbb{R}$ and $z \in A(\tau)$ there exists a complete trajectory ψ through z at τ (i.e., $\psi(\tau) = z$) such that $\psi(t) \in A(t)$ for all $t \in \mathbb{R}$.
- A is weakly positively invariant if for every $\tau \leq t$ and $z \in A(\tau)$ we have that $U_{\mathcal{G}}(t, \tau, z) \cap A(t) \neq \varnothing$.

It is obvious that A is invariant if and only if it is both positively and negatively invariant and that if A is quasi-invariant, then A is negatively invariant. It is also well-known [24] that if A is invariant, then it is quasi-invariant.

A family $A = \{A(t)\}_{t\in\mathbb{R}}$ is said to be closed (compact, bounded) if every set $A\,(t)$ is closed (compact, bounded).

For compact families of sets of strict multivalued processes we will check that negatively invariance together with weakly positively invariance implies quasi-invariance. This fact is proved in a similar way to the autonomous case [16, 19, 22].

Lemma 8.1 *Let* $(C1)$*–*$(C4)$ *hold and let* $A = \{A(t)\}_{t\in\mathbb{R}}$ *be compact. If* A *is weakly positively invariant and negatively invariant, then it is quasi-invariant.*

Proof First, we prove that for every $\tau \in \mathbb{R}$, $x_0 \in A(\tau)$ there exists $\varphi \in \mathcal{G}\,(\tau)$ such that $\varphi\,(\tau) = x_0$ and $\varphi(t) \in A\,(t)$ for all $t \geq \tau$. We observe that for this statement it is only necessary to assume that A is closed.

There is an $x_{11} \in U_{\mathcal{G}}(\tau + 1, \tau, x_0) \cap A\,(\tau + 1)$ and a map $\varphi_1 \in \mathcal{G}(\tau)$ such that $\varphi_1(\tau) = x_0$ and $\varphi_1(\tau + 1) = x_{11}$. In the same way, we choose

$$x_{21} \in U_{\mathcal{G}}(\tau + \frac{1}{2}, \tau, x_0) \cap A(\tau + \frac{1}{2}),\ x_{22} \in U_{\mathcal{G}}(\tau + 1, \tau + \frac{1}{2}, x_{21}) \cap A(\tau + 1),$$

so by $(C3)$ there exists $\varphi_2 \in \mathcal{G}$ such that $\varphi_2(\tau) = x_0$, $\varphi_2(\tau + \frac{1}{2}) = x_{21}$ and $\varphi_2(\tau + 1) = x_{22}$.

Repeating this procedure several times we obtain a map $\varphi_{n+1} \in \mathcal{G}\,(\tau)$ such that $\varphi_{n+1}(\tau) = x_0$ and $\varphi_{n+1}(t) \in A\,(t)$ for $t = \tau + \frac{j}{2^n}, j = 1, 2, \ldots, 2^n$.

From $(C4)$ there exist $\varphi^0 \in \mathcal{G}(\tau)$ and a subsequence of φ_n such that $\varphi^0(t) = \lim_{k\to\infty} \varphi_{n_k}(t)$, for all $t \geq 0$. Thus, since A is closed, $\varphi^0(t) \in A(t)$ for all $t \in [0, 1]$

which are binary fractions. As A is closed and φ^0 is continuous, it follows that $\varphi^0(t) \in A(t)$, for all $t \in [0, 1]$.

We repeat the same proof in order to define a sequence $\varphi^j \in \mathscr{G}(\tau+j), j \in \mathbb{N}$, such that $\varphi^j(\tau+j) = \varphi^{j-1}(\tau+j)$ and $\varphi^j(t) \in A(t)$ for all $t \in [\tau+j, \tau+j+1]$. Using $(C3)$ we concatenate these maps and obtain $\varphi \in \mathscr{G}(\tau)$ such that $\varphi(\tau) = x_0$ and $\varphi(t) \in A$, for all $t \geq \tau$, as desired.

Second, arguing in a similar way we can check that if A is compact, then for every $\tau \in \mathbb{R}, x_0 \in A(\tau)$ there exists a complete trajectory ψ such that $\phi(\tau) = x_0$ and $\phi(t) \in A(t)$ for all $t \leq \tau$.

Concatenating the maps ϕ and φ we obtain a complete trajectory ψ such that $\psi(\tau) = x_0$ and $\psi(t) \in A(t)$ for all $t \in \mathbb{R}$.

Let us recall now the concept of pullback attractor.

Definition 8.2 The family $K = \{K(t)\}_{t\in\mathbb{R}}$ is called pullback attracting for $U_{\mathscr{G}}$ if

$$dist(U_{\mathscr{G}}(t,s,B), K(t)) \to 0, \text{ as } s \to -\infty, \text{ for all } B \in \mathscr{B}(X),\ t \in \mathbb{R}.$$

That is, if it pullback attracts any bounded set at any time $t \in \mathbb{R}$.

Definition 8.3 The family $\mathscr{A} = \{\mathscr{A}(t)\}_{t\in\mathbb{R}}$ is said to be a global pullback attractor for $U_{\mathscr{G}}$ if:

(1) $\mathscr{A}$ is compact;
(2) $\mathscr{A}$ is pullback attracting;
(3) $\mathscr{A}$ is negatively invariant;
(4) $\mathscr{A}$ is minimal, that is, if $\widehat{\mathscr{A}} = \{\widehat{\mathscr{A}}(t)\}_{t\in\mathbb{R}}$ is a closed pullback attracting family, then $\mathscr{A}(t) \subset \widehat{\mathscr{A}}(t)$ for all $t \in \mathbb{R}$.

An essential property for proving the existence of a pullback attractor is the so-called pullback asymptotic compactness.

Definition 8.4 $U_{\mathscr{G}}$ is called pullback asymptotically compact at time t if for all $B \in \mathscr{B}(X)$ each sequence $\{\xi_j\}_{j\in\mathbb{N}}$ such that $\xi_j \in U_{\mathscr{G}}(t, \tau_j, B)$, where $\tau_j \to -\infty$, has a convergent subsequence. If this property is satisfied for each time $t \in \mathbb{R}$, then we say that $\mathscr{G}$ is pullback asymptotically compact.

Theorem 6 and Lemma 8 in [4] imply the following lemma.

Lemma 8.2 *Let $(C1)$–$(C2)$ hold. If $\mathscr{G}$ is pullback asymptotically compact, then for any $B \in \mathscr{B}(X)$ and $t \in \mathbb{R}$ the ω-limit set $\omega(t, B)$ is non-empty, compact and pullback attracts B at time t, that is,*

$$dist(U_{\mathscr{G}}(t,s,B), \omega(t,B)) \to 0, \text{ as } s \to -\infty.$$

Moreover, $\mathscr{G}$ is pullback asymptotically compact if and only if for each $t \in \mathbb{R}$ and $B \in \mathscr{B}(X)$ there exists a compact set $D(t, B)$ satisfying

$$dist\,(U_{\mathscr{G}}(t,s,B), D(t,B)) \to 0 \text{ as } s \to -\infty.$$

Remark 8.1 There exist other definitions which are equivalent to pullback asymptotic compactness. See [13, 24] for more details.

In [24, Theorem 12.5] the ω-limit set was proved to satisfy an additional invariance property.

Lemma 8.3 *Let* $(C1)$–$(C2)$, $(C4)$ *hold. If* $\mathscr{G}$ *is pullback asymptotically compact, then for any* $B \in \mathscr{B}(X)$ *the family of sets* $\{\omega(t, B)\}_{t\in\mathbb{R}}$ *is quasi-invariant. If, moreover,* $(C3)$ *holds and* $U_{\mathscr{G}}(t, r, \omega(r, B)) \subset B$ *for all* $r \leq t$, *then* $\{\omega(t, B)\}_{t\in\mathbb{R}}$ *is invariant.*

Let us consider now sufficient and necessary conditions for the existence of a pullback attractor.

Condition $(C4)$ implies easily that for all $(t, s) \in \mathbb{R}_d$ the map $x \mapsto U_{\mathscr{G}}(t, s, x)$ has closed graph (see Theorem 12.3 in [24] for more details). Hence, we obtain from Theorem 18 in [4] the following result, which provides a sufficient condition for the existence of a pullback attractor.

Theorem 8.1 *Let* $(C1)$–$(C2)$, $(C4)$ *hold. If there exists a pullback attracting family of compact sets* $D(t)$, *then the family of sets* $\mathscr{A} = \{\mathscr{A}(t)\}_{t\in\mathbb{R}}$ *defined by*

$$\mathscr{A}(t) = \overline{\bigcup_{B\in\mathscr{B}(X)} \omega(t, B)} \tag{8.2}$$

is a global pullback attractor for $U_{\mathscr{G}}$. *Moreover, the sets* $\mathscr{A}(t)$ *are compact and* $\mathscr{A}(t) \subset D(t)$ *for all* $t \in \mathbb{R}$.

Definition 8.5 The family of sets $\{K(t)\}_{t\in\mathbb{R}}$ is called backwards bounded if for some τ the set $K_\tau = \cup_{t\leq\tau} K(t)$ is bounded.

In [13, Proposition 4.3] and [9, Lemma 5] it was proved that if the pullback attractor is backwards bounded and $U_{\mathscr{G}}$ is strict, then it is invariant. Since conditions $(C1)$–$(C4)$ imply that the semiflow $U_{\mathscr{G}}$ is strict, we have the following result.

Lemma 8.4 *Let* $(C1) - (C4)$ *hold. If* $U_{\mathscr{G}}$ *possesses a backwards bounded pullback attractor* $\mathscr{A} = \{\mathscr{A}(t)\}_{t\in\mathbb{R}}$, *then* $\mathscr{A}$ *is invariant.*

When we use pullback asymptotic compactness in order to prove the existence of a pullback attractor, we need to add some dissipative assumptions.

Definition 8.6 The family $\{A(t)\}_{t\in\mathbb{R}}$ pullback absorbs bounded subsets of X if for each $t \in \mathbb{R}, B \in \mathscr{B}(X)$ there exists $T = T(t, B) \leq t$ such that $U_{\mathscr{G}}(t, \tau, B) \subset A(t)$, for all $\tau \leq T$.

Definition 8.7 $U_{\mathscr{G}}$ is called pullback bounded dissipative if there exists a family $\mathscr{B}_0 := \{B_0(t)\}_{t\in\mathbb{R}}$ with $B_0(t) \in \mathscr{B}(X)$ for any $t \in \mathbb{R}$ which pullback absorbs bounded subsets of X. $\mathscr{B}_0$ is said to be pullback absorbing. It is said to be monotonically pullback bounded dissipative if, in addition, $B_0(s) \subset B_0(t)$ for every $s \leq t$.

In [13, Theorem 3.6 and Proposition 4.2] the authors proved that if $U_{\mathcal{G}}$ is a multivalued process such that the graph of the map $x \mapsto U_{\mathcal{G}}(t,s,x)$ is closed, then pullback asymptotically compactness and monotonically pullback dissipativeness are necessary and sufficient conditions for the existence of the unique backwards bounded pullback attractor $A = \{A(s) : s \in \mathbb{R}\}$. We extend this result by showing that the pullback attractor is characterized in this case by formula (8.2) but without the closure.

Theorem 8.2 *Let* $(C1)$–$(C2)$, $(C4)$ *hold. Then* $U_{\mathcal{G}}$ *is pullback asymptotically compact and monotonically pullback bounded dissipative if and only if it possesses the unique backwards bounded global pullback attractor* $\mathscr{A} = \{\mathscr{A}(s) : s \in \mathbb{R}\}$ *defined by*

$$\mathscr{A}(t) = \bigcup_{B \in \mathscr{B}(X)} \omega(t,B). \tag{8.3}$$

If $(C3)$ *is also satisfied, then* $\mathscr{A}$ *is invariant.*

Proof In view of [13, Theorem 3.6 and Proposition 4.2] it only remains to prove the equality (8.3). From the proof of Theorem 3.6 in [13] we know that $\mathscr{A}(t) = \overline{\bigcup_{B\in\mathscr{B}(X)} \omega(t,B)} = \omega(t, B_0(t))$. Hence,

$$\mathscr{A}(t) := \omega(t, B_0(t)) \subset \bigcup_{B\in\mathscr{B}(X)} \omega(t,B) \subset \overline{\bigcup_{B\in\mathscr{B}(X)} \omega(t,B)} = \mathscr{A}(t),$$

so (8.3) follows.

The last statement is a consequence of Lemma 8.4.

Remark 8.2 It is interesting to know whether it is possible to obtain the existence of the pullback attractor assuming that $U_{\mathcal{G}}$ is just pullback bounded dissipative. We will give an answer to this question in the next section.

Finally, let us consider the characterization of the dynamics inside the pullback attractor using complete trajectories. In [9] it is shown that backwards bounded pullback attractors can be characterized by the union of all backwards bounded complete trajectories. We recall that a complete trajectory ψ is said to be bounded if the set $\cup_{t\in\mathbb{R}} \psi(t)$ is bounded.

Theorem 8.3 ([9]) *Let either* $(C1)$–$(C2)$, $(C4)$ *or* $(C1)$–$(C3)$ *hold. If* $U_{\mathcal{G}}$ *possesses the backwards bounded global pullback attractor* $\mathscr{A} = \{\mathscr{A}(s) : s \in \mathbb{R}\}$, *then*

$$\mathscr{A}(t) = \{\psi(t) : \psi \text{ is a backwards bounded complete trajectory}\}. \tag{8.4}$$

If we do not assume that the attractor is backwards bounded, then we can only obtain that every backwards bounded complete trajectory belongs to it.

Theorem 8.4 *Let* $(C1)$–$(C4)$ *be satisfied and let* $U_{\mathcal{G}}$ *possess the global pullback attractor* $\mathscr{A} = \{\mathscr{A}(s) : s \in \mathbb{R}\}$. *The following statements hold:*

(1) If $\psi : \mathbb{R} \to X$ is a bounded complete trajectory, then $\psi(s) \in \mathscr{A}(s)$ for all $s \in \mathbb{R}$.
(2) If, moreover, $\mathscr{A}$ is invariant, then for each $z \in \mathscr{A}(t)$ there exist a complete trajectory ψ_z such that $\psi_z(t) = z$ and $\psi_z(s) \in \mathscr{A}(s)$ for all $s \in \mathbb{R}$.

Proof Let $\psi : \mathbb{R} \to X$ be a bounded complete trajectory. Consider the set $B := \bigcup_{s\in\mathbb{R}} \psi(s) \in B(X)$. Then for any $s \in \mathbb{R}$ and $\varepsilon > 0$ there exists $T = T(s, B) < s$ such that $U_{\mathscr{G}}(s, \ell, B) \subset O_\varepsilon(\mathscr{A}(s))$ for all $\ell < T$. Hence,

$$\psi(s) \in U_{\mathscr{G}}(s, s-t, \psi(s-t)) \subset U_{\mathscr{G}}(s, s-t, B) \subset O_\varepsilon(\mathscr{A}(s))$$

for t large enough (i.e., $s - t < T$). Then, $\psi(s) \in \mathscr{A}(s)$.

The second statement is a consequence of Lemma 8.1.

Remark 8.3 Conditions $(C3)$–$(C4)$ are not necessary for the first statement.

8.3 Pullback Attraction of Families of Sets

In this section we will consider the theory of pullback attractors which attract certain families of sets instead of bounded sets. We will recall first the theory of existence of such attractors, which was developed in [6, 7] for multivalued processes. After that, we will study their characterization using complete trajectories.

Let $\mathscr{D}$ be a class of families of non-empty sets $D = \{D(t) : t \in \mathbb{R}\}$. We will say that the class $\mathscr{D}$ is inclusion-closed if $D \in \mathscr{D}$ and $\emptyset \neq D'(t) \subset D(t)$, for all $t \in \mathbb{R}$, imply that $D' = \{D'(t) : t \in \mathbb{R}\}$ belongs to $\mathscr{D}$.

Definition 8.8 The family $\mathscr{A} = \{\mathscr{A}(t) : t \in \mathbb{R}\}$ is said to be a global pullback $\mathscr{D}$-attractor for $U_{\mathscr{G}}$ if it satisfies:

1. $\mathscr{A}(t)$ is compact for any $t \in \mathbb{R}$;
2. $\mathscr{A}$ is pullback $\mathscr{D}$-attracting, i.e.

$$\lim_{\tau\to-\infty} dist_X(U_{\mathscr{G}}(t, \tau, D(\tau)), \mathscr{A}(t)) = 0 \ \forall t \in \mathbb{R},$$

   for all $D \in \mathscr{D}$;
3. $\mathscr{A}$ is negatively invariant.

$\mathscr{A}$ is said to be a strict global pullback D-attractor if it is also invariant.

In this framework, the theorem stating the existence of a global pullback $\mathscr{D}$-attractor is similar to the corresponding one in the autonomous case, as unlike the situation in the previous section we do not need to assume that the absorbing family is backwards bounded.

Definition 8.9 We say that a family of non-empty sets $\mathscr{B}_0 = \{B_0(t){:}t \in \mathbb{R}\}$ is pullback $\mathscr{D}$-absorbing if for every $D \in \mathscr{D}$ and every $t \in \mathbb{R}$, there exists $\tau(t, D) \leq t$ such that

$$U_{\mathscr{G}}(t,\tau,D(\tau)) \subset B_0(t) \textit{ for all } \tau \leq \tau(t,D).$$

Definition 8.10 The multivalued process $U_{\mathscr{G}}$ is asymptotically compact with respect to a family $\widehat{B} = \{B(t) : t \in \mathbb{R}\}$ if for all $t \in \mathbb{R}$ and every sequence $\tau_n \leq t$ tending to $-\infty$, any sequence $y_n \in U_{\mathscr{G}}(t, \tau_n, B(\tau_n))$ is relatively compact.

We say that $U_{\mathscr{G}}$ is upper semicontinuous if for all $t \geq \tau$ the mapping $U_{\mathscr{G}}(t,\tau,\cdot)$ is upper-semicontinuous, i.e., for any $x_0 \in X$ and for every neighborhood $\mathscr{O}$ in X of the set $U_{\mathscr{G}}(t,\tau,x_0)$, there exists $\delta > 0$ such that $U_{\mathscr{G}}(t,\tau,y) \subset \mathscr{O}$ whenever $\rho(x_0,y) < \delta$.

Condition $(C4)$ implies easily that $U_{\mathscr{G}}(t,\tau,\cdot)$ is upper-semicontinuous and has closed values. Therefore, the following result is a slight modification of Theorem 3.3 in [7].

Theorem 8.5 *Let* $(C1)$–$(C2)$, $(C4)$ *hold. Assume that there exists a pullback* $\mathscr{D}$-*absorbing family* $\mathscr{B}_0 = \{B_0(t) : t \in \mathbb{R}\}$ *and that* $U_{\mathscr{G}}$ *is asymptotically compact with respect to* $\mathscr{B}_0$. *Then, the set* $\mathscr{A}$ *given by*

$$\mathscr{A}(t) := \overline{\bigcup_{D\in\mathscr{D}} \omega(t,D)} \subset \omega(t,\mathscr{B}_0), \tag{8.5}$$

where $\omega(t,D) = \bigcap_{s\leq t} \overline{\bigcup_{\tau\leq s} U_{\mathscr{G}}(t,\tau,D(\tau))}$, *is a global pullback* $\mathscr{D}$-*attractor for* $U_{\mathscr{G}}$. $\mathscr{A}$ *is the minimal closed pullback* $\mathscr{D}$-*attracting family. If* $\mathscr{B}_0 \in \mathscr{D}$, *then*

$$\mathscr{A}(t) := \omega(t,\mathscr{B}_0). \tag{8.6}$$

Moreover, suppose that $\mathscr{D}$ *is inclusion closed,* $\mathscr{B}_0 \in \mathscr{D}$, *and that* $B(t)$ *is closed in* X *for any* $t \in \mathbb{R}$. *Then* $\mathscr{A} \in \mathscr{D}$ *and is the unique global pullback* $\mathscr{D}$-*attractor with this property. In addition, if* $(C3)$ *is also satisfied, then* $\mathscr{A}$ *is invariant.*

Proof In view of Lemma 3.2 in [7] the family $\omega(t,\mathscr{B}_0)$ is non-empty, compact, negatively invariant and pullback attracts $\mathscr{B}_0$. It is also proved in [7, p. 33] that $\omega(t,\mathscr{B}_0)$ pullback attracts every $D \in \mathscr{D}$. Hence, $U_{\mathscr{G}}$ is asymptotically compact with respect to every $D \in \mathscr{D}$. Thus, using again Lemma 3.2 in [7] the family $\{\omega(t,D)\}_{t\in\mathbb{R}}$ pullback attracts D and is negatively invariant and compact. Moreover, $\{\omega(t,D)\}_{t\in\mathbb{R}}$ is the minimal closed family that pullback attracts D. Indeed, let $\{A(t)\}_{t\in\mathbb{R}}$ be a closed family pullback attracting D. Since for any $y \in \omega(t,D)$ there exists a sequence $y_n \in U_{\mathscr{G}}(t,\tau_n,D(\tau_n))$, $\tau_n \to -\infty$, such that $y_n \to y$, we have

$$dist(y,A(t)) \leq \rho(y,y_n) + dist(y_n,A(t)) \to 0,$$

so $y \in A(t)$. Then it follows that $\omega(t,D) \subset \omega(t,\mathscr{B}_0)$.

Therefore, $\mathscr{A}$ is the minimal closed pullback $\mathscr{D}$-attracting family and $\mathscr{A}(t) \subset \omega(t,\mathscr{B}_0)$. It is clear that the sets $\mathscr{A}(t)$ are compact. It remains to prove that $\mathscr{A}$ is negatively invariant. Indeed, let $y \in \mathscr{A}(t)$ and $y_n \in \omega(t,D_n)$ be such that $y_n \to y$. For any $\tau < t$ there are $x_n \in \omega(\tau,D_n)$ satisfying $y_n \in U_{\mathscr{G}}(t,\tau,x_n)$. Passing to a subsequence we can assume that $x_n \to x \in \mathscr{A}(\tau)$ and then $y \in U_{\mathscr{G}}(t,\tau,x)$, as the graph of the map $x \mapsto U_{\mathscr{G}}(t,\tau,x)$ is closed. Thus, $\mathscr{A}(t) \subset U_{\mathscr{G}}(t,\tau,\mathscr{A}(\tau))$.

If $\mathscr{B}_0 \in \mathscr{D}$, then $\omega(t,\mathscr{B}_0) \subset \mathscr{A}(t)$, which implies (8.6).

The other statements follow from [7, Theorem 3.3] or [6, Theorem 3.4].

Further, let us study the structure of the pullback attractor. More precisely, we will prove analogous results as in Theorem 8.3, the main difference being that the attractor will be described now as the union of all complete trajectories which belong to the class $\mathscr{D}$.

Theorem 8.6 *Assume that (C1)–(C2), (C4) hold, $\mathscr{D}$ is inclusion closed and that $U_{\mathscr{G}}$ possesses the global pullback $\mathscr{D}$-attractor $\mathscr{A}$, which belong to $\mathscr{D}$. Then*

$$\mathscr{A}(t) = \{\psi(t) : \psi \text{ is a complete trajectory and } \psi \in \mathscr{D}\}.$$

Proof First, let $\psi \in \mathscr{D}$ be a complete trajectory. Then

$$\psi(t) \in U_{\mathscr{G}}(t,s,\psi(s)), \text{for any} s \leq t.$$

Since $dist(U_{\mathscr{G}}(t,s,\psi(s)),\mathscr{A}(t)) \to 0$ as $s \to -\infty$, we obtain that $\psi(t) \in \mathscr{A}(t)$ for any $t \in \mathbb{R}$.

Second, let $z \in \mathscr{A}(t)$, $t \in \mathbb{R}$ be arbitrary. Since $\mathscr{A}$ is negatively invariant, for an arbitrary sequence $s_n \to -\infty$ we have $z \in \mathscr{A}(t) \subset U(t,s_n,\mathscr{A}(s_n))$, so there is $\varphi_n \in \mathscr{G}(s_n)$ such that $z = \varphi_n(t)$ and $\varphi_n(s_n) \in \mathscr{A}(s_n)$. Condition $(C2)$ implies that $v_n^0 = \varphi_n|_{[t,\infty)} \in \mathscr{G}(t)$. By $(C4)$, passing to a subsequence, $v_n^0(r) \to v^0(r)$, for all $r \geq t$, where $v^0 \in \mathscr{G}(t)$, $v^0(t) = z$. Since $v^0(r) = \lim_{n\to\infty} \varphi_n(r)$ and $\varphi_n(r) \in U_{\mathscr{G}}(r,s_n,\mathscr{A}(s_n))$, we obtain that $v^0(r) \in \omega(r,\mathscr{A}) \subset \mathscr{A}(r)$ for any $r \geq t$.

Let now $v_n^1 = \varphi_n|_{[t-1,\infty)} \in \mathscr{G}(t-1)$. Since

$$v_n^1(t-1) = \varphi_n(t-1) \in U_{\mathscr{G}}(t-1,s_n,\mathscr{A}(s_n)),$$

passing to a subsequence $v_n^1(t-1) \to z_{-1}$. Therefore, repeating the same argument as before we obtain a map $v^1 \in \mathscr{G}(t-1)$ such that, up to a subsequence, $v_n^1(r) \to v^1(r)$ for all $r \geq t-1$. Also, $v^1(r) \in \mathscr{A}(r)$, for any $r \geq t-1$, and $v^1(r) = v^0(r)$ if $r \geq t$. In particular, $v^1(t) = z$.

Arguing as in the previous cases we define a sequence of functions $v^j \in \mathscr{G}(t-j)$, $j \in \mathbb{Z}^+$, such that $v^j(r) \in \mathscr{A}(r)$, for any $r \geq t-j$, $v^j(r) = v^{j-1}(r)$, for $r \geq t-j+1$, and $v^j(t) = z$.

Let $\psi(\cdot)$ be the function which takes the common value of the functions $v^j(\cdot)$ for all $r \in \mathbb{R}$. It follows that $\psi(\cdot)$ is a complete trajectory and $\psi(t) = z$. Moreover, since $\psi(r) \in \mathscr{A}(r)$, for any $r \in \mathbb{R}$, $\mathscr{A} \in \mathscr{D}$ and $\mathscr{D}$ is inclusion closed, we get that $\psi \in \mathscr{D}$.

Theorem 8.7 *Assume that (C1)–(C3) hold, $\mathscr{D}$ is inclusion closed and that $U_{\mathscr{G}}$ possesses the global pullback $\mathscr{D}$-attractor $\mathscr{A}$, which belong to $\mathscr{D}$. Then*

$$\mathscr{A}(t) = \{\psi(t) : \psi \text{ is a complete trajectory and } \psi \in \mathscr{D}\}.$$

Proof We know from the proof of Theorem 8.6 that $\psi(t) \in \mathscr{A}(t)$ for any complete trajectory ψ such that $\psi \in \mathscr{D}$.

Let $z \in \mathscr{A}(t)$. By $(C1)$ there exists $\varphi^0 \in \mathscr{G}(t)$ such that $\varphi^0(t) = z$. The pullback attractor $\mathscr{A}$ is invariant. Indeed, by $(C3)$ for any $s \leq r$ we have

$$U_{\mathscr{G}}(r, s, \mathscr{A}(s)) \subset U_{\mathscr{G}}(r, s, U_{\mathscr{G}}(s, \tau, \mathscr{A}(\tau))) \subset U_{\mathscr{G}}(r, \tau, \mathscr{A}(\tau)) \to \mathscr{A}(r), \quad (8.7)$$

as $\tau \to -\infty$.

Hence, $\mathscr{A}(r) = U_{\mathscr{G}}(r, t, \mathscr{A}(t))$, which implies that $\varphi^0(r) \in \mathscr{A}(r)$ for any $r \geq t$. Further, $z \in \mathscr{A}(t) \subset U_{\mathscr{G}}(t, t-1, \mathscr{A}(t-1))$ implies the existence of $v^1 \in \mathscr{G}(t-1)$ satisfying $v^1(r) \in \mathscr{A}(r)$, for all $r \geq t-1$, and $v^1(t) = z$. In view of $(C3)$ concatenating v^1 and φ^0 we obtain a function $\varphi^1 \in \mathscr{G}(t-1)$ such that $\varphi^1(r) \in \mathscr{A}(r)$, for all $r \geq t-1$, $\varphi^1(t) = z$ and $\varphi^1(r) = \varphi^0(r)$ for $r \geq t$. Then, we define inductively a sequence of functions $\varphi^j \in \mathscr{G}(t-j), j \in \mathbb{Z}^+$, such that $\varphi^j(r) \in \mathscr{A}(r)$, for all $r \geq t-j$, $\varphi^j(t) = z$ and $\varphi^j(r) = \varphi^{j-1}(r)$ if $r \geq t-j+1$. Let ψ be the function defined by the common value of φ^j at any point $t \in \mathbb{R}$, which is a complete trajectory satisfying $\psi(t) = z$ and $\psi(r) \in \mathscr{A}(r)$ for any $r \in \mathbb{R}$. Since $\mathscr{D}$ is inclusion closed and $\mathscr{A} \in \mathscr{D}$, we obtain that $\psi \in \mathscr{D}$.

Remark 8.4 As far as we know, this characterization of the pullback attractor is new even in the case where the map $U_{\mathscr{G}}$ is single-valued.

An interesting question appears when the multivalued semiflow posseses both a pullback $\mathscr{D}$-attractor and a pullback attractor in the sense of Definition 8.3. We will denote these attractors by $\mathscr{A}_{\mathscr{D}}$ and $\mathscr{A}$, respectively. Namely, what is the relationship between them?

Lemma 8.5 *Let* $(C1)$–$(C2)$ *and let every family of the type* $\mathscr{B} = \{B(t) \equiv B \in \mathscr{B}(X) : t \in \mathbb{R}\}$ *belong to* $\mathscr{D}$ *(that is, any family of fixed bounded sets belong to* $\mathscr{D}$*). Assume that* $U_{\mathscr{G}}$ *possesses both a pullback* $\mathscr{D}$*-attractor* $\mathscr{A}_{\mathscr{D}}$ *and a pullback attractor* $\mathscr{A}$*. Then*

$$\mathscr{A}(t) \subset \mathscr{A}_{\mathscr{D}}(t) \text{ for all } t \in \mathbb{R}. \quad (8.8)$$

If, moreover, $\mathscr{A}_{\mathscr{D}}$ *is backwards bounded, then*

$$\mathscr{A}(t) = \mathscr{A}_{\mathscr{D}}(t) \text{ for all } t \in \mathbb{R}. \quad (8.9)$$

Proof Since the families of fixed bounded sets belong to $\mathscr{D}$, $\mathscr{A}_{\mathscr{D}}$ is a closed pullback attracting family in the sense of Definition 8.2. The minimality of $\mathscr{A}$ implies (8.8).

Let $t \in \mathbb{R}$ be arbitrary. If $\mathscr{A}_{\mathscr{D}}$ is backwards bounded, then

$$\mathscr{A}_{\mathscr{D}}(t) \subset U_{\mathscr{G}}(t, s, \mathscr{A}_{\mathscr{D}}(s)) \subset U_{\mathscr{G}}(t, s, A_{\tau}), \text{ for any } s \leq \tau,$$

where $\tau \le t$ is such that $A_\tau = \cup_{r\le\tau}\mathscr{A}_{\mathscr{D}}(r)$ is bounded. From

$$dist\left(U_{\mathscr{G}}\left(t,s,A_\tau\right), A\left(t\right)\right) \to 0,\ as\ s \to -\infty,$$

we have that $\mathscr{A}_{\mathscr{D}}(t) \subset \mathscr{A}(t)$. Thus, (8.9) is proved.

Remark 8.5 In the single-valued case, an answer to this problem was given in [20]. In the multivalued framework properties (8.8), (8.9) have been proved in [21] using similar conditions to those in Theorem 8.5.

As commented before, it is an interesting question whether one can prove the existence of a pullback attractor (in the sense of Definition 8.2) assuming just that $U_{\mathscr{G}}$ is pullback bounded dissipative. In [20] this problem was solved in the single-valued framework by modifying the pullback compactness condition. Using Theorem 8.5 we will prove a similar result in the multivalued case. We note that this result was already stated in [21].

Theorem 8.8 *Let (C1)–(C2), (C4) hold. If $U_{\mathscr{G}}$ is pullback bounded dissipative and asymptotically compact with respect to the absorbing family $\mathscr{B}_0$, then it possesses the global pullback attractor $\mathscr{A} = \{\mathscr{A}(s) : s \in \mathbb{R}\}$ defined by (8.5).*

If (C3) is also satisfied and $\mathscr{A}_{\mathscr{D}}$ is backwards bounded, then $\mathscr{A}$ is invariant and (8.4) holds.

Proof Consider the class of families $\mathscr{D}$ consisting of bounded sets, that is, $D \in \mathscr{D}$ if and only if $D = \{D(t) \equiv B \in \mathscr{B}(X) : t \in \mathbb{R}\}$. The existence of the pullback attractor follows from Theorem 8.5.

The second part is a consequence of Lemma 8.4 and Theorem 8.3.

8.4 Application to a Reaction-Diffusion Equation

Let us consider the following reaction-diffusion problem

$$\begin{cases} \frac{\partial u}{\partial t} - \Delta u = f(x,u) + h(t)\ \ in\ \ \Omega \times (\tau,+\infty), \\ u = 0\ \ on\ \ \partial\Omega \times (\tau,+\infty), \\ u(x,\tau) = u_\tau(x),\ \ x \in \Omega, \end{cases} \tag{8.10}$$

where $\tau \in \mathbb{R}$, $u_\tau \in L^2(\Omega)$, $h \in L^2_{loc}(\mathbb{R}; H^{-1}(\Omega))$, $f : \Omega \times \mathbb{R} \to \mathbb{R}$ is a measurable function such that $f(x,\cdot) \in C(\mathbb{R})$ for almost every $x \in \Omega$, and f satisfies that there exist constants $\alpha_1 > 0$, $\alpha_2 > 0$, $p \ge 2$ and positive functions $C_1(x), C_2(x) \in L^1(\Omega)$ such that

$$|f(x,s)|^{\frac{p}{p-1}} \le \alpha_1 |s|^p + C_1(x)\ \ \forall s \in \mathbb{R}, x \in \Omega, \tag{8.11}$$

$$f(x,s)s \le -\alpha_2 |s|^p + C_2(x)\ \forall s \in \mathbb{R}, x \in \Omega. \tag{8.12}$$

Here, $\Omega \subset \mathbb{R}^N$ is a nonempty open set, not necessarily bounded, satisfying the Poincaré inequality, i.e., there exists a constant $\lambda_1 > 0$ such that

$$\int_\Omega |u(x)|^2 \, dx \leq \lambda_1^{-1} \int_\Omega |\nabla u(x)|^2 \, dx \;\; \forall u \in H_0^1(\Omega). \tag{8.13}$$

We assume also that $h = \sum_{i=1}^N \frac{\partial h_i}{\partial x_i}$, where $h_i \in L^2_{loc}(\mathbb{R}; L^2(\Omega))$ are such that

$$\sum_{i=1}^N \int_{-\infty}^t e^{\lambda_1 s} |h_i(s)|^2 \, ds < +\infty \;\; \forall t \in \mathbb{R}. \tag{8.14}$$

By $|\cdot|$, $\|\cdot\|$, $\|\cdot\|_*$ we denote the norms in $L^2(\Omega)$, $H_0^1(\Omega)$ and $H^{-1}(\Omega)$, respectively. We will use $(\cdot,\cdot)$ to denote the scalar product in either $L^2(\Omega)$ or $[L^2(\Omega)]^N$, and $\langle \cdot,\cdot \rangle$ to denote the duality pairing between $H^{-1}(\Omega) + L^q(\Omega)$ and $H_0^1(\Omega) \cap L^p(\Omega)$, where $\frac{1}{p} + \frac{1}{q} = 1$.

A weak solution of (8.10) is a function $u : (\tau, +\infty) \to L^p(\Omega) \cap H_0^1(\Omega)$ such that $u \in L^p(\tau, T; L^p(\Omega)) \cap L^2(\tau, T; H_0^1(\Omega))$, for all $T > \tau$, and

$$(u(t), w) + \int_\tau^t (\nabla u(s), \nabla w) \, ds = (u_\tau, w) + \int_\tau^t \langle f(x, u(s)) + h(s), w \rangle \, ds, \tag{8.15}$$

for all $t \geq \tau$, $w \in L^p(\Omega) \cap H_0^1(\Omega)$.

It follows from [12, p. 285] that any weak solution u to problem (8.10) satisfies $u \in C([\tau, +\infty); L^2(\Omega))$. Moreover the function $t \mapsto |u(t)|^2$ is absolutely continuous on every interval $[\tau, T]$ and $\frac{d}{dt}|u(t)|^2 = 2\left\langle \frac{du}{dt}, u \right\rangle$ for a.a. $t \in (\tau, T)$. Hence, it satisfies the energy equality

$$|u(t)|^2 + 2\int_\tau^t \|u(s)\|^2 \, ds = |u_\tau|^2 + 2\int_\tau^t \langle f(x, u(s)) + h(s), u(s) \rangle \, ds \;\; \forall t \geq \tau.$$

It is well-known [1, Theorem 2] that for all $\tau \in \mathbb{R}$, $u_\tau \in L^2(\Omega)$ there exists at least one weak solution u to (8.10). For each $\tau \in \mathbb{R}$ we define then the space $W_\tau = C([\tau, \infty); L^2(\Omega))$ and

$$\mathcal{G}(\tau) = \{u \in W_\tau : u \text{ is a weak solution to } (8.10)\}.$$

Hence, condition $(C1)$ is satisfied. It also follows from the proof of Lemma 11 in [1] that $(C2)$–$(C3)$ hold. Therefore, the map $U_\mathcal{G}$ defined by (8.1) is a strict multivalued semiflow. In addition, Proposition 16 in [1] implies that $(C4)$ is satisfied, so $\mathcal{G}$ is a exact generalized process.

We shall apply the results of Sects. 8.2 and 8.3 in order to obtain that the pullback attractor of $U_\mathcal{G}$ is described as the union of a certain type of complete trajectories.

We begin with the second case, that is, when we study a global pullback $\mathscr{D}$-attractor for a suitable class of families of sets $\mathscr{D}$.

Let $\mathscr{R}_{\lambda_1}$ be the set of all functions $r : \mathbb{R} \to (0, +\infty)$ such that

$$\lim_{t \to -\infty} e^{\lambda_1 t} r^2(t) = 0,$$

where λ_1 is the constant in the Poincaré inequality (8.13). Denote by $\mathscr{D}$ the class of all families $D = \{D(t) : t \in \mathbb{R}\}, D(t) \in \mathscr{P}(L^2(\Omega))$, such that $D(t) \subset \overline{B}(0, r_D(t))$ for some $r_D \in \mathscr{R}_{\lambda_1}$, where $\overline{B}(0, r_D(t))$ is a closed ball in $L^2(\Omega)$ centered at zero with radius $r_D(t)$. The class $\mathscr{D}$ is obviously inclusion-closed.

We recall some results from [1].

Lemma 8.6 ([1, Lemma 12]) *The balls $B_0(t) = \overline{B}_{L^2(\Omega)}(0, R_{\lambda_1}(t))$, where $R_{\lambda_1}(t)$ is the nonnegative number given by*

$$R_{\lambda_1}^2(t) = 2e^{-\lambda_1 t} \sum_{i=1}^{N} \int_{-\infty}^{t} e^{\lambda_1 s} |h_i(s)|^2 \, ds + 2\lambda_1^{-1} \|C_2\|_{L^1(\Omega)} + 1, \tag{8.16}$$

form a family $\mathscr{B}_0 \in \mathscr{D}$ which is pullback $\mathscr{D}$-absorbing for $U_{\mathscr{G}}$.

Lemma 8.7 ([1, Lemma 18]) *$U_{\mathscr{G}}$ is asymptotically compact with respect to any family $D \in \mathscr{D}$.*

Remark 8.6 In [1] this lemma is stated only with respect to the absorbing family $\mathscr{B}_0$, but in fact the proof works for an arbitrary $D \in \mathscr{D}$.

Theorem 8.9 ([1, Theorem 19]) *The multivalued process $U_{\mathscr{G}}$ possesses a unique global pullback $\mathscr{D}$-attractor $\mathscr{A}_{\mathscr{D}} = \{\mathscr{A}_{\mathscr{D}}(t) : t \in \mathbb{R}\}$ belonging to $\mathscr{D}$, which is given by*

$$\mathscr{A}_{\mathscr{D}}(t) := \bigcap_{s \leq t} \overline{\bigcup_{\tau \leq s} U_{\mathscr{G}}(t, \tau, B_0(\tau))}, \tag{8.17}$$

where $\mathscr{B}_0 = \{B(t) : t \in \mathbb{R}\}$ is defined in Lemma 8.6. Moreover, $\mathscr{A}$ is invariant.

Now, using Theorem 8.9 and either Theorem 8.6 or 8.7 we obtain the characterization of the pullback $\mathscr{D}$-attractor.

Theorem 8.10 *The global pullback $\mathscr{D}$-attractor $\mathscr{A}_{\mathscr{D}}$ is given by*

$$\mathscr{A}_{\mathscr{D}}(t) = \{\psi(t) : \psi \text{ is a complete trajectory and } \psi \in \mathscr{D}\}.$$

In other words, $\mathscr{A}_{\mathscr{D}}$ is the union of all complete trajectories ψ such that

$$\lim_{t \to -\infty} e^{\lambda_1 t} \|\psi(t)\|^2 = 0.$$

Further, let us consider the first case, that is, when we study the attraction of bounded sets.

We observe that for any bounded set B, the associated constant family $\widehat{B} = \{B(t) : t \in \mathbb{R}\}$, where $B(t) \equiv B$, belongs to $\mathscr{D}$. Therefore, Lemmas 8.6 and 8.7 imply in particular that $U_{\mathscr{G}}$ is pullback asymptotically compact and bounded dissipative in the sense of Definitions 8.4 and 8.7. In order to obtain that $U_{\mathscr{G}}$ is monotonically bounded dissipative we need an extra assumption.

Lemma 8.8 *Assume that*

$$\sup_{t\leq t_0} e^{-\lambda_1 t}\sum_{i=1}^{N}\int_{-\infty}^{t} e^{\lambda_1 s}\left|h_i(s)\right|^2 ds < \infty \text{ for any } t_0 \in \mathbb{R}. \tag{8.18}$$

Then $U_{\mathscr{G}}$ is monotonically bounded dissipative.

Proof In view of (8.18) the absorbing family $\mathscr{B}_0$ satisfies that the sets $\cup_{t\leq t_0} B_0(t)$ are bounded for any $t_0 \in \mathbb{R}$. Hence, the results follows from Proposition 3.4 in [13].

Applying Lemma 8.8 and Theorems 8.2 and 8.3 we obtain the following result.

Theorem 8.11 *Assume that (8.18) holds. Then the multivalued semiflow $U_{\mathscr{G}}$ possesses the unique backwards bounded global pullback attractor $\mathscr{A} = \{\mathscr{A}(s) : s \in \mathbb{R}\}$ defined by*

$$\mathscr{A}(t) = \bigcup_{B\in\mathscr{B}(X)} \omega(t, B).$$

Moreover,

$$\mathscr{A}(t) = \{\psi(t) : \psi \text{ is a backwards bounded complete trajectory}\}. \tag{8.19}$$

We can also prove that the pullback attractors $\mathscr{A}_{\mathscr{D}}$ and $\mathscr{A}$ coincide.

Lemma 8.9 *If (8.18) holds, then $\mathscr{A}_{\mathscr{D}}$ is backwards bounded and $\mathscr{A} = \mathscr{A}_{\mathscr{D}}$. In addition,*

$$\mathscr{A}_{\mathscr{D}}(t) = \{\psi(t) : \psi \text{ is a backwards bounded complete trajectory}\}. \tag{8.20}$$

Proof We have seen in the proof of Lemma 8.8 that the sets $\cup_{t\leq t_0} B_0(t)$ are bounded for any $t_0 \in \mathbb{R}$. In view of (8.17), $\mathscr{B}_0 \in \mathscr{D}$ and the closedness of $B_0(t)$ we get $\mathscr{A}_{\mathscr{D}}(t) \subset B_0(t)$, so $\mathscr{A}_{\mathscr{D}}$ is backwards bounded. Since the families of fixed bounded sets belong to $\mathscr{D}$, the equality $\mathscr{A} = \mathscr{A}_{\mathscr{D}}$ follows from Lemma 8.5. Finally, (8.20) is a consequence of (8.19).

Acknowledgments The first author has been supported by a CNPq scholarship, process 202645/2014-2 (Brazil).

The second author has been partially supported by the Spanish Ministry of Economy and Competitiveness and FEDER, projects MTM2015-63723-P and MTM2012-31698, and by Junta de Andalucía (Spain), project P12-FQM-1492.

The authors would like to thank the anonymous referee for her/his helpful remarks.

References

1. Anguiano, M., Caraballo, T., Real, J., Valero, J.: Pullback attractors for reaction-diffusion equations in some unbounded domains with an H^{-1}-valued non-autonomous forcing term and without uniqueness of solutions. Discret. Contin. Dyn. Syst. Ser. B **14**, 307–326 (2010)
2. Ball, J.M.: On the asymptotic behavior of generalized process with applications to nonlinear evolution equations. J. Differ. Equ. **27**, 224–265 (1978)
3. Caraballo, T., Kloeden, P.E., Marín-Rubio, P.: Weak pullback attractors of setvalued processes. J. Math. Anal. Appl. **288**, 692–707 (2003)
4. Caraballo, T., Langa, J.A., Melnik, V.S., Valero, J.: Pullback attractors of nonautonomous and stochastic multivalued dynamical systems. Set-Valued Anal. **11**, 153–201 (2003)
5. Caraballo, T., Lukaszewicz, G., Real, J.: Pullback attractors for asymptotically compact nonautonomous dynamical systems. Nonlinear Anal. **64**, 484–498 (2006)
6. Caraballo, T., Garrido-Atienza, M.J., Schmalfuss, B., Valero, J.: Non-autonomous and random attractors for delay random semilinear equations without uniqueness. Discret. Contin. Dyn. Syst. **21**, 415–443 (2008)
7. Caraballo, T., Kloeden, P.E.: Non-autonomous attractors for integro-differential evolution equations. Discret. Contin. Dyn. Syst. Ser. S **2**, 17–36 (2009)
8. Caraballo, T., Carvalho, A.N., Langa, J.A., Rivero, F.: Existence of pullback attractors for pullback asymptotically compact processes. Nonlinear Anal. **72**, 1967–1976 (2010)
9. Caraballo, T., Langa, J.A., Valero, J.: Structure of the pullback attractor for a non-autonomous scalar differential inclusion. Discret. Contin. Dyn. Syst. Ser. S, to appear
10. Carvalho, A.N., Langa, J.A., Robinson, J.C.: On the continuity of pullback attractors for evolution processes. Nonlinear Anal. **71**, 1812–1824 (2009)
11. Carvalho, A.N., Langa, J.A., Robinson, J.C.: Attractors for Infinite-Dimensional Nonautonomous Dynamical Systems. Springer, New-York (2013)
12. Chepyzhov, V.V., Vishik, M.I.: Attractors for Equations of Mathematical Physics. American Mathematical Society, Providence (2002)
13. Coti Zelati, M., Kalita, P.: Minimality properties of set-valued processes and their pullback attractors. SIAM J. Math. Anal. **47**, 1530–1561 (2015)
14. Crauel, H., Flandoli, F.: Attractors for random dynamical systems. Probab. Theory Relat. Fields **100**, 365–393 (1994)
15. Crauel, H., Debussche, A., Flandoli, F.: Random attractors. J. Dyn. Differ. Equ. **9**, 307–341 (1997)
16. da Costa, H., Valero, J.: Morse decompositions and Lyapunov functions for dynamically gradient multivalued semiflows. Nonlinear Dyn. **84**, 19–34 (2016)
17. Kapustyan, O.V., Kasyanov, P.O., Valero, J.: Pullback attractors for a class of extremal solutions of the 3D Navier-Stokes system. J. Math. Anal. Appl. **373**, 535–547 (2011)
18. Kloeden, P.E., Schmalfuss, B.: Asymptotic behaviour of non-autonomous difference inclusions. Syst. Control Lett. **33**, 275–280 (1998)
19. Li, D.: On dynamical stability in general dynamical systems. J. Math. Anal. Appl. **263**, 455–478 (2001)
20. Marín-Rubio, P., Real, J.: On the relation between two different concepts of pullback attractors for non-autonomous dynamical systems. Nonlinear Anal. **71**, 3956–3963 (2009)

21. Marín-Rubio, P., Real, J.: Pullback attractors for 2D-Navier-Stokes equations with delays in continuous and sub-linear operators. Discret. Contin. Dyn. Syst. **26**, 989–1006 (2010)
22. Roxin, E.: Stability in general control systems. J. Differ. Equ. **1**, 115–150 (1965)
23. Schmalfuss, B.: Attractors for the non-autonomous dynamical systems. In: Fiedler, B., Gröger, K., Sprekels, J. (eds.) Proceedings of Equadiff, Berlin, vol. 99, pp. 684–689. World Scientific, Singapore (2000)
24. Simsen, J., Capelato, E.: Some properties for exact generalized processes. In: Sadovnichiy, V.A., Zgurovsky, M.Z. (eds.) Continuous and Distributed Systems II, pp. 209–219. Springer, Berlin (2015)

Chapter 9
Global Attractors for Discontinuous Dynamical Systems with Multi-valued Impulsive Perturbations

Oleksiy V. Kapustyan and Iryna V. Romaniuk

Abstract In this work, we consider impulsive infinite-dimensional dynamical systems generated by parabolic equations with continuous bounded right-hand side $\varepsilon F(y)$ and with impulsive multi-valued perturbations. Moments of impulses are not fixed and defined by moments of intersection of solutions with some subset of the phase space. We find an explicit formula in the case $\varepsilon = 0$ and prove that for sufficiently small value of the parameter $\varepsilon > 0$ the corresponding nonlinear system also has a global attractor.

9.1 Introduction

An autonomous evolution system is called discontinuous (or impulsive) dynamical system (DS) if its trajectories have jumps at moments of intersection with certain surface of the phase space [19]. Unlike systems with impulses at fixed moments of time, the behavior of impulsive DS is far from complete understanding. Some aspects of the qualitative behavior of impulsive finite-dimensional DS such as stability of solutions or properties of ω-limit sets have been studied by many authors [4, 13, 14, 16, 18, 19]. For infinite-dimensional dissipative systems, one of the most powerful tools of investigation of their limit behavior is the theory of global attractors [20]. Lack of continuous dependence on initial data in impulsive DS required a new concept of global attractor for such systems. The first approach was proposed in [2, 3]. The

O.V. Kapustyan (✉)
Taras Shevchenko National University of Kyiv,
Volodymyrska Str., 64, Kyiv 01601, Ukraine
e-mail: kapustyanav@gmail.com

O.V. Kapustyan
Institute for Applied System Analysis, National Technical University of Ukraine
"Kyiv Polytechnic Institute", Kyiv, Ukraine

I.V. Romaniuk
Taras Shevchenko National University of Kyiv,
Akademika Glushkova Avenue, 4-e, Kyiv 03127, Ukraine
e-mail: romanjuk.iv@gmail.com

V.A. Sadovnichiy and M.Z. Zgurovsky (eds.), *Advances in Dynamical Systems and Control*, Studies in Systems, Decision and Control 69,
DOI 10.1007/978-3-319-40673-2_9

key point of these papers was to keep invariance property in definition of attractor. For this purpose, the authors considered either systems with finite number of impulses or systems satisfying very special conditions ("tube condition") which required detailed information about the character of intersection of a given set by the trajectories of the given nonlinear system. These conditions were formulated in an abstract form and could not be effectively tested without explicit formulas of solutions. The second approach was proposed in [10, 11]. It was based on the notion of global attractor for nonautonomous systems [5, 8, 9], in particular for the systems with impulsive effects at fixed moments of time [6, 17]. For such systems, it is natural to require minimality property in the definition of global attractor instead of invariance. This approach allowed us to give necessary and sufficient conditions of existence of global attractors for impulsive DS with infinite number of impulsive points under natural assumptions on parameters. In particular, some model examples of existence and nonexistence of global attractor were considered. In this chapter, we extend results of [10, 11] on wider classes of impulsive DS. More precisely, we consider impulsive infinite-dimensional dynamical systems generated by parabolic equations with continuous bounded right-hand side $\varepsilon F(y)$ and with impulsive multi-valued perturbations. For description of such systems, the theory of global attractors of multi-valued DS is used [7, 15, 21]. We find an explicit formula of attractor in the case $\varepsilon = 0$ and prove that for sufficiently small value of the parameter $\varepsilon > 0$ the corresponding multi-valued DS also has a global attractor.

9.2 Construction of Impulsive DS with Multi-valued Impulsive Perturbation

Let (X, ρ) be a metric space, $P(X)$ ($\beta(X)$) be a set of all nonempty (nonempty bounded) subsets of X.

Definition 9.1 ([7]) A multi-valued map $G : R_+ \times X \to P(X)$ is called multi-valued DS (MDS), if

(1) $\forall x \in X \; G(0, x) = x$;
(2) $\forall x \in X \; \forall t, s \geq 0 \; G(t + s, x) \subseteq G(t, G(s, x))$.

The following definition is borrowed from the attractors theory of nonautonomous processes [5, 9, 17]

Definition 9.2 A subset $\Theta \subset X$ is called a global attractor of MDS G, if

(1) Θ is a compact set ;
(2) Θ is uniformly attracting set, i.e.,

$$\forall B \in \beta(X) \;\; dist(G(t, B), \Theta) \to 0, \;\; t \to \infty;$$

(3) Θ is minimal among closed uniformly attracting sets.

Remark 9.1 In the definition of MDS, we assume no conditions of continuity for the map $x \to G(t, x)$. If the MDS G also has global attractor in the classical sense [7], i.e., if there exists a compact uniformly attracting set $\Theta_1 \subset X$ and $\forall t \geq 0$ $\Theta_1 \subset G(t, \Theta_1)$, then $\Theta = \Theta_1$.

Following result guarantees existence criteria of global attractors for dissipative MDS.

Lemma 1 ([17, 21]) *Assume that MDS G satisfies dissipativity condition:*

$$\exists B_0 \in \beta(X) \quad \forall B \in \beta(X) \quad \exists T = T(B) > 0 \quad \forall t \geq T \quad G(t, B) \subset B_0. \tag{9.1}$$

Then, the following conditions are equivalent:

(1) MDS G has a global attractor Θ;
(2) MDS G is asymptotically compact, i.e.,

$$\forall t_n \nearrow \infty \ \forall B \in \beta(X) \ \forall \xi_n \in G(t_n, B) \text{ the sequence } \{\xi_n\} \text{ is precompact in } X. \tag{9.2}$$

Moreover,

$$\Theta = \omega(B_0) := \bigcap_{\tau>0} \overline{\bigcup_{t \geq \tau} G(t, B_0)}. \tag{9.3}$$

We consider MDS G generated by the following impulsive problem

$$\frac{du}{dt} = L(u), \ u \notin M, \tag{9.4}$$

$$\Delta u|_{u \in M} \in Iu - u, \tag{9.5}$$

where (9.4) is some evolution autonomous problem, which generates continuous semigroup $V : R_+ \times X \mapsto X$ in the phase space X, $M \subset X$ is impulsive set, $I : M \mapsto P(X)$ is impulsive map, $\Delta u|_{t=\tau} = u(\tau + 0) - u(\tau - 0)$.

The phase point $u(t)$ moves along trajectories of (9.4) and when it reaches the set M, it jumps to a new position $u^+ \in Iu$. We shall consider right continuous impulsive trajectories. For constructing of such trajectories, we assume the following conditions:

$$M \cap I(M) = \emptyset; \tag{9.6}$$

$$\forall x \in M \ \exists \tau = \tau(x) \quad \forall t \in (0, \tau) \ V(t, x) \notin M. \tag{9.7}$$

We define

$$\forall x \in X \ M^+(x) = \Big(\bigcup_{t>0} V(t, x)\Big) \cap M.$$

If $M^+(x) \neq \varnothing$, then [1], there exists a moment of time $s := \phi(x) > 0$ such that

$$\begin{cases} V(t,x) \notin M \ \ \forall t \in (0,s); \\ V(s,x) \in M. \end{cases}$$

Then, the impulsive trajectory which started from the point can be constructed according to the classical scheme [13] with slight "multi-valued" modifications as follows.

If $M^+(x) = \varnothing$, then $\varphi(t) = V(t,x)\ \forall t \geq 0$.

If $M^+(x) \neq \varnothing$, then for $s_0 := \phi(x) > 0$, $x_1 := V(s_0,x) \in M$, $x_1^+ \in Ix_1$ we define φ on $[0, s_0]$ by the following rule:

$$\varphi(t) = \begin{cases} V(t,x), & t \in [0,s_0); \\ x_1^+, & t = s_0. \end{cases}$$

If $M^+(x_1^+) = \varnothing$, then $\varphi(t) = V(t - s_0, x_1^+)\ \forall t \geq s_0$.

If $M^+(x_1^+) \neq \varnothing$, then for $s_1 := \phi(x_1^+) > 0$, $x_2 := V(s_1, x_1^+) \in M$, $x_2^+ \in Ix_2$ we define φ on $[s_0, s_0 + s_1]$ by the following rule:

$$\varphi(t) = \begin{cases} V(t - s_0, x_1^+), & t \in [s_0, s_0 + s_1); \\ x_2^+, & t = s_0 + s_1. \end{cases}$$

Repeating this procedure, we obtain impulsive trajectory with finite or infinite number of impulsive points $\{x_n^+\}_{n\geq 1} \subset X$ and corresponding moments of time $\{s_n\}_{n\geq 0} \subset (0, +\infty)$.

If we put

$$t_0 := 0,\ t_{n+1} := \sum_{k=0}^{n} s_k,\ n \geq 0,$$

then for the case of infinite number of impulses we obtain $\forall n \geq 0\ \forall t \in [t_n, t_{n+1}]$

$$\varphi(t) = \begin{cases} V(t - t_n, x_n^+), & t \in [t_n, t_{n+1}); \\ x_{n+1}^+, & t = t_{n+1}. \end{cases}$$

By K_x, we denote the set of all impulsive trajectories which start from the point x.

We assume the following conditions:

$$\forall x \in X \text{ every trajectory } \varphi \in K_x \text{ is defined on } [0, +\infty), \tag{9.8}$$

i.e., for every impulsive trajectory, the number of impulsive points is either no more than finite or $\sum_{k=0}^{\infty} s_k = \infty$.

We put

$$\forall x \in X \; \forall t \geq 0 \; G(t, x) = \{\varphi(t) | \varphi \in K_x\}. \tag{9.9}$$

It is easy to show that $G : R_+ \times X \to P(X)$ satisfies conditions of Definition 1. So, (9.9) defines a MDS which will be called *impulsive MDS.*

In all further arguments, we shall say that the problem (9.4), (9.5) *generates an impulsive MDS* (according to the rule (9.9) if the conditions (9.6)–(9.8) are satisfied.

9.3 The Main Results

Consider a triple $V \subset H \subset V^*$ of Hilbert spaces with compact and dense embedding. Denote by $\|\cdot\|$ and $(\cdot, \cdot)$, respectively, the norm and scalar product in H. Let $\|\cdot\|_V$ be a norm in V and

$$\exists \alpha > 0 \; \forall u \in V \; \|u\|^2 \leq \alpha \|u\|_V^2.$$

Consider a linear continuous self-adjoint operator $A : V \to V^*$ such that

$$\exists \beta > 0 \;\; \forall u \in V \;\; \langle Au, u \rangle \geq \beta \|u\|_V^2.$$

We consider the problem

$$\frac{dy}{dt} = -Ay, \; t > 0. \tag{9.10}$$

The corresponding semigroup $V : R_+ \times H \mapsto H$ is defined by the formula

$$\forall \; y_0 = \sum_{i=1}^{\infty} c_i \psi_i \in H \quad V(t, y_0) = y(t) = \sum_{i=1}^{\infty} c_i e^{-\lambda_i t} \psi_i,$$

where $\{\psi_i\}$, $\{\lambda_i\}$ are solutions of the spectral problem

$$\forall i \geq 1 \;\; A\psi_i = \lambda_i \psi_i, \;\; 0 < \lambda_1 \leq \lambda_2 \leq \ldots, \;\; \lambda_i \to \infty, \; i \to \infty.$$

The semigroup V has trivial global attractor $\{0\}$. As it was shown in [10] that an arbitrary small impulsive perturbation can destroy global attractor. More precisely, let us consider impulsive parameters

$$M = \{y \in H | \;\; \|y\| = \varepsilon\}, \; Iy = (1 + \mu)y, \; \varepsilon > 0, \; \mu > 0. \tag{9.11}$$

Lemma 2 ([10, 11]) *For every $\varepsilon > 0$, $\mu > 0$ the problem (9.10), (9.11) generates an impulsive DS, which is dissipative but does not possess global attractor in the phase space H.*

In [10], existence of global attractor was proved for the impulsive parameters

$$M = \left\{ y \in H \mid (y, \psi_1) = a \right\}, \quad I : M \mapsto H,$$

$$I\Big(\sum_{i=1}^{\infty} c_i \psi_i\Big) = (1+\mu) c_1 \psi_1 + \sum_{i=2}^{\infty} c_i \psi_i$$

in both linear and weakly nonlinear cases. In the present chapter, our aim was to extend this result to the following case: for fixed $p \geq 1$, $\{\alpha_i\}_{i=1}^{p} \subset (0, +\infty)$, $a > 0$, $\mu > 0$,

$$M = \left\{ y = \sum_{i=1}^{\infty} c_i \psi_i \in H \mid \forall i = \overline{1, p} \ \ c_i \geq 0, \ \sum_{i=1}^{p} \alpha_i c_i = a \right\}, \tag{9.12}$$

$$I : M \to P(H) \ and \ for \ \ y = \sum_{i=1}^{\infty} c_i \psi_i \in M$$

$$Iy = \left\{ \sum_{i=1}^{p} c_i' \psi_i + \sum_{i=p+1}^{\infty} c_i \psi_i \mid \forall i = \overline{1, p} \ \ c_i' \geq 0, \ \sum_{i=1}^{p} \alpha_i c_i' = a(1+\mu) \right\}. \tag{9.13}$$

The following Lemma can be proved by direct calculations with the help of explicit formula of semigroup V.

Lemma 3 *For every $p \geq 1$, $\{\alpha_i\}_{i=1}^{p} \subset (0, +\infty)$, $a > 0$, $\mu > 0$, the impulsive problem (9.10), (9.12), (9.13) generates an impulsive MDS $G : R_+ \times H \mapsto P(H)$, which has global attractor Θ.*

Moreover,

$$\forall t \geq 0 \ \ G(t, \Theta \setminus M) \subset \Theta \setminus M, \tag{9.14}$$

and the following equality takes place:

$$\Theta = \left\{ \sum_{i=1}^{p} c_i e^{-\lambda_i \tau} \psi_i \mid \tau \in [0, \overline{\tau}], \ \ c_i \geq 0, \ \sum_{i=1}^{p} \alpha_i c_i e^{-\lambda_i \overline{\tau}} = a, \sum_{i=1}^{p} \alpha_i c_i = a(1+\mu) \right\}. \tag{9.15}$$

Remark 9.2 Note that if in (9.15) numbers $c_i \geq 0$ are such that $\sum_{i=1}^{p} \alpha_i c_i = a(1+\mu)$, then the moment of time $\overline{\tau}$ is uniquely determined by the equality $\sum_{i=1}^{p} \alpha_i c_i e^{-\lambda_i \overline{\tau}} = a$.

The main result of the paper was to prove the existence of global attractor for weakly nonlinear case when there is no explicit formula of solutions.

We consider the following nonlinear problem

$$\frac{dy}{dt} + Ay = \varepsilon \cdot f(y), \ t > 0, \tag{9.16}$$

where $\varepsilon > 0$ is a small parameter and Lipschitz-continuous nonlinear term $f : H \mapsto H$ satisfies the following assumption:

$$\exists C > 0 \ \ \forall y \in H \ \ \|f(y)\| \le C. \tag{9.17}$$

It is well known that under such conditions for every $y_0 \in H, \varepsilon > 0$ there exists a unique (mild) solution y of (9.16) with $y(0) = y_0$. Therefore, the problem (9.16) generates the continuous semigroup $V_\varepsilon : R_+ \times H \to H$.

Moreover, if $\varepsilon_n \to \varepsilon_0$, $y^{(n)}(t) = V_{\varepsilon_n}(t, y_0^{(n)})$, $y(t) = V_{\varepsilon_0}(t, y_0)$, then $\forall T > 0$ we have the following regularity result [21]:

$$\text{if } y_0^{(n)} \to y_0 \text{ in } H_w, \text{ then } \forall \tau > 0 \ y^{(n)} \to y \text{ in } C([0,T]; H_w) \cap C([\tau, T]; H), \tag{9.18}$$

$$\text{if } y_0^{(n)} \to y_0 \text{ in } H, \text{ then } y^{(n)} \to y \text{ in } C([0,T]; H), \tag{9.19}$$

where H_w is the space H with weak topology.

The main result of the paper is the following theorem.

Theorem 9.1 *For sufficiently small $\varepsilon > 0$, the impulsive problem (9.16), (9.12), (9.13) generates an impulsive MDS $G_\varepsilon : R_+ \times H \mapsto P(H)$, which has a global attractor Θ_ε.*

Moreover,

$$dist(\Theta_\varepsilon, \Theta) \to 0, \ \varepsilon \to 0, \tag{9.20}$$

where Θ is given by (9.15).

Remark 9.3 In all further arguments, the phrase "*for sufficiently small ε*" means that there exists $\varepsilon_1 > 0$ which depends only on the parameters of the problem (9.16), (9.12), (9.13) such that some property fulfilled for every $\varepsilon \in [0, \varepsilon_1]$.

Proof First of all, we must verify conditions (9.6)–(9.8). Let us consider some properties of solutions of (9.16). For every solution y and for a.a. $t > 0$, we have

$$\frac{1}{2}\frac{d}{dt}\|y(t)\|^2 + \langle Ay(t), y(t)\rangle = \varepsilon(f(y(t)), y(t)). \tag{9.21}$$

Then for sufficiently small ε from (9.17) and Gronwall Lemma, we deduce

$$\forall t \ge s \ge 0 \ \ \|y(t)\|^2 \le \|y(s)\|^2 e^{-\frac{\beta}{\alpha}(t-s)} + 1. \tag{9.22}$$

Every solution y also satisfies the following equality: $\forall i \geq 1\ \ \forall t \geq 0$

$$(y(t), \psi_i) = e^{-\lambda_i t}(y_0, \psi_i) + \varepsilon \int_0^t e^{-\lambda_i (t-\tau)}(f(y(\tau)), \psi_i) d\tau \tag{9.23}$$

From the definition of the set M and the map I, we immediately obtain (9.6). To verify (9.7), we take an arbitrary $y_0 \in M$, an arbitrary solution y of (9.16) with $y(0) = y_0$ and consider the function

$$g_\varepsilon(t) = \sum_{i=1}^p \alpha_i e^{-\lambda_i t}(y_0, \psi_i) + \varepsilon \int_0^t \sum_{i=1}^p \alpha_i e^{-\lambda_i (t-\tau)}(f(y(\tau)), \psi_i) d\tau.$$

As for some $\tau_0 = \tau_0(y_0) > 0$, we have

$$\forall t \in (0, \tau_0)\ \ g_0(t) < a - \frac{a\lambda_1}{2} t,$$

then for $t \in (0, \tau_0)$

$$g_\varepsilon(t) < a - \frac{a\lambda_1}{2} t + \varepsilon C \cdot \sum_{i=1}^p \alpha_i t. \tag{9.24}$$

So for sufficiently small $\varepsilon > 0$ we obtain

$$\exists \tau = \tau(\varepsilon, y_0)\ \ \forall t \in (0, \tau)\ \ g_\varepsilon(t) \neq a. \tag{9.25}$$

Let us prove property (9.8). It is obvious if y does not intersect M. To investigate the other situation, we take an arbitrary solution y with $y(0) = y_0 \in IM$. First of all let us show that y intersects M. For this aim we consider the function

$$F(\varepsilon, t) = \sum_{i=1}^p \alpha_i e^{-\lambda_i t}(y_0, \psi_i) - a + \varepsilon \int_0^t \sum_{i=1}^p \alpha_i e^{-\lambda_i (t-\tau)}(f(y(\tau)), \psi_i) d\tau.$$

As

$$g_0(0) = a(1 + \mu),\ g_0'(t) = -\sum_{i=1}^p \alpha_i \lambda_i e^{-\lambda_i t}(y_0, \psi_i) < 0\ \forall\, t \geq 0,\ \lim_{t\to\infty} g_0(t) = 0,$$

so there exists $s_0 > 0$ such that $F(0, s_0) = 0$. Due to conditions on f, the function

$$(-1, 1) \times (0, +\infty) \ni (\varepsilon, t) \mapsto F(\varepsilon, t)$$

is continuous on the first variable and has continuous derivative on the second variable. Moreover, the following estimates take place

$$|F(\varepsilon, s_0) - F(0, s_0)| \le \varepsilon C_1,$$

$$\left| F(\varepsilon, t') - F(\varepsilon, t'') - \frac{\partial F}{\partial t}(0, s_0)(t' - t'') \right| \le C_2 \left(|t_0 - t'| + |t_0 - t''| + \varepsilon \right) \left| t' - t'' \right|,$$

where positive constants C_1, C_2 depend only on parameters of the problem (9.16). Therefore, from the Implicit Value Theorem for sufficiently small ε there exists $s_\varepsilon = s_\varepsilon(y_0) > 0$ such that $F(\varepsilon, s_\varepsilon) = 0$. The last equality means that $y(s_\varepsilon) \in M$. Without loss of generality, we can assume that

$$\forall t \in (0, s_\varepsilon) \;\; y(t) \notin M, \quad y(s_\varepsilon) \in M.$$

Let us give an estimation for s_ε. Using (9.23), we get

$$a = \sum_{i=1}^{p} \alpha_i e^{-\lambda_i s_\varepsilon}(y_0, \psi_i) + \varepsilon \int_0^{s_\varepsilon} \sum_{i=1}^{p} \alpha_i e^{-\lambda_i (s_\varepsilon - \tau)}(f(y(\tau)), \psi_i) d\tau \le$$

$$\sum_{i=1}^{p} \alpha_i \|y_0\| \cdot e^{-\lambda_1 s_\varepsilon} + \varepsilon C \sum_{i=1}^{p} \frac{\alpha_i}{\lambda_i}.$$

So for sufficiently small ε, we have

$$s_\varepsilon \le \frac{1}{\lambda_1} \ln \frac{2\|y_0\| \sum_{i=1}^{p} \alpha_i}{a}. \tag{9.26}$$

Using again (9.23), we obtain

$$a = \sum_{i=1}^{p} \alpha_i e^{-\lambda_i s_\varepsilon}(y_0, \psi_i) + \varepsilon \int_0^{s_\varepsilon} \sum_{i=1}^{p} \alpha_i e^{-\lambda_i (s_\varepsilon - \tau)}(f(y(\tau)), \psi_i) d\tau \ge$$

$$e^{-\lambda_p s_\varepsilon} a(1 + \mu) - \frac{a\mu}{2}.$$

Therefore,

$$s_\varepsilon \ge \frac{1}{\lambda_p} \ln(1 + \mu'), \quad \mu' = \frac{\mu}{2 + \mu}. \tag{9.27}$$

From (9.27), we get the required property.

It is important to note that the previous arguments guarantee that from every initial point $y(0) = y_0 \in IM$ starts at least one trajectory with infinite number of impulsive perturbations.

Properties (9.6)–(9.8) guarantee that for sufficiently small ε formula

$$G_\varepsilon(t, y_0) = \{y(t) \mid y(\cdot) \in K^\varepsilon_{y_0}\} \tag{9.28}$$

generates an impulsive MDS, where $K^{\varepsilon}_{y_0}$ is a set of all solutions of (9.16), (9.12), (9.13) with initial point y_0.

Let us prove the dissipativity condition for impulsive MDS (9.28). If $y \in K^{\varepsilon}_{y_0}$, $\|y_0\| \leq R$ does not intersect M then from (9.22)

$$\|y(t)\| \leq \sqrt{2} \quad \forall t \geq T = \frac{2\alpha}{\beta} \ln R. \tag{9.29}$$

If for some $\tau > 0$ $y(t) \notin M$ $\forall t \in (0, \tau)$, $y(\tau) \in M$ then from (9.26) follows

$$\tau \leq \frac{1}{\lambda_1} \ln \frac{2R \sum\limits_{i=1}^{p} \alpha_i}{a} \tag{9.30}$$

Thus, it is enough to prove the following property for sufficiently small ε:

$$\exists R_0 > 0 \;\; \forall R > 0 \;\; \exists T = T(R) > 0 \;\; \forall y_0 \in IM, \;\; \|y_0\| \leq R,$$

$$\forall y \in K^{\varepsilon}_{y_0} \quad \forall t \geq T \quad \|y(t)\| \leq R_0. \tag{9.31}$$

Without loss of generality in all further arguments, we assume that if $y_0 \in IM$ then $y \in K^{\varepsilon}_{y_0}$ has an infinite number of impulsive points.

So for given $y \in K^{\varepsilon}_{y_0}$ with $y_0 \in IM$, $\|y_0\| \leq R$ from (9.27), there are $\{s_i\}_{i=0}^{\infty}$ such that $y(\cdot)$ has jumps at the moments $\{s_0, s_0 + s_1, \ldots\}$ with impulsive points $\{y_i^+\}_{i=1}^{\infty}$ and $\forall i \geq 0$ $s_i \geq \frac{1}{\lambda_p} \ln(1 + \mu')$.

Let

$$V_{\varepsilon}(s_0, y_0) = y(s_0 - 0) = \sum_{i=1}^{p} c_i \psi_i + \sum_{i=p+1}^{\infty} c_i \psi_i,$$

$$\|y(s_0 - 0)\|^2 = \sum_{i=1}^{\infty} c_i^2 \leq \|y_0\|^2 e^{-\delta s_0} + 1, \quad \delta = \frac{\beta}{\alpha} > 0,$$

$$y(s_0) = y_1^+ = \sum_{i=1}^{p} c_i^{'} \psi_i + \sum_{i=p+1}^{\infty} c_i \psi_i.$$

Using inequality

$$\forall\, i = \overline{1, p} \; c_i^{'} \leq \frac{a(1+\mu)}{\varkappa}, \quad \varkappa := \min_{1 \leq i \leq p} \alpha_i > 0,$$

we get for $k \geq 1$

$$\|y(\sum_{i=0}^{k} s_i - 0)\|^2 \le \|y_0\|^2 e^{-\delta \sum_{i=0}^{k} s_i} + p\frac{(1+\mu)^2}{\varkappa^2} a^2 (e^{-\delta s_k} + \ldots + e^{-\delta(s_k+\ldots+s_1)}) + \tag{9.32}$$

$$e^{-\delta s_k} + \ldots + e^{-\delta(s_k+\ldots+s_1)} + 1$$

$$\|y_{k+1}^{+}\|^2 \le \|y_0\|^2 e^{-\delta \sum_{i=0}^{k} s_i} + (p\frac{(1+\mu)^2}{\varkappa^2} a^2 + 1)(e^{-\delta s_k} + \ldots + e^{-\delta(s_k+\ldots+s_1)} + 1) \tag{9.33}$$

Using (9.27), from (9.32), (9.33) we get

$$\exists T = T(R) \ \forall t \ge T \ \ \|y(t)\|^2 \le 1 + \frac{(p\frac{(1+\mu)^2}{\varkappa^2} a^2 + 1)}{1 - (1+\mu')^{-\frac{\beta}{\alpha \cdot \lambda_p}}} := R_0 \tag{9.34}$$

Finally, let us prove that G_ε is asymptotically compact. Let $\{y_0^{(n)}\}$ be an arbitrary-bounded sequence of initial data, $\|y_0^{(n)}\| \le R$, $\xi_n \in G_\varepsilon(t_n, y_0^{(n)})$, $t_n \nearrow +\infty$. Then, $\xi_n = y_n(t_n)$, where $y_n \in K^{\varepsilon}_{y_0^{(n)}}$. If y_n does not intersect M, then $\forall\ t \ge 0\ y_n(t) = V_\varepsilon(t, y_0^{(n)})$. So

$$\xi_n = y_n(t_n) = V_\varepsilon(1, y_n(t_n - 1)).$$

From (9.22) we obtain

$$\|y_n(t_n - 1)\| \le \sqrt{2} \quad \forall n \ge N(R).$$

Therefore from (9.18) the sequence $\{\xi_n\}$ is precompact in H. If the function y_n intersects M at the first time at a point τ_n, then from (9.26) sequence τ_n is bounded and $\{V_\varepsilon(\tau_n, y_0^{(n)})\}$ is also bounded in H. So from the inequality

$$\forall\ y \in M\ \forall y^+ \in Iy\ \|y^+\|^2 \le pa^2 \frac{(1+\mu)^2}{\varkappa^2} + \|y\|^2, \tag{9.35}$$

it will be enough to prove the precompactness of the sequence $\{\xi_n\} \subset H$, where

$$\xi_n \in \tilde{G}_\varepsilon(t_n, z_n), \quad t_n \nearrow \infty, \quad z_n \in IM, \quad \|z_n\| \le R.$$

Let $\xi_n = y_n(t_n)$, $y_n \in K^{\varepsilon}_{z_n}$, $\{T_{i+1}^{(n)} = \sum_{k=0}^{i} s_k^{(n)}\}_{i=0}^{\infty}$ be the impulse moments for $y_n(\cdot)$, $\{\eta_i^{(n)+}\}_{i=1}^{\infty}$ be the corresponding impulsive points.

Firstly, we want to prove the precompactness of $\{\eta_i^{(n)+}\}$. Due to boundness of f, we deduce from the Uniform Gronwall Lemma [20] that $\forall r > 0\ \forall\ y_0 \in H$

$$\beta \|V_\varepsilon(r, y_0)\|_V^2 \le \frac{\|y(0)\|^2 + 1}{r} + C^2 r + \frac{\alpha}{\beta} C^2. \tag{9.36}$$

For the sequence y_n from (9.22) and (9.34), we deduce that there exists $c = c(R) > 0$ such that

$$\forall t \geq 0 \ \ \forall n \geq 1 \ \ \|y_n(t)\| \leq c(R). \tag{9.37}$$

Using the inequality

$$\forall\, y \in M \cap V \ \forall\, y^+ \in Iy \ \|y^+\|_V^2 \leq p\lambda_p a^2 \frac{(1+\mu)^2}{\varkappa^2} + \|y\|_V^2,$$

from (9.26), (9.27), and (9.36), we deduce $\forall i \geq 1 \ \ \forall n \geq 1$

$$\|\eta_i^{(n)+}\|_V^2 = \|y_n(T_i^n)\|_V^2 \leq p\lambda_p a^2 \frac{(1+\mu)^2}{\varkappa^2} +$$

$$+\frac{1}{\beta}\Big(\frac{(c^2(R)+1)}{ln(1+\mu')}\lambda_p + c_1^2 \cdot \frac{1}{\lambda_1} ln \frac{2c(R)\sum\limits_{i=1}^{p}\alpha_i}{a} + \frac{\alpha}{\beta}c_1^2\Big) \tag{9.38}$$

As the embedding $V \subset H$ is compact, we obtain the required precompactness of $\{\eta_i^{(n)+}\}$ in H. As for every sufficiently large n, there exists a number $i = i(n) \geq 1$, $i(n) \to \infty, \ \ n \to \infty$ such that

$$t_n \in [T_{i(n)}^{(n)}, T_{i(n)+1}^{(n)}),$$

so (9.19) provides precompactness of $\{\xi_n = y_n(t_n)\}$. Therefore, according to Lemma 1 and dissipativity estimate (9.34), impulsive MDS G_ε has global attractor $\Theta_\varepsilon = \omega_\varepsilon(B_0)$, where dissipativity set B_0 does not depend on ε. Let us prove the limit equality (9.20). For this purpose, it is enough to prove that for $\varepsilon_n \to 0$, $\xi^{(n)} \in \Theta_{\varepsilon_n}$ on some subsequence

$$\xi^{(n)} \to \xi \in \Theta \text{ in } H, \ n \to \infty. \tag{9.39}$$

There exist sequences $\{t_n \nearrow \infty\}$, $\{z_n\} \subset B_0$, $y_n \in K_{z_n}^{\varepsilon_n}$ such that

$$\forall\, n \geq 1 \ \ \|\xi^{(n)} - y_n(t_n)\| \leq \frac{1}{n}.$$

From previous arguments, we have for $\xi_n = y_n(t_n)$

$$\xi_n = V_{\varepsilon_n}(\tau_n, \eta_n^+),$$

where

$$\tau_n = t_n - T_{i(n)}^{(n)+}, \ \eta_n^+ = \eta_{i(n)}^{(n)+}, \ i(n) \to \infty, \ n \to \infty.$$

Moreover, from (9.19) we can claim that

$$\tau_n \to \tau \in [0, \bar{\tau}],\ \eta_n^+ \to \eta,\ \xi_n \to \xi = V(\tau, \eta) \text{ in } H.$$

Using (9.23), (9.27) and "nonimpulsive" character of coordinates $c_j(t),\ j \geq p+1$ of every impulsive trajectory, we deduce

$$\forall\, j \geq p+1\ \ (\eta_n^+, \psi_j) \to 0,\ n \to \infty.$$

Therefore, $\xi \in \Theta$ and theorem is proved.

Remark 9.4 It is also possible to prove invariance property (9.14) for the global attractor Θ_ε. It will be done in our forthcoming paper.

References

1. Bonotto, E.M.: Flows of characteristic 0+ in impulsive semidynamical systems. J. Math. Anal. Appl. **332**, 81–96 (2007)
2. Bonotto, E.M., Demuner, D.P.: Attractors of impulsive dissipative semidynamical systems. Bull. Sci. Math. **137**, 617–642 (2013)
3. Bonotto, E.M., Bortolan, M.C., Carvalho, A.N., Czaja, R.: Global attractors for impulsive dynamical systems - a precompact approach. J. Diff. Eqn. **259**, 2602–2625 (2015)
4. Ciesielski, K.: On stability in impulsive dynamical systems. Bull. Pol. Acad. Sci. Math. **52**, 81–91 (2004)
5. Chepyzhov, V. V., Vishik, M. I.: Attractors of equations of mathematical physics, AMS **49**, 363 (2002)
6. Iovane, G., Kapustyan, O.V., Valero, J.: Asymptotic behaviour of reaction-diffusion equations with non-damped impulsive effects. Nonlinear Anal. **68**, 2516–2530 (2008)
7. Kapustyan, A.V., Mel'nik, V.S.: On global attractors of multivalued semidynamical systems and their approximations. Doklady Academii Nauk. **366**(4), 445–448 (1999)
8. Kapustyan, O.V., Melnik, V.S., Valero, J.: A weak attractor and properties of solutions for the three-dimensional Benard problem. Discret. Contin. Dyn. Syst. **18**, 449–481 (2007)
9. Kapustyan, O.V., Kasyanov, P.O., Valero, J.: Pullback attractors for some class of extremal solutions of 3D Navier-Stokes system. J. Math. Anal. Appl. **373**, 535–547 (2011)
10. Kapustyan, O.V., Perestyuk, M.O.: Existence of global attractors for impulsive dynamical systems, Reports of the NAS of Ukraine. Mathematics **12**, 13–18 (2015)
11. Kapustyan, O.V., Perestyuk, M.O.: Global attractors for impulsive infinite-dimensional systems. Ukr. Math. J. **68**(4), 517–528 (2016)
12. Kasyanov, P.O.: Multivalued dynamics of solutions of autonomous differential-operator inclusion with pseudomonotone nonlinearity. Cybern. Syst. Anal. **47**(5), 800–811 (2011)
13. Kaul, S.K.: On impulsive semidynamical systems. J. Math. Anal. Appl. **150**(1), 120–128 (1990)
14. Kaul, S.K.: Stability and asymptotic stability in impulsive semidynamical systems. J. Appl. Math. Stoch. Anal. **7**(4), 509–523 (1994)
15. Melnik, V.S., Valero, J.: On attractors of multi-valued semi-flows and differential inclusions. Set-Valued Anal. **6**, 83–111 (1998)
16. Perestjuk, Y.M.: Discontinuous oscillations in an impulsive system. J. Math. Sci. **194**(4), 404–413 (2013)
17. Perestyuk, M.O., Kapustyan, O.V.: Long-time behavior of evolution inclusion with non-damped impulsive effects. Mem. Differ. Equ. Math. Phys. **56**, 89–113 (2012)

18. Rozko, V.: Stability in terms of Lyapunov of discontinuous dynamic systems. Differ. Uravn. **11**(6), 1005–1012 (1975)
19. Samoilenko, A.M., Perestyuk, N.A.: Impulsive Differential Equations. World Scientific, Singapore (1995)
20. Temam, R.: Infinite-Dimensional Dynamical Systems in Mechanics and Physics, Springer, Berlin (1988)
21. Zgurovsky, M.Z., Kasyanov, P.O., Kapustyan, O.V., Valero, J., Zadoianchuk, N.V.: Evolution Inclusions and Variation Inequalities for Earth Data Processing III. Long-Time Behavior of Evolution Inclusions Solutions in Earth Data Analysis. Springer, Berlin (2012)

Chapter 10
A Random Model for Immune Response to Virus in Fluctuating Environments

Yusuke Asai, Tomás Caraballo, Xiaoying Han and Peter E. Kloeden

Abstract In this work, we study a model for virus dynamics with a random immune response and a random production rate of susceptible cells from cell proliferation. In traditional models for virus dynamics, the rate at which the viruses are cleared by the immune system is constant, and the rate at which susceptible cells are provided is constant or a function depending on the population of all cells. However, the human body in general is never stationary, and thus, these rates can barely be constant. Here, we assume that the human body is a random environment and models the rates by random processes, which result in a system of random differential equations. We then analyze the long-term behavior of the random system, in particular the existence and geometric structure of the random attractor, by using the theory of random dynamical systems. Numerical simulations are provided to illustrate the theoretical result.

Y. Asai
Department of Hygiene, Graduate School of Medicine, Hokkaido University, Sapporo 060-8638, Japan
e-mail: yusuke.asai@med.hokudai.ac.jp

T. Caraballo (✉)
Departamento de Ecuaciones Diferenciales y Análisis Numérico, Universidad de Sevilla, Apdo. de Correos 1160, 41080 Sevilla, Spain
e-mail: caraball@us.es

X. Han
Department of Mathematics and Statistics, Auburn University, 221 Parker Hall, Auburn, AL 36849, USA
e-mail: xzh0003@auburn.edu

P.E. Kloeden
School of Mathematics and Statistics, Huazhong University of Science & Technology, Wuhan 430074, China
e-mail: kloeden@mathematik.uni-kl.de

P.E. Kloeden
Felix-Klein-Zentrum Für Mathematik, TU Kaiserslautern, 67663 Kaiserslautern, Germany

V.A. Sadovnichiy and M.Z. Zgurovsky (eds.), *Advances in Dynamical Systems and Control*, Studies in Systems, Decision and Control 69,
DOI 10.1007/978-3-319-40673-2_10

10.1 Introduction

Basic models for virus dynamics were introduced in the classic text by May and Nowak [12]. The assumption for simplest models is that the body is modeled as a "well-stirred" chemostat containing the virus and two kinds of cells, uninfected but susceptible cells and cells infected by virus. In a chemostat, microorganisms grow by feeding on nutrients in the culture vessel and are flushed out to the collecting vessel. Similarly in the human body, the virus grows from dead infected cells and is cleared by the immune system. Modeling chemostats by systems of non-autonomous or random differential equations is fully justified (see, e.g., [6, 7]), as the environment for a chemostat usually varies in time (either deterministically or randomly). Using the argument that the human body also varies in time, we will model the virus dynamics by a system of random differential equations in this work.

Denote by v the population size of free virus, x the population size of uninfected cells (food for virus), and y the population size of infected cells. First, uninfected cells are produced by cell proliferation at a constant rate Λ, live for an average lifetime, and die at an average death rate γ_1. Second, virus infects susceptible cells to produce infected cells, with an "efficiency," β. Since cells are infected by contact with viruses, the infection can be modeled as a simple mass action reaction

$$x + v \xrightarrow{\beta} y.$$

Third, infected cells die at an average rate γ_2 and release new viruses at a rate κ. At the same time, these viruses are cleared by the immune system at a rate α. Then, we arrive at the basic model of virus dynamics:

$$\frac{\mathrm{d}x(t)}{\mathrm{d}t} = \Lambda - \gamma_1 x - \beta x v, \tag{10.1}$$

$$\frac{\mathrm{d}y(t)}{\mathrm{d}t} = \beta x v - \gamma_2 y, \tag{10.2}$$

$$\frac{\mathrm{d}v(t)}{\mathrm{d}t} = \kappa y - \alpha v. \tag{10.3}$$

The ordinary differential equation system (10.1)–(10.3) can be used to describe the dynamics of various types of virus and healthy and infected cells, but with limitations. First, the model assumes that the contribution of the immune response (to the death of infected cells or free virus and to reducing the rate of infection of new cells) is constant over time. Second, the dynamics of the susceptible cell population assumes a constant production rate from a pool of precursors. These assumptions may be justified for stationary environments, within a short term of time span. However, in the long term, the human body is never a stationary environment—it varies over time in principle, and hence, system (10.1)–(10.3) is not adequate to explain the real dynamic of virus and the immune response.

In this work, we will assume that the human body is a random environment that varies randomly with respect to time. Due to this random variation, the contribution of the immune response and the production rate of susceptible cells from cell proliferation will also fluctuate randomly with respect to time. More precisely, we assume that parameters Λ and α are perturbed by real noise, i.e., $\Lambda = \Lambda(\theta_t\omega)$ and $\alpha = \alpha(\theta_t\omega)$ are continuous and essentially bounded:

$$\Lambda(\theta_t\omega) \in \lambda \cdot [1 - \delta_1, 1 + \delta_1], \quad \lambda > 0, \quad 0 < \delta_1 < 1, \tag{10.4}$$

$$\alpha(\theta_t\omega) \in a \cdot [1 - \delta_2, 1 + \delta_2], \quad a > 0, \quad 0 < \delta_2 < 1. \tag{10.5}$$

Then, system (10.1)–(10.3) becomes

$$\frac{\mathrm{d}x(t,\omega)}{\mathrm{d}t} = \Lambda(\theta_t\omega) - \gamma_1 x - \beta x v, \tag{10.6}$$

$$\frac{\mathrm{d}y(t,\omega)}{\mathrm{d}t} = \beta x v - \gamma_2 y, \tag{10.7}$$

$$\frac{\mathrm{d}v(t,\omega)}{\mathrm{d}t} = \kappa y - \alpha(\theta_t\omega) v, \tag{10.8}$$

where γ_1, γ_2, β, κ are positive constants and $\Lambda(\theta_t\omega)$ and $\alpha(\theta_t\omega)$ are defined as in (10.4) and (10.5), respectively.

Bounded noise can be modeled in various ways. For example in [2], given a stochastic process Z_t such as Ornstein–Uhlenbeck (OU) process, the stochastic process

$$\zeta(Z_t) := \zeta_0\Big(1 - 2\varepsilon\frac{Z_t}{1 + Z_t^2}\Big), \tag{10.9}$$

where ζ_0 and ε are positive constants with $\varepsilon \in (0, 1)$, takes values in the interval $\zeta_0[1 - \varepsilon, 1 + \varepsilon]$ and tends to peak around $\zeta_0(1 \pm \varepsilon)$. It is thus suitable for a noisy switching scenario. In another example, the stochastic process

$$\eta(Z_t) := \eta_0\Big(1 - \frac{2\varepsilon}{\pi}\arctan Z_t\Big), \tag{10.10}$$

where η_0 and ε are positive constants with $\varepsilon \in (0, 1)$, takes values in the interval $\eta_0[1 - \varepsilon, 1 + \varepsilon]$ and is centered on η_0. In the theory of random dynamical systems, the driving noise process $Z_t(\omega)$ is replaced by a canonical driving system $\theta_t\omega$. This simplification allows a better understanding of the pathwise approach to model noise: A system influenced by stochastic processes for each single realization ω can be interpreted as wandering along a path $\theta_t\omega$ in Ω and thus may provide additional statistical information to the modeler.

In this paper, we will study the properties of solutions to (10.6)–(10.8). In particular, we are interested in the long-term behavior of solutions to (10.6)–(10.8), characterized by a global random attractor. The rest of the paper is organized as follows. In Sect. 10.2, we provide preliminaries on the theory of random dynamical systems. In Sect. 10.3, we prove the existence and uniqueness of a positive bounded solution to (10.6)–(10.8) and show that the solution generates a random dynamical system. In Sect. 10.4, we prove the existence and uniqueness of a global random attractor to the random dynamical system generated by the solution to (10.6)–(10.8) and also investigate the conditions under which the global random attractor consists of a singleton axial solution (endemic), or non-trivial component sets (pandemic). Numerical simulations are provided in Sect. 6, to illustrate the conditions for the endemic and pandemic of system (10.6)–(10.8).

10.2 Preliminaries on Random Dynamical Systems

In this section, we first present some concepts (from [1]) related to general random dynamical systems (RDSs) and random attractors that we require in the sequel. Our situation is, in fact, somewhat simpler, but to facilitate the reader's access to the literature, we give more general definitions here.

Let $(X, \|\cdot\|_X)$ be a separable Banach space, and let $(\Omega, \mathscr{F}, \P)$ be a probability space where $\mathscr{F}$ is the $\sigma-$algebra of measurable subsets of Ω (called "events") and $\P$ is the probability measure. To connect the state ω in the probability space Ω at time 0 with its state after a time of t elapses, we define a flow $\theta = \{\theta_t\}_{t\in\mathbb{R}}$ on Ω with each θ_t being a mapping $\theta_t : \Omega \to \Omega$ that satisfies

(1) $\theta_0 = \mathrm{Id}_\Omega$,
(2) $\theta_s \circ \theta_t = \theta_{s+t}$ for all $s, t \in \mathbb{R}$,
(3) the mapping $(t, \omega) \mapsto \theta_t\omega$ is measurable, and
(4) the probability measure $\P$ is preserved by θ_t, i.e., $\theta_t\P = \P$.

This setup establishes a time-dependent family θ that tracks the noise, and $(\Omega, \mathscr{F}, \P, \theta)$ is called a *metric dynamical system* [1].

Definition 10.1 A stochastic process $\{S(t, \omega)\}_{t\geq 0, \omega\in\Omega}$ is said to be a continuous RDS over $(\Omega, \mathscr{F}, \mathbb{P}, (\theta_t)_{t\in\mathbb{R}})$ with state space X if $S : [0, +\infty) \times \Omega \times X \to X$ is $(\mathscr{B}[0, +\infty) \times \mathscr{F} \times \mathscr{B}(X), \mathscr{B}(X))$-measurable and for each $\omega \in \Omega$,

(1) the mapping $S(t, \omega) : X \to X, x \mapsto S(t, \omega)x$ is continuous for every $t \geq 0$;
(2) $S(0, \omega)$ is the identity operator on X;
(3) (cocycle property) $S(t + s, \omega) = S(t, \theta_s\omega)S(s, \omega)$ for all $s, t \geq 0$.

Definition 10.2 (1) A set-valued mapping $B : \omega \to 2^X\backslash\emptyset$ is said to be a random set if the mapping $\omega \mapsto \mathrm{dist}_X(x, B(\omega))$ is measurable for any $x \in X$.
(2) A random set $B(\omega)$ is said to be bounded if $B(\omega)$ is bounded for a.e. $\omega \in \Omega$; a random set $B(\omega)$ is said to be compact if $B(\omega)$ is compact for a.e. $\omega \in \Omega$; a random set is said to be closed if $B(\omega)$ is closed for a.e. $\omega \in \Omega$.

(3) A bounded random set $B(\omega) \subset X$ is said to be tempered with respect to $(\theta_t)_{t\in\mathbb{R}}$ if for a.e. $\omega \in \Omega$,

$$\lim_{t\to\infty} e^{-\beta t} \sup_{x\in B(\theta_{-t}\omega)} \|x\|_X = 0, \quad \text{for all } \beta > 0;$$

a random variable $\omega \mapsto r(\omega) \in \mathbb{R}$ is said to be tempered with respect to $(\theta_t)_{t\in\mathbb{R}}$ if for a.e. $\omega \in \Omega$,

$$\lim_{t\to\infty} e^{-\beta t} \sup_{t\in\mathbb{R}} |r(\theta_{-t}\omega)| = 0, \quad \text{for all } \beta > 0.$$

In what follows, we use $\mathscr{D}(X)$ to denote the set of all tempered random sets of X.

Definition 10.3 A random set $K(\omega) \subset X$ is called a random absorbing set in $\mathscr{D}(X)$ if for any $B \in \mathscr{D}(X)$ and a.e. $\omega \in \Omega$, there exists $T_B(\omega) > 0$ such that

$$S(t, \theta_{-t}\omega)B(\theta_{-t}\omega) \subset K(\omega), \quad \forall t \geq T_B(\omega).$$

Definition 10.4 Let $\{S(t,\omega)\}_{t\geq 0,\omega\in\Omega}$ be an RDS over $(\Omega, \mathscr{F}, \mathbb{P}, (\theta_t)_{t\in\mathbb{R}})$ with state space X, and let $\mathscr{A}(\omega)(\subset X)$ be a random set. Then, $\mathscr{A}(\omega)$ is called a global random $\mathscr{D}$ attractor (or pullback Δ attractor) for $\{S(t,\omega)\}_{t\geq 0,\omega\in\Omega}$ if $\omega \mapsto \mathscr{A}(\omega)$ satisfies

(1) (random compactness) $\mathscr{A}(\omega)$ is a compact set of X for a.e. $\omega \in \Omega$;
(2) (invariance) for a.e. $\omega \in \Omega$ and all $t \geq 0$, it holds

$$S(t,\omega)\mathscr{A}(\omega) = \mathscr{A}(\theta_t\omega);$$

(3) (attracting property) for any $B \in \mathscr{D}(X)$ and a.e. $\omega \in \Omega$,

$$\lim_{t\to\infty} \text{dist}_X(S(t,\theta_{-t}\omega)B(\theta_{-t}\omega), \mathscr{A}(\omega)) = 0,$$

where

$$\text{dist}_X(G, H) = \sup_{g\in G} \inf_{h\in H} \|g - h\|_X$$

is the Hausdorff semimetric for $G, H \subseteq X$.

Proposition 10.1 ([5, 9, 10]) *Let $B \in \mathscr{D}(X)$ be an absorbing set for the continuous random dynamical system $\{S(t,\omega)\}_{t\geq 0,\omega\in\Omega}$ which is closed and satisfies the asymptotic compactness condition for a.e. $\omega \in \Omega$, i.e., each sequence $x_n \in S(t_n, \theta_{-t_n}, B(\theta_{-t_n}\omega))$ has a convergent subsequence in X when $t_n \to \infty$. Then, the cocycle S has a unique global random attractor with component subsets*

$$\mathscr{A}(\omega) = \bigcap_{\tau\geq t_B(\omega)} \overline{\bigcup_{t\geq\tau} S(t,\theta_{-t}\omega)B(\theta_{-t}\omega)}.$$

If the pullback absorbing set is positively invariant, i.e., $S(t,\omega)B(\omega) \subset B(\theta_t\omega)$ for all $t \geq 0$, then

$$\mathscr{A}(\omega) = \bigcap_{t\geq 0} S(t, \theta_{-t}\omega)B(\theta_{-t}\omega).$$

For state space $X = \mathbb{R}^d$ as in this paper, the asymptotic compactness follows trivially. Note that the random attractor is pathwise attracting in the pullback sense, but need not be pathwise attracting in the forward sense, although it is forward attracting in probability, due to some possible large deviations (see, e.g., Arnold [1].

When the cocycle mapping is strictly uniformly contracting [8, 11], i.e., there exists $K > 0$ such that

$$\|S(t,\omega)x_0 - S(t,\omega)y_0\|_X \leq e^{-Kt}\|x_0 - y_0\|_X$$

for all $t \geq 0$, $\omega \in \Omega$, and $x_0, y_0 \in X$, then the random attractor consists of singleton subsets $\mathscr{A}(\omega) = \{A(\omega)\}$. It is thus essentially a single stochastic process with sample paths $A(\theta_t\omega)$ for all $t \in \mathbb{R}$. The proof uses a Cauchy sequence rather than compactness argument. In this case, the random attractor is pathwise attracting in both the pullback and forward senses.

10.3 Properties of Solutions

In this section, we will prove the existence, uniqueness, and boundedness of positive solutions to (10.6)–(10.8). In addition, we prove that the solution generates a random dynamical system. Denote by

$$\mathbb{R}^3_+ = \{(x, y, v) \in \mathbb{R}^3 : x \geq 0, y \geq 0, v \geq 0\},$$

and for simplicity, we write $\mathbf{u}(t,\omega) = (x(t,\omega), y(t,\omega), v(t,\omega))^T$.

Theorem 10.1 *For any $\omega \in \Omega$, $t_0 \in \mathbb{R}$, and initial data $\boldsymbol{u}_0 = (x(t_0), y(t_0), v(t_0))^T \in \mathbb{R}^3_+$, system (10.6)–(10.8) has a unique nonnegative bounded solution $\boldsymbol{u}(\cdot; t_0, \omega, \boldsymbol{u}_0) \in \mathscr{C}\left([t_0,\infty), \mathbb{R}^3_+\right)$, with $\boldsymbol{u}(t_0; t_0, \omega, \boldsymbol{u}_0) = \boldsymbol{u}_0$. Moreover, the solution generates a random dynamical system $\varphi(t,\omega)(\cdot)$ defined as*

$$\varphi(t,\omega)\boldsymbol{u}_0 = \boldsymbol{u}(t; 0, \omega, \boldsymbol{u}_0), \quad \forall t \geq 0,\ \boldsymbol{u}_0 \in \mathbb{R}^3_+,\ \omega \in \Omega.$$

Proof Write

$$L(\theta_t\omega) = \begin{pmatrix} -\gamma_1 & 0 & 0 \\ 0 & -\gamma_2 & 0 \\ 0 & \kappa & -\alpha(\theta_t\omega) \end{pmatrix} \quad \text{and} \quad f(\theta_t\omega, \mathbf{u}) = \begin{pmatrix} \Lambda(\theta_t\omega) - \beta xv \\ \beta xv \\ 0 \end{pmatrix},$$

then Eqs. (10.6)–(10.8) become

$$\frac{du(t,\omega)}{dt} = L(\theta_t\omega)\mathbf{u} + f(\theta_t\omega, \mathbf{u}). \tag{10.11}$$

First, since $\alpha(\theta_t\omega)$ is bounded, the operator L generates an evolution system on $\mathbb{R}^3$. Second, since $\Lambda(\theta_t\omega)$ is continuous with respect to t, function f is continuous with respect to t and locally Lipschitz with respect to $\mathbf{u}$. Hence, system (10.11) has a unique local solution $\mathbf{u}(\cdot; t_0, \omega, \mathbf{u}_0) \in \mathscr{C}\left([t_0, T), \mathbb{R}^3\right)$.

By continuity of solutions, each solution has to take value 0 before it reaches a negative value. Notice that

$$\left.\frac{dx(t,\omega)}{dt}\right|_{x=0,y\geq 0,v\geq 0} = \Lambda(\theta_t\omega) > 0,$$
$$\left.\frac{dy(t,\omega)}{dt}\right|_{x\geq 0,y=0,v\geq 0} = \beta xv \geq 0,$$
$$\left.\frac{dv(t,\omega)}{dt}\right|_{x\geq 0,y\geq 0,v=0} = \kappa y \geq 0,$$

we have $x(t)$ strictly increasing at $x = 0$, $y(t)$ and $v(t)$ and non-decreasing at $y = 0$ and $v = 0$, respectively. This implies that $\mathbf{u}(t) \in \mathbb{R}^3_+$ for $t \in [t_0, T)$.

For $\mathbf{u}(t) \in \mathbb{R}^3_+$, define

$$\|\mathbf{u}(t)\|_1 := x(t) + y(t) + v(t).$$

Let $s(t) = 2\kappa x(t) + 2\kappa y(t) + \gamma_2 v(t)$, then

$$\|\mathbf{u}(t)\|_1 \leq \frac{s(t)}{\min\{2\kappa, \gamma_2\}}.$$

On the other hand by (10.6)–(10.8), we have

$$\begin{aligned}\frac{ds(t,\omega)}{dt} &= 2\kappa\Lambda(\theta_t\omega) - 2\kappa\gamma_1 x - \kappa\gamma_2 y - \gamma_2\alpha(\theta_t\omega)v\\ &\leq 2\kappa\lambda(1+\delta_1) - 2\kappa\gamma_1 x - \kappa\gamma_2 y - \gamma_2 a(1-\delta_2)v\\ &\leq 2\kappa\lambda(1+\delta_1) - \mu_1 s(t),\end{aligned} \tag{10.12}$$

where

$$\mu_1 = \min\{\gamma_1, \gamma_2/2, a(1-\delta_2)\} > 0. \tag{10.13}$$

For $s(t_0) \geq 2\kappa\lambda(1+\delta_1)/\mu_1$, $s(t)$ will be non-increasing for $t \geq t_0$, and thus, $s(t) \leq s(t_0)$. Otherwise, for $s(t_0) \leq 2\kappa\lambda(1+\delta_1)/\mu_1$, $s(t)$ will stay $\leq 2\kappa\lambda(1+\delta_1)/\mu_1$. In summary,

$$0 \leq \|\mathbf{u}\|_1 \leq \frac{s(t)}{\min\{2\kappa, \gamma_2\}} \leq \frac{\max\{2\kappa x(t_0) + 2\kappa y(t_0) + \gamma_2 v(t_0), 2\kappa\lambda(1+\delta_1)/\mu_1\}}{\mu_2},$$

where

$$\mu_2 = \min\{2\kappa, \gamma_2\}. \tag{10.14}$$

This implies that system (10.11) has a unique positive and bounded global solution $\mathbf{u}(\cdot; t_0, \omega, \mathbf{u}_0) \in \mathbb{R}^3_+$.

It is straightforward to check that

$$\mathbf{u}(t + t_0; t_0, \omega, \mathbf{u}_0) = \mathbf{u}(t; 0, \theta_{t_0}\omega, \mathbf{u}_0)$$

for all $t_0 \in \mathbb{R}, t \geq t_0, \omega \in \Omega$, and $\mathbf{u}_0 \in \mathbb{R}^3_+$. This allows us to define a mapping $\varphi(t, \omega)(\cdot)$:

$$\varphi(t, \omega)\mathbf{u}_0 = \mathbf{u}(t; 0, \omega, \mathbf{u}_0), \quad \forall t \geq 0, \ \mathbf{u}_0 \in \mathbb{R}^3_+, \ \omega \in \Omega. \tag{10.15}$$

From now on, we will simply write $\mathbf{u}(t; \omega, \mathbf{u}_0)$ instead of $\mathbf{u}(t; 0, \omega, \mathbf{u}_0)$.

For any $\mathbf{u}_0 \in \mathbb{R}^3_+$, solution $\mathbf{u}(\cdot; \omega, \mathbf{u}_0) \in \mathbb{R}^3_+$ for $t \in [0, \infty)$. Since function $f(\mathbf{u}, \theta_t\omega) = f(\mathbf{u}, t, \omega)$ is continuous in $\mathbf{u}$, t, and is measurable in ω, $\mathbf{u} : [0, \infty) \times \Omega \times \mathbb{R}^3_+ \to \mathbb{R}^3_+$, $(t; \omega, \mathbf{u}_0) \mapsto \mathbf{u}(t; \omega, \mathbf{u}_0)$ is $(\mathscr{B}[0, \infty) \times \mathscr{F}_0 \times \mathscr{B}(\mathbb{R}^3_+), \mathscr{B}(\mathbb{R}^3_+))$-measurable. It then follows directly that (10.11) generates a continuous random dynamical system $\varphi(t, \omega)(\cdot)$ defined by (10.15). This completes the proof.

10.4 Existence and Geometric Structure of Global Random Attractors

In this section, we will first prove the existence of a global random attractor for the random dynamical system $\{\varphi(t, \omega)\}_{t \geq 0, \omega \in \Omega}$. In addition, we will investigate the geometric structure of this random attractor.

Theorem 10.2 *The random dynamical system generated by system* (10.11) *possesses a unique global random attractor* $\mathscr{A} = \{A(\omega) : \omega \in \Omega\}$.

Proof We first prove that for $\omega \in \Omega$, there exists a tempered bounded closed random absorbing set $K(\omega) \in \Delta(\mathbb{R}^3_+)$ of the random dynamical system $\{\varphi(t, \omega)\}_{t \geq 0, \omega \in \Omega}$ such that for any $B \in \Delta(\mathbb{R}^3_+)$ and each $\omega \in \Omega$, there exists $T_B(\omega) > 0$ yielding

$$\varphi(t, \theta_{-t}\omega)B(\theta_{-t}\omega) \subset K(\omega) \quad \forall t \geq T_B(\omega).$$

In fact, recall that $\mathbf{u}(t; \omega, \mathbf{u}_0) = \varphi(t, \omega)\mathbf{u}_0$ denotes the solution of system (10.11) satisfying $\mathbf{u}(0; \omega, \mathbf{u}_0) = \mathbf{u}_0$. Then, for any $\mathbf{u}_0 := \mathbf{u}_0(\theta_{-t}\omega) \in B(\theta_{-t}\omega)$,

$$\|\varphi(t,\theta_{-t}\omega)\mathbf{u}_0\|_1 = \|\mathbf{u}(t;\theta_{-t}\omega,\mathbf{u}_0(\theta_{-t}\omega))\|_1 \le \frac{1}{\mu_2}\cdot s(t;\theta_{-t}\omega,s_0(\theta_{-t}\omega)).$$

Using inequality (10.12) and substituting ω by $\theta_{-t}\omega$, we obtain

$$\begin{aligned} s(t;\theta_{-t}\omega,s_0(\theta_{-t}\omega))\ s_0 &\le e^{-\mu_1 t} + \frac{2\kappa\lambda(1+\delta_1)}{\mu_1} \\ &\le e^{-\mu_1 t} \sup_{(x,y,v)\in B(\theta_{-t}\omega)} (2\kappa x + 2\kappa y + \gamma_2 v) + \frac{2\kappa\lambda(1+\delta_1)}{\mu_1}. \end{aligned}$$

Therefore, for any $\epsilon > 0$, and $\mathbf{u}_0 \in B(\theta_{-t}\omega)$, there exists $T_B(\omega)$ such that when $t > T_B$,

$$\begin{aligned} \|\varphi(t,\theta_{-t}\omega)\mathbf{u}_0\|_1 &\le \frac{1}{\mu_2}\cdot s(t;\theta_{-t}\omega,s_0(\theta_{-t}\omega)) \\ &\le \frac{1}{\mu_2}\cdot\frac{2\kappa\lambda(1+\delta_1)}{\mu_1} + \varepsilon, \end{aligned}$$

Define

$$K_\epsilon(\omega) = \overline{\left\{(x,y,v)\in\mathbb{R}^3_+ : x+y+v \le \frac{1}{\mu_2}\cdot\frac{2\kappa\lambda(1+\delta_1)}{\mu_1} + \epsilon\right\}}. \tag{10.16}$$

Then, $K_\epsilon(\omega)$ is positively invariant and absorbing in $\mathbb{R}^3_+$.

It follows directly from Proposition 10.1 that the random dynamical system generated by system (10.6)–(10.8) possesses a random attractor $\mathscr{A} = \{A(\omega) : \omega \in \Omega\}$, consisting of non-empty compact random subsets of $\mathbb{R}^3_+$ contained in $K_\epsilon(\omega)$. This completes the proof.

Next, we will investigate the details of the random attractor $\mathscr{A}$.

Theorem 10.3 *The random pullback attractor $\mathscr{A} = \{A(\omega) : \omega \in \Omega\}$ for the random dynamical system generated by system* (10.6)–(10.8) *has singleton component sets $A(\omega) = \{(x^*(\omega), 0, 0)\}$ for every $\omega \in \Omega$, provided that*

$$\frac{\kappa}{\gamma_2} \le 1 \quad \textit{and} \quad \frac{\beta\lambda(1+\delta_1)}{\mu_1 a(1-\delta_2)} < 1. \tag{10.17}$$

Proof Summing (10.7) and (10.8), we obtain

$$\frac{\mathrm{d}(y+v)}{\mathrm{d}t} = -(\gamma_2-\kappa)y - (\alpha(\theta_t\omega) - \beta x)v.$$

Recall that due to (10.16), for any $\epsilon > 0$, there exists $T_B(\omega)$ such that when $t > T_B$,

$$x(t) \le \|\mathbf{u}(t)\|_1 \le \frac{1}{\mu_2}\cdot\frac{2\kappa\lambda(1+\delta_1)}{\mu_1} + \epsilon.$$

By definition of μ_2 in (10.14), we have that $2\kappa/\mu_2 \leq 1$. Then, picking ϵ small enough, we have

$$\alpha(\theta_t\omega) - \beta x > \alpha(1-\delta_2) - \beta \cdot \frac{1}{\mu_2} \cdot \frac{2\kappa\lambda(1+\delta_1)}{\mu_1}$$
$$\geq \alpha(1-\delta_2) - \beta \cdot \frac{\lambda(1+\delta_1)}{\mu_1} > 0,$$

which implies that $y + v$ decreases to 0 as t approaches ∞.

Letting $y = v = 0$ in Eq. (10.6), we obtain

$$\frac{\mathrm{d}x}{\mathrm{d}t} = \Lambda(\theta_t\omega) - \gamma_1 x. \tag{10.18}$$

Solving Eq. (10.18) gives

$$x(t;\omega, x_0) = x_0 e^{-\gamma_1 t} + \int_0^t \Lambda(\theta_s\omega)e^{\gamma_1(t-s)}\mathrm{d}s,$$

and consequently,

$$x(t;\theta_{-t}\omega, x_0) = x_0 e^{-\gamma_1 t} + \int_{-t}^0 \Lambda(\theta_s\omega)e^{-\gamma_1 s}\mathrm{d}s \stackrel{t\to\infty}{\longrightarrow} \int_{-\infty}^0 \Lambda(\theta_s\omega)e^{-\gamma_1 s}\mathrm{d}s := x^*(\omega).$$

This completes the proof.

Theorem 10.3 implies that $(x^*(\theta_t\omega), 0, 0)$ is asymptotically stable as $t \to \infty$, i.e., endemic occurs when the parameters satisfy (10.17). We next investigate the conditions under which epidemic occurs.

Theorem 10.4 *The random pullback attractor* $\mathscr{A} = \{A(\omega) : \omega \in \Omega\}$ *for the random dynamical system generated by system* (10.6)–(10.8) *possesses non-trivial component sets which include* $(x^*(\omega), 0, 0)$ *and strictly positive points, provided that*

$$\frac{\beta\lambda(1+\delta_1)}{\mu_1 a(1+\delta_2)} > \frac{\gamma_2}{\kappa}. \tag{10.19}$$

Proof First, notice that the Eq. (10.7) is deterministic and implies that the surface $y = \frac{\beta}{\gamma_2}xv$ is invariant. The dynamics of x and v restricted on this invariant surface satisfy

$$\frac{\mathrm{d}x(t,\omega)}{\mathrm{d}t} = \Lambda(\theta_t\omega) - \gamma_1 x - \beta x v, \tag{10.20}$$
$$\frac{\mathrm{d}v(t,\omega)}{\mathrm{d}t} = \frac{\kappa\beta}{\gamma_2}xv - \alpha(\theta_t\omega)v. \tag{10.21}$$

Define the region Γ_ε by

$$\Gamma_\varepsilon := \left\{(x, v) \in \mathbb{R}^2_+ : x \geq \frac{a(1+\delta_2)\gamma_2}{\kappa\beta} + \varepsilon, v \geq \varepsilon, \frac{\kappa}{\gamma_2}x(t) + v(t) \leq \frac{\kappa\lambda}{\mu_1\gamma_2}(1+\delta_1) + \varepsilon\right\}.$$

For any $(x, v) \in \Gamma_\varepsilon$, we have

$$\frac{\mathrm{d}v}{\mathrm{d}t} = \left(\frac{\kappa\beta}{\gamma_2}x - \alpha(\theta_t\omega)\right)v > \left(\frac{\kappa\beta}{\gamma_2} \cdot \frac{a(1+\delta_2)\gamma_2}{\kappa\beta} - a(1+\delta_2)\right)v \geq 0.$$

On the other hand, we have

$$\begin{aligned}\frac{\mathrm{d}}{\mathrm{d}t}\left(\frac{\kappa}{\gamma_2}x(t) + v(t)\right) &= \frac{\kappa}{\gamma_2}\Lambda(\theta_t\omega) - \gamma_1\frac{\kappa}{\gamma_2}x - \alpha(\theta_t\omega)v \\ &\leq \frac{\kappa\lambda}{\gamma_2}(1+\delta_1) - \gamma_1\frac{\kappa}{\gamma_2}x - a(1-\delta_2)v \\ &\leq \frac{\kappa\lambda}{\gamma_2}(1+\delta_1) - \mu_1\left(\frac{\kappa}{\gamma_2}x(t) + v(t)\right),\end{aligned}$$

where μ_1 is as defined in (10.13). This implies that

$$\frac{\kappa}{\gamma_2}x(t) + v(t) \leq \frac{\kappa\lambda}{\mu_1\gamma_2}(1+\delta_1) + \varepsilon$$

for t large enough. Assumption (10.19) ensures that Γ_ε is a non-empty compact positive invariant absorbing set, which then ensures the existence of a non-trivial pullback attractor $\mathscr{A}_\varepsilon = \{A_\varepsilon(t) : t \in \mathbb{R}\}$ in Γ_ε. This completes the proof.

10.5 Numerical Simulations

In this section, we will simulate the system (10.6)–(10.8) numerically and verify that conditions (10.17) and (10.19) give rise to an endemic state (all infected cells and viruses are cleared) and a pandemic state (susceptible cells, infected cells, and viruses coexist) of system (10.6)–(10.8), respectively.

First, we transform the system (10.6)–(10.8) and two OU processes $Z_1(t)$ and $Z_2(t)$ into a system of random ordinary differential equation (RODE)–stochastic ordinary differential equation (SODE) pair [2, 4]:

$$d\begin{pmatrix} x(t) \\ y(t) \\ v(t) \\ Z_1(t) \\ Z_2(t) \end{pmatrix} = \begin{pmatrix} \Lambda(Z_1) - \gamma_1 x - \beta x v \\ \beta x v - \gamma_2 y \\ \kappa y - \alpha(Z_2)v \\ \theta_{11} - \theta_{12}Z_1 \\ \theta_{21} - \theta_{22}Z_2 \end{pmatrix} dt + \begin{pmatrix} 0 \\ 0 \\ 0 \\ \theta_{13} \\ \theta_{23} \end{pmatrix} dW_t.$$

The OU processes $Z_1(t)$ and $Z_2(t)$ can be generated independently, and we solve only the RODE part, i.e., x, y, and v compartments, of the RODE–SODE system. The system is assumed to be stiff, and the implicit 1.5-order RODE–Taylor scheme in [2] is applied here.

In the following simulation, we suppose that the cell proliferation rate $\Lambda(Z_1)$ has a switching effect and the loss rate of viruses $\alpha(Z_2)$ is distributed in a finite interval. They are randomized by the Eqs. (10.9) and (10.10), respectively, and given by

$$\Lambda(Z_1) = \lambda\left(1 - 2\delta_1\frac{Z_1}{1+Z_1^2}\right),$$
$$\alpha(Z_2) = a\left(1 - \frac{2\delta_2}{\pi}\arctan Z_2\right),$$

which satisfy (10.4) and (10.5).

Initial conditions for x, y, and v compartments are set as $x_0 = 2\times10^5$, $y_0 = 1\times10^5$, and $v_0 = 1\times10^6$. The coefficients for the OU processes are fixed to $\theta_{11} = 1, \theta_{12} = 3, \theta_{13} = 0.8, \theta_{21} = 0, \theta_{22} = 1$, and $\theta_{23} = 0.5$ for all examples. We will choose different set of parameters that satisfy assumption (10.17) or assumption (10.19).

Example 1 In this example, we set the parameters to be $\gamma_1 = 0.25$, $\gamma_2 = 0.5$, $\beta = 1\times10^{-5}$, $\lambda = 4\times10^4$, $a = 3$, $\delta_1 = 0.45$, $\delta_2 = 0.2$, and $\kappa = 0.2$. Assumption (10.17) is satisfied by this set of parameters. Figure 10.1 shows that the y and v compartments go to zero after enough amount of time and only x compartment remains nonzero, which means that the endemic state is achieved for parameters satisfying (10.17).

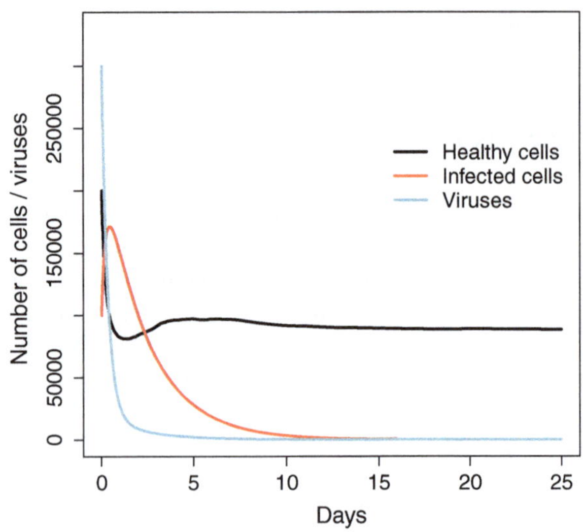

Fig. 10.1 With parameters $\gamma_1 = 0.25$, $\gamma_2 = 0.5$, $\beta = 1\times10^{-5}$, $\lambda = 4\times10^4$, $a = 3$, $\delta_1 = 0.45$, $\delta_2 = 0.2$, and $\kappa = 0.2$ satisfying assumption (10.17), both infected cells and viruses are cleared; only healthy cells remain

Example 2 In this example, we set the parameters to be $\gamma_1 = 0.25$, $\gamma_2 = 0.5$, $\beta = 1 \times 10^{-5}$, $\lambda = 4 \times 10^4$, $a = 3$, $\delta_1 = 0.45$, $\delta_2 = 0.2$, and $\kappa = 2$. Assumption (10.19) is satisfied by this set of parameters. Figure 10.2 shows that x, y, and v all remain nonzero for a time long enough, which means that the pandemic state is achieved for parameters satisfying (10.19).

Notice that the only parameter that has different values in Example 1 and Example 2 is κ. This implies that the rate at which virus is generated by dead susceptible cells is critical. A series of numerical simulations with different parameters were done to support this argument, among which we picked one more example to present

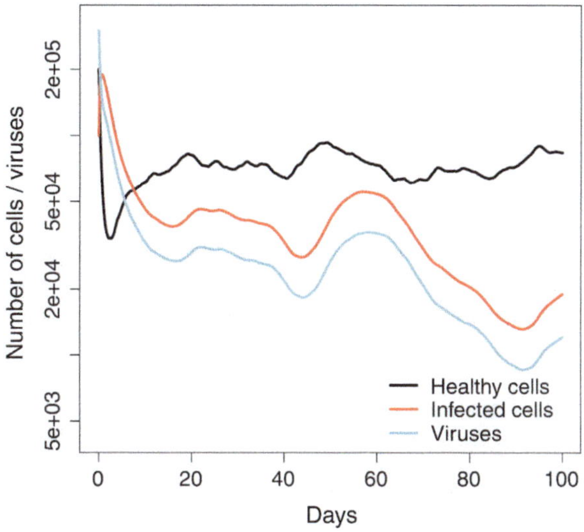

Fig. 10.2 With parameters $\gamma_1 = 0.25$, $\gamma_2 = 0.5$, $\beta = 1 \times 10^{-5}$, $\lambda = 4 \times 10^4$, $a = 3$, $\delta_1 = 0.45$, $\delta_2 = 0.2$, and $\kappa = 2$ satisfying assumption (10.19), infected cells, susceptible cells, and viruses coexist

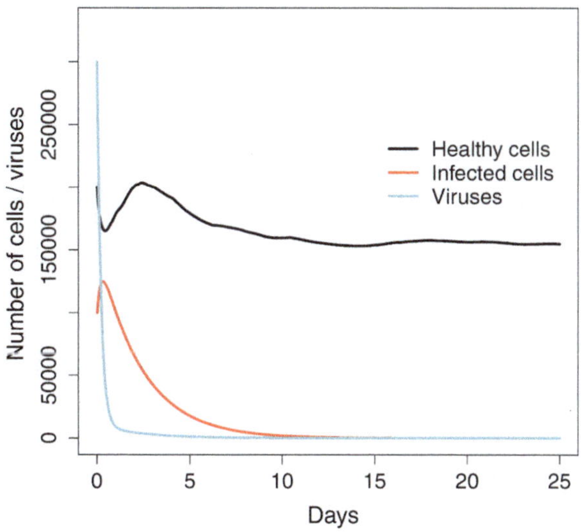

Fig. 10.3 With parameters $\gamma_1 = 0.4$, $\gamma_2 = 0.5$, $\beta = 5 \times 10^{-5}$, $\lambda = 10^5$, $a = 5$, $\delta_1 = 0.4$, $\delta_2 = 0.2$, and $\kappa = 0.3$ satisfying assumption (10.17), both infected cells and viruses are cleared; only healthy cells remain

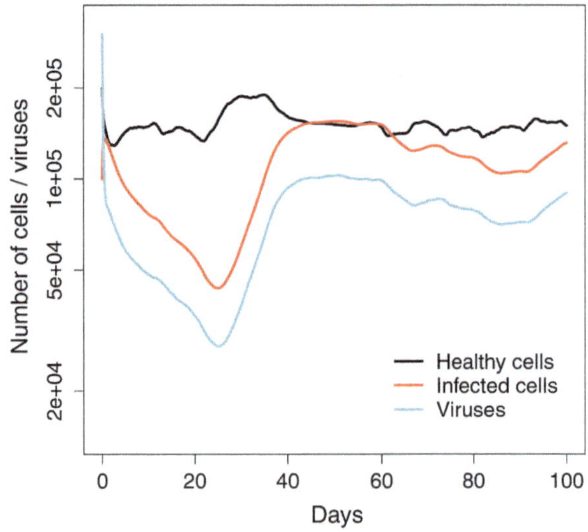

Fig. 10.4 With parameters $\gamma_1 = 0.4$, $\gamma_2 = 0.5$, $\beta = 5 \times 10^{-5}$, $\lambda = 10^5$, $a = 5$, $\delta_1 = 0.4$, $\delta_2 = 0.2$ and $\kappa = 3$ satisfying assumption (10.19), infected cells, susceptible cells and viruses coexist

below. In the following example, the parameters are chosen to be $\gamma_1 = 0.4$, $\gamma_2 = 0.5$, $\beta = 5 \times 10^{-5}$, $\lambda = 10^5$, $a = 5$, $\delta_1 = 0.4$, and $\delta_2 = 0.2$. When $\kappa = 0.3$, assumption (10.17) is satisfied and we obtain an endemic state (see Fig. 10.3). When $\kappa = 3$, assumption (10.19) is satisfied, and we obtain a pandemic state (see Fig. 10.4).

Acknowledgments This work has been partially supported by the Chinese NSF Grant No. 1157112, the Spanish Ministerio de Economía y Competitividad project MTM2015-63723-P and the Consejería de Innovación, Ciencia y Empresa (Junta de Andalucía) under grant 2010/FQM314, and Proyecto de Excelencia P12-FQM-1492.

References

1. Arnold, L.: Random Dynamical Systems. Springer, Berlin (1998)
2. Asai, Y., Kloeden, P.E.: Numerical schemes for random ODEs via stochastic differential equations. Commun. Appl. Anal. **17**, 521–528 (2013)
3. Asai, Y., Herrmann, E., Kloeden, P.E.: Stable integration of stiff random ordinary differential equations. J. Stoch. Anal. Appl. **31**, 293–313 (2013)
4. Asai, Y., Kloeden, P.E.: Multi-step methods for random ODEs driven by Itô diffusions. J. Comput. Appl. Math **294**, 210–224 (2016)
5. Bates, P.W., Lisei, H., Lu, K.: Attractors for stochastic lattice dynamical systems. Stoch. Dyn. **6**, 1–21 (2006)
6. Caraballo, T., Han, X., Kloeden, P.E.: Chemostats with random inputs and wall growth. Math. Meth. Appl. Sci. **38**, 3538–3550 (2015)
7. Caraballo, T., Han, X., Kloeden, P.E.: Nonautonomous chemostats with varible delays. SIAM J. Math. Anal. **47**, 2178–2199 (2015)
8. Caraballo, T., Kloeden, P.E., Schmalfuss, B.: Exponentially stable stationary solutions for stochastic evolution equations and their perturbation. Appl. Math. Optim. **50**, 183–207 (2004)

9. Caraballo, T., Lukaszewicz, G., Real, J.: Pullback attractors for asymptotically compact nonautonomous dynamical systems. Nonlinear Anal. TMA **6**, 484–498 (2006)
10. Flandoli, F., Schmalfuss, B.: Random attractors for the 3D stochastic Navier-Stokes equation with multiplicative noise. Stoch. Stoch. Rep. **59**(1–2), 21–45 (1996)
11. Kloeden, P.E., Lorenz, T.: Pullback incremental attraction. Nonautonomous & Random Dynamical Systems 53–60 (2013). doi:10.2478/msds-2013-0004
12. May, R., Nowak, M.: Virus Dynamics: Mathematical Principles of Immunology and Virology. Oxford University Press, Oxford (2001)
13. Perelson, A.S., Ribeiro, R.M.: Modeling the within-host dynamics of HIV infection. BMC Biol. **11**, 96 (2013)

[illegible]

Chapter 11
Some Aspects Concerning the Dynamics of Stochastic Chemostats

Tomás Caraballo, María J. Garrido-Atienza and Javier López-de-la-Cruz

Abstract In this paper, we study a simple chemostat model influenced by white noise which makes this kind of models more realistic. We use the theory of random attractors and, to that end, we first perform a change of variable using the Ornstein–Uhlenbeck process, transforming our stochastic model into a system of differential equations with random coefficients. After proving that this random system possesses a unique solution for any initial value, we analyze the existence of random attractors. Finally, we illustrate our results with some numerical simulations.

11.1 Introduction

Modeling chemostats is a really interesting and important problem with special interest in mathematical biology, since they can be used to study recombinant problems in genetically altered microorganisms [12, 13], waste water treatment [9, 17] and play an important role in theoretical ecology [2, 8, 11, 16, 21–23, 25]. Derivation and analysis of chemostat models are well documented in [18, 19, 24] and references therein.

Two standard assumptions for simple chemostat models are as follows: (1) the availability of the nutrient and its supply rate are fixed and (2) the tendency of the microorganisms to adhere to surfaces is not taken into account. However, these are very strong restrictions as the real world is non-autonomous and stochastic, and this justifies the analysis of stochastic chemostat models.

T. Caraballo · M.J. Garrido-Atienza (✉) · J. López-de-la-Cruz
Departamento de Ecuaciones Diferenciales y Análisis Numérico, Universidad de Sevilla, Apdo. de Correos 1160, 41080 Sevilla, Spain
e-mail: mgarrido@us.es

T. Caraballo
e-mail: caraball@us.es

J. López-de-la-Cruz
e-mail: jlopez78@us.es

V.A. Sadovnichiy and M.Z. Zgurovsky (eds.), *Advances in Dynamical Systems and Control*, Studies in Systems, Decision and Control 69,
DOI 10.1007/978-3-319-40673-2_11

Let us first consider one of the simplest chemostat models,

$$\frac{dS}{dt} = (S^0 - S)D - \frac{mSx}{a+S}, \tag{11.1}$$

$$\frac{dx}{dt} = x\left(\frac{mS}{a+S} - D\right), \tag{11.2}$$

where $S(t)$ and $x(t)$ denote concentrations of the nutrient and the microbial biomass, respectively; S^0 denotes the volumetric dilution rate, a is the half-saturation constant, D is the dilution rate, and m is the maximal consumption rate of the nutrient and also the maximal specific growth rate of microorganisms. We notice that all parameters are positive and we use a function Holling type-II as functional response of the microorganism describing how the nutrient is consumed by the species (see [20] for more details and biological explanations about this model).

However, we can consider a more realistic model by introducing a white noise in one of the parameters; therefore, we replace the dilution rate D by $D + \alpha\dot{W}(t)$, where $W(t)$ is a white noise, i.e., is a Brownian motion, and $\alpha \geq 0$ represents the intensity of noise. Then, system (11.1) and (11.2) is replaced by the following system of stochastic differential equations

$$dS = \left[(S^0 - S)D - \frac{mSx}{a+S}\right]dt + \alpha(S^0 - S)dW(t), \tag{11.3}$$

$$dx = x\left(\frac{mS}{a+S} - D\right)dt - \alpha x dW(t). \tag{11.4}$$

System (11.3) and (11.4) has been analyzed in [26] by using the classic techniques from stochastic analysis and some stability results are provided there. However, as in our opinion there are some unclear points in the analysis carried out in [26], our aim in this paper is to use an alternative approach to this problem, specifically the theory of random dynamical systems, which will allow us to partially improve the results in [26]. In addition, we will provide some results which hold with probability one while those from [26] are said to hold in probability.

System (11.3) and (11.4) is understood in the Itô sense. Then, we first consider its equivalent Stratonovich formulation which is given by

$$dS = \left[(S^0 - S)\left(D + \frac{\alpha^2}{2}\right) - \frac{mSx}{a+S}\right]dt + \alpha(S^0 - S) \circ dW(t), \tag{11.5}$$

$$dx = x\left(\frac{mS}{a+S} - D + \frac{\alpha^2}{2}\right)dt - \alpha x \circ dW(t). \tag{11.6}$$

In Sect. 11.2, we recall some basic results on random dynamical systems. In Sect. 11.3, we start with the study of equilibria and we prove a result related to the existence and uniqueness of global solution of (11.5) and (11.6), by using the so-called Ornstein–Uhlenbeck process. Then, we define a random dynamical system and prove the existence of a random attractor for system (11.5) and (11.6) giving an explicit expression for it. Finally, in Sect. 11.3.5 we show some numerical simulations with different values of α and we can see what happens when α increases.

11.2 Random Dynamical Systems

In this section, we present some basic results related to random dynamical systems (RDSs) and random attractors which will be necessary for our analysis. For more detailed information about RDSs and their importance, see [1].

Let $(X, \|\cdot\|_X)$ be a separable Banach space and let $(\Omega, \mathscr{F}, \mathbb{P})$ be a probability space where $\mathscr{F}$ is the σ-algebra of measurable subsets of Ω (called "events") and $\mathbb{P}$ is the probability measure. To connect the state ω in the probability space Ω at time 0 with its state after a time of t elapses, we define a flow $\theta = \{\theta_t\}_{t\in\mathbb{R}}$ on Ω with each θ_t being a mapping $\theta_t : \Omega \to \Omega$ that satisfies

(1) $\theta_0 = \mathrm{Id}_\Omega$,
(2) $\theta_s \circ \theta_t = \theta_{s+t}$ for all $s, t \in \mathbb{R}$,
(3) the mapping $(t, \omega) \mapsto \theta_t\omega$ is measurable,
(4) the probability measure $\mathbb{P}$ is preserved by θ_t, i.e., $\theta_t\mathbb{P} = \mathbb{P}$.

This setup establishes a time-dependent family θ that tracks the noise, and $(\Omega, \mathscr{F}, \mathbb{P}, \theta)$ is called a *metric dynamical system* [1].

Definition 11.1 A stochastic process $\{\varphi(t, \omega)\}_{t\geq 0, \omega\in\Omega}$ is said to be a continuous RDS over $(\Omega, \mathscr{F}, \mathbb{P}, \{\theta_t\}_{t\in\mathbb{R}})$ with state space X if $\varphi : [0, +\infty) \times \Omega \times X \to X$ is $(\mathscr{B}[0, +\infty) \times \mathscr{F} \times \mathscr{B}(X), \mathscr{B}(X))$- measurable, and for each $\omega \in \Omega$,

(i) the mapping $\varphi(t, \omega) : X \to X, x \mapsto \varphi(t, \omega)x$ is continuous for every $t \geq 0$,
(ii) $\varphi(0, \omega)$ is the identity operator on X,
(iii) (cocycle property) $\varphi(t + s, \omega) = \varphi(t, \theta_s\omega)\varphi(s, \omega)$ for all $s, t \geq 0$.

Definition 11.2 Let $(\Omega, \mathscr{F}, \mathbb{P})$ be a probability space. A random set K is a measurable subset of $X \times \Omega$ with respect to the product σ-algebra $\mathscr{B}(X) \times \mathscr{F}$.

The ω-section of a random set K is defined by

$$K(\omega) = \{x : (x, \omega) \in K\}, \quad \omega \in \Omega.$$

In the case that a set $K \subset X \times \Omega$ has closed or compact ω-sections, it is a random set as soon as the mapping $\omega \mapsto d(x, K(\omega))$ is measurable (from Ω to $[0, \infty)$) for every $x \in X$, see [7]. Then, K will be said to be a closed or a compact, respectively, random set. It will be assumed that closed random sets satisfy $K(\omega) \neq \emptyset$ for all or at least for $\mathbb{P}$-almost all $\omega \in \Omega$.

Remark 11.1 It should be noted that in the literature very often random sets are defined provided that $\omega \mapsto d(x, K(\omega))$ is measurable for every $x \in X$. Obviously, this is satisfied, for instance, when $K(\omega) = N$ for all ω, where N is some non-measurable subset of X, and also when $K = (U \times F) \cup (\overline{U} \times F^c)$ for some open set $U \subset X$ and $F \notin \mathscr{F}$. In both cases, $\omega \mapsto d(x, K(\omega))$ is constant, hence measurable, for every $x \in X$. However, both cases give $K \subset X \times \Omega$ which is not an element of the product σ-algebra $\mathscr{B}(X) \times \mathscr{F}$.

Definition 11.3 A bounded random set $K(\omega) \subset X$ is said to be tempered with respect to $\{\theta_t\}_{t\in\mathbb{R}}$ if for a.e. $\omega \in \Omega$,

$$\lim_{t\to\infty} e^{-\beta t} \sup_{x\in K(\theta_{-t}\omega)} \|x\|_X = 0, \quad \text{for all } \beta > 0;$$

a random variable $\omega \mapsto r(\omega) \in \mathbb{R}$ is said to be tempered with respect to $\{\theta_t\}_{t\in\mathbb{R}}$ if for a.e. $\omega \in \Omega$,

$$\lim_{t\to\infty} e^{-\beta t} \sup_{t\in\mathbb{R}} |r(\theta_{-t}\omega)| = 0, \quad \text{for all } \beta > 0.$$

In what follows we use $\mathscr{D}(X)$ to denote the set of all tempered random sets of X.

Definition 11.4 A random set $B(\omega) \subset X$ is called a random absorbing set in $\mathscr{D}(X)$ if for any $D \in \mathscr{D}(X)$ and a.e. $\omega \in \Omega$, there exists $T_D(\omega) > 0$ such that

$$\varphi(t, \theta_{-t}\omega)D(\theta_{-t}\omega) \subset B(\omega), \quad \forall t \geq T_D(\omega).$$

Definition 11.5 Let $\{\varphi(t, \omega)\}_{t\geq 0,\omega\in\Omega}$ be an RDS over $(\Omega, \mathscr{F}, \mathbb{P}, \{\theta_t\}_{t\in\mathbb{R}})$ with state space X and let $A(\omega)(\subset X)$ be a random set. Then, $\mathscr{A} = \{A(\omega)\}_{\omega\in\Omega}$ is called a global random $\mathscr{D}$-attractor (or pullback $\mathscr{D}$-attractor) for $\{\varphi(t, \omega)\}_{t\geq 0,\omega\in\Omega}$ if

(i) (compactness) $A(\omega)$ is a compact set of X for any $\omega \in \Omega$;
(ii) (invariance) for any $\omega \in \Omega$ and all $t \geq 0$, it holds

$$\varphi(t, \omega)A(\omega) = A(\theta_t\omega);$$

(iii) (attracting property) for any $D \in \mathscr{D}(X)$ and a.e. $\omega \in \Omega$,

$$\lim_{t\to\infty} \text{dist}_X(\varphi(t, \theta_{-t}\omega)D(\theta_{-t}\omega), A(\omega)) = 0,$$

where

$$\text{dist}_X(G, H) = \sup_{g\in G} \inf_{h\in H} \|g - h\|_X$$

is the Hausdorff semi-metric for $G, H \subseteq X$.

Proposition 11.1 ([4, 10]) *Let $B \in \mathscr{D}(X)$ be a closed absorbing set for the continuous random dynamical system $\{\varphi(t, \omega)\}_{t\geq 0,\omega\in\Omega}$ that satisfies the asymptotic compactness condition for a.e. $\omega \in \Omega$, i.e., each sequence $x_n \in \varphi(t_n, \theta_{-t_n}\omega)B(\theta_{-t_n}\omega)$ has a convergent subsequence in X when $t_n \to \infty$. Then, φ has a unique global random attractor $\mathscr{A} = \{A(\omega)\}_{\omega\in\Omega}$ with component subsets*

$$A(\omega) = \bigcap_{\tau \geq T_B(\omega)} \overline{\bigcup_{t \geq \tau} \varphi(t, \theta_{-t}\omega) B(\theta_{-t}\omega)}.$$

If the pullback absorbing set is positively invariant, i.e., $\varphi(t, \omega)B(\omega) \subset B(\theta_t\omega)$ *for all* $t \geq 0$*, then*

$$A(\omega) = \bigcap_{t \geq 0} \overline{\varphi(t, \theta_{-t}\omega) B(\theta_{-t}\omega)}.$$

Remark 11.2 When the state space $X = \mathbb{R}^d$ as in this paper, the asymptotic compactness follows trivially. Note that the random attractor is path-wise attracting in the pullback sense, but does not need to be path-wise attracting in the forward sense, although it is forward attracting in probability, due to some possible large deviations, see e.g. [1].

The next result ensures when two random dynamical systems are conjugated (see also [3, 6]).

Lemma 11.1 *Let* φ_u *be a random dynamical system on* X*. Suppose that the mapping* $T : \Omega \times X \to X$ *possesses the following properties: for fixed* $\omega \in \Omega$, $T(\omega, \cdot)$ *is a homeomorphism on* X*, and for* $x \in X$*, the mappings* $T(\cdot, x)$, $T^{-1}(\cdot, x)$ *are measurable. Then, the mapping*

$$(t, \omega, x) \to \varphi_v(t, \omega)x := T^{-1}(\theta_t\omega, \varphi_u(t, \omega)T(\omega, x))$$

is a (conjugated) random dynamical system.

11.3 Random Chemostat

In this section, we will investigate the stochastic system (11.5) and (11.6). To this end, we first transform it into differential equations with random coefficients and without white noise.

Let W be a two-sided Wiener process. Kolmogorov's theorem ensures that W has a continuous version that we will denote by ω, whose canonical interpretation is as follows: let Ω be defined by

$$\Omega = \{\omega \in \mathscr{C}(\mathbb{R}, \mathbb{R}) : \omega(0) = 0\} = \mathscr{C}_0(\mathbb{R}, \mathbb{R}),$$

$\mathscr{F}$ be the Borel σ-algebra on Ω generated by the compact open topology (see [1] for details) and $\mathbb{P}$ the corresponding Wiener measure on $\mathscr{F}$. We consider the Wiener shift flow given by

$$\theta_t\omega(\cdot) = \omega(\cdot + t) - \omega(t), \quad t \in \mathbb{R},$$

then $(\Omega, \mathscr{F}, \mathbb{P}, \{\theta_t\}_{t\in\mathbb{R}})$ is a metric dynamical system. Now let us introduce the following Ornstein–Uhlenbeck process on $(\Omega, \mathscr{F}, \mathbb{P}, \{\theta_t\}_{t\in\mathbb{R}})$

$$z^*(\theta_t\omega) = -\int_{-\infty}^{0} e^s\theta_t\omega(s)ds, \quad t \in \mathbb{R}, \quad \omega \in \Omega,$$

which solves the following Langevin equation [1, 5]

$$dz + zdt = d\omega(t), \quad t \in \mathbb{R}.$$

Proposition 11.2 ([1, 5]) *There exists a θ_t-invariant set $\tilde{\Omega} \in \mathscr{F}$ of Ω of full $\mathbb{P}$ measure such that for $\omega \in \tilde{\Omega}$, we have*

(i) the random variable $|z^*(\omega)|$ is tempered.
(ii) the mapping

$$(t, \omega) \to z^*(\theta_t\omega) = -\int_{-\infty}^{0} e^s\omega(t+s)\mathrm{d}s + \omega(t)$$

is a stationary solution of (11.7) *with continuous trajectories;*
(iii) in addition, for any $\omega \in \tilde{\Omega}$:

$$\lim_{t\to\pm\infty} \frac{|z^*(\theta_t\omega)|}{t} = 0;$$
$$\lim_{t\to\pm\infty} \frac{1}{t}\int_0^t z^*(\theta_s\omega)ds = 0;$$
$$\lim_{t\to\pm\infty} \frac{1}{t}\int_0^t |z^*(\theta_s\omega)|ds = \mathbb{E}[z^*] < \infty.$$

In what follows we will consider the restriction of the Wiener shift θ to the set $\tilde{\Omega}$, and we restrict accordingly the metric dynamical system to this set, that is also a metric dynamical system, see [6]. For simplicity, we will still denote the restricted metric dynamical system by the old symbols $(\Omega, \mathscr{F}, \mathbb{P}, \{\theta_t\}_{t\in\mathbb{R}})$.

11.3.1 Stochastic Chemostat Becomes a Random Chemostat

In what follows we use the Ornstein–Uhlenbeck process to transform (11.5) and (11.6) into a random system. Let us note that analyzing the equilibria we obtain that the only one is the axial equilibrium $(S^0, 0)$ and then we define two new variables σ and κ by

$$\sigma(t) = (S(t) - S^0)e^{\alpha z^*(\theta_t\omega)}, \tag{11.7}$$

$$\kappa(t) = x(t)e^{\alpha z^*(\theta_t\omega)}. \tag{11.8}$$

For the sake of simplicity, we will write z^* instead of $z^*(\theta_t\omega)$, and σ and κ instead of $\sigma(t)$ and $\kappa(t)$.

On the one hand, by differentiation, we have

$$\begin{aligned}
d\sigma &= e^{\alpha z^*}dS + (S - S^0)e^{\alpha z^*}\alpha dz^* \\
&= \left\{\left[(S^0 - S)\left(D + \frac{\alpha^2}{2}\right) - \frac{mSx}{a+S}\right]dt + \alpha(S^0 - S)\circ dW(t)\right\}e^{\alpha z^*} \\
&\quad + (S - S^0)e^{\alpha z^*}\alpha\left\{-z^*dt + dW(t)\right\} \\
&= (S^0 - S)\left(D + \frac{\alpha^2}{2}\right)e^{\alpha z^*}dt - \frac{mSx}{a+S}e^{\alpha z^*}dt + \alpha(S^0 - S)e^{\alpha z^*}\circ dW(t) \\
&\quad - (S - S^0)\alpha e^{\alpha z^*}z^*dt + (S - S^0)e^{\alpha z^*}\alpha \circ dW(t) \\
&= \left[-\left(D + \frac{\alpha^2}{2}\right)\sigma - \frac{mS\kappa}{a+S} - \alpha\sigma z^*\right]dt \\
&= \left[-\left(D + \frac{\alpha^2}{2}\right)\sigma - \frac{m(S^0 + \sigma e^{-\alpha z^*})}{a + S^0 + \sigma e^{-\alpha z^*}}\kappa - \alpha\sigma z^*\right]dt.
\end{aligned}$$

On the other hand,

$$\begin{aligned}
d\kappa &= e^{\alpha z^*}dx + xe^{\alpha z^*}\alpha dz^* \\
&= \left[x\left(\frac{mS}{a+S} - D + \frac{\alpha^2}{2}\right)dt - \alpha x \circ dW(t)\right]e^{\alpha z^*} + \alpha xe^{\alpha z^*}\left[-z^*dt + dW(t)\right] \\
&= \frac{xmS}{a+S}e^{\alpha z^*}dt + x\left(-D + \frac{\alpha^2}{2}\right)e^{\alpha z^*}dt - \alpha xe^{\alpha z^*}\circ dW(t) \\
&\quad - \alpha xz^*e^{\alpha z^*}dt + \alpha xe^{\alpha z^*}\circ dW(t) \\
&= \left[\frac{m(S^0 + \sigma e^{-\alpha z^*})}{a + S^0 + \sigma e^{-\alpha z^*}}\kappa - \left(D - \frac{\alpha^2}{2}\right)\kappa - \alpha z^*\kappa\right]dt.
\end{aligned}$$

Thus, we have obtained the following random system

$$\frac{d\sigma}{dt} = -(\bar{D} + \alpha z^*)\sigma - \frac{m(S^0 + \sigma e^{-\alpha z^*})}{a + S^0 + \sigma e^{-\alpha z^*}}\kappa, \tag{11.9}$$

$$\frac{d\kappa}{dt} = -(\widetilde{D} + \alpha z^*)\kappa + \frac{m(S^0 + \sigma e^{-\alpha z^*})}{a + S^0 + \sigma e^{-\alpha z^*}}\kappa, \tag{11.10}$$

where $\bar{D} := D + \frac{\alpha^2}{2}$ and $\widetilde{D} := D - \frac{\alpha^2}{2}$.

11.3.2 Random Chemostat Generates an RDS

Next we prove that the random chemostat system (11.9) and (11.10) generates an RDS. From now on, we denote $\mathscr{X} := \{(x, y) \in \mathbb{R}^2 : x \in \mathbb{R}, y \geq 0\}$, the upper half-plane.

Lemma 11.2 *Assume that*

$$D \geq \frac{\alpha^2}{2}, \qquad \tilde{\lambda} := \frac{\tilde{D}a}{m - \tilde{D}} \geq S^0. \tag{11.11}$$

Then for any $\omega \in \Omega$ and any initial value $u_0 := (\sigma_0, \kappa_0) \in \mathscr{X}$, where $\sigma_0 := \sigma(0)$ and $\kappa_0 := \kappa(0)$, system (11.9) and (11.10) possesses a unique global solution $u(\cdot; \omega, u_0) := (\sigma(\cdot; \omega, u_0), \kappa(\cdot; \omega, u_0)) \in \mathscr{C}^1([0, +\infty), \mathscr{X})$ with $u(0; \omega, u_0) = u_0$. Moreover, the solution mapping generates a random dynamical system $\varphi_u : \mathbb{R}^+ \times \Omega \times \mathscr{X} \to \mathscr{X}$ defined as

$$\varphi_u(t, \omega)u_0 = u(t; \omega, u_0), \quad \forall t \in \mathbb{R}^+, \; u_0 \in \mathscr{X}, \; \omega \in \Omega.$$

Proof Observe that we can rewrite one of the terms in the previous equations as

$$\frac{m(S^0 + \sigma e^{-\alpha z^*})}{a + S^0 + \sigma e^{-\alpha z^*}}\kappa = \frac{m(S^0 + \sigma e^{-\alpha z^*} + a - a)}{a + S^0 + \sigma e^{-\alpha z^*}}\kappa = m\kappa - \frac{ma\kappa}{a + S^0 + \sigma e^{-\alpha z^*}}$$

and therefore, system (11.9) and (11.10) turns into

$$\frac{d\sigma}{dt} = -(\bar{D} + \alpha z^*)\sigma - m\kappa + \frac{ma}{a + S^0 + \sigma e^{-\alpha z^*}}\kappa, \tag{11.12}$$

$$\frac{d\kappa}{dt} = -(\tilde{D} + \alpha z^*)\kappa + m\kappa - \frac{ma}{a + S^0 + \sigma e^{-\alpha z^*}}\kappa. \tag{11.13}$$

Denoting $u(\cdot; \omega, u_0) := (\sigma(\cdot; \omega, u_0), \kappa(\cdot; \omega, u_0))$, system (11.12) and (11.13) can be rewritten as

$$\frac{du}{dt} = L(\theta_t\omega) \cdot u + F(u, \theta_t\omega),$$

where

$$L(\theta_t\omega) = \begin{pmatrix} -(\bar{D} + \alpha z^*) & -m \\ 0 & -(\tilde{D} + \alpha z^*) + m \end{pmatrix}$$

and $F : \mathscr{X} \times [0, +\infty) \longrightarrow \mathbb{R}^2$ is given by

$$F(\xi, \theta_t\omega) = \begin{pmatrix} \dfrac{ma}{a + S^0 + \xi_1 e^{-\alpha z^*}}\xi_2 \\ \dfrac{-ma}{a + S^0 + \xi_1 e^{-\alpha z^*}}\xi_2 \end{pmatrix},$$

where $\xi = (\xi_1, \xi_2) \in \mathscr{X}$.

Since $z^*(\theta_t\omega)$ is continuous, L generates an evolution system on $\mathbb{R}^2$. Moreover, we notice that

$$\frac{\partial}{\partial \xi_2}\left[\pm\frac{am}{a + S^0 + \xi_1 e^{-\alpha z^*}}\xi_2\right] = \pm\frac{am}{a + S^0 + \xi_1 e^{-\alpha z^*}}$$

and

$$\frac{\partial}{\partial \xi_1}\left[\pm\frac{am}{a + S^0 + \xi_1 e^{-\alpha z^*}}\xi_2\right] = \mp\frac{ame^{-\alpha z^*}}{(a + S^0 + \xi_1 e^{-\alpha z^*})^2}\xi_2$$

so $F(\cdot, \theta_t\omega) \in \mathscr{C}(\mathscr{X} \times [0, +\infty); \mathbb{R}^2)$ and is continuously differentiable with respect to the variables (ξ_1, ξ_2), which implies that it is locally Lipschitz with respect to $(\xi_1, \xi_2) \in \mathscr{X}$.

Therefore, thanks to classical results from the theory of ordinary differential equations, system (11.12) and (11.13) possesses a unique local solution. Let us check now that in fact this solution is a global one. In order to do that we split our analysis into two different cases: first, we assume $\sigma(t) \geq 0$ for all $t \geq 0$. Thus, from (11.9) and (11.10)

$$\begin{aligned} \frac{d}{dt}(\sigma + \kappa) &= -\bar{D}\sigma - \alpha z^*\sigma - \widetilde{D}\kappa - \alpha z^*\kappa \\ &\leq -\widetilde{D}\sigma - \alpha z^*\sigma - \widetilde{D}\kappa - \alpha z^*\kappa \\ &= -(\widetilde{D} + \alpha z^*)(\sigma + \kappa). \end{aligned}$$

Hence

$$\sigma(t) + \kappa(t) \leq (\sigma(0) + \kappa(0))e^{-\widetilde{D}t - \alpha\int_0^t z^*(\theta_s\omega)ds},$$

so $\sigma + \kappa$ tends to zero when t goes to infinity since $D \geq \frac{\alpha^2}{2}$, i.e., $\widetilde{D} \geq 0$.

Moreover, since $S^0 + \sigma e^{-\alpha z^*} = S \geq 0$, we have

$$\begin{aligned} \frac{d\sigma}{dt} &= -(\bar{D} + \alpha z^*)\sigma - \frac{m(S^0 + \sigma e^{-\alpha z^*})}{a + S^0 + \sigma e^{-\alpha z^*}}\kappa \\ &\leq -(\bar{D} + \alpha z^*)\sigma \end{aligned}$$

and solving this differential equation, we obtain

$$\sigma(t) \leq \sigma(0)e^{-\bar{D}t+\alpha\int_0^t z^*(\theta_s\omega)ds},$$

which implies that σ always tends to zero when t goes to infinity, because $\bar{D} \geq 0$.

Summing up, we have

$$0 \leq \sigma(t) \longrightarrow 0, \quad \text{when} \quad t \uparrow +\infty,$$

$$0 \leq \sigma(t) + \kappa(t) \longrightarrow 0, \quad \text{when} \quad t \uparrow +\infty,$$

since $D \geq \frac{\alpha^2}{2}$, so we have

$$0 \leq \kappa(t) = (\sigma(t) + \kappa(t)) - \sigma(t) \longrightarrow 0, \quad \text{when} \quad t \uparrow +\infty.$$

In particular, σ and κ are bounded.

Now, we assume there exists some $\widetilde{t} \geq 0$ such that $\sigma(\widetilde{t}) < 0$. In this case, there exists t^* such that $\sigma(t^*) = 0$ and then

$$\begin{aligned}\frac{d\sigma}{dt}(t^*) &= \left[-(\bar{D} + \alpha z^*)\sigma - \frac{m(S^0 + \sigma e^{-\alpha z^*})}{a + S^0 + \sigma e^{-\alpha z^*}}\kappa\right](t^*) \\ &= -\frac{mS^0}{a + S^0}\kappa(t^*) < 0.\end{aligned}$$

Therefore, we have $\sigma(t) < 0$ for all $t > t^*$, and from (11.7), we get that $S(t) < S^0$, for all $t > t^*$.

Now, since the mapping $f(S) := \frac{mS}{a+S}$ is an increasing function, then $f(S(t)) < f(S^0)$, for all $t > t^*$, i.e., we have

$$\frac{mS}{a + S} < \frac{mS^0}{a + S^0}.$$

Hence, from (11.9) and (11.10)

$$\begin{aligned}\frac{d\sigma}{dt} &= -(\bar{D} + \alpha z^*)\sigma - \frac{m(S^0 + \sigma e^{-\alpha z^*})}{a + S^0 + \sigma e^{-\alpha z^*}}\kappa \\ &\leq -(\bar{D} + \alpha z^*)\sigma\end{aligned}$$

and for $t > t^*$

$$\begin{aligned} \frac{d\kappa}{dt} &= -(\widetilde{D} + \alpha z^*)\kappa + \frac{m(S^0 + \sigma e^{-\alpha z^*})}{a + S^0 + \sigma e^{-\alpha z^*}}\kappa \\ &= -(\widetilde{D} + \alpha z^*)\kappa + \frac{mS}{a + S}\kappa \\ &< -(\widetilde{D} + \alpha z^*)\kappa + \frac{mS^0}{a + S^0}\kappa \\ &= -\left(\widetilde{D} - \frac{mS^0}{a + S^0} + \alpha z^*\right)\kappa, \end{aligned} \tag{11.14}$$

thus

$$\sigma(t) \leq \sigma(0)e^{-\bar{D}t - \alpha\int_0^t z^*(\theta_s\omega)ds}$$

and for $t > t^*$

$$\kappa(t) < \kappa(0)e^{-\left(\widetilde{D} - \frac{mS^0}{a+S^0}\right)t - \alpha\int_0^t z^*(\theta_s\omega)ds}.$$

Summing up, in this second case σ and κ also keep bounded because of the assumption $\tilde{\lambda} \geq S^0$.

Therefore, the unique local solution to system (11.12) and (11.13) can be extended to a unique global solution.

Notice that, although σ remains negative, it will never make vanish the denominator $a + S^0 + \sigma e^{-\alpha z^*}$. Indeed, if we suppose that there exists $\bar{t} > t^* > 0$ such that

$$a + S^0 + \sigma(\bar{t})e^{-\alpha z^*(\theta_{\bar{t}}\omega)} = 0,$$

then for every $M > 0$ given, there exists $t_M \in (t^*, \bar{t})$ such that

$$\frac{m(S^0 + \sigma(t)e^{-\alpha z^*(\theta_t\omega)})}{a + S^0 + \sigma(t)e^{-\alpha z^*(\theta_t\omega)}} \geq M$$

for all $t \in (t_M, \bar{t}]$.

Hence, κ satisfies the following differential inequality

$$\frac{d\kappa}{dt} \geq -(\widetilde{D} - M + \alpha z^*)\kappa, \tag{11.15}$$

thus, if we choose $M > \frac{mS^0}{a+S^0}$ and evaluate the solution of (11.15) that starts in t_M in the instant $\bar{t}$, we obtain

$$\kappa(\bar{t}) > \kappa(t_M)e^{-\left(\widetilde{D} - \frac{mS^0}{a+S^0}\right)(\bar{t} - t_M) - \alpha\int_{t_M}^{\bar{t}} z^*(\theta_s\omega)ds}. \tag{11.16}$$

On the other hand, by solving (11.14) and evaluating it in $t = \bar{t}$, we have

$$\kappa(\bar{t}) \leq \kappa(t_M) e^{-\left(\tilde{D} - \frac{mS^0}{a+S^0}\right)(\bar{t}-t_M) - \alpha \int_{t_M}^{\bar{t}} z^*(\theta_s \omega) ds}, \tag{11.17}$$

which clearly contradicts (11.16).

As a consequence, we deduce that for all $t \in \mathbb{R}$

$$\sigma(t) > -(a + S^0) e^{\alpha z^*(\theta_t \omega)}.$$

Now we would like to check that this global solution belongs to the set $\mathscr{X}$ for any $t \in \mathbb{R}^+$. If there exists $t \in \mathbb{R}^+$ such that $\kappa(t) = 0$, assuming $\sigma(0) > 0$, we have

$$\begin{aligned} \frac{d\sigma}{dt}(t) &= \left[-(\bar{D} + \alpha z^*)\sigma - \frac{m(S^0 + \sigma e^{-\alpha z^*})}{a + S^0 + \sigma e^{-\alpha z^*}} \kappa \right](t) \\ &= -(\bar{D} + \alpha z^*)\sigma(t), \end{aligned}$$

and therefore,

$$\sigma(t) = \sigma(0) e^{-\bar{D}t - \alpha \int_0^t z^*(\theta_s \omega) ds},$$

which, since $\bar{D} \geq 0$, implies that

$$\lim_{t \uparrow +\infty} \sigma(t) = 0 \quad \text{and} \quad \lim_{t \downarrow -\infty} \sigma(t) = +\infty.$$

Similarly, assuming $\kappa(t) = 0$ and $\sigma(0) < 0$, we obtain

$$\lim_{t \uparrow +\infty} \sigma(t) = 0 \quad \text{and} \quad \lim_{t \downarrow -\infty} \sigma(t) = -\infty.$$

By the previous analysis, we deduce that for any initial data $u_0 \in \mathscr{X}$, the solution $u(t)$ remains in $\mathscr{X}$.

Now we can define the mapping $\varphi_u : \mathbb{R}^+ \times \Omega \times \mathscr{X} \to \mathscr{X}$ given by

$$\varphi_u(t, \omega)u_0 := u(t; \omega, u_0), \quad \forall t \geq 0, \; u_0 \in \mathscr{X}, \; \omega \in \Omega.$$

Since the function F is continuous in u, t, and is measurable in ω, we obtain the $(\mathscr{B}[0, +\infty) \times \mathscr{F} \times \mathscr{B}(\mathscr{X}), \mathscr{B}(\mathscr{X}))$-measurability of the previous mapping. Items (i), (ii), and (iii) in Definition 11.1 follow easily by the definition of φ_u.

11.3.3 Existence of the Random Attractor

Now, we study the existence of a random attractor, describing it explicitly.

Lemma 11.3 *Under the assumption (11.11), there exists a tempered compact random absorbing set $B_\varepsilon(\omega) \in \mathscr{D}(\mathscr{X})$, for all $\varepsilon > 0$, of the random dynamical system $\{\varphi_u(t,\omega)\}_{t\geq 0,\,\omega\in\Omega}$, that is, for any $D \in \mathscr{D}(\mathscr{X})$ and each $\omega \in \Omega$, there exists $T_D(\omega) > 0$ such that*

$$\varphi_u(t,\theta_{-t}\omega)D(\theta_{-t}\omega) \subset B_\varepsilon(\omega) \quad \forall t \geq T_D(\omega).$$

Proof Recall that $\varphi_u(t,\omega)u_0 = u(t;\omega,u_0)$ denotes the solution of system (11.12) and (11.13), satisfying $u(0;\omega,u_0) = u_0$, where $u_0 := u_0(\theta_{-t}\omega) \in D(\theta_{-t}\omega)$.

First we assume that $\sigma(t) \geq 0$ for all $t \geq 0$ and define $\|\cdot\|_1$ as

$$\begin{aligned}\|\varphi_u(t,\theta_{-t}\omega)u_0\|_1 &= \|u(t;\theta_{-t}\omega,u_0(\theta_{-t}\omega))\|_1\\ &= \sigma(t;\theta_{-t}\omega,u_0(\theta_{-t}\omega)) + \kappa(t;\theta_{-t}\omega,u_0(\theta_{-t}\omega)).\end{aligned}$$

Note that

$$\begin{aligned}&\sigma(t;\theta_{-t}\omega,u_0(\theta_{-t}\omega)) + \kappa(t;\theta_{-t}\omega,u_0(\theta_{-t}\omega))\\ &\leq \sup_{(\sigma_0,\kappa_0)\in D(\theta_{-t}\omega)} \{\sigma_0+\kappa_0\} e^{-\widetilde{D}t-\alpha\int_0^t z^*(\theta_s\theta_{-t}\omega)ds}\\ &= \sup_{(\sigma_0,\kappa_0)\in D(\theta_{-t}\omega)} \{\sigma_0+\kappa_0\} e^{-\widetilde{D}t-\alpha\int_{-t}^0 z^*(\theta_s\omega)ds}.\end{aligned}$$

Therefore, thanks to the temperedness of $D(\omega)$ and (11.11), there exists $T_D(\omega)$ such that $\|u(t;\theta_{-t}\omega,u_0(\theta_{-t}\omega))\|_1 \leq \varepsilon$, for all $\varepsilon > 0, u_0 \in D(\theta_{-t}\omega)$, when $t > T_D(\omega)$.

Define

$$B_\varepsilon^1(\omega) := \{(\sigma,\kappa) \in \mathscr{X} \,:\, 0 \leq \sigma+\kappa \leq \varepsilon\},$$

then $B_\varepsilon^1(\omega)$ is absorbing in $\mathscr{X}$.

Now we assume that there exists some $\widetilde{t} \geq 0$ such that $\sigma(\widetilde{t}) < 0$. In this case, we proved that $\sigma(t) < 0$ for all $t \geq \widetilde{t}$. We now get

$$\begin{aligned}\sigma(t;\theta_{-t}\omega,u_0(\theta_{-t}\omega)) &\leq \sup_{(\sigma_0,\kappa_0)\in D(\theta_{-t}\omega)} \{\sigma_0\} e^{-\bar{D}t-\alpha\int_0^t z^*(\theta_s\theta_{-t}\omega)ds}\\ &= \sup_{(\sigma_0,\kappa_0)\in D(\theta_{-t}\omega)} \{\sigma_0\} e^{-\bar{D}t-\alpha\int_{-t}^0 z^*(\theta_s\omega)ds}\end{aligned}$$

and

$$\kappa(t;\theta_{-t}\omega,u_0(\theta_{-t}\omega)) \le \sup_{(\sigma_0,\kappa_0)\in D(\theta_{-t}\omega)} \{\kappa_0\} e^{-\left(\tilde{D}-\frac{mS^0}{a+S^0}\right)t-\alpha\int_0^t z^*(\theta_s\theta_{-t}\omega)ds}$$
$$= \sup_{(\sigma_0,\kappa_0)\in D(\theta_{-t}\omega)} \{\kappa_0\} e^{-\left(\tilde{D}-\frac{mS^0}{a+S^0}\right)t-\alpha\int_{-t}^0 z^*(\theta_s\omega)ds}.$$

Therefore, thanks to the temperedness of $D(\omega)$ and (11.11), there exists $T_D(\omega)$ such that

$$\sigma(t;\theta_{-t}\omega,u_0(\theta_{-t}\omega)) \le \varepsilon \quad \text{and} \quad \kappa(t;\theta_{-t}\omega,u_0(\theta_{-t}\omega)) \le \varepsilon$$

for all $\varepsilon > 0$, $u_0 \in D(\theta_{-t}\omega)$, when $t > T_D(\omega)$.

On the other hand, from (11.9) and (11.10) we always have

$$\frac{d(\sigma+\kappa)}{dt} \ge -(\bar{D}+\alpha z^*)(\sigma+\kappa),$$

thus

$$(\sigma+\kappa)(t) \ge (\sigma+\kappa)(0)e^{-\bar{D}t-\alpha\int_0^t z^*(\theta_s\omega)ds}$$

which tends to zero when t goes to infinity since $\bar{D} \ge 0$.

Hence, $\sigma+\kappa \ge 0$ iff $\sigma \ge -\kappa$, thus

$$\sigma(t;\theta_{-t}\omega,\sigma_0(\theta_{-t}\omega)) \ge -\varepsilon$$

for all $\varepsilon > 0$, $u_0 \in D(\theta_{-t}\omega)$, when $t > T_D(\omega)$.

We define

$$B_\varepsilon^2(\omega) := \{(\sigma,\kappa) \in \mathscr{X} \ : \ -\varepsilon \le \sigma \le \varepsilon,\ 0 \le \kappa \le \varepsilon\},$$

then $B_\varepsilon^2(\omega)$ is absorbing in $\mathscr{X}$.

In conclusion, considering

$$B_\varepsilon(\omega) = B_\varepsilon^1(\omega) \cup B_\varepsilon^2(\omega) = B_\varepsilon^2(\omega),$$

it follows directly from Proposition 11.1 that the random dynamical system generated by the system (11.12) and (11.13) possesses a unique random attractor given by

$$\mathscr{A} = \{A(\omega)\}_{\omega\in\Omega} \subset B_\varepsilon(\omega), \quad \text{for all } \varepsilon > 0,$$

thus

$$\mathscr{A} = \{A(\omega)\}_{\omega\in\Omega} = \{(0,0)\}.$$

11.3.4 Existence of the Random Attractor for the Stochastic Chemostat System

We have proved that the system (11.9) and (11.10) has a unique global solution $u(t;\omega,u_0)$ which remains in $\mathscr{X}$ for all $u_0 \in \mathscr{X}$ and generates the RDS φ_u.

Now, we define a mapping

$$T : \Omega \times \mathscr{X} \longrightarrow \mathscr{X}$$

as follows

$$T(\omega,\zeta) = T(\omega,(\zeta_1,\zeta_2)) = \begin{pmatrix} T_1(\omega,\zeta_1) \\ T_2(\omega,\zeta_2) \end{pmatrix} = \begin{pmatrix} (\zeta_1 - S^0)e^{\alpha z^*(\omega)} \\ \zeta_2 e^{\alpha z^*(\omega)} \end{pmatrix}$$

whose inverse is given by

$$T^{-1}(\omega,\zeta) = \begin{pmatrix} S^0 + \zeta_1 e^{-\alpha z^*(\omega)} \\ \zeta_2 e^{-\alpha z^*(\omega)} \end{pmatrix}.$$

We know that $v(t) = (S(t), x(t))$ and $u(t) = (\sigma(t), \kappa(t))$ are related by (11.7) and (11.8). Since T is a homeomorphism, thanks to Lemma 11.1 we obtain a conjugated RDS given by

$$\begin{aligned} \varphi_v(t,\omega)v_0 &:= T^{-1}(\theta_t\omega, \varphi_u(t,\omega)T(\omega,v_0)) \\ &= T^{-1}\left(\theta_t\omega, \varphi_u(t,\omega)\begin{pmatrix} (S(0) - S^0)e^{\alpha z^*(\omega)} \\ x(0)e^{\alpha z^*(\omega)} \end{pmatrix}\right) \\ &= T^{-1}(\theta_t\omega, \varphi_u(t,\omega)u_0) \\ &= T^{-1}(\theta_t\omega, u(t;\omega,u_0)) \\ &= \begin{pmatrix} S^0 + \sigma(t)e^{-\alpha z^*(\theta_t\omega)} \\ \kappa(t)e^{-\alpha z^*(\theta_t\omega)} \end{pmatrix} \\ &= v(t;\omega,v_0) \end{aligned}$$

which means that φ_v is an RDS for our original stochastic system (11.5) and (11.6).

Moreover, the global random attractor of the random system (11.9) and (11.10)

$$\mathscr{A} = \{A(\omega)\}_{\omega\in\Omega} = \{(0,0)\}$$

becomes

$$\widetilde{\mathscr{A}} = \{\widetilde{A}(\omega)\}_{\omega\in\Omega} = \{(S^0, 0)\},$$

the global random attractor of the stochastic system (11.5) and (11.6).

11.3.5 Numerical Simulations and Final Comments

To confirm the results above, in this section we show some numerical simulations for (11.3) and (11.4). We use the Euler–Maruyama method [14] considering an initial value $(S_0, x_0) = (5, 10)$, $S^0 = 1$, $D = 3$, $a = 0.6$, $m = 3$ and the following numerical scheme:

$$S_j = S_{j-1} + f(x_{j-1}, S_{j-1})\Delta t + g(x_{j-1}, S_{j-1}) \cdot (W(\tau_j) - W(\tau_{j-1})),$$
$$x_j = x_{j-1} + \widetilde{f}(x_{j-1}, S_{j-1})\Delta t + \widetilde{g}(x_{j-1}, S_{j-1}) \cdot (W(\tau_j) - W(\tau_{j-1})),$$

where we define functions f, g, $\widetilde{f}$, and $\widetilde{g}$ as

$$\begin{aligned} f(x_{j-1}, S_{j-1}) &= \left[(S^0 - S_{j-1})D - \frac{mS_{j-1}x_{j-1}}{a + S_{j-1}} \right], \\ g(x_{j-1}, S_{j-1}) &= \alpha(S^0 - S_{j-1}), \\ \widetilde{f}(x_{j-1}, S_{j-1}) &= x_{j-1}\left(\frac{mS_{j-1}}{a + S_{j-1}} - D \right), \\ \widetilde{g}(x_{j-1}, S_{j-1}) &= \alpha x_{j-1}, \end{aligned}$$

and

$$W(\tau_j) - W(\tau_{j-1}) = \sum_{k=jR-R+1}^{jR} dW_k,$$

where R is a non-negative integer number and dW_k are $\mathscr{N}(0, 1)$-distributed independent random variables which can be generated numerically by pseudorandom number generators.

From now on, the red lines in the pictures represent the stochastic solutions of system (11.3) and (11.4) and the blue ones the deterministic solutions of the same system.

By the previous sections, we know that system (11.3) and (11.4) possesses a random attractor given by $\tilde{\mathscr{A}} = \{(S^0, 0)\}$ as long as (11.11) is satisfied. For the following different values of α, we obtain the following values of $\tilde{\lambda}$:

(a) **Case** $\alpha = 0.1$:

$$\tilde{\lambda} := \frac{\widetilde{D}a}{m - \widetilde{D}} = 359.4 \geq 1 = S^0.$$

(b) **Case** $\alpha = 0.5$:

$$\tilde{\lambda} := \frac{\widetilde{D}a}{m - \widetilde{D}} = 13.8 \geq 1 = S^0.$$

(c) **Case** $\alpha = 1$:

$$\widetilde{\lambda} := \frac{\widetilde{D}a}{m - \widetilde{D}} = 3 \geq 1 = S^0.$$

(d) **Case** $\alpha = 1.5$:

$$\widetilde{\lambda} := \frac{\widetilde{D}a}{m - \widetilde{D}} = 1 \geq 1 = S^0.$$

Summing up, in all the above cases $\widetilde{\lambda} \geq S^0$ and $D \geq \frac{\alpha^2}{2}$ hold; hence, the solutions of system (11.3) and (11.4) for the previous values of the parameters go to $(S^0, 0) = (1, 0)$, the random attractor.

The following pictures show what we expected from the theory and numerical computing and we also can observe what happens when the intensity of noise increases (Figs. 11.1 and 11.2).

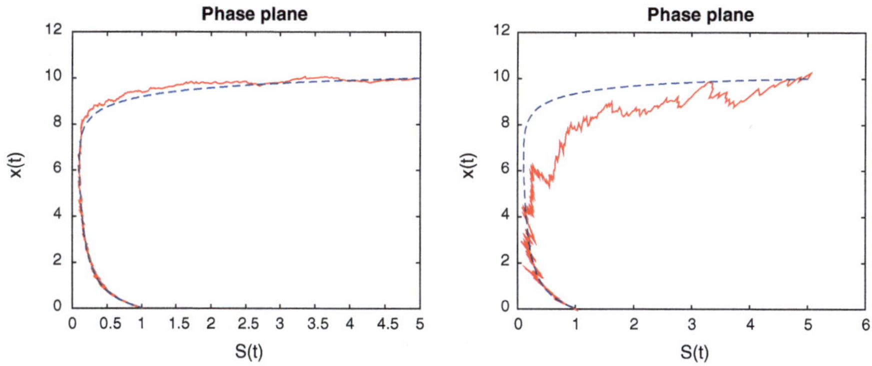

Fig. 11.1 $\alpha = 0.1$ on the *left* and $\alpha = 0.5$ on the *right*

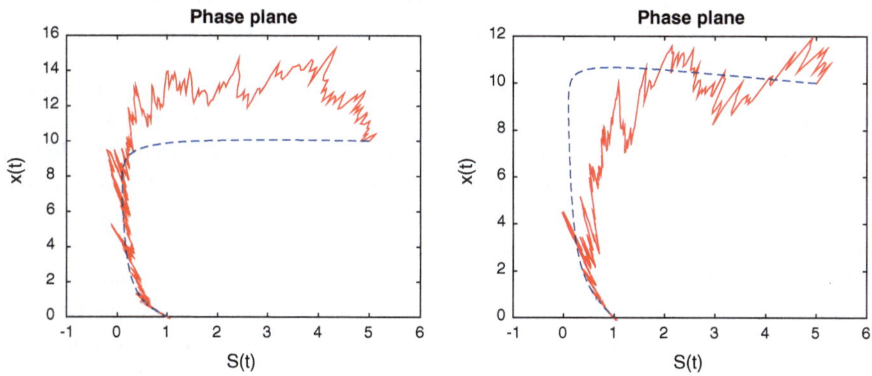

Fig. 11.2 $\alpha = 1$ on the *left* and $\alpha = 1.5$ on the *right*

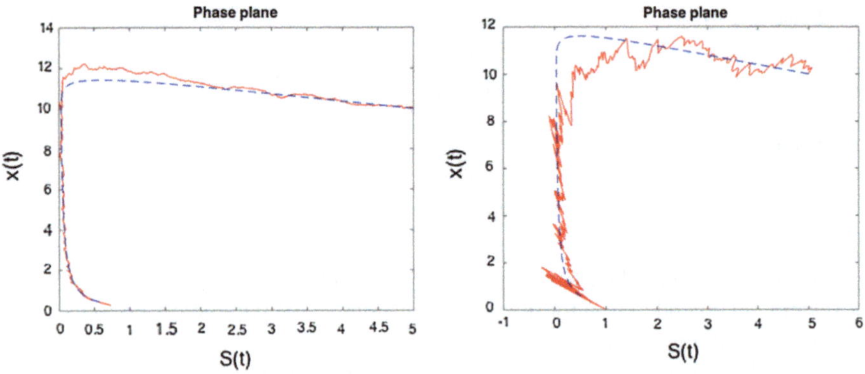

Fig. 11.3 $\alpha = 0.1$ on the *left* and $\alpha = 0.5$ on the *right*

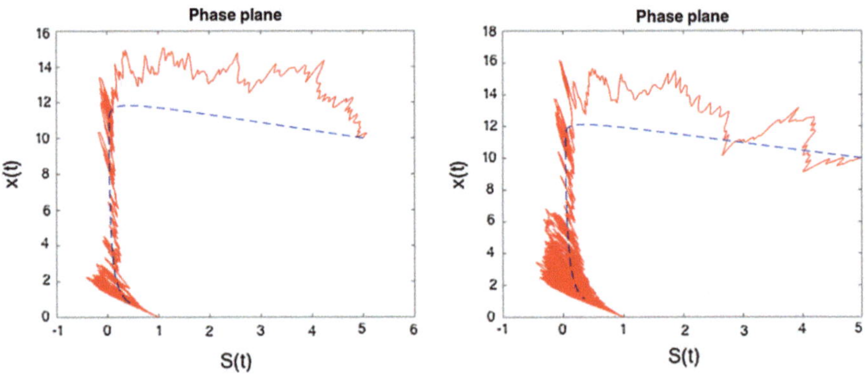

Fig. 11.4 $\alpha = 0.7$ on the *left* and $\alpha = 0.9$ on the *right*

However, the other pictures show what happens if $\widetilde{\lambda} < S^0$ holds true. In this case, $D = 1.5$ instead of $D = 3$ as in the previous cases (Figs. 11.3 and 11.4).

Remark 11.3 We would like to mention that the fact that the substrate S (or its corresponding σ) may take negative values does not produce any mathematical inconsistence in our analysis, in other words, our mathematical analysis is accurate to handle the mathematical problem. However, from a biological point of view, this may reflect some troubles and suggest that either the fact of perturbing the dilution rate with an additive noise may not be a realistic situation, or that we should try to use a some kind of switching system to model our real chemostat in such a way that when the dilution may be negative, we use a different equation to model the system. This will lead us to a different analysis in some subsequent papers by considering a different kind of randomness or stochasticity in this parameter or designing a different model for our problem.

On the other hand, it could also be considered a noisy term in each equation of the deterministic model in the same fashion as in the paper by Imhof and Walcher [15], which ensures the positivity of both the nutrient and biomass, although does not preserve the washout equilibrium from the deterministic to the stochastic model. We are currently interested on this kind of chemostat models and we will analyze them in future papers.

Acknowledgments This paper was partially supported by FEDER and Ministerio de Economía y Competitividad under Grant MTM2015-63723-P and Junta de Andalucía under Proyecto de Excelencia P12-FQM-1492. We also would like to thank Alain Rapaport and Stefanie Sonner for the nice discussions that we had with them during the final writing of the paper. Thanks to their helpful suggestions we were able to improve the preliminary version of this paper. Finally, we are really grateful to the referee for the kind comments and useful suggestions which helped us to improve the paper.

References

1. Arnold, L.: Random Dynamical Systems. Springer, Berlin (1998)
2. Bungay, H.R., Bungay, M.L.: Microbial interactions in continuous culture. Adv. Appl. Microbiol. **10**, 269–290 (1968)
3. Caraballo, T., Kloeden, P.E., Schmalfuß, B.: Exponentially stable stationary solutions for stochastic evolution equations and their perturbation. Appl. Math. Optim. **50**, 183–207 (2004)
4. Caraballo, T., Lukaszewicz, G., Real, J.: Pullback attractors for asymptotically compact nonautonomous dynamical systems. Nonlinear Anal. TMA **6**, 484–498 (2006)
5. Caraballo, T., Lu, K.: Attractors for stochastic lattice dynamical systems with a multiplicative noise. Front. Math. China **3**(3), 317–335 (2008)
6. Caraballo, T., Garrido-Atienza, M.J., Schmalfuß, B., Valero, J.: Asymptotic behaviour of a stochastic semilinear dissipative functional equation without uniqueness of solutions. Discret. Contin. Dyn. Syst. Ser. B **14**(2), 439–455 (2010)
7. Crauel, H.: Random Probability Measures on Polish Spaces. Taylor & Francis, London (2002)
8. Cunningham, A., Nisbet, R.M.: Transients and Oscillations in Continuous Cultures. Mathematics in Microbiology. Academic Press, London (1983)
9. D'ans, G., Kokotovic, P.V., Gottlieb, D.: A nonlinear regulator problem for a model of biological waste treatment. IEEE Trans. Autom. Control **AC–16**, 341–347 (1971)
10. Flandoli, F., Schmalfuß, B.: Random attractors for the 3D stochastic Navier–Stokes equation with multiplicative noise. Stoch. Stoch. Rep. **59**(1–2), 21–45 (1996)
11. Fredrickson, A.G., Stephanopoulos, G.: Microbial competition. Science **213**(4511), 972–979 (1981)
12. Freter, R.: Mechanisms that Control the Microflora in the Large Intestine. In: Hentges, D.J. (ed.) Human Intestinal Microflora in Health and Disease, pp. 33–54. Academic Press, New York (1983)
13. Freter, R.: An understanding of colonization of the large intestine requires mathematical analysis. Microecol. Ther. **16**, 147–155 (1986)
14. Higham, D.J.: An algorithmic introduction to numerical simulation of stochastic differential equations. SIAM Rev. **43**(3), 525–546 (2001)
15. Imhof, L., Walcher, S.: Exclusion and persistence in deterministic and stochastic chemostat models. J. Differ. Equ. **217**, 26–53 (2005)
16. Jannash, H.W., Mateles, R.T.: Experimental bacterial ecology studies in continuous culture. Adv. Microb. Physiol. **11**, 165–212 (1974)
17. La Riviere, J.W.M.: Microbial ecology of liquid waste. Adv. Microb. Ecol. **1**, 215–259 (1977)

18. Smith, H.L.: Monotone Dynamical Systems: an Introduction to the Theory of Competitive and Cooperative Systems, Mathematical Surveys and Monographs 41. American Mathematical Society, Providence (1995)
19. Smith, H.L., Waltman, P.: The Theory of the Chemostat: Dynamics of Microbial Competition. Cambridge University Press, Cambridge (1995)
20. Sree Hari Rao, V., Raja Sekhara Rao, P.: Dynamic Models and Control of Biological Systems. Springer, Heidelberg (2009)
21. Taylor, P.A., Williams, J.L.: Theoretical studies on the coexistence of competing species under contunous flow conditions. Can. J. Microbiol. **21**, 90–98 (1975)
22. Veldcamp, H.: Ecological studies with the chemostat. Adv. Microb. Ecol. **1**, 59–95 (1977)
23. Waltman, P.: Competition Models in Population Biology, CBMS-NSF Regional Conference Series in Applied Mathematics 45. Society for Industrial and Applied Mathematics, Philadelphia (1983)
24. Waltman, P.: Coexistence in chemostat-like model. Rocky Mt. J. Math. **20**, 777–807 (1990)
25. Waltman, P., Hubbel, S.P., Hsu, S.B.: Theoretical and Experimental Investigations of Microbial Competition in Continuous Culture. Modeling and Differential Equations in Biology (Conference on southern Illinois Univ. Carbonadle, III., 1978), Lecture Notes in Pure and Applied Mathematics, vol. 58, pp. 107–152. Dekker, New York (1980)
26. Xu, C., Yuan, S., Zhang, T.: Asymptotic Behaviour of a Chemostat Model with Stochastic Perturbation on the Dilution Rate. Abstract and Applied Analysis. Hindawi Publishing Corporation (2013)

Chapter 12
Higher-Order Allen–Cahn Models with Logarithmic Nonlinear Terms

Laurence Cherfils, Alain Miranville and Shuiran Peng

Abstract Our aim in this chapter was to study higher-order (in space) Allen–Cahn models with logarithmic nonlinear terms. In particular, we obtain well-posedness results, as well as the existence of the global attractor.

12.1 Introduction

The Allen–Cahn equation describes the ordering of atoms during the phase separation of a binary alloy (see [1]) and reads

$$\frac{\partial u}{\partial t} - \alpha \Delta u + f(u) = 0, \ \alpha > 0. \tag{12.1}$$

We studied in [9] generalizations of (12.1) of the form

$$\frac{\partial u}{\partial t} + P(-\Delta)u + f(u) = 0, \tag{12.2}$$

where $P(s) = \sum_{i=1}^{k} a_i s^i$, $a_k > 0$, $k \geq 1$; in particular, when $k = 1$, we recover the Allen–Cahn equation (12.1) and, when $k = 2$, the model contains the Swift–Hohenberg equation (see [30, 32]).

L. Cherfils
Université de La Rochelle, Laboratoire Mathématiques, Image et Applications,
Avenue Michel Crépeau, 17042 La Rochelle Cedex, France
e-mail: lcherfil@univ-lr.fr

A. Miranville (✉) · S. Peng
Université de Poitiers, Laboratoire de Mathématiques et Applications,
UMR CNRS 7348 - SP2MI, Boulevard Marie et Pierre Curie - Téléport 2,
86962 Chasseneuil Futuroscope Cedex, France
e-mail: Alain.Miranville@math.univ-poitiers.fr

S. Peng
e-mail: Shuiran.Peng@math.univ-poitiers.fr

V.A. Sadovnichiy and M.Z. Zgurovsky (eds.), *Advances in Dynamical Systems and Control*, Studies in Systems, Decision and Control 69,
DOI 10.1007/978-3-319-40673-2_12

Such higher-order (in space) terms were proposed in [4] in the context of phase transition models and in the isotropic limit of more general higher-order terms. We can note that a second-order term in phase separation is obtained by the truncation of higher-order ones (see [6]); it can also be seen as a first-order approximation of a (spatially) nonlocal term accounting for long-ranged interactions (see [16, 17]). In particular, nonlocal models have been much studied recently, see, e.g., [20, 22] and the references therein; it is interesting to note that, from a mathematical point of view, the picture is more complete when compared to the original (approximated) ones (see the aforementioned references).

In [9], we considered regular nonlinear terms (a typical choice is the usual cubic nonlinear term $f(s) = s^3 - s$). It is, however, important to note that, in phase separation, such a regular nonlinear term actually is an approximation of thermodynamically relevant logarithmic ones of the form $f(s) = -\lambda_1 s + \frac{\lambda_2}{2} \ln \frac{1+s}{1-s}$, $s \in (-1, 1)$, $0 < \lambda_2 < \lambda_1$, which follow from a mean-field model (see [6, 10]; in particular, the logarithmic terms correspond to the entropy of mixing); note that, as far as the Swift–Hohenberg equation is concerned, it is not clear whether logarithmic nonlinear terms are relevant.

The study of the classical Allen–Cahn equation (12.1) (i.e., $k = 1$ in (12.2)) with logarithmic nonlinear terms is well established (see, e.g., [28]). However, when $k \geq 2$ in (12.2), the situation is much more involved and we are not able to prove the existence of a solution in a classical sense (meaning in a classical weak/variational sense). Nevertheless, we are able to prove the existence of a (weaker) variational solution. This notion of a variational solution was introduced in [35] for the Cahn–Hilliard equation with singular nonlinear terms and dynamic boundary conditions and is based on a variational inequality (see also [21] for a different, though related, approach based on duality techniques). It was also applied with success in other situations in [8, 11, 30, 32].

Our aim in this chapter was to study the well-posedness of (12.2) with a logarithmic nonlinear term in the variational sense mentioned above. We also prove the dissipativity of the corresponding solution operator, as well as the existence of the global attractor.

12.2 Setting of the Problem

We consider the following initial and boundary value problem in a bounded and regular domain $\Omega \subset \mathbb{R}^n$, $n = 1, 2$ or 3, with boundary Γ:

$$\frac{\partial u}{\partial t} + P(-\Delta)u + f(u) = 0, \tag{12.3}$$

$$u = \Delta u = \cdots = \Delta^{k-1} u = 0 \text{ on } \Gamma, \tag{12.4}$$

$$u|_{t=0} = u_0. \tag{12.5}$$

We assume that the polynomial P is defined by

$$P(s) = \sum_{i=1}^{k} a_i s^i, \ a_k > 0, \ k \geq 1, \ s \in \mathbb{R}. \tag{12.6}$$

As far as the nonlinear term f is concerned, we assume that

$$f(s) = -\lambda_1 s + \frac{\lambda_2}{2} \ln \frac{1+s}{1-s}, \ s \in (-1, 1), \ 0 < \lambda_2 < \lambda_1. \tag{12.7}$$

In particular, it is not difficult to show that it satisfies the following properties:

$$f \in \mathscr{C}^{\infty}(-1, 1), \ f(0) = 0, \tag{12.8}$$

$$\lim_{s \to \pm 1} f(s) = \pm\infty, \ \lim_{s \to \pm 1} f'(s) = +\infty, \tag{12.9}$$

$$f' \geq -\lambda_1, \tag{12.10}$$

$$-c_1 \leq F(s), \ F(s) + \frac{1}{2}|f(s)| \leq f(s)s + c_2, \ c_1, \ c_2 \geq 0, \ s \in (-1, 1), \tag{12.11}$$

where $F(s) = \int_0^s f(\xi)\, d\xi$. We can also note that F is bounded on $(-1, 1)$; indeed, there holds

$$F(s) = -\frac{\lambda_1}{2} s^2 + \frac{\lambda_2}{2}((1+s)\ln(1+s) + (1-s)\ln(1-s)). \tag{12.12}$$

Remark 12.1 We can note that all properties above easily follow from the explicit expression of f. Actually, (12.10) and (12.11) follow from (12.8) and (12.9). The only difficulty here is to prove that $F(s) \leq f(s)s + c$, $c \geq 0$, $s \in (-1, 1)$. To do so, it suffices to study the variations of the function $s \mapsto f(s)s - F(s) + \frac{\lambda_1}{2}s^2$, whose derivative has, owing to (12.10), the sign of s. We can thus consider more general singular nonlinear terms only satisfying (12.8) and (12.9). Indeed, the boundedness of F is not necessary and just allows us to consider more general initial data.

Setting

$$F(s) = -\frac{\lambda_1}{2} s^2 + F_1(s),$$

we introduce the following approximated functions $F_{1,N} \in \mathscr{C}^4(\mathbb{R})$, $N \in \mathbb{N}$:

$$F_{1,N}(s) = \begin{cases} \sum_{i=0}^{4} \frac{1}{i!} F_1^{(i)}(1-\frac{1}{N})(s-1+\frac{1}{N})^i, \ s \geq 1-\frac{1}{N}, \\ F_1(s), \ |s| \leq 1-\frac{1}{N}, \\ \sum_{i=0}^{4} \frac{1}{i!} F_1^{(i)}(-1+\frac{1}{N})(s+1-\frac{1}{N})^i, \ s \leq -1+\frac{1}{N}. \end{cases} \tag{12.13}$$

Setting $F_N(s) = -\frac{\lambda_1}{2}s^2 + F_{1,N}(s), f_{1,N} = F'_{1,N}$ and $f_N = F'_N$, there holds

$$f_N \in \mathscr{C}^3(\mathbb{R}), \ f_N(0) = 0, \tag{12.14}$$

$$f'_{1,N} \geq 0, \ f'_N \geq -\lambda_1, \tag{12.15}$$

$$F_N \geq -c_1, \tag{12.16}$$

$$F_N(s) \geq c_3 s^4 - c_4, \ c_3 > 0, \ c_4 \geq 0, \ s \in \mathbb{R}, \tag{12.17}$$

$$f_N(s)s \geq c_5(F_N(s) + |f_N(s)|) - c_6, \ c_5 > 0, \ c_6 \geq 0, \ s \in \mathbb{R}. \tag{12.18}$$

Furthermore, all constants can be chosen independently of N. These properties follow from the fact that we have similar properties for the original singular nonlinear term and from the explicit expression of $F_{1,N}$; we refer the reader to [14, 33, 35] for more details. We can also note that F_N is bounded, independently of N, in the neighborhood of ± 1.

We then consider the approximated problems

$$\frac{\partial u^N}{\partial t} + P(-\Delta)u^N + f_N(u^N) = 0, \tag{12.19}$$

$$u^N = \Delta u^N = \cdots = \Delta^{k-1}u^N = 0 \text{ on } \Gamma, \tag{12.20}$$

$$u^N|_{t=0} = u_0. \tag{12.21}$$

The existence, uniqueness, and regularity of the solution u^N to (12.19)–(12.21) were proved in [9].

Notation

We denote by $((\cdot,\cdot))$ the usual L^2-scalar product, with associated norm $\|\cdot\|$. More generally, $\|\cdot\|_X$ denotes the norm on the Banach space X.

We then consider the operator $-\Delta$ associated with Dirichlet boundary conditions; it is a strictly positive, self-adjoint, and unbounded linear operator with compact inverse $(-\Delta)^{-1}$, with domain $H^2(\Omega) \cap H_0^1(\Omega)$. In particular, this allows us (see, e.g., [42]) to define the operators $(-\Delta)^m$, $m \in \mathbb{R}$ (being understood that, when $m = 0$, then $(-\Delta)^0$ is the identity operator). For $m \in \mathbb{N}$, $(-\Delta)^m$ has for domain $\{v \in H^{2m}(\Omega), \ v = \Delta v = \cdots = \Delta^{m-1}v = 0 \text{ on } \Gamma\}$. We set, for $m \in \mathbb{N}$,

$$\dot{H}^m(\Omega) = \{v \in H^m(\Omega),\ v = \Delta v = \cdots = \Delta^{[\frac{m-1}{2}]} v = 0 \text{ on } \Gamma\},$$

where $[\cdot]$ denotes the integer part. This space, endowed with the usual H^m-norm, is a closed subspace of $H^m(\Omega)$. Furthermore, $v \mapsto \|(-\Delta)^{\frac{m}{2}} v\|$ is a norm on $\dot{H}^m(\Omega)$ which is equivalent to the usual H^m-norm.

Throughout the chapter, the same letters c, c', and c'' denote (generally positive) constants which may vary from line to line and are independent of N. Similarly, the same letter Q denotes (positive) monotone increasing and continuous functions which may vary from line to line and are independent of N.

12.3 A Priori Estimates

Our aim in this section was to derive uniform (with respect to N) a priori estimates on u^N which will allow us, in the next section, to pass to the limit $N \to +\infty$ and prove the existence of a solution to the original singular problem, in a suitable setting (i.e., as mentioned in the introduction, based on a proper variational inequality).

Though formal, these a priori estimates can be fully justified in view of the regularity results obtained in [9].

We assume from now on that $-1 < u_0(x) < 1$ a.e. $x \in \Omega$.

Remark 12.2 For a more general singular nonlinear term f, we would need a stronger separation property from the singular values ± 1, namely $\|u_0\|_{L^\infty(\Omega)} < 1$.

We multiply (12.19) by $\frac{\partial u^N}{\partial t}$ and have, integrating over Ω and by parts,

$$\frac{d}{dt}\Big(\sum_{i=1}^{k} a_i \|(-\Delta)^{\frac{i}{2}} u^N\|^2 + 2\int_\Omega F_N(u^N)\, dx\Big) + 2\Big\|\frac{\partial u^N}{\partial t}\Big\|^2 = 0. \tag{12.22}$$

We then multiply (12.19) by u^N to obtain

$$\frac{1}{2}\frac{d}{dt}\|u^N\|^2 + \sum_{i=1}^{k} a_i \|(-\Delta)^{\frac{i}{2}} u^N\|^2 + ((f_N(u^N), u^N)) = 0. \tag{12.23}$$

Employing the interpolation inequality

$$\|(-\Delta)^{\frac{i}{2}} v\| \le c(i) \|(-\Delta)^{\frac{m}{2}} v\|^{\frac{i}{m}} \|v\|^{1-\frac{i}{m}}, \tag{12.24}$$

$$v \in \dot{H}^m(\Omega),\ i \in \{1, \ldots, m-1\},\ m \in \mathbb{N},\ m \ge 2,$$

from which it follows that, for $i \in \{1, \ldots, k-1\}$ and $k \ge 2$,

$$\|(-\Delta)^{\frac{i}{2}} u^N\|^2 \le \varepsilon \|(-\Delta)^{\frac{k}{2}} u^N\|^2 + c(i,\varepsilon)\|u^N\|^2,\ \forall \varepsilon > 0, \tag{12.25}$$

Equations (12.18), (12.23) and (12.25) yield

$$\frac{d}{dt}\|u^N\|^2+c(\|u^N\|^2_{H^k(\Omega)}+\int_\Omega F_N(u^N)\,dx+\|f_N(u^N)\|_{L^1(\Omega)})\le c'(\|u^N\|^2+1),\ c>0. \tag{12.26}$$

Noting finally that

$$\|u^N\|^2\le\varepsilon\|u^N\|^4_{L^4(\Omega)}+c(\varepsilon),\ \forall\varepsilon>0, \tag{12.27}$$

we deduce from (12.17) and (12.26)–(12.27) that

$$\frac{d}{dt}\|u^N\|^2+c(\|u^N\|^2_{H^k(\Omega)}+\int_\Omega F_N(u^N)\,dx+\|f_N(u^N)\|_{L^1(\Omega)})\le c',\ c>0. \tag{12.28}$$

Summing (12.22) and (12.28), we find, noting that $\sum_{i=1}^k a_i\|(-\Delta)^{\frac{i}{2}}u^N\|^2\le c\|u^N\|^2_{H^k(\Omega)}$, a differential inequality of the form

$$\frac{dE_{1,N}}{dt}+c(E_{1,N}+\|f_N(u^N)\|_{L^1(\Omega)}+\|\frac{\partial u^N}{\partial t}\|^2)\le c',\ c>0, \tag{12.29}$$

where

$$E_{1,N}=\sum_{i=1}^k a_i\|(-\Delta)^{\frac{i}{2}}u^N\|^2+2\int_\Omega F_N(u^N)\,dx+\|u^N\|^2$$

satisfies

$$E_{1,N}\ge c(\|u^N\|^2_{H^k(\Omega)}+\int_\Omega F_N(u^N)\,dx)-c',\ c>0. \tag{12.30}$$

Indeed, it follows from the interpolation inequality (12.24) that

$$E_{1,N}\ge c(\|u^N\|^2_{H^k(\Omega)}+\int_\Omega F_N(u^N)\,dx)-c'\|u^N\|^2-c''$$

and we conclude by employing (12.17) and (12.27).

We then multiply (12.19) by $-\Delta u^N$ and have, owing to (12.15),

$$\frac{d}{dt}\|\nabla u^N\|^2 + 2\sum_{i=1}^{k} a_i\|(-\Delta)^{\frac{i+1}{2}} u^N\|^2 \leq 2\lambda_1\|\nabla u^N\|^2. \tag{12.31}$$

Summing (12.29) and δ_1 times (12.31), where $\delta_1 > 0$ is small enough, we obtain, employing once more the interpolation inequality (12.24), a differential inequality of the form

$$\frac{dE_{2,N}}{dt} + c(E_{2,N} + \|u^N\|^2_{H^{k+1}(\Omega)} + \|f_N(u^N)\|_{L^1(\Omega)} + \|\frac{\partial u^N}{\partial t}\|^2) \leq c',\ c > 0, \tag{12.32}$$

where

$$E_{2,N} = E_{1,N} + \delta_1\|\nabla u^N\|^2$$

satisfies

$$E_{2,N} \geq c(\|u^N\|^2_{H^k(\Omega)} + \int_\Omega F_N(u^N)\,dx) - c',\ c > 0. \tag{12.33}$$

In particular, it follows from (12.32) and (12.33) and Gronwall's lemma that

$$\|u^N(t)\|^2_{H^k(\Omega)} \leq ce^{-c't}(\|u_0\|^2_{H^k(\Omega)} + \int_\Omega F_N(u_0)\,dx) + c'',\ c' > 0,\ t \geq 0, \tag{12.34}$$

and

$$\int_t^{t+r} (\|u^N\|^2_{H^{k+1}(\Omega)} + \|\frac{\partial u^N}{\partial t}\|^2)\,ds \tag{12.35}$$

$$\leq ce^{-c't}(\|u_0\|^2_{H^k(\Omega)} + \int_\Omega F_N(u_0)\,dx) + c''(r),\ c' > 0,\ t \geq 0,$$

$r > 0$ given. Actually, noting that $F_N(u_0)$ is bounded (independently of N and u_0), there holds

$$\|u^N(t)\|^2_{H^k(\Omega)} \leq ce^{-c't}\|u_0\|^2_{H^k(\Omega)} + c'',\ c' > 0,\ t \geq 0, \tag{12.36}$$

and

$$\int_t^{t+r} (\|u^N\|^2_{H^{k+1}(\Omega)} + \|\frac{\partial u^N}{\partial t}\|^2)\,ds \tag{12.37}$$

$$\leq ce^{-c't}\|u_0\|^2_{H^k(\Omega)} + c''(r),\ c' > 0,\ t \geq 0,$$

$r > 0$ given.

We now differentiate (12.3) with respect to time to find

$$\frac{\partial}{\partial t}\frac{\partial u^N}{\partial t} + P(-\Delta)\frac{\partial u^N}{\partial t} + f_N'(u^N)\frac{\partial u^N}{\partial t} = 0, \tag{12.38}$$

$$\frac{\partial u^N}{\partial t} = \Delta\frac{\partial u^N}{\partial t} = \cdots = \Delta^{k-1}\frac{\partial u^N}{\partial t} = 0 \text{ on } \Gamma, \tag{12.39}$$

$$\frac{\partial u^N}{\partial t}|_{t=0} = -P(-\Delta)u_0 - f_N(u_0). \tag{12.40}$$

Multiplying (12.38) by $\frac{\partial u^N}{\partial t}$, we have, employing (12.15) and the interpolation inequality (12.24),

$$\frac{d}{dt}\|\frac{\partial u^N}{\partial t}\|^2 \leq c\|\frac{\partial u^N}{\partial t}\|^2. \tag{12.41}$$

It then follows from (12.37), say, for $r = 1$, and the uniform Gronwall's lemma (see, e.g., [42]) that

$$\|\frac{\partial u^N}{\partial t}(t)\|^2 \leq ce^{-c't}\|u_0\|^2_{H^k(\Omega)} + c'',\ c > 0,\ t \geq 1. \tag{12.42}$$

Remark 12.3 (i) Actually, it follows from the uniform Gronwall's lemma that

$$\|\frac{\partial u^N}{\partial t}(t+r)\|^2 \leq \frac{c(r)}{r}e^{-c't}\|u_0\|^2_{H^k(\Omega)} + c''(r),\ c' > 0,\ t \geq 0, \tag{12.43}$$

$r > 0$ given.

(ii) We assume that $\|u_0\|_{L^\infty(\Omega)} < 1$. We can note that, if $u_0 \in H^{2k}(\Omega)$, then $\frac{\partial u^N}{\partial t}(0) \in L^2(\Omega)$ and it follows from the continuity of f and the continuous embedding $H^{2k}(\Omega) \subset \mathscr{C}(\overline{\Omega})$ that, for N large enough (note that $f_{1,N}$, coincides with $f_1 = F_1'$ when $|s| \leq 1 - \frac{1}{N}$),

$$\|\frac{\partial u^N}{\partial t}(0)\| \leq Q(\|u_0\|_{H^{2k}(\Omega)}). \tag{12.44}$$

It then follows from (12.41) and Gronwall's lemma that

$$\|\frac{\partial u^N}{\partial t}(t)\| \leq e^{ct}Q(\|u_0\|_{H^{2k}(\Omega)}),\ t \geq 0. \tag{12.45}$$

Collecting (12.42) and (12.45) (for $t \in [0, 1]$), we finally deduce that

$$\|\frac{\partial u^N}{\partial t}(t)\| \leq e^{-ct}Q(\|u_0\|_{H^{2k}(\Omega)}) + c', \ c > 0, \ t \geq 0. \tag{12.46}$$

We finally rewrite (12.19) as an elliptic equation, for $t > 0$ fixed,

$$P(-\Delta)u^N + f_N(u^N) = -\frac{\partial u^N}{\partial t}, \ u^N = \Delta u^N = \cdots = \Delta^{k-1}u^N = 0 \text{ on } \Gamma. \tag{12.47}$$

Multiplying (12.47) by $-\Delta u^N$, we find, owing to (12.15) and employing the interpolation inequality (12.24),

$$\|u^N\|^2_{H^{k+1}(\Omega)} \leq c(\|\frac{\partial u^N}{\partial t}\|^2 + \|u^N\|^2_{H^1(\Omega)}),$$

which yields, owing to (12.36) and (12.42),

$$\|u^N(t)\|^2_{H^{k+1}(\Omega)} \leq ce^{-c't}\|u_0\|^2_{H^k(\Omega)} + c'', \ c' > 0, \ t \geq 1. \tag{12.48}$$

Remark 12.4 We assume that $\|u_0\|_{L^\infty(\Omega)} < 1$. There also holds, owing to (12.46) and for N large enough,

$$\|u^N(t)\|^2_{H^{k+1}(\Omega)} \leq e^{-ct}Q(\|u_0\|_{H^{2k}(\Omega)}) + c', \ c > 0, \ t \geq 0. \tag{12.49}$$

Of course, we have a similar H^{2k}-estimate on u^N (see [9]), but, in that case, the constants and the function Q a priori depend on N.

12.4 The Dissipative Semigroup

We assume in this section that $k \geq 2$. For $k = 1$, i.e., for the classical Allen–Cahn equation, one can prove the existence (and the uniqueness) of a classical (strong) solution u, owing to the fact that u is strictly separated from the singular values ± 1, meaning that we essentially have to deal with a regular (and even bounded) nonlinear term (see [28]).

Our main aim was to prove the existence (and uniqueness) of solutions to (12.3)–(12.5) in a suitable sense, namely based on a variational inequality.

To do so, we first derive a variational inequality from (12.3). In this regard, we multiply this equation by $u - v$, where $v = v(x)$ is smooth enough and satisfies $v = \Delta v = \cdots = \Delta^{k-1}v = 0$ on Γ. We then have, recalling that $f(s) = f_1(s) - \lambda_1 s$, $s \in (-1, 1)$,

$$((\frac{\partial u}{\partial t}, u - v)) + \sum_{i=1}^{k} a_i(((-\Delta)^{\frac{i}{2}} u, (-\Delta)^{\frac{i}{2}}(u - v)))$$

$$+((f_1(u), u - v)) - \lambda_1((u, u - v)) = 0.$$

Noting that f_1 is monotone increasing, this yields the variational inequality

$$((\frac{\partial u}{\partial t}, u - v)) + \sum_{i=1}^{k} a_i(((-\Delta)^{\frac{i}{2}} u, (-\Delta)^{\frac{i}{2}}(u - v))) \tag{12.50}$$

$$+((f_1(v), u - v)) - \lambda_1((u, u - v)) \leq 0,$$

i.e., the nonlinear term now acts on the test functions rather than on the solutions.
Based on this, we give the following definition (see also [35]):

Definition 12.1 We assume that $u_0 \in \dot{H}^k(\Omega)$, with $-1 < u_0(x) < 1$ a.e. $x \in \Omega$. Then, $u = u(t, x)$ is a variational solution to (12.3)–(12.5) if, for all $T > 0$,

(i) $-1 < u(t, x) < 1$ a.e. (t, x),
(ii) $u \in \mathscr{C}([0, T]; L^2(\Omega)) \cap L^\infty(0, T; \dot{H}^k(\Omega)) \cap L^2(0, T; \dot{H}^{k+1}(\Omega))$,
(iii) $\frac{\partial u}{\partial t} \in L^2(0, T; L^2(\Omega))$,
(iv) $f_1(u) \in L^1((0, T) \times \Omega)$,
(v) $u(0) = u_0$,
(vi) the variational inequality (12.50) is satisfied for every $t > 0$ and every test function $v = v(x)$ such that $v \in \dot{H}^k(\Omega)$, with $f_1(v) \in L^1(\Omega)$.

We first prove the uniqueness of variational solutions. To do so, we need to define as admissible test functions the solutions themselves; i.e., we need to define admissible time-dependent test functions. More precisely, we call admissible any function $v = v(t, x)$ such that $v \in \mathscr{C}([0, T]; L^2(\Omega)) \cap L^\infty(0, T; \dot{H}^k(\Omega)) \cap L^2(0, T; \dot{H}^{k+1}(\Omega))$, $f_1(v) \in L^1((0, T) \times \Omega)$ and $\frac{\partial v}{\partial t} \in L^2(0, T; L^2(\Omega))$, $\forall T > 0$.

Next, we write (12.50) for $v = v(t, \cdot)$, for almost every $t > 0$. Noting that, owing to the regularity assumptions on u and v, all terms are L^1 with respect to time, we can integrate with respect to time to obtain

$$\int_s^t [((\frac{\partial u}{\partial t}, u - v)) + \sum_{i=1}^{k} a_i(((-\Delta)^{\frac{i}{2}} u, (-\Delta)^{\frac{i}{2}}(u - v))) \tag{12.51}$$

$$+((f_1(v), u - v)) - \lambda_1((u, u - v))]\, d\xi \leq 0,$$

for all $0 < s < t$ and for every admissible test function $v = v(t, x)$. In particular, since $H^k(\Omega) \subset \mathscr{C}(\overline{\Omega})$, $k \geq 2$, it follows from the above regularity that $((f_1(u), u - v)) \in L^1(0, T)$, $\forall T > 0$.

Remark 12.5 We can replace (12.50) by (12.51) in Definition 12.1, (vi).

We will actually need a second variational inequality. To do so, let $w = w(t, x)$ be an admissible test function and set

$$v_\eta = (1 - \eta)u + \eta w, \ \eta \in (0, 1].$$

Noting that

$$f_1''(s)\,\mathrm{sgn}(s) \geq 0, \ s \in (-1, 1), \tag{12.52}$$

it follows that $|f_1|$ is convex, so that

$$|f_1(v_\eta)| \leq |f_1(u)| + |f_1(w)|. \tag{12.53}$$

This yields that $f_1(v_\eta) \in L^1((0, T) \times \Omega)$ and v_η is an admissible test function. Taking $v = v_\eta$ in (12.51) and dividing by η, we find

$$\int_s^t [((\frac{\partial u}{\partial t}, u - w)) + \sum_{i=1}^k a_i(((-\Delta)^{\frac{i}{2}} u, (-\Delta)^{\frac{i}{2}} (u - w)))$$
$$+((f_1(v_\eta), u - w)) - \lambda_1((u, u - w))]\, d\xi \leq 0.$$

Passing finally to the limit $\eta \to 0$ and employing Lebesgue's dominated convergence theorem (see (12.53)), we have

$$\int_s^t [((\frac{\partial u}{\partial t}, u - w)) + \sum_{i=1}^k a_i(((-\Delta)^{\frac{i}{2}} u, (-\Delta)^{\frac{i}{2}} (u - w))) \tag{12.54}$$
$$+((f_1(u), u - w)) - \lambda_1((u, u - w))]\, d\xi \leq 0,$$

for all $0 < s < t$ and for every test function $w = w(t, x)$.

Let now u_1 and u_2 be two variational solutions with initial data $u_{1,0}$ and $u_{2,0}$, respectively. We take $u = u_1$ and $v = u_2$ in (12.51) and $u = u_2$ and $w = u_1$ in (12.54) and sum the two resulting inequalities. We obtain, after simplifications (recall that f_1 is monotone increasing) and noting that all terms are absolutely continuous from $[0, T]$ onto $L^2(\Omega)$,

$$\frac{1}{2}\|u_1(t) - u_2(t)\|^2 - \frac{1}{2}\|u_1(s) - u_2(s)\|^2 \tag{12.55}$$
$$+\int_s^t (\sum_{i=1}^k a_i\|(-\Delta)^{\frac{i}{2}}(u_1 - u_2)\|^2 - \lambda_1\|u_1 - u_2\|^2)\, d\xi \leq 0.$$

Employing the interpolation inequality (12.24), we deduce that

$$\frac{1}{2}\|u_1(t)-u_2(t)\|^2-\frac{1}{2}\|u_1(s)-u_2(s)\|^2\leq c\int_s^t\|u_1-u_2\|^2\,d\xi,$$

so that, employing Gronwall's lemma,

$$\|u_1(t)-u_2(t)\|\leq e^{c(t-s)}\|u_1(s)-u_2(s)\|,$$

where the constant c is independent of t, s, u_1, and u_2. Passing finally to the limit $s\to 0$, we find

$$\|u_1(t)-u_2(t)\|\leq e^{ct}\|u_{1,0}-u_{2,0}\|,\ t\geq 0, \tag{12.56}$$

hence the uniqueness, as well as the continuous dependence with respect to the initial data in the L^2-norm.

We now have the

Theorem 12.1 *We assume that $u_0\in\dot{H}^k(\Omega)$, with $-1<u_0<1$ a.e. $x\in\Omega$. Then,* (12.3)–(12.5) *possesses a unique variational solution u.*

Proof There remains to prove the existence of a variational solution. To do so, we consider the solution u_N to the approximated problem (12.19)–(12.21) (as already mentioned, the existence, uniqueness, and regularity of u^N are known). Furthermore, proceeding as above, it is easy to see that u_N satisfies a variational inequality which is analogous to (12.51), namely

$$\int_s^t[((\frac{\partial u^N}{\partial t},u^N-v))+\sum_{i=1}^k a_i(((-\Delta)^{\frac{i}{2}}u^N,(-\Delta)^{\frac{i}{2}}(u^N-v))) \tag{12.57}$$

$$+((f_{1,N}(v),u^N-v))-\lambda_1((u^N,u^N-v))]\,d\xi\leq 0,$$

for all $0<s<t$ and for every admissible test function $v=v(t,x)$.

It then follows from the uniform (with respect to N) a priori estimates derived in the previous section (which are fully justified at this stage) that, up to a subsequence, u_N converges to a limit function u such that, $\forall T>0$,

$$u^N\to u \text{ in } L^\infty(0,T;H^k(\Omega)) \text{ weak}-\star \text{ and in } L^2(0,T;H^{k+1}(\Omega)) \text{ weak},$$

$$\frac{\partial u^N}{\partial t}\to\frac{\partial u}{\partial t} \text{ in } L^2(0,T;L^2(\Omega)) \text{ weak},$$

$$u^N\to u \text{ in } \mathscr{C}([0,T];H^{k-\varepsilon}(\Omega)),\ L^2(0,T;H^{k+1-\varepsilon}(\Omega)) \text{ and a.e. in } (0,T)\times\Omega,\ \varepsilon>0.$$

Our aim was to pass to the limit in (12.57). We can note that the above convergences allow us to pass to the limit in all terms in (12.57), except in the nonlinear term $\int_s^t ((f_{1,N}(v), u^N - v))\, d\xi$. To pass to the limit in the nonlinear term, we can note that, by construction,

$$|f_{1,N}(v)| \leq |f_1(v)|$$

and we are in a position to use Lebesgue's dominated convergence theorem (recall that if v is an admissible test function, then $f_1(v) \in L^1((0,T) \times \Omega)$; also note that u and v belong to $L^\infty((0,T) \times \Omega)$).

We now need to prove the separation property (i). To do so, we note that, owing to (12.29) and (12.30), $f_{1,N}(u^N)$ is uniformly (with respect to N) bounded in $L^1((0,T) \times \Omega)$. Then, owing to the explicit expression of $f_{1,N}$, we have

$$\text{meas}\{(t,x) \in (0,T) \times \Omega,\ |u^M(t,x)| > 1 - \frac{1}{N}\} \leq \frac{c}{f_1(1-\frac{1}{N})},\ M \geq N, \tag{12.58}$$

where the constant c is independent of $M \geq N$ and N (note that f_1 and $f_{1,N}$ are odd functions). Indeed, there holds

$$\int_0^T \int_\Omega |f_{1,M}(u^M)|\, dx\, dt \geq \int_{E_{N,M}} |f_{1,M}(u^M)|\, dx\, dt \geq c'\text{meas}(E_{N,M}) f_1(1 - \frac{1}{N}),$$

where

$$E_{N,M} = \{(t,x) \in (0,T) \times \Omega,\ |u^M(t,x)| > 1 - \frac{1}{N}\},$$

the constant c' being independent of N and M. Passing to the limit $M \to +\infty$ (employing Fatou's Lemma) and then $N \to +\infty$ (noting that $f_1(1 - \frac{1}{N}) \to +\infty$ as $N \to +\infty$) in (12.58), it follows that

$$\text{meas}\{(t,x) \in (0,T) \times \Omega,\ |u(t,x)| \geq 1\} = 0, \tag{12.59}$$

hence the separation property.

In order to complete the proof of existence, there remains to prove (iv). To do so, we note that it follows from the almost everywhere convergence of u^N to u, the separation property (i), and the explicit expression of $f_{1,N}$ again that

$$f_{1,N}(u^N) \to f_1(u) \text{ a.e. in } (0,T) \times \Omega.$$

Then, we deduce from Fatou's lemma that

$$\|f(u)\|_{L^1((0,T)\times\Omega)} \leq \liminf \|f_N(u^N)\|_{L^1((0,T)\times\Omega)} < +\infty,$$

which finishes the proof of existence.

Remark 12.6 A natural question is whether a solution in the sense of Definition 12.1 is a classical variational solution (i.e., it satisfies a variational equality instead of a variational inequality). To prove this, one solution is to obtain a uniform (with respect to N) bound on $f_{1,N}(u^N)$ in $L^p((0,T)\times\Omega)$, for some $p > 1$ (and not just for $p = 1$). Unfortunately, we have not been able to derive such an estimate when $k \geq 2$ so that the question of whether a variational solution is a classical (variational) one is an open problem.

It follows from Theorem 12.1 that we can define the family of operators $S(t) : \Phi \to \Phi$, $u_0 \mapsto u(t)$, $t \geq 0$, where

$$\Phi = \{v \in \dot{H}^k(\Omega),\ -1 < v(x) < 1 \text{ a.e. } x \in \Omega\}.$$

This family of operators forms a semigroup (i.e., $S(0) = I$ (identity operator) and $S(t+\tau) = S(t)\circ S(\tau)$, $t,\ \tau \geq 0$) which is, owing to (12.56), continuous in the L^2 topology. Furthermore, it follows from (12.36) (which also holds in the limit $N \to +\infty$) that this semigroup is dissipative, in the sense that it possesses a bounded absorbing set $\mathscr{B}_0 \subset \Phi$ (i.e., $\forall B \subset \Phi$ bounded, $\exists t_0 = t_0(B) \geq 0$ such that $t \geq t_0 \implies S(t)B \subset \mathscr{B}_0$).

It then follows from (12.56) that we can actually extend (in a unique way and by continuity) $S(t)$ to the closure of Φ in the L^2-topology, namely

$$S(t) : \Phi_1 \to \Phi_1,\ t \geq 0,$$

where

$$\Phi_1 = \{v \in L^\infty(\Omega),\ \|v\|_{L^\infty(\Omega)} \leq 1\}.$$

It also follows from the a priori estimates derived in the previous section that $S(t)$ instantaneously regularizes, i.e.,

$$S(t) : \Phi_1 \to \Phi,\ t > 0,$$

and that it possesses a bounded absorbing set $\mathscr{B}_1$ which is compact in $L^2(\Omega)$ and bounded in $H^{k+1}(\Omega)$. We thus deduce from standard results (see, e.g., [34, 42]) that we have the

Theorem 12.2 *The semigroup $S(t)$ possesses the global attractor $\mathscr{A}$ which is compact in $L^2(\Omega)$ and bounded in $H^{k+1}(\Omega)$.*

Remark 12.7 We recall that the global attractor $\mathscr{A}$ is the smallest (for the inclusion) compact set of the phase space which is invariant by the flow (i.e., $S(t)\mathscr{A} = \mathscr{A}$, $\forall t \geq 0$) and attracts all bounded sets of initial data as time goes to infinity; it thus appears as a suitable object in view of the study of the asymptotic behavior of the system. We refer the reader to, e.g., [34, 42], for more details and discussions on this.

Remark 12.8 An important question is whether the global attractor $\mathscr{A}$ has finite dimension, in the sense of covering dimensions such as the Hausdorff and the fractal dimensions. The finite-dimensionality means, very roughly speaking, that even though the initial phase space has infinite dimension, the reduced dynamics can be described by a finite number of parameters (we refer the interested reader to, e.g., [34, 42], for discussions on this subject). When $k = 1$, i.e., for the classical Allen–Cahn equation, this can easily be established, owing again to the strict separation from the singular values ± 1 (see, e.g., [28]). However, when $k \geq 2$, the situation is much more involved and one idea could be to proceed as in [35]. This will be addressed elsewhere.

Remark 12.9 We can adapt the above analysis to the higher-order Cahn–Hilliard model

$$(-\Delta)^{-1}\frac{\partial u}{\partial t} + P(-\Delta)u + f(u) = 0, \tag{12.60}$$

$$u = \Delta u = \cdots = \Delta^{k-1}u = 0 \text{ on } \Gamma, \tag{12.61}$$

$$u|_{t=0} = u_0, \tag{12.62}$$

where P and f are as above. In particular, for $k = 1$, we recover the classical Cahn–Hilliard equation which describes phase separation processes (spinodal decomposition and coarsening) in binary alloys (see [5, 6] and the review papers [10, 36] for more details). When $k = 2$, the model contains sixth-order Cahn–Hilliard models. We can note that there is currently a strong interest in the study of sixth-order Cahn–Hilliard equations. Such equations arise in situations such as strong anisotropy effects being taken into account in phase separation processes (see [43]), atomistic models of crystal growth (see [2, 3, 13, 15]), the description of growing crystalline surfaces with small slopes which undergo faceting (see [41]), oil–water–surfactant mixtures (see [18, 19]), and mixtures of polymer molecules (see [12]). We refer the reader to [7, 23–27, 29–32, 37–40, 44–46] for the mathematical and numerical analysis of such models.

References

1. Allen, S.M., Cahn, J.W.: A microscopic theory for antiphase boundary motion and its application to antiphase domain coarsening. Acta Metall. **27**,1085–1095 (1979)
2. Berry, J., Elder, K.R., Grant, M.: Simulation of an atomistic dynamic field theory for monatomic liquids: freezing and glass formation. Phys. Rev. E **77**, 061506 (2008)
3. Berry, J., Grant, M., Elder, K.R.: Diffusive atomistic dynamics of edge dislocations in two dimensions. Phys. Rev. E **73**, 031609 (2006)
4. Caginalp, G., Esenturk, E.: Anisotropic phase field equations of arbitrary order. Discret. Contin. Dyn. Syst. S **4**, 311–350 (2011)
5. Cahn, J.W.: On spinodal decomposition. Acta Metall. **9**, 795–801 (1961)
6. Cahn, J.W., Hilliard, J.E.: Free energy of a nonuniform system I. Interfacial free energy. J. Chem. Phys. **2**, 258–267 (1958)
7. Chen, F., Shen, J.: Efficient energy stable schemes with spectral discretization in space for anisotropic Cahn-Hilliard systems. Commun. Comput. Phys. **13**, 1189–1208 (2013)
8. Cherfils, L., Gatti, S., Miranville, A.: A variational approach to a Cahn-Hilliard model in a domain with nonpermeable walls. J. Math. Sci. **189**, 604–636 (2013)
9. Cherfils, L., Miranville, A., Peng, S.: Higher-order models in phase separation. J. Appl. Anal. Comput. (to appear)
10. Cherfils, L., Miranville, A., Zelik, S.: The Cahn-Hilliard equation with logarithmic potentials. Milan J. Math. **79**, 561–596 (2011)
11. Conti, M., Gatti, S., Miranville, A.: Attractors for a Caginalp model with a logarithmic potential and coupled dynamic boundary conditions. Anal. Appl. **11**, 1350024 (2013)
12. de Gennes, P.G.: Dynamics of fluctuations and spinodal decomposition in polymer blends. J. Chem. Phys. **72**, 4756–4763 (1980)
13. Emmerich, H., Löwen, H., Wittkowski, R., Gruhn, T., Tóth, G.I., Tegze, G., Gránásy, L.: Phase-field-crystal models for condensed matter dynamics on atomic length and diffusive time scales: an overview. Adv. Phys. **61**, 665–743 (2012)
14. Frigeri, S., Grasselli, M.: Nonlocal Cahn-Hilliard-Navier-Stokes systems with singular potentials. Dyn. PDE **9**, 273–304 (2012)
15. Galenko, P., Danilov, D., Lebedev, V.: Phase-field-crystal and Swift-Hohenberg equations with fast dynamics. Phys. Rev. E **79**, 051110 (2009)
16. Giacomin, G., Lebowitz, J.L.: Phase segregation dynamics in particle systems with long range interaction I. Macroscopic limits. J. Stat. Phys. **87**, 37–61 (1997)
17. Giacomin, G., Lebowitz, J.L.: Phase segregation dynamics in particle systems with long range interaction II. Interface motion. SIAM J. Appl. Math. **58**, 1707–1729 (1998)
18. Gompper, G., Kraus, M.: Ginzburg-Landau theory of ternary amphiphilic systems. I. Gaussian interface fluctuations. Phys. Rev. E **47**, 4289–4300 (1993)
19. Gompper, G., Kraus, M.: Ginzburg-Landau theory of ternary amphiphilic systems. II. Monte Carlo simulations. Phys. Rev. E **47** 4301–4312 (1993)
20. Grasselli, M.: Finite-dimensional global attractor for a nonlocal phase-field system. Istit. Lombardo Accad. Sci. Lett. Rend. A **146**, 3–22 (2012)
21. Grasselli, M., Miranville, A., Schimperna, G.: The Caginalp phase-field system with coupled dynamic boundary conditions and singular potentials. Discret. Contin. Dyn. Syst. **28**, 67–98 (2010)
22. Grasselli, M., Schimperna, G.: Nonlocal phase-field systems with general potentials. Discret. Contin. Dyn. Syst. A **33**, 5089–5106 (2013)
23. Grasselli, M., Wu, H.: Well-posedness and longtime behavior for the modified phase-field crystal equation. Math. Model. Methods Appl. Sci. **24**, 2743–2783 (2014)
24. Grasselli, M., Wu, H.: Robust exponential attractors for the modified phase-field crystal equation. Discret. Contin. Dyn. Syst. **35**, 2539–2564 (2015)
25. Hu, Z., Wise, S.M., Wang, C., Lowengrub, J.S.: Stable finite difference, nonlinear multigrid simulation of the phase field crystal equation. J. Comput. Phys. **228**, 5323–5339 (2009)

26. Korzec, M., Nayar, P., Rybka, P.: Global weak solutions to a sixth order Cahn-Hilliard type equation. SIAM J. Math. Anal. **44**, 3369–3387 (2012)
27. Korzec, M., Rybka, P.: On a higher order convective Cahn-Hilliard type equation. SIAM J. Appl. Math. **72**, 1343–1360 (2012)
28. Miranville, A.: Some mathematical models in phase transition. Discret. Contin. Dyn. Syst. S **7**, 271–306 (2014)
29. Miranville, A.: Asymptotic behavior of a sixth-order Cahn-Hilliard system. Central Eur. J. Math. **12**, 141–154 (2014)
30. Miranville, A.: Sixth-order Cahn-Hilliard equations with logarithmic nonlinear terms. Appl. Anal. **94**, 2133–2146 (2015)
31. Miranville, A.: Sixth-order Cahn-Hilliard systems with dynamic boundary conditions. Math. Methods Appl. Sci. **38**, 1127–1145 (2015)
32. Miranville, A.: On the phase-field-crystal model with logarithmic nonlinear terms. RACSAM (to appear)
33. Miranville, A., Zelik, S.: Robust exponential attractors for Cahn-Hilliard type equations with singular potentials. Math. Methods Appl. Sci. **27**, 545–582 (2004)
34. Miranville, A., Zelik, S.: Attractors for dissipative partial differential equations in bounded and unbounded domains. In: Dafermos, C.M., Pokorny, M. (eds.) Handbook of Differential Equations, Evolutionary Partial Differential Equations, vol. 4, pp. 103–200. Elsevier, Amsterdam (2008)
35. Miranville, A., Zelik, S.: The Cahn-Hilliard equation with singular potentials and dynamic boundary conditions. Discret. Cont. Dyn. Syst. **28**, 275–310 (2010)
36. Novick-Cohen, A.: The Cahn-Hilliard equation. In: Dafermos, C.M., Pokorny, M. (eds.) Handbook of Differential Equations, Evolutionary Partial Differential Equations, pp. 201–228. Elsevier, Amsterdam (2008)
37. Pawlow, I., Schimperna, G.: On a Cahn-Hilliard model with nonlinear diffusion. SIAM J. Math. Anal. **45**, 31–63 (2013)
38. Pawlow, I., Schimperna, G.: A Cahn-Hilliard equation with singular diffusion. J. Diff. Equ. **254**, 779–803 (2013)
39. Pawlow, I., Zajaczkowski, W.: A sixth order Cahn-Hilliard type equation arising in oil-water-surfactant mixtures. Commun. Pure Appl. Anal. **10**, 1823–1847 (2011)
40. Pawlow, I., Zajaczkowski, W.: On a class of sixth order viscous Cahn-Hilliard type equations. Discret. Contin. Dyn. Syst. S **6**, 517–546 (2013)
41. Savina, T.V., Golovin, A.A., Davis, S.H., Nepomnyashchy, A.A., Voorhees, P.W.: Faceting of a growing crystal surface by surface diffusion. Phys. Rev. E **67**, 021606 (2003)
42. Temam, R.: Infinite-dimensional dynamical systems in mechanics and physics. Applied Mathematical Sciences, vol. 68, 2nd edn. Springer, New York (1997)
43. Torabi, S., Lowengrub, J., Voigt, A., Wise, S.: A new phase-field model for strongly anisotropic systems. Proc. R. Soc. A **465**, 1337–1359 (2009)
44. Wang, C., Wise, S.M.: Global smooth solutions of the modified phase field crystal equation. Methods Appl. Anal. **17**, 191–212 (2010)
45. Wang, C., Wise, S.M.: An energy stable and convergent finite difference scheme for the modified phase field crystal equation. SIAM J. Numer. Anal. **49**, 945–969 (2011)
46. Wise, S.M., Wang, C., Lowengrub, J.S.: An energy stable and convergent finite difference scheme for the phase field crystal equation. SIAM J. Numer. Anal. **47**, 2269–2288 (2009)

Chapter 13
Uniform Global Attractor for Nonautonomous Reaction–Diffusion Equations with Carathéodory's Nonlinearity

Nataliia V. Gorban and Liliia S. Paliichuk

Abstract We consider nonautonomous reaction–diffusion system with Carathéodory's nonlinearity. We investigate the long-time dynamics of all globally defined weak solutions under the standard sign and polynomial growth conditions. We obtain new topological properties of solutions, in particular flattening property, prove the existence of uniform global attractor for multivalued semiflow generated by considered problem.

13.1 Introduction and Statement of the Problem

Let $N, M = 1, 2, \ldots$. In a bounded domain $\Omega \subset \mathbf{R}^N$ with sufficiently smooth boundary $\partial\Omega$, we consider the following problem:

$$\begin{cases} u_t = a\Delta u - f(x,t,u), & x \in \Omega,\ t > 0, \\ u|_{\partial\Omega} = 0, \end{cases} \tag{13.1}$$

where $u = u(x,t) = (u^{(1)}(x,t), \ldots, u^{(M)}(x,t))$ is unknown vector function, a is real $M \times M$ matrix, $f = f(x,t,u) = (f^{(1)}(x,t,u), \ldots, f^{(M)}(x,t,u))$ is given interaction function.

Note that Problem (13.1) is a nonautonomous reaction–diffusion system. There are a lot of papers on qualitative behavior of solutions for evolution systems of reaction–diffusion type. This is due to theoretical and applied importance of such objects. The partial cases of reaction–diffusion problem are Kolmogorov–Petrovsky–Piskunov equations (the problem on the gene diffusion) [1], models of Belousov–Zhabotinsky reaction [2, 3], Gause–Vitta models [4, 5], and Selkov model for glycolysis [6, 7].

N.V. Gorban · L.S. Paliichuk (✉)
Institute for Applied System Analysis, National Technical University of Ukraine "Kyiv Polytechnic Institute", Peremogy ave., 37, Kyiv 03056, Ukraine
e-mail: lili262808@gmail.com

N.V. Gorban
e-mail: nataliia.v.gorban@gmail.com

V.A. Sadovnichiy and M.Z. Zgurovsky (eds.), *Advances in Dynamical Systems and Control*, Studies in Systems, Decision and Control 69,
DOI 10.1007/978-3-319-40673-2_13

Reaction–diffusion equations are actively used for modeling various biological and chemical processes.

Remark that existence and properties of global attractors for autonomous reaction–diffusion equations with smooth interaction functions are well-known results (see [8, 9]). The autonomous equations and inclusions without uniqueness are investigated in [10–15]. In [16, 17] for autonomous reaction–diffusion inclusion of subgradient type, the existence of Lyapunov function is obtained, the structure of global attractor is studied, and the application to climatology model is considered. For nonautonomous equations of such type with almost periodic interaction functions, the results on trajectory attractors are obtained in [18]. In [19], the existence of uniform trajectory attractor for nonautonomous Problem (13.1) with Carathéodory's nonlinearity is proved. In this chapter, we prove the existence of uniform global attractor for Problem (13.1).

Remark 13.1 Let $\gamma \geq 1$ and $\mathscr{Y}$ be a real separable Banach space. We consider the Fréchet space $L_\gamma^{\mathrm{loc}}(\mathbb{R}_+; \mathscr{Y})$ of all locally integrable functions with values in $\mathscr{Y}$, i.e., $\varphi \in L_\gamma^{\mathrm{loc}}(\mathbb{R}_+; \mathscr{Y})$ if and only if for any finite interval $[\tau, T] \subset \mathbb{R}_+$ the restriction of φ on $[\tau, T]$ belongs to the space $L_\gamma(\tau, T; \mathscr{Y})$ [19].

Definition 13.1 ([19]) A function $\varphi \in L_1^{\mathrm{loc}}(\mathbb{R}_+; L_1(\Omega)$ is called a translation uniform integrable one in $L_1^{\mathrm{loc}}(\mathbb{R}_+; L_1(\Omega))$, if

$$\lim_{K\to+\infty} \sup_{t\geq 0} \int_t^{t+1} \int_\Omega |\varphi(x,s)| \chi_{\{|\varphi(x,s)|\geq K\}} dx ds = 0.$$

Remark 13.2 A function $\varphi \in L_1^{\mathrm{loc}}(\mathbb{R}_+; L_1(\Omega))$ is a translation uniform integrable one in $L_1^{\mathrm{loc}}(\mathbb{R}_+; L_1(\Omega))$ if and only if for every sequence of elements $\{\tau_n\}_{n\geq 1} \subset \mathbb{R}_+$ the sequence $\{\varphi(\cdot + \tau_n)\}_{n\geq 1}$ contains a subsequence which converges weakly in $L_1^{\mathrm{loc}}(\mathbb{R}_+; L_1(\Omega))$.

The following condition

$$\sup_{t\geq 0} \int_t^{t+1} \|\varphi(s)\|_{\mathscr{E}}^\gamma ds < +\infty$$

is the sufficient condition for the translation uniform integrability of function φ; see [19].

The Main Assumptions on Parameters of Problem (13.1)

Assumption (A) There exists a positive constant d such that $\frac{1}{2}(a + a^*) \geq dI$, where I is the identity $M \times M$ matrix, a^* is a transposed matrix for a.

Assumption (B) The interaction function $f = (f^{(1)}, \ldots, f^{(M)}) : \Omega \times \mathbb{R}_+ \times \mathbb{R}^M \to \mathbb{R}^M$ satisfies the standard Carathéodory's conditions, i.e., $(x, t, y) \to$

$f(x, t, y)$ is continuous map in $y \in \mathbb{R}^M$ for a.e. $(x, t) \in \Omega \times \mathbb{R}_+$, and it is measurable map in $(x, t) \in \Omega \times \mathbb{R}_+$ for any $y \in \mathbb{R}^M$.

Assumption (C) There exist a translation uniform integrable in $L_1^{\text{loc}}(\mathbb{R}_+; L_1(\Omega))$ function $c_1 : \Omega \times \mathbb{R}_+ \to \mathbb{R}_+$ and a constant $c_2 > 0$ such that

$$\sum_{i=1}^{M} \left| f^{(i)}(x, t, y) \right|^{q_i} \leq c_1(x, t) + c_2 \sum_{i=1}^{M} \left| y^{(i)} \right|^{p_i}$$

for any $y = (y^{(1)}, \dots, y^{(M)}) \in \mathbb{R}^M$ and a.e. $(x, t) \in \Omega \times \mathbb{R}_+$, where $p_i \geq 2$ and $q_i > 1$ are such that $\frac{1}{p_i} + \frac{1}{q_i} = 1$ for any $i = 1, 2, \dots, M$.

Assumption (D) There exist a constant $\alpha > 0$ and a translation uniform integrable in $L_1^{\text{loc}}(\mathbb{R}_+; L_1(\Omega))$ function $\beta : \Omega \times \mathbb{R}_+ \to \mathbb{R}_+$ such that

$$\sum_{i=1}^{M} f^{(i)}(x, t, y) y^{(i)} \geq \alpha \sum_{i=1}^{M} \left| y^{(i)} \right|^{p_i} - \beta(x, t)$$

for any $y = (y^{(1)}, \dots, y^{(M)}) \in \mathbb{R}^M$ and a.e. $(x, t) \in \Omega \times \mathbb{R}_+$.

Consider the evolution triple (V, H, V^*), where $H = (L_2(\Omega))^M$, $V = (H_0^1(\Omega))^M$, and $V^* = (H^{-1}(\Omega))^M$ with standard respective inner products and norms $(\cdot, \cdot)_H$ and $\|\cdot\|_H$, $(\cdot, \cdot)_V$ and $\|\cdot\|_V$, and $(\cdot, \cdot)_{V^*}$ and $\|\cdot\|_{V^*}$.

Let $0 \leq \tau < T < +\infty$. Denote

$$\begin{gathered}
\mathbf{L_p}(\Omega) := L_{p_1}(\Omega) \times \ldots \times L_{p_M}(\Omega), \quad \mathbf{L_q}(\Omega) := L_{q_1}(\Omega) \times \ldots \times L_{q_M}(\Omega), \\
\mathbf{L_p}(\tau, T; \mathbf{L_p}(\Omega)) := L_{p_1}(\tau, T; L_{p_1}(\Omega)) \times \ldots \times L_{p_M}(\tau, T; L_{p_M}(\Omega)), \\
\mathbf{L_q}(\tau, T; \mathbf{L_q}(\Omega)) := L_{q_1}(\tau, T; L_{q_1}(\Omega)) \times \ldots \times L_{q_M}(\tau, T; L_{q_M}(\Omega)),
\end{gathered}$$

where $\mathbf{p} = (p_1, p_2, \dots, p_M)$ and $\mathbf{q} = (q_1, q_2, \dots, q_M)$.

Definition 13.2 A function $u = u(x, t) \in \mathbf{L}_2(\tau, T; V) \cap \mathbf{L_p}(\tau, T; \mathbf{L_p}(\Omega))$ is called a *weak solution* of Problem (13.1) on $[\tau, T]$ if for any function $\varphi = \varphi(x) \in (C_0^\infty(\Omega))^M$ the following equality holds

$$\frac{d}{dt} \int_\Omega u(x, t) \cdot \varphi(x) dx + \int_\Omega \{a \nabla u(x, t) \cdot \nabla \varphi(x) + f(x, t, u(x, t)) \cdot \varphi(x)\} dx = 0$$

in the sense of scalar distributions on (τ, T).

Conditions (A)–(D) guarantee the existence of at least one weak solution on arbitrary interval (τ, T), $0 \leq \tau < T < \infty$, with initial condition $u(\tau) = u_\tau$, $u_\tau \in H$ [20, pp. 283–284]. But the uniqueness is not provided.

The main goal of this paper is to investigate the uniform long-time behavior of all globally defined weak solutions for Problem (13.1) with initial data $u_\tau \in H$ under listed above assumptions, in particular to prove the existence of uniform global attractor for all globally defined weak solutions of Problem (13.1).

13.2 Auxiliaries

Let $0 \le \tau < T < \infty, u^{(\tau)} \in H$. Denote by $\mathscr{D}_{\tau,T}(u^{(\tau)})$ the family of all weak solutions on $[\tau, T]$ with initial data $u(\tau) = u^{(\tau)}$; that is,

$$\mathscr{D}_{\tau,T}(u^{(\tau)}) = \{u(\cdot)\,|\, u \text{ is a weak solution of Problem (13.1) on } [\tau, T],\ u(\tau) = u^{(\tau)}\}.$$

Remark that $\mathscr{D}_{\tau,T}(u^{(\tau)}) \neq \emptyset$ and $\mathscr{D}_{\tau,T}(u^{(\tau)}) \subset W_{\tau,T}$ where $u^{(\tau)} \in H$. Moreover, the concatenation of weak solutions of Problem (13.1) is a weak solution too, i.e., if $0 \le \tau < t < T$, $u^{(\tau)} \in H$, $u(\cdot) \in \mathscr{D}_{\tau,t}(u^{(\tau)})$, and $v(\cdot) \in \mathscr{D}_{t,T}(u(t))$, then

$$z(s) = \begin{cases} u(s),\ s \in [\tau, t], \\ v(s),\ s \in [t, T] \end{cases}$$

belongs to $\mathscr{D}_{\tau,T}(u^{(\tau)})$ (cf. [21, pp. 55–56]).

Listed above properties of solutions and Grönwall's lemma provide that for any finite time interval $[\tau, T] \subset \mathbb{R}_+$ each weak solution u of Problem (13.1) on $[\tau, T]$ satisfies estimates

$$\begin{array}{l} \|u(t)\|_H^2 - 2\int_\tau^t \int_\Omega \beta(x,\xi)dxd\xi + 2\alpha \sum_{i=1}^M \int_s^t \|u^{(i)}(\xi)\|_{L_{p_i}(\Omega)}^{p_i} d\xi \\ \quad +2d\int_s^t \|u(\xi)\|_V^2 d\xi \le \|u(s)\|_H^2 - 2\int_\tau^s \int_\Omega \beta(x,\xi)dxd\xi, \end{array} \tag{13.2}$$

$$\|u(t)\|_H^2 \le \|u(s)\|_H^2 e^{-2d\lambda_1(t-s)} + 2\int_s^t \int_\Omega \beta(x,\xi) e^{-2d\lambda_1(t-\xi)} dxd\xi \tag{13.3}$$

for any $t, s \in [\tau, T], t \ge s$, where λ_1 is the first eigenvalue of the scalar operator $-\Delta$ with Dirichlet boundary conditions (cf. [20, p. 285], [21, p. 56], [22] and references therein).

Any weak solution u of Problem (13.1) on a finite time interval $[\tau, T] \subset \mathbb{R}_+$ can be extended to a global one, defined on $[\tau, +\infty)$. For arbitrary $\tau \ge 0$ and $u^{(\tau)} \in H$ denote by $\mathscr{D}_\tau(u^{(\tau)})$ the set of all weak solutions (defined on $[\tau, +\infty)$) of Problem (13.1) with initial data $u(\tau) = u^{(\tau)}$. Consider the family of all weak solutions of Problem (13.1) defined on the semi-infinite time interval $[\tau, +\infty)$:

$$\mathscr{K}_\tau^+ = \cup_{u^{(\tau)} \in H} \mathscr{D}_\tau(u^{(\tau)}).$$

Consider the Fréchet space $C^{\text{loc}}(\mathbf{R}_+; H)$ [23, p. 918]. We denote $T(h)u(\cdot) = u_h(\cdot)$, where $u_h(t) = u(t+h)$ for any $u \in C^{\text{loc}}(\mathbf{R}_+; H)$ and $t, h \ge 0$ [24].

Remark that in the autonomous case the set $\mathscr{K}^+ := \mathscr{K}_0^+$ is *translation semi-invariant*, i.e., $T(h)\mathscr{K}^+ \subseteq \mathscr{K}^+$ for any $h \ge 0$. Such autonomous problems were investigated in [20, Chap. XIII], [25–28], [21, Chap. 2] and references therein; see also [29]. In the nonautonomous case, we have that $T(h)\mathscr{K}_0^+ \nsubseteq \mathscr{K}_0^+$. So, we consider *a united trajectory space* [19] of the following form:

$$\mathcal{K}_{\cup}^{+} := \bigcup_{\tau \geq 0} \left\{ u(\cdot + \tau) \in C^{\mathrm{loc}}(\mathbb{R}_+; H) \, : \, u(\cdot) \in \mathcal{K}_{\tau}^{+} \right\}.$$

Then $T(h)\{u(\cdot + \tau) : u \in \mathcal{K}_{\tau}^{+}\} \subseteq \{u(\cdot + \tau + h) : u \in \mathcal{K}_{\tau+h}^{+}\}$ for any $\tau, h \geq 0$. So, $T(h)\mathcal{K}_{\cup}^{+} \subseteq \mathcal{K}_{\cup}^{+}$ for any $h \geq 0$. Then, we consider *an extended united trajectory space* for Problem (13.1):

$$\mathcal{K}_{C^{\mathrm{loc}}(\mathbf{R}_+; H)}^{+} = \mathrm{cl}_{C^{\mathrm{loc}}(\mathbf{R}_+; H)} \left[\mathcal{K}_{\cup}^{+} \right], \tag{13.4}$$

where $\mathrm{cl}_{C^{\mathrm{loc}}(\mathbf{R}_+; H)}[\,\cdot\,]$ is the closure in $C^{\mathrm{loc}}(\mathbf{R}_+; H)$. Note that

$$T(h)\mathcal{K}_{C^{\mathrm{loc}}(\mathbf{R}_+; H)}^{+} \subseteq \mathcal{K}_{C^{\mathrm{loc}}(\mathbf{R}_+)}^{+} \text{ for any } h \geq 0 \tag{13.5}$$

(cf. [19, 23, 25]).

The following theorem characterizes the compactness properties of shifted solutions for Problem (13.1) in the induced topology from $C^{\mathrm{loc}}(\mathbb{R}_+; H)$.

Theorem 13.1 ([19, Theorem 4.1]) *Let Assumptions (A)–(D) hold. If* $\{u_n\}_{n \geq 1} \subset \mathcal{K}_{C^{\mathrm{loc}}(\mathbb{R}_+; H)}^{+}$ *is an arbitrary sequence, which is bounded in* $L_\infty(\mathbf{R}_+; H)$*, then there exist a subsequence* $\{u_{n_k}\}_{k \geq 1} \subseteq \{u_n\}_{n \geq 1}$ *and an element* $u \in \mathcal{K}_{C^{\mathrm{loc}}(\mathbb{R}_+; H)}^{+}$ *such that*

$$\|\Pi_{\tau,T} u_{n_k} - \Pi_{\tau,T} u\|_{C([\tau,T]; H)} \to 0, \quad k \to +\infty,$$

for any finite time interval $[\tau, T] \subset (0, +\infty)$*. Moreover, for any* $u \in \mathcal{K}_{C^{\mathrm{loc}}(\mathbb{R}_+; H)}^{+}$ *the following estimate holds:*

$$\|u(t)\|_H^2 \leq \|u(0)\|_H^2 e^{-c_3 t} + c_4$$

for any $t \geq 0$*, where positive constants* c_3 *and* c_4 *do not depend on* $u \in \mathcal{K}_{C^{\mathrm{loc}}(\mathbb{R}_+; H)}^{+}$ *and* $t \geq 0$*.*

Let us define the multivalued map $G : \mathbb{R}_+ \times H \to 2^H \setminus \{\emptyset\}$ as

$$G(t, u_0) = \{u(t) \in H \mid u(\cdot) \in \mathcal{K}_{C^{\mathrm{loc}}(\mathbf{R}_+; H)}^{+} \, : \, u(0) = u_0\}. \tag{13.6}$$

Then the multivalued map G is a multivalued semiflow (see [21], (13.4) and (13.5)).

Definition 13.3 (*see* [15, 21]) The set $\Theta \subset X$ is called a uniform global attractor for multivalued semiflow G from (13.6) if the following conditions hold:

- Θ is an attracting set for G, that is for arbitrary bounded nonempty set $B \subset H$

  $$dist_H(G(t, B), \Theta) \to 0 \text{ as } t \to +\infty,$$

  where $dist_H(A, B) = \sup_{x \in A} \inf_{y \in B} ||x - y||_H$ for any non-empty sets $A, B \subset H$.

- Θ is the minimal attracting set, that is $\Theta \subset cl_H Y$ for arbitrary attracting set $Y \subset H$;
- $\Theta \subset G(t, \Theta)$ for all $t \geq 0$.

Uniform global attractor is invariant if $\Theta = G(t, \Theta)$ for all $t \geq 0$.

Definition 13.4 ([30, Definition 2.7]) Multivalued semiflow $\mathcal{G} : \mathbf{R}_+ \times H \to 2^H \setminus \emptyset$ satisfies the flattening property if for arbitrary bounded set $B \subset H$ and $\varepsilon > 0$ there exist $t_0(B, \varepsilon)$ and finite-dimensional subspace E of H such that for bounded projector $P : H \to E$ the set $P(\bigcup_{t>t_0} \mathcal{G}(t, B))$ is bounded in H, and

$$(I - P)(\bigcup_{t>t_0} \mathcal{G}(t, B)) \subset B(0, \varepsilon).$$

The following lemma provides the sufficient condition for justice of flattening property for multivalued semiflow G.

Lemma 13.1 ([30, Lemmas 2.4, 2.6], [31, p. 35]) *Let $\mathcal{G}$ be an asymptotically compact multivalued semiflow in H, that is for arbitrary sequence $\{\varphi_n\}_{n\geq 1} \subset \mathcal{G}$ with $\{\varphi_n(0)\}_{n\geq 1}$ bounded, and for any sequence $\{t_n\}_{n\geq 1}$: $t_n \to +\infty$ as $n \to \infty$, the sequence $\{\varphi_n(t_n)\}_{n\geq 1}$ has a convergent subsequence. Then for $\mathcal{G}$ the flattening property holds.*

13.3 Main Results

The main result of this note has the following formulation:

Theorem 13.2 *Let Assumptions (A)–(D) hold. Then the multivalued semiflow G, defined in (13.6), has a compact uniform global attractor Θ in the phase space H.*

Proof From [21], we have that the following conditions are sufficient for the existence of a compact uniform global attractor for the multivalued semiflow G: for each $t \geq 0$, the mapping $H \ni u \mapsto G(t, u)$ has a closed graph; G is asymptotically compact multivalued semiflow; there exists $R_0 > 0$ such that $\forall R > 0\ \exists T \geq 0$ (depended on R) such that $\forall t \geq T$

$$G(t, \{u \in H \mid \rho(u, 0) \leq R\}) \subset B_0 = \{u \in H \mid \rho(u, 0) \leq R_0\}. \tag{13.7}$$

The first condition follows from (13.6). Theorem 13.1 and assumptions (C), (D) and estimates (13.2), (13.3) provide the asymptotically compactness of multivalued semiflow G and the fulfillment of (13.7).

The following theorem implies that dynamics of all weak solutions of studied problem is finite-dimensional within a small parameter.

Theorem 13.3 *Let Assumptions (A)–(D) on the parameters of Problem (13.1) hold. Then the multivalued semiflow G satisfies the flattening condition.*

Proof The statement of the theorem directly follows from Lemma 13.1 and proof of Theorem 13.2.

Remark 13.3 All statements of Theorems 13.2 and 13.3 hold for function $f(x, t, u)$ equals to the sum of interaction function $f_1(x, t, u)$, satisfying Assumptions (A)–(D), and an external force $g \in L_2^{loc}(\mathbb{R}_+; V^*)$, which satisfies

$$\sup_{t\geq 0} \int_t^{t+1} \|g(s)\|_{V*}^2 ds < +\infty.$$

The proofs are similar with some standard technical modifications.

As applications we may consider Fitz–Hugh–Nagumo system (signal transmission across axons), complex Ginzburg–Landau equation (theory of superconductivity), Lotka–Volterra system with diffusion (ecology models), Belousov–Zhabotinsky system (chemical dynamics) and many other reaction–diffusion-type systems [32], whose dynamics are well studied in autonomous case [9, 20], and in nonautonomous case, when all coefficients are uniformly continuous on time variable (see [20, 21] and references therein).

Acknowledgments This work was partially supported by the Ukrainian State Fund for Fundamental Researches under grant GP/F61/017 and by the National Academy of Sciences of Ukraine under grant 2284.

References

1. Kolmogorov, A.N., Petrovsky, I.H., Piskunov, N.S.: Investigation of the equation of diffusion combined with increasing of the substance and its application to a biological problem. Bull. Mosc. State Univ. Ser. A: Math. Mech. **1**(6), 1–25 (1937)
2. Prigogine, I.: From being to becoming: time and complexity in the physical sciences. Philos. Sci. **51**(2), 355–357 (1984)
3. Field, R.: Experimental and Mechanistic Characterization of Bromate-Ion-Driven Chemical Oscillations and Traveling Waves in Closed Systems. Oscillations and Traveling Waves in Chemical Systems. Wiley-Interscience, New York (1985)
4. Svirezhev, Y.M.: Nonlinear Waves, Dissipative Structures and Catastrophes in Ecology. Nauka, Moscow (1987) (in Russian)
5. Svirezhev, Y.M., Logofet, D.O.: Stability of Biological Communities. Mir, Moscow (1983)
6. Pearson, J.E.: Complex patterns in a simple system. Science **261**, 189–192 (1993)
7. Kyoung, J.L., McCormic, W.D., Ouyang, Q., Swinney, H.L.: Pattern formation by interacting chemical fronts. Science **261**, 192–194 (1993)
8. Babin, A.V., Vishik, M.I.: Attractors of Evolution Equations. Nauka, Moscow (1989)
9. Temam, R.: Infinite-Dimensional Dynamical Systems in Mechanics and Physics. Appl. Math. Sci. Springer, New York (1988)
10. Kapustyan, O.V., Kas'yanov, P.O.: Global attractor for a nonautonomous inclusion with discontinuous right-hand side. Ukr. Math. J. **55**(11), 1765–1776 (2003). doi:10.1023/B:UKMA.0000027041.12041.e8

11. Morillas, F., Valero, J.: Attractors for reaction-diffusion equation in R^n with continuous nonlinearity. Asymptot. Anal. **44**(1–2), 111–130 (2005)
12. Wang, B.: Attractors for reaction-diffusion equations in unbounded domains. Phys. D **128**, 41–52 (1999)
13. Gorban, N.V., Kasyanov, P.O.: On regularity of all weak solutions and their attractors for reaction-diffusion inclusion in unbounded domain. Continuous and distributed systems. Theory and applications. Ser. Solid Mech. Appl. **211**, 205–220 (2014). doi:10.1007/978-3-319-03146-0_15
14. Kasyanov, P.O., Toscano, L., Zadoianchuk, N.V.: Regularity of weak solutions and their attractors for a parabolic feedback control problem. Set-Valued Var. Anal. **21**(2), 271–282 (2013). doi:10.1007/s11228-013-0233-8
15. Kapustyan, O.V., Mel'nik, V.S., Valero, J., Yasinsky, V.V.: Global Attractors for Multivalued Dynamical Systems. Naukova Dumka, Kyiv (2008)
16. Gluzman, M.O., Gorban, N.V., Kasyanov, P.O.: Lyapunov functions for differential inclusions and applications in physics, biology, and climatology. Continuous and distributed systems II. Theory and applications. Ser. Stud. Syst. Decis. Control. **30**, 233–243 (2015). doi:10.1007/978-3-319-19075-4_14
17. Gluzman, M.O., Gorban, N.V., Kasyanov, P.O.: Lyapunov type functions for classes of autonomous parabolic feedback control problems and applications. Appl. Math. Lett. **39**, 19–21 (2015). doi:10.1016/j.aml.2014.08.006
18. Chepyzhov, V., Vishik, M.: Attractors of non-autonomous dynamical systems and their dimension. J. Math. Pures Appl. **73**(3), 279–333 (1994)
19. Gorban, N.V., Kapustyan, O.V., Kasyanov, P.O.: Uniform Trajectory attractor for non-autonomous reaction-diffusion equations with Caratheodory's nonlinearity. Nonlinear Anal. Theory Methods Appl. **98**, 13–26 (2014). doi:10.1016/j.na.2013.12.004
20. Chepyzhov, V.V., Vishik, M.I.: Attractors for Equations of Mathematical Physics. American Mathematical Society, Providence (2002)
21. Zgurovsky, M.Z., Kasyanov, P.O., Kapustyan, O.V., Valero J., Zadoianchuk, N.V.: Evolution Inclusions and Variation Inequalities for Earth Data Processing III. Springer, Berlin (2012). doi:10.1007/978-3-642-28512-7
22. Valero, J., Kapustyan, A.V.: On the connectedness and asymptotic behaviour of solutions of reaction-diffusion systems. J. Math. Anal. Appl. **323**(1), 614–633 (2006)
23. Chepyzhov, V.V., Vishik, M.I.: Evolution equations and their trajectory attractors. J. Math. Pures Appl. **76**, 913–964 (1997)
24. Vishik, M., Chepyzhov, V.: Trajectory and global attractors of three-dimensional Navier-Stokes systems. Math. Notes **71**(1–2), 177–193 (2002)
25. Vishik, M.I., Zelik, S.V., Chepyzhov, V.V.: Strong trajectory attractor for a dissipative reaction-diffusion system. Dokl. Math. **82**(3), 869–873 (2010)
26. Melnik, V.S., Valero, J.: On attractors of multivalued semi-flows and generalized differential equations. Set-Valued Anal. **6**(1), 83–111 (1998)
27. Kasyanov, P.O.: Multivalued dynamics of solutions of autonomous operator differential equations with pseudomonotone nonlinearity. Math. Notes **92**, 205–218 (2012). doi:10.1134/S0001434612070231
28. Kasyanov, P.O.: Multivalued dynamics of solutions of an autonomous differential-operator inclusion with pseudomonotone nonlinearity. Cybern. Syst. Anal. **47**, 800–811 (2011). doi:10.1007/s10559-011-9359-6
29. Balibrea, F., Caraballo, T., Kloeden, P.E., Valero, J.: Recent developments in dynamical systems: three perspectives. Int. J. Bifurc. Chaos **20**(9), 2591–2636 (2010)
30. Kalita, P., Łukaszewicz, G.: Global attractors for multivalued semiflows with weak continuity properties. Nonlinear Anal. **101**, 124–143 (2014)
31. Ball, J.M.: Global attractors for damped semilinear wave equations. DCDS **10**, 31–52 (2004)
32. Smoller, J.: Shock Waves and Reaction-Diffusion Equations (Grundlehren der mathematischen Wissenschaften). Springer, New York (1983)

Chapter 14
Some Problems Connected with the Thue–Morse and Fibonacci Sequences

Francisco Balibrea

Abstract We deal with outstanding properties of the Thue–Morse sequence and consider some of its appearances in combinatorial, symbolic, and physical problems. In particular, we consider the solution of a problem in semigroups theory, and additionally, we consider a system of difference equations associated with a transmission of waves problem, studied in Avishai and Berend, Phys Rev B 45:6873–688, 1991, [1], and Avishai and Berend, Phys Rev B 45:2717–2774, 1992 [2]. Such system has interesting properties from the dynamical point of view, particularly what concerns to periodic points and asymptotic behaviors of most of the non-periodic orbits. Additionally, we state similar problems using Fibonacci and Rudin–Shapiro sequences.

14.1 Introduction

Given two symbols a and b, the Thue–Morse sequence denoted by $(T - M)$ associated with them is a non-periodic sequence given by

$$abbabaabbaababbabaababbaabbabaab....$$

Usually, it is represented in the literature by the following sequence of $0's$ and $1's$

$$\mathbf{t} = (t_n)_{n\geq 0} = 0110100110010...$$

Such sequence is an ubiquitous mathematical object. It comes up in algebra, number theory, combinatorics, topology, and other areas.

$(T - M)$ appeared for the first time in a paper of Eugene Prouhet in 1851 devoted to problems in number theory [19]. However, Prouhet did not mention the sequence

F. Balibrea (✉)
Facultad de Matemáticas, Campus de Espinardo, Universidad de Murcia, 30100 Murcia, Spain
e-mail: balibrea@um.es

V.A. Sadovnichiy and M.Z. Zgurovsky (eds.), *Advances in Dynamical Systems and Control*, Studies in Systems, Decision and Control 69,
DOI 10.1007/978-3-319-40673-2_14

explicitly; this was made by Axel Thue in 1906 [22], who used in combinatorics on words composed of $0's$ and $1's$. The sequence was only brought to mathematical worldwide attention by Marston Morse in 1921 ([15]) in his construction of geodesics in surfaces of negative curvature, and in turn, it was the beginning of a fruitful part of the theory of dynamical systems, called the symbolic dynamics.

$(T - M)$ was rediscovered independently many times, not always by professional research mathematicians. For example, Max Euwe, a chess grandmaster, who held the world championship title from 1935 to 1937, and mathematics teacher, discovered it in 1929 in an application to chess. Using its overlap property, he showed how to circumvent a rule aimed at preventing infinitely protracted games by declaring repetition of moves a draw ([10]).

14.1.1 (T − M) and Some Definitions and Properties

Depending on the problem to solve or simply what properties of the sequence we want to take into account, there has been obtained different approaches for introducing the sequence. In this subsection, we will consider only the approach needed in the developing of the rest of the paper.

Given the *alphabet* {0, 1}, we will call as *word* to a finite sequence of symbols from the alphabet and *letter* to any member of it. If the word is composed of infinite symbols, then it is called an *infinite word* or simply a sequence of symbols. In this paper, we will refer always to words with a finite number of symbols.

Define a sequence of words of 0's and 1's as follows:

$$a_0 = 0, \qquad b_0 = 1$$

$$a_1 = a_0 b_0, \qquad b = b_0 a_0$$

$$\ldots.$$

$$a_{n+1} = a_n b_n, \qquad b_{n+1} = b_n a_n$$

$$a_{n+1} = a_n b_n$$

where the words $a_n b_n$ and $b_n a_n$ are composed of symbols $0's$ and $1's$ in number 2^{n+1} For example, we find

$$a_0 = 0$$

$$a_1 = 01$$

$$a_2 = 0110$$

$$a_3 = 01101001$$

$$a_4 = 0110100110010110$$

$$....$$

and so on. Given two words A *and* B, we denote by AB their *concatenation*; that is, we get a word of more number of symbols by adding to the last symbol of A the symbols of B.

Given any word A, we will denote by A' the word which is composed of the same symbols that A but taken in a inverse order. Immediately, we observe that

$$a_n = a_n', \qquad b_n = b_n', \quad (n \ is \ even)$$

$$a_n = b_n', \qquad b_n = a_n', \quad (n \ is \ odd)$$

A *morphism* on words is a map h that satisfies the identity $h(XY) = h(X)h(Y)$ for all words $XandY$.

Define the Thue–Morse morphism as $\mu(0) = 01$ and $\mu(1) = 10$. Then, the successive iterations of the morphism are as follows:

$$\mu(0) = 01$$

$$\mu^2(0) = \mu(\mu(0)) = 0110$$

$$\mu^3(0) = 01101001$$

$$\mu^4(0) = 0110100110010110...$$

and so on.

Then, it can be proved immediately by induction on n that $\mu^n(0) = X_n$ and $\mu^n(1) = \bar{X}_n$. This process leads to a sequence

$$b = (b_i)_{i=0}^{\infty} = lim_{n\to\infty}\mu^n$$

obtained simply as a concatenation of symbols. In the literature, there others methods of introducing $(T - M)$ (see, e.g., [5]).

An *overlap* on the set of words is a word of the form $aXaXa$ where a is a letter and X is a word. As examples, using all letters of English or Spanish languages, the words *alfalfa* and *entente* (both exist in the two languages) are examples of overlap words. A word is called *overlap-free* if it contains no word that is an overlap.

The $(T - M)$ is an example of a infinite overlap-free word. It was proved in [16] in the result,

Theorem 14.1 *$(M - T)$ does not contain any word B of the form $D\bar{D}d$ holding $D = \bar{D}$ and when d is the initial symbol of D.*

where $\bar{D}$ will be defined at the end of the paragraph. From it may applications have been made. Here, we present two of them, one concerning the solution of a chess problem (solved by Euwe in [10]) and another giving an answer to a problem in semigroup theory.

Another interesting appearance of $(M - T)$ is connected with the notion of uniform recurrence on shift of two symbols and the non-trivial construction of examples of them. Let us denote by $((\Sigma^2, d), \sigma))$ (d is a metric) such shift, where σ is the map *shift* acting on $s \in \Sigma^2$); that is, $(\sigma(s))_n = s_{n+1}$ for all $n = 0, 1, 2, \ldots$. $(T - M)$ is a *uniform recurrent point* with respect to the shift map. The proof of this statement can be done seeing that all in the $(T - M)$ are *syndetic* (see [6] for definitions on recurrence, uniform recurrence, and proofs therein).

On next subsection, we will give a report of one old and interesting problem on the construction of semigroups holding some conditions.

14.1.2 On the Solution of a Problem on Semigroups

In [5], R.P. Dilworth stated that some problems in semigroups theory could be solved using the $(T - M)$. Let S be a semigroup of elements where the operation between $a, b \in S$ (a product) is denoted by ab and where with e we denote the zero element, that is, $ea = ae = e$ for all $a \in S$. If A and B are subsets of S, we denote by AB the set of all products ab taking $a \in A$ and $b \in B$. The semigroup S is said to be *nilpotent* if there exist an integer m such that $S^m = e$. It is wondered if it is possible to construct a non-nilpotent semigroup S generated by three elements (which means that there are $a, b, andc$ in S in such a way that all members of S can be obtained by a finite number of operations of them), and in such a way that $w^2 = e$ for all $w \in S$. It can be done if it is possible to construct an infinite sequence of the three elements containing no word of the form BB where B is a word.

We prove that using the $(T - M)$, it is possible to solve such problem. The proof is just an application of Theorem 14.1 and using some words considered in [17] (see again [16]). The construction is contained in [16], and here, for the aim of completeness and interest, we include it with some little variants. Such proof is constructive and purely combinatorial.

First, we use a process called in [17], *association* in which, given a symbolic sequence it is determined another sequence holding additional properties. In this subsection, $(T - M)$ will be represented by the indexed sequence of symbols $t_0t_1t_2\ldots$ and let denote by B_i the word of two symbols from the sequence starting in the index i. Since $(T - M)$ has two generating symbols 0 and 1, there are at most four different words B_i of two symbols, 00, 01, 10, and 11 which appear in $(T - M)$. Now take the indexed sequence

$$B_0 B_1 B_2 .....(4)$$

which has four generating symbols. Take as S the symbolic sequence of which (4) is an indexed representation. If we denote the $2 - words$ given above by 1, 2, 3, and 4, a simple computation shows that $B_0 B_1 B_2 ...$ begins as follows

$$2432, 3124, 3123, 2432, 3123, 2431$$

Now let us suppose that in the indexed representation, a word D with initial index $i + 1$ occurs and prove the following statement, where if

$$D = t_{i+1} t_{i+2} ... t_{i+\omega}$$

then we will denote by $\bar{D}$, the word

$$t_{i+\omega+1} t_{i+\omega+2} ... t_{i+2\omega}$$

Proposition 14.1 *The symbolic sequence S contains no word of the form $B\bar{B}$ holding $B = \bar{B}$.*

Proof If the statement were not true, it would follow from the construction of S that $(T - M)$ would contain a word of the form $D\bar{D}d$ where $D = \bar{D}$ and d would be the initial symbol of the word D. But this is not possible according to Theorem 14.1.

Now let U be the symbolic sequence obtained from S just changing the index 4 by the index 1. The result is that U would contain the word

$$213231213123213231232131$$

Such symbolic trajectory U allows us to prove the following result which in fact is an answer to the main question in this section

Theorem 14.2 *The symbolic trajectory U has three generators and contains no word of the form $E\bar{E}$ holding $E = \bar{E}$*

Proof It is immediate by the construction of U from S that it has three generators.

Let us suppose that U would contain a word of the form $E\bar{E}$ with $E = \bar{E}$. This would imply the existence of a word CC in S and that $C = \bar{C}$ when 4 be replaced by 1. Suppose that in the indexed representation of S, C would have the representation of $B_{i+1} B_{i+2} ... B_{i+\omega}$ and $\bar{C}$ the representation of $B_{i+\omega+1} B_{i+\omega+2} ... B_{i+2\omega}$. Since the index 1 in S corresponds to 00 in $(T - M)$ and 4 in S corresponds to 11 in $(T - M)$, while $(T - M)$ contains neither the words 000 or 111, it follows that the index 1 in S must be preceded by the index 3, while 4 must be preceded by 2 and followed by 3.

Let us ω denote the number of symbols of a word. If $\omega = 1$, from the hypothesis $E = \bar{E}$, it would follow that either $C = \bar{C}$ or $C\bar{C} = 14$ or $C\bar{C} = 41$. But the former proposition implies that the first possibility is impossible, and since 1 is followed by 2 and preceded by 3 in S, the second and third cases are also impossible. Therefore, we cannot have $\omega = 1$.

If $\omega > 1$, let j be index holding $i + 1 < j < i + \omega$. Using the hypothesis $E = \bar{E}$, we claim that it must be $B_{i+\omega} = B_{i+2\omega}$ unless $B_{i+\omega} = 4$ and $B_{i+2\omega} = 1$ or viceversa. But $B_i = 4$ implies $B_{ii+3} = 3$ and therefore $B_{i+1+\omega} = 3$, and consequently, $B_{i+\omega} = 4 = B_i$. If $B_i = 1$, then $B_{i+1} = 2$, and then, $B_{i+1+\omega} = 2$ and $B_{i+\omega} = 1 = B_i$. Thus we have

$$B_i = B_{i+\omega}, \qquad for \quad i + 1 \le j \le i + \omega$$

But this implies that $C = \bar{C}$ which is impossible according to Proposition 14.1.

The assumption that U contains a word of the form $E\bar{E}$ holding $E = \bar{E}$ leads to a contradiction and the proof is complete.

14.2 A Problem on Transmission of Waves

In [1, 2], Y. Avishai and D. Berend considered the transmission of a wave described by a parameter denoted by $|t_N|$ (in modulus) and reflection $|r_N|$ of a plane wave (number wave given by $k > 0$) through a one-dimensional array of N δ-function potentials having equal strengths ν placed on a Thue–Morse chain sequence x_n with distances d_1 and d_2 when $N \to \infty$.

By means of number theoretical theory and analytic methods, such authors obtain the following interesting results which describes the physics of the problem.

(1) For any k, if ν is large enough, the sequence of reflection coefficients $(|r_N|)_{N=1}^{\infty}$ has a subsequence that converges exponentially to unity.
(2) If k is an integer multiple of $\frac{\pi}{|d_1 - d_2|}$, then there exists a threshold value ν_0 for the values of ν, such that for $\nu \ge \nu_0$ is $|r_N|_{N\to\infty} = 1$. If $\nu < \nu_0$, then it is $|r_n| \ne 1$. In fact, something more can be said. Is $lim \quad sup_{N\to\infty}|r_N| < 1$ and $lim \quad inf_{N\to\infty}|r_N| = 0$.
(3) For other values of k, it is claimed that if k is not a multiple of $\frac{\pi}{|d_1 - d_2|}$ always the sequence $(|r_N|)_{N=1}^{\infty}$ has a subsequence tending to unity independently of ν except for a set of measure zero.
(4) After numerical simulations, it seems that if we test that the above sequence has a subsequence converging to unity, then the whole sequence is converging to unity.

The central problem to be solved is the case when we are dealing with quasicrystals and trying to decide whether a one-dimensional array behaves as conductor ($|r_N|_{N\to\infty} \ne 1$) or an insulator ($|r_N|_{N\to\infty} = 1$). More specifically, there is or not a curve in the (ν, k) parameter space separating the conductor and insulator domains.

Given a one-dimensional array of $N - \delta$-function potentials

$$V(x) = v \sum_{n=1}^{N} \delta(x - x_n)$$

where $\nu > 0$ and x_n is given assuming that

$x_{n+1} - x_n = y_n$ takes only two positive values d_1 or d_2 depending on if $\xi_n = 0\, or\ 1$, where $\xi_n = [1 + (-1)^{s(n)}]/2$ and $s(n)$is the number of ones in the binary expansion of n.

A plane wave with momentum k, given by e^{-ikx} (coming from right) will have refection and transmission amplitudes r_N and t_N, respectively, when crossing the array. When $N = 1$, we have

$$r_1 = \frac{\nu}{2ik - \nu}, \qquad t_1 = \frac{2ik}{2ik - \nu}$$

holding unity and continuity at the point x_0 conditions

$$|r_1|^2 + |t|^2 = 1, \quad t_1 r_1^\star + t *_1 \star = 0 \qquad (1)$$

where $a^\star$ means the conjugate of the complex number a and

$$t_1 = 1 + r_1$$

The unitary condition (1) is held for any N. If $N > 1$, the reflection and transmission amplitudes are determined for the following recursion. First, we introduce additional notation

$$a_n = \frac{1}{t_n}, \quad b_n = \frac{r_n}{t_n},$$

$$A_1 = \begin{pmatrix} 1/t_1 & -r_1/t_1 \\ r_1/t_1 & (t_1^2 - r_1^2)/t_1 \end{pmatrix}$$

$$\Lambda_n = \begin{pmatrix} e^{-iky_n} & 0 \\ 0 & e^{iky-n} \end{pmatrix}$$

Then $D_n = A_1 \Lambda_n$ with

$$det(A_1) = det(\Lambda_n) = det(D_n) = 1$$

This D_n is the transfer matrix at the site n. Taking into account the product of n transfer matrices, we define

$$M_n = \Lambda_n A \Lambda_{n-1} ... A \Lambda_1 A$$

Then, we obtain

$$(a_{n+1} \qquad b_{n+1}) = A_1 \Lambda_n (a_n \qquad b_n)$$

The *conductance* of the initial system (one-dimensional array) is given by $lim_{N\to\infty}|t_N|^2 = \frac{1}{|a_N|^2}$. This is equivalent to obtain the limit of $|r_N|^2 = |\frac{b_N}{a_N}|^2$. This leads to the following criterium. When $|t_N| \to 0$ (or equivalently $|r_N \to 1$), we say that the systems behaves like an insulator. When $|t_N|$ does not converges to 0, then the system may conduct. At this point, what it is really interesting is to find out for what values of the parameters momentum k and the strength ν, the system is an insulator or a conductor.

The matrices A_1, Λ_n, and M_n belong to the multiplicative group of 2, defined by

$$SU(1,1) = \begin{pmatrix} \alpha & \beta \\ \beta^\star & \alpha^\star \end{pmatrix}$$

where $\alpha, \beta \in C$, $|\alpha|^2 - |\beta|^2 = 1$.

Given the sequence $(y_n)_{n=1}^{\infty}$, our main problem is deciding the values of ν and k for which we have $|r_N|_{N\to\infty} \to 1$. Before considering the Thue–Morse sequence and independently of the values reached by y_n (only that there are two values) in [1, 2], it is proved that for every k multiple of $\frac{\pi}{(d_1-d_2)}$, there is a threshold value ν_0 such that

(1)

$$|r_N| \to 1 \Leftrightarrow \nu \geq \nu_0 \quad (both\ \ positives)$$

$$\nu_0 = 2ktg\frac{k}{2} \quad if \ \ sin\ \ k < 0, \qquad -2ktg\frac{k}{2} \quad sin\ \ k < 0$$

(2) When $|r_N|_{N\to 1}$, then the sequence r_N lies on a circle of diameter $\frac{q}{|sin\ \ k-q\ \ cos\ \ k|}$ (<1) passing through the origin and additionally

$$lim\ \ sup|r_N| < 1; \qquad lim\ \ inf|r_N| = 0$$

This result covers the values of k which are integer multiples of $mm = \frac{\pi}{(d_1-d_2)}$. Further, we will assume that is not a multiple of m. We state

$$\Phi = kd_1, \qquad \Psi = kd_2$$

and define recurrently two sequences of matrices $(P_n)_{n=0}^{\infty}$ and $(Q_n)_{n=0}^{\infty}$ belonging to $SU(1,1)$

$$P_0 = \begin{pmatrix} e^{-i\Phi} & 0 \\ 0 & e^{i\Phi} \end{pmatrix},$$

$$Q_0 = \begin{pmatrix} e^{-i\Psi} & 0 \\ 0 & e^{i\Psi} \end{pmatrix}$$

$$P_n = Q_{n-1}P_{n-1}, \qquad Q_n = P_{n-1}Q_{n-1} \qquad (1)$$

where A denotes the transfer matrix introduced before. We obtain by a direct calculation that

$$M_{2^n} = P_n \qquad n \geq 0.$$

In [2], it is made some numerical simulations. As a consequence, he claimed that the behavior of last subsequence $(M_{2^n})_{n=0}^{\infty}$ can be taken as the behavior of the whole sequence. To see the behavior of the sequence $(P_n)_{n=1}^{\infty}$, we are considering what is called the trace map. Let us denote by $\chi_n = tr(P_n), n \geq 0$. It is immediate that $tr(Q_n) = \chi_n, n \geq 1$.

Our main purpose now is to find out for which the values of k and ν the sequence of norms $(||P_n||)_{n=1}^{\infty}$ converges to infinity and for which not. Instead, in order to simplify computations, we will use the sequence of traces $(\chi_n)_{n=1}^{\infty}$. In this case if $|\chi_n|_{n\to\infty}$, then $||P_n||_{n\to\infty}$. If $|\chi_n|_{n\to\infty}$, then all results can appear concerning the other sequence (for a discussion in this point, see [2]).

Also in [2], it is proved that if $(P_n)_{n=0}^{\infty}$ and $(Q_n)_{n=0}^{\infty}$ are any two sequences of matrices from $SU(1, 1)$, holding (1) and $\chi_n = tr(P_n)$, then

$$\chi_{n+2} = \chi_n^2(\chi_{n+1} - 2) + 2, \qquad n \geq 1, \qquad (2)$$

To study the sequence $(\chi_n)_{n=1}^{\infty}$, we will consider the unfolding of the nonlinear difference equation (2) by the planar transformation $H : R^2 \to R^2$

$$H(x, y) = (y, x^2y - 2x^2 + 2)$$

Using the map, $\gamma : R^2 \to R^2$ given by

$$\gamma(x, y) = (x^2, y)$$

which is a semiconjugacy, it is immediate that $\gamma \circ H = T \circ \gamma$. Further, we will deal with the difference equation given by T which we will call the Thue–Morse difference equation since it is generated using the idea of such a sequence and properties we are remarked before.

The Thue–Morse sequence is also connected with the description of quasicrystals. In place of every 0, we can image an atom placed in a small square and when have an 1 another different atom. If we construct an infinite array of squares, we have an image of an 1D-quasicrystal.

14.2.1 Dynamics of the Thue–Morse System

The previous problem on transmission leads to the following system of difference equations

$$x_{n+1} = x_n(4 - x_n - y_n)$$

$$y_{n+1} = x_n y_n$$

The system can be seen as a two-dimensional dynamical system given by the pair (R^2, T) where

$$T(x, y) = (x(4 - x - y), xy)$$

It is easy to test that systems

$$S(x, y) = ((y - 2)^2, xy)$$

and

$$B(x, y) = (xy, (x - 2)^2)$$

are topologically conjugate in R^2 to $T(x, y)$; that is, there exist bijections in R^2, Φ and Ψ such that

$$\Phi \circ T = S \circ \Phi$$

and

$$\Phi \circ T = B \circ \Psi$$

These conjugations are interesting since the dynamical properties of the systems are kept by topological conjugation, in particular existence of periodic orbits, density of them, transitivity, etc. In the paper, we are dealing with the Thue–Morse transformation given in the form $T(x, y) = (x(4 - x - y), xy)$ from R^2 into itself which it is a nonlinear transformation.

We will call the *trajectory* of a point $P \in R^2$ the sequence $(T^n(P))_{n=0}^{\infty}$ where with $T^n = T \circ (T^{n-1})$ and $T^0(P) = P$ for all points in R^2. We will call *orbit* of P the set of points of the trajectory of P.

The most interesting part of the dynamics is concentrated in the interior of the triangle Δ obtained connecting the three points $(0, 0)$, $(4, 0)$, $and (0, 4)$. The line l connecting $(4, 0)$ and $(0, 4)$ is given by $x + y = 4$, and we will denote by Γ_1 the

Fig. 14.1 The partition of Δ

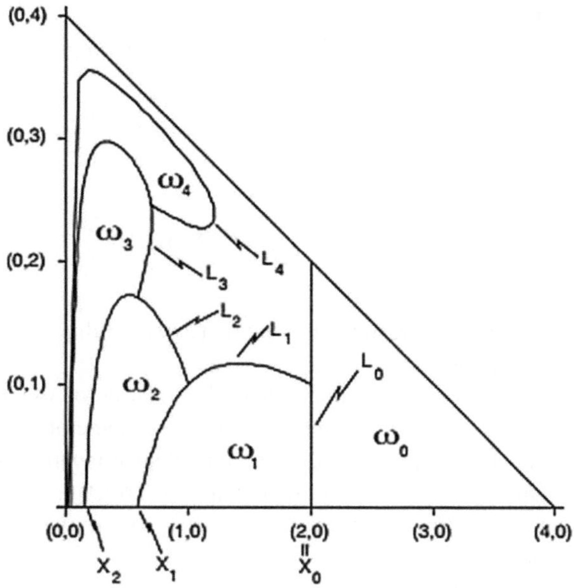

segment connecting these two points, by Γ_2 the segment connecting (0, 0) and (0, 4), and γ the segment connecting (0, 0) and (4, 0).

Inside Δ, there is a sequence of domains whose interiors are pairwise disjoint and such that the terms of the orbit of a point stating in any of them follow the sequence of domains $\omega_0, \omega_1, \omega_2, ..., \omega_n$.... It means that if a starting point P belongs to ω_i, then $T(P)$ belongs to ω_{i+1} and so on. See Fig. 14.2, where their boundaries inside Δ are denoted by L_i and by X_i their intersections with the axis $y = 0$ (see Fig. 14.1).

The dynamics is easily understood if we split Δ into two sets

$$\Delta = \Delta_l \cup \omega_0$$

where $\Delta_l = \{(x, y) : 0 < x < 2\}$ and $\omega_0 = \{(x, y) : 2 < 4\}$. Since every point in $int\Delta$ has two preimages, the map T is not invertible in it, but it is easy to see that the restriction to $int\Delta_l$ and $int\omega_0$ it is. In fact, can be obtained explicit expressions of the inverse maps of such restrictions (see [3]). Of interest is the segment $\{(2, y) : 0 < y < 2\}$ which is the channel of communication of the inside of Δ and the point (0, 0) through the boundary of Δ.

Since all points outside Γ_1 but belonging to the line l are preimages of (0, 0), and there are no preimages of (3, 0) inside Δ since all of them belong only to γ. The following decomposition of Δ is very effective.

Proposition 14.2 (see [3])
The triangle Δ can be decomposed into the following way

$$\Delta = (\bigcup_{n=0}^{\infty} F^{-n}(0,0)|\Delta) \cup (\bigcup_{n=0}^{\infty} F^{-n}(1,2)) \cup (\bigcup_{n=0}^{\infty} F^{-n}(3,0))\cup$$

$$\cup(I(\bigcup_{n=0}^{\infty} F^{-n}(3,0)) \cup \bigcup_{n=0}^{\infty} F^{-n}(0,0))) \cup R$$

Because of this result, it is clear that any periodic orbit of T, if they exist, must belong to the ser R.

Taking pieces of curve from boundaries of the ω-sets mentioned above, it is possible to obtain invariant sets with the shape of a spiral. In Fig. 14.2, we joint the points $(1, 2)$ and $(0, 0)$ through one of such invariant spirals.

In Fig. 14.3, we have made with the Program R, a representation of three orbits starting in three relevant part of $int D$ which give us an idea of the complexity of orbits.

It seems that there exists a arrow strip parallel to γ which attracts initially to almost points from $int\Delta$, but when the iterates of the point are closed to it, then it is an repelling effect. The topological and geometrical structure of such a string is not yet known.

The dynamics on $\partial\Delta$ (boundary of Δ) and $int\Delta$ is easy to known.

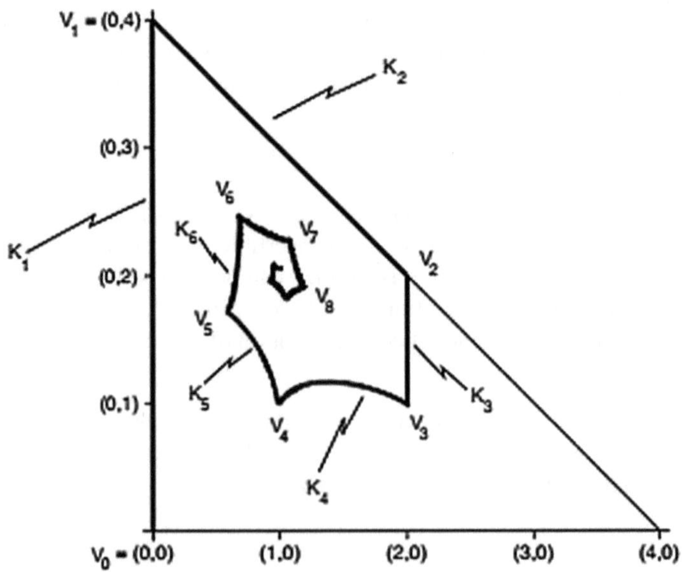

Fig. 14.2 Invariant set as a spiral

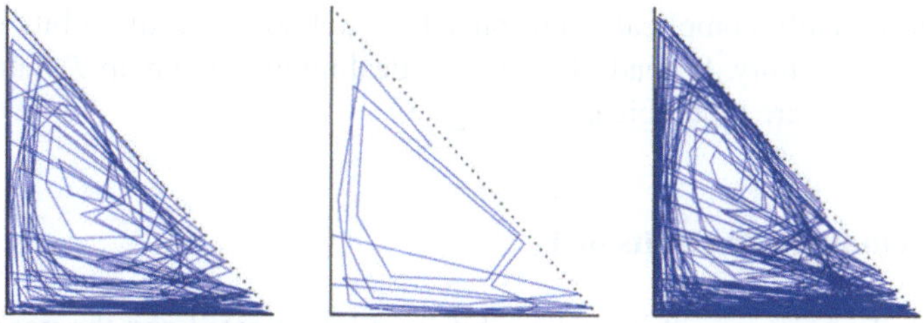

Fig. 14.3 On left we have of orbit of (0.5, 3.2), on center that of (1.5, 2.2), and on right that of (3.5, 0.2). The three after 10.000 iterations

Proposition 14.3 *The map $T|\partial\Delta$ verifies:*

(1)

$$T(\partial\Delta) = \partial\Delta$$

(2)

$$T(int\Delta) = int\Delta$$

Proof 1 is immediate. To prove 2, we take into account that points (x, y) in $int\Delta$ fulfills the conditions, $x > 0$, $y > 0$, $4 - x - y > 0$. If (X, Y) is the image of (x, y), we have $X = x(4 - x - y) > 0$, $Y = xy > 0$ and $4 - X - y = (x - 2)^2 > 0$ except for the points of γ.

14.2.2 Sharkovskii's Program

In 1993 in a conference in Oberwolfach, A. Sharkovskii motivated by the former cited paper [2] proposed investigation on the two-dimensional dynamical system

$$S(x, y) = ((y - 2)^2, xy)$$

and stated a program of research with the following questions:

(1) Are the periodic points of S dense in Δ?
(2) Is $H|\Delta$ transitive?
(3) Is Γ_1 an attractor of Δ in Milnor's sense?
(4) Does there exist a point P such that $\omega_S(P)$ (the ω-limit set of the point P under S) be unbounded but holding $\omega_S(P) \cap \Gamma_1 \neq \emptyset$

The aim of the rest of the paper was to deal with the above problems and with other stated in the literature (see [3, 7, 12, 20]). We have answered to part of Sharkovskii's questions and complete the knowledge of the dynamics of the map $T(x, y)$ outside Δ. We have completed the analysis with graphical simulations. The dynamics inside the triangle is really complicated but outside is not so, since the orbits of almost all points go to infinity. Instead of S, we are dealing with the map T, but we have obtained that they are topologically conjugate.

14.2.2.1 On Periodic Orbits of T

On [3, 12], the existence of periodic orbits was investigated and the possibility of the existence or not of forcing patterns of periodic orbits in the line of Sharkovskiĭ results was stated. This is the case in the restriction of T to the segment γ, but this nothing new since in γ, T behaves as like one-dimensional.

We are summarizing the above-referred papers and complete some aspects of them concerning periodic orbits.

By elementary algebraic computations, it is easy to see that inside Δ, there is only one fixed point, $(1, 2)$. At the boundary of Δ, we have two fixed points, $(0, 0)$ and $(3, 0)$. Outside the triangle, we have no fixed point. There are no two-periodic points, neither three periodic points.

Using the algebraic method of *resultant* (see [3]), we obtain that the interior point

$$(1 - \sqrt{2}/2, 1 + \sqrt{2}/2)$$

is periodic of period 4. By a numerical approach in [3], it proved the existence of a unique periodic points of period 5 and by direct calculus that

$$(1, (3 + \sqrt{5})/2))$$

is a periodic orbit of period six.

Using an adapted symbolic dynamics to this problem, P. Malicky has shown that for $n \geq 4$, there is point in int(Δ) of period n. It remains open if such points are unique or not.

The key point of Malicky's proof is to prove that given a saddle periodic point P in γ, there exists in the interior of Δ a periodic point having the same itinerary. It is interesting to have a criterium to prove the existence of saddle points in γ. In fact, let $P = [4sin^2(k\pi/(2^n + (-)1), 0]$, where $n > 0$ and k are integer numbers. If

$$1 \leq k \leq \frac{\sqrt{2}(2^n + (-)1)}{\pi 2^{\sqrt{2n+1/4}}},$$

then P is a saddle fixed point of T^n (see [11–13]). The following Fig. 14.4 allows us to see the saddle nature of the internal periodic points in the triangle. In such figure, it is represented with several colors from red to yellow the sum of the distances between two consecutive points the orbits starting in points of the triangle according with a technique of representation introduced in [9].

The restriction $T|[0, 4]$ is the logistic parabola $p(x) = x(4 - x)$. We will recall some properties concerning periodic points. Let $I = [0, 1]$ be the unit interval and

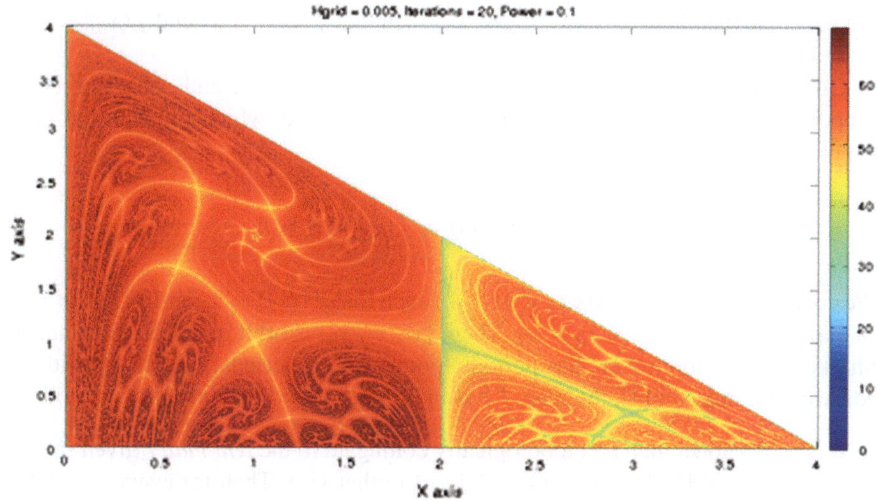

Fig. 14.4 graphical representations of saddle points inside Δ

$f(x) = 4x(1-x)$ (equivalent to $p(x)$) be the logistic map. The map f is onto and two to one for all points in I except for $x = \frac{1}{2}$.

Lemma 14.1 *For every $p \in N$, the map f has a periodic orbit of period p*

Proof We claim that the map f has in I a periodic point of period 3. In fact,

$$f(0) = f(1) = 0, \qquad f(\frac{1}{2}) = 1;$$

therefore, $f^3(0) = 0$. We also know that $f'(\frac{1}{2}) = 0$.

Let $x_0 = (1 - \sqrt{\frac{1}{2}})/2$, thus, $0 > x_0 < \frac{1}{2}$ and $4x_0(1-x_0) = \frac{1}{2}$; that is, $f(x_0) = \frac{1}{2}$. And therefore

$$f^3(x_0) = f^2(\frac{1}{2}) = f(1) = 0,$$

then $0 < x_1 < x_0$ and $f(x_1) = x_0$

Let

$$x_1 = \frac{1}{2}(1 - \sqrt{\frac{1+\sqrt{1/2}}{2}})$$

On the other hand, $(f^3)'(x) = f'(f^2(x))f'(f(x))f'(x)$. Using the above property, we have

$$f^3(x_1) = f^2(x_0) = f'(\frac{1}{2}) = 1 > x_1$$

We conclude that there exists $x_2 \in (x_1, x_0)$ such that $f^3(x_2) = x_2$ and $f(x_2) = x_1 > x_2$ and $f^2(x_2) = f(x_1) = x_0 > x_2$.

On other hand is $(f^3)'(x) = f'((f^2(x))f'((f(x))f'(x)$, and using the above property, we have $f'(f^2(x_1) = f'(f(x_0)) = f'(\frac{1}{2}) = 0$, which implies that $(f^3)'(x_1 = 0$. Then, we obtain that

$$f^3(x_1) = f^2(x_0) = f(\frac{1}{2}) = 1 > x_1$$

We conclude that there exists $x_2 \in (x_1, x_0)$ such that $f^3(x_2) = x_2$ and $f(x_2) = x_1 > x_2$ and $f^2(x_2) = f(x_1) = x_0 > x_2$. Then, $\{x_1, x_2, x_0\}$ is orbit of period three and by Sharkovskii's theorem [21] has periodic points for all periods. According with the construction made in [4], the map f has in $[x_1, x_3]$ periodic points of all periods.

It is well known that f is topologically conjugate to the *tent map t* given in I by $t(x) = 2x$ if $x \in [0, \frac{1}{2}]$, and $t(x) = 2(1 - x)$ otherwise. Then for every $n \geq 1$ and for all $0 \leq k \leq 2^n$ is

$$t^n(\frac{k}{2^n}) = 0$$

if k is even and equal to 1 otherwise. It proves that in $[\frac{k-1}{2^n}, \frac{k}{2^n}]$, t has precisely one fixed point. Then, we have proved that

Lemma 14.2 *The set of periodic points of f is dense in I.*

In order to study with more detail the existence of periodic points in the interior of Δ, it is suitable to consider another chart. For this, we introduce a linear change of variable denoted by $\Phi(x, y) = (4 - x - y, y) = (u, v)$. Now, we introduce a new map G by

$$G(u, v) = (4 - u - v)(u, v)$$

This new map, G, is again onto on Δ which is also invariant, that is, $G(\Delta) = \Delta$. It is easy to see that

Lemma 14.3 *(1) The set of fixed points of G is*

$$\{(0, 0)\} \cup \{(u, v) \in \Delta : u + v = 3\}$$

(2)

$$G^{-1}(\{(0, 0)\}) = \{(0, 0)\} \cup \{(u, v) \in \Delta : u + v = 4\}$$

(3) The maps T and G are related by the formulas $F \circ \Phi = G$ *for every* $(x, y) \in R^2$

Now, we have the interesting result on the periodic points of maps T and G

Theorem 14.3 *The map* $G = T \circ \Phi$ *has periodic orbits of all orders. Moreover, its sets of periodic points is dense in* Δ.

Proof Let (x_0, y_0) a point of Γ, that is, $0 \leq x_0 \leq 4$ and $y_0 = 4 - x_0$. It will be denoted by $S(x_0)$ the segment joining the points $(0, 0)$ and (x_0, y_0), that is

$$S(x_0) = t(x_0, y_0)) : 0 \leq t \leq 1$$

If we compute now $G(t(x_0, y_0)$, we have

$$G(t(x_0, y_0)) = (4 - tx_0 - ty_0) = (4 - tx_0 - ty_0) = (4 - tx_0 - 4t + 4t_0)(tx_0, t(4 - x_0)) =$$

$$= (4(1 - t)(tx_0, t(4 - x_0)) = (4t(1 - t)x_0, 4t(1 - t)(4 - x_0)) = (4t(1 - t)x_0, 4t(1 - t)y_0))$$

Therefore, the point (x_0, y_0) belongs to $S(x_0)$ since $4t(1 - t) \leq 1$. The map $f(t) = 4t(1 - t)$ has periodic orbits of all periods, and as a consequence, it is possible to choose orbits of all periods. Given a periodic orbit of $f(t)$, $(t_1, t_2, ..., t_n)$, we compute the points $t_i(x_0, y_0)$ for $i = 1, ..., n$ which all belong to $S(x_0)$ and form a periodic orbit of period n which is contained in $S(x_0)$. Therefore in such a segment, there periodic points of G of all periods. The procedure for finding them is changing the value x_0.

Remark 14.1 Since the maps T and G are not topologically conjugate, the existence of periodic orbits for G does not imply they are automatically transmitted to T. For example, it is the case with periodic orbits of period two and three which appear for map G but not for T. In fact, we claim that the set of periodic points in $\int D$ is not dense, although currently we are not able to propose an easy argument of it.

14.2.2.2 Dynamics on $\partial \Delta$ and Axes $x = 0$ and $y = 0$

All points of the form $(0, y)$ with $y \in R$ are preimages of $(0, 0)$. The images of points of the segment $\gamma = \{(x, 0)\}$ with $0 \leq x \leq 4$ remains in γ and $T|\gamma = (f(x), 0)$ with $f(x) = x(4 - x)$ and $T(\gamma) = \gamma$.

In the axis $y = 0$, all points outside γ_2 transform into point with negative abscise. It is $T(4, \infty) = (-\infty, 0)$ and $T(-\infty, 0) = (-\infty, 0)$, that is

$$A = T((4, \infty) \cup T(-\infty, 0)) = (-\infty, 0),$$

and it is immediate to test that $lim_{n \to \infty} f^n(x) = -\infty$ for every $x \in A$.

Every point (x, y) belonging to the line $x + y = 4$ is transformed into points $(0, Y)$ where $Y = xy$ and the sign of X depends on signs of x and y. It is evident that $T(\partial \Delta) = \partial \Delta$. Therefore, the sets $\partial \Delta$ and $(-\infty, 0)$ are invariant by T.

It is immediate to see that T is not invertible in Δ since $(0, 0)$ has many preimages, but restricted to the sets

$$\Delta_- = \{(x, y) : 0 < x < 2\}$$

and

$$\Delta_+ = \{(x, y) : 2 < x < 4\}$$

It is invertible and the open segment $\alpha = \{(2, y)\}$ with $0 < y < 2$ is composed of preimages of third order of $(0, 0)$, that is, $T^3(2, y) = (0, 0)$ for every point in α which it is a channel of communication of the inside and the boundary of Δ. In fact, the set of preimages of α is dense in Δ. In particular, all points belonging to Γ are eventually fixed to $(0, 0)$ of order two. The point $(0, 0)$ has also in γ infinitely many preimages of all orders.

14.2.2.3 Outside the Triangle Δ

Let us consider a partition of R^2 into open domains shown in Fig. 14.3. Such domains have as boundaries the line $\{(x, y) : x + y = 4\}$ and/or segments of the axes. The next result states the description of all orbits starting in interior of such domains.

Theorem 14.4 *The behavior of the domains with respect to orbits starting in their points is as follows.*

(1) The first iteration of all points belonging to Domain 1, belongs to Domain 3, except the points $\{(2, y) : y > 2\}$ *whose first iterate belongs to the set* $\{(x, y) : x + y = 4, x < 0\}$ *and consequently their third orbit is the point* $(0, 0)$. *It means there is a supply of iterations from points outside the triangle to its boundary*

(2) All points in Domain 2 *are transformed in points belonging to Domain* 5 *while all points from Domain 6 are transformed in points of Domain* 4.

(3) All points of Domain 3 *are transformed in points of Domain* 4 *and viceversa. This means that the orbit of every point from Domain* 3 *and* 4 *oscillates around the axis* $y = 0$ *and*

$$lim_{n\to\infty}||T^n(x, y)|| = \infty$$

(4) Domain 6 *is* T*-invariant and the orbit of every point verifies also*

$$lim_{n\to\infty}||T^n(x, y)|| = \infty$$

Proof What it is important is the position of points in the plane. For points (x, y), up the line l is $4 - x - y < 0$. For points in l is $4 - x - y = 0$ and down l is $4 - x - y > 0$.

It remains open the description of the dynamics of the map outside Δ. Analytically and graphically it is proved that outside the triangle there are no periodic points. Using Fig. 14.3, we appreciate the movement of points starting in each of the regions. In particular, it is proved that Domain 6 is invariant and there are or not periodic points.

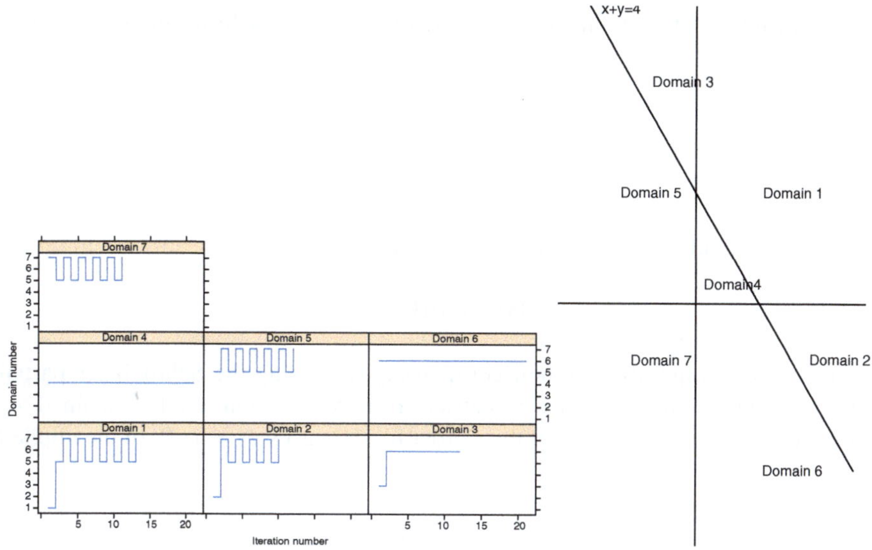

Fig. 14.5 The number of points that remain in the different domains of the plane after taking initial points in different domains

We claim that there are some periodic points, a non-wandering set of points and the rest tending to infinite when $n \to \infty$. The set

$$\Delta_1 = \{(x, y) : x, y \geq 0, \qquad x + y \geq 4\}$$

has no wandering points and the set

$$\Delta_2 = \{(x, y) : x, y \geq 0, \qquad x + y \leq 4\}$$

is also T-invariant and can contain non-wandering points. Using the former arguments, the answer to problems 3 and 4 from Sharkovskii is negative, (see in [14] for the definition of attractor on Milnor sense).

According to numerical experiments, it seems that most points go to infinite when $n \to \infty$ (Fig. 14.5).

14.2.3 A Fibonacci System

In [1], Y. Avishai and D. Berend considered a similar problem than what we have considered before for the case of a Thue–Morse chain. The definitions of reflection and transmission coefficients are the same than in Thue–Morse problem. Now, we wonder for the values of k and ν such that $|r_N|_{N\to\infty} = 1$ or there is no convergence.

For the introduction of the Fibonacci sequence, we use the following substitutions rules:

$$0 \rightarrow 01$$

$$1 \rightarrow 0$$

and with these rules, we obtain the mentioned sequence.

$$(00101001001....)$$

with it we construct also a Fibonacci quasicrystal. Using the Schrodinger partial differential equation, trace maps of matrices from $SO(1, 1)$ in [1], it is obtained the following difference equations of third order for the traces $\chi_n = tr P_n$ of the implied matrices

$$\chi_{n+3} = \chi_{n+1}\chi_{n+2} - \chi_n$$

whose unfolding

$$F(x, y, z) = (y, z, yz - x)$$

is a nonlinear dynamical system from R^3 into itself. There is a few work made on it. We claim that the orbits of all points outside a bounded set in R^3 are converging to infinity. In this bounded set is seem to concentrate a complicate dynamics.

Similar problems are open to be considered for the case of Rudin–Shapiro systems (see [8, 18]).

Remark 14.2 Many thanks to Carlos Lopesino allowing us to reproduce Fig. 14.4. The research has been supported by the Proyecto MTM2014-51891-P from Spanish MINECO and from Research Project 19294/PI/14 supported by the Fundacion Seneca—the Regional Agency of Science and Technology of the Regional Government of Murcia (Spain) in the setting of PCTIRM 2011–2014.

References

1. Avishai, Y., Berend, D.: Transmission through Fibonacci chain. Phys. Rev. B **45**, 6873–6880 (1991)
2. Avishai, Y., Berend, D.: Transmission through a Thue-Morse chain. Phys. Rev. B **45**, 2717–2774 (1992)
3. Balibrea, F., García-Guirao, J.L., Lampart, M., Llibre, J.: Dynamics of a Lotka-Volterra map. Fundam. Math. **191**, 265–279 (2006)
4. Du, B.-S.: A simple proof of Sharkovsky's theorem. Am. Math. Monthly **111**, 595–599 (2004)
5. Dilworth, R.P.: The arithmetical theory of Birkhoff lattices. Trans. Am. Math. Soc. **8**, 286–299 (1941)

6. Furstenberg, H.: Recurrence in Ergodic Theory and Combinatorial Number Theory. Princeton University Press, Princeton (1981)
7. García-Guirao, J.L., Lampart, M.: Transitivity of a Lotka-Volterra map. Discret. Contin. Dyn. Syst. Ser B **9**(1), 75–82 (2008)
8. Kohmoto, M., Kadanoff, L.P., Tang, C.: Localization problem in one dimension:mapping and scape. Phys. Rev. Lett. **50**, 1860–1872 (1983)
9. Lopesino, C., Balibrea-Iniesta, F., Wiggins, S., Mancho, A.M.: Lagrangian descriptors for two dimensional area preserving autonomous and non autonomous maps. Commun. Nonlinear Sci. Number Simul. **27**(1–3), 152–166 (2015)
10. Max, E.: Mengentheoretische Betrachtungen ber das Schachspiel. Proc. Koninklijke Nederlandske Akademie van Wetenschappen **32**, 633–642 (1929)
11. Maličk P.: On a number of interior points of a Lotka-Volterra map. Acta Univ. Matthiae Belli Ser. Math. **19**, 21–30 (2011)
12. Maličk, P.: Interior periodic points of a Lotka-Volterra map. J. Differ. Equ. App. **18**(4), 553–567 (2012)
13. Maličk, P.: Modified Lotka-Volterra maps and their interior periodic points. Proceedings ESSAIM-2014 (2014)
14. Milnor, J.: On the concept of attractor. Commun. Math. Phys. **99**, 177–195 (1985)
15. Morse, M.: Recurrent geodesics on a surface of negative curvature. Trans. Am. Math. Soc. **22**, 84–100 (1921)
16. Morse, M.: Unending chess, symbolic dynamics and a problem in semigroups. Duke Math. J. **11**, 1–7 (1944)
17. Morse, M., Hedlund, A.: Symbolic dynamics. Am. J. Math. **60**, 815–866 (1938)
18. Ostlund, S., Pandit, R., Rand, D., Schnellnhuber, H.J., Siggia, E.D.: One dimensional Schredinger equation. Phys. Rev. Lett. **60**, 1873–1876 (1983)
19. Prouhet, E.: Memoire sur quelques relations entre les puissances des nombres. C. R. Acad. Sci. Paris Ser. I **33**, 225 (1851)
20. Swirszcz, G.: On a certain map of the triangle. Fundam. Math. **155**, 45–57 (1998)
21. Sharkovskii, A.N.: Low dimensional dynamics, Tagengsbericht 20/1993. Proceedings of Mathematisches Forschungsinstitut Oberwolfach, vol. 17 (1993)
22. Thue, A.: Uber die gegenseitige lafe gleicher teite gewisser Zeicheruresihen. Kra. Vidensk. Selsk., Skrifter 1, Mat.-Nat.Kl., Chistiana, Nr. 10 (1912)

Chapter 15
Existence of Chaos in a Restricted Oligopoly Model with Investment Periods

Jose S. Cánovas

Abstract The aim of this paper is to give a proof of the existence of chaos in an oligopoly model stated in Cánovas, Panchuk, and Puu, Math. Comput. Simul. 117, pp 20–38 (2015) [14]. For that, we consider a restricted case of the model and prove that for a wide range of parameter values, the topological entropy is positive, implying that the model exhibits topological chaos. In addition, we discuss whether this topological chaos is also physically observable, that is, can be shown in a computer simulation.

15.1 Introduction

In [14], an oligopoly model with investment periods was introduced. In oligopoly models, several firms compete in a market where some economic rules are fixed (see, e.g., [6, 33] or [27]). The model studied in this chapter, which will be introduced in the next section, is quite hard to analyze from an analytically point of view. However, numerically, it has shown the existence of clusterization of firms that make their investments at the same time, and moreover, in a periodic way. In this case, there are some invariant subsets in which the model can adopt a simplified form, and the restricted model can be analyzed mathematically, proving for instance, in a rigorous way, the existence of chaos on it. This is the main aim of this paper.

The paper is organized as follows. The next section will be devoted to introduce the model in a concise way. We refer the reader to [14] for a complete model description. Then, we will make a short introduction of the dynamical systems techniques that we are going to use to describe the model, and finally, we apply the above-mentioned techniques to study our model.

J.S. Cánovas (✉)
Departamento de Matemática Aplicada y Estadística, Universidad Politécnica de Cartagena, C/ Doctor Fleming sn, 30202 Cartagena, Spain
e-mail: Jose.Canovas@upct.es

V.A. Sadovnichiy and M.Z. Zgurovsky (eds.), *Advances in Dynamical Systems and Control*, Studies in Systems, Decision and Control 69,
DOI 10.1007/978-3-319-40673-2_15

15.2 The Model

In our model, $n+1$ firms compete in a market in which firms produce similar goods. Each firm i produces a quantity q_i and possesses a capital k_i. The capital decreases on time, so the model has an investment period, long run, where the capital is renovated, and a production period called short run. The model is constructed in such a way each firm profit Π_i is maximized in both long run and short run. If Q denotes the total output of all the firms, and $Q_i = Q - q_i$ denotes the residual supply for firm i, the model has the form for the short run

$$\begin{aligned} q_i(t+1) &= F_{w,\theta}(q_i(t), Q_i(t), k_i(t)), \\ k_i(t+1) &= k_i(t), \\ T_i(t+1) &= T_i(t) - \kappa^{q_i(t)-\sqrt{\frac{r}{c}}k_i(t)}, \end{aligned}$$

where

$$F_{w,\theta}(q_i, Q_i, k_i) := \begin{cases} \theta k_i \frac{\sqrt{\frac{Q_i}{w}}-Q_i}{k_i+\sqrt{\frac{Q_i}{w}}} + (1-\theta)\, q_i, & Q_i \le \frac{1}{w}, \\ (1-\theta)\, q_i, & Q_i > \frac{1}{w}, \end{cases}$$

and for the long run

$$\begin{aligned} q_i(t+1) &= G_{c,\theta}(q_i(t), Q_i(t)), \\ k_i(t+1) &= \frac{\sqrt{c}}{\sqrt{r}} G_{c,\theta}(q_i(t), Q_i(t)), \\ T_i(t+1) &= T. \end{aligned}$$

where

$$G_{c,\theta}(q_i, Q_i) := \begin{cases} \theta \left(\sqrt{\frac{Q_i}{c}} - Q_i \right) + (1-\theta)\, q_i, & Q_i \le \frac{1}{c}, \\ (1-\theta)\, q_i, & Q_i > \frac{1}{c}. \end{cases}$$

Of course, we must explain the meaning of the variables involved in the model, and some relation among them. The model is discrete, and sequences are denoted by $q(t)$, $t \in \mathbb{N} \cup \{0\}$. The maps $F_{w,\theta}$ and $G_{c,\theta}$ are called reaction functions. T_i represents the lifetime of capital, while T denotes its maximum length. Denote capital rent and wage rate as r and w, respectively, and the long-run unit cost is $c = \left(\sqrt{r} + \sqrt{w}\right)^2$, which are the economic constants of the model. The parameter θ is chosen to construct the model with adaptative expectations. When $\theta = 1$, we receive naive expectations, in which firms expect to win the maximum value of the previous period, given by the reaction functions. The parameter κ must be greater than zero, and it is related with the life of capital. Therefore, the model depends on too many parameters. In this paper, we will only consider the naive expectation case, that is, we fix $\theta = 1$.

As it can be checked in [14], we may assume without loss of generality that $w = 1$, and so the number of parameters is reduced in one. On the other hand, all the numerical simulations made in [14] establish that firms do invest in a periodic way, and there is a clusterization process in the firm investment. This is important because in the case of firm clusterization, we can study a simplified model instead the original one.

Here, we consider the limited case of firm clusterization; that is, we suppose that all the firms do invest at the same time. On the other hand, initially we consider that capital is zero in two periods of time, and then, firms play the short run just once. In other words, we are going to consider the simplified model in which we alternate the long run and the short run, that is,

$$\begin{aligned} q_i(t+1) &= F_1(Q_i(t), k_i(t)), \\ k_i(t+1) &= k_i(t), \end{aligned}$$

where

$$F_1(Q_i, k_i) := \begin{cases} k_i \frac{\sqrt{Q_i} - Q_i}{k_i + \sqrt{Q_i}}, & Q_i \leq 1, \\ 0, & Q_i > 1, \end{cases}$$

and for the long run

$$\begin{aligned} q_i(t+1) &= G_c(Q_i(t)), \\ k_i(t+1) &= \frac{\sqrt{c}}{\sqrt{r}} G_c(Q_i(t)). \end{aligned}$$

where

$$G_c(Q_i) := \begin{cases} \sqrt{\frac{Q_i}{c}} - Q_i, & Q_i \leq \frac{1}{c}, \\ 0, & Q_i > \frac{1}{c}, \end{cases}$$

and T_i is either 0 or 1 for $i = 1, 2, \ldots, n+1$.

In any case, when the map G_c is evaluated, all the future outputs depend on Q_i and no longer on the initial value of k_i; that is, the second iteration is given by

$$q_i(t+2) = \max \left\{ 0, \frac{\sqrt{c}}{\sqrt{r}} G_c(Q_i(t)) \frac{\sqrt{\sum_{j \neq i} G_c(Q_j(t))} - \sum_{j \neq i} G_c(Q_j(t))}{\frac{\sqrt{c}}{\sqrt{r}} \sqrt{\sum_{j \neq i} G_c(Q_j(t)) + G_c(Q_i(t))}} \right\}, \tag{15.1}$$

and

$$k_i(t+2) = \frac{\sqrt{c}}{\sqrt{r}} G_c(Q_i(t)). \tag{15.2}$$

Obviously, neither the variable Q_i nor k_i depend on k_i, so we can avoid Eq. (15.2) and concentrate our efforts in analyzing Eq. (15.1). It is easy to see that the diagonal

$$\Delta = \{(q, q, \ldots, q) \in \mathbb{R}^{n+1} : q \geq 0\}$$

is invariant by the system, and on it, the Eq. (15.1) reads as

$$\begin{aligned} q(t+2) &= \frac{\sqrt{c}}{\sqrt{r}} G_c(nq(t)) \frac{\sqrt{nG_c(nq(t))} - nG_c(nq(t))}{\frac{\sqrt{c}}{\sqrt{r}}\sqrt{nG_c(nq(t))} + G_c(nq(t))} \\ &= \sqrt{c}G_c(nq(t)) \frac{\sqrt{nG_c(nq(t))} - nG_c(nq()t)}{\sqrt{c}\sqrt{nG_c(nq(t))} + \sqrt{r}G_c(nq(t))}, \end{aligned}$$

where $G_c(nq) := G_c(q, q, \ldots, q) = \max\{0, \sqrt{nq/c} - nq\}$.

For a proper analysis of the dynamics of the difference equation

$$q(t+2) = f(q(t)) = \max\{0, \varphi(q(t))\},$$

where

$$\varphi(q) = \sqrt{c}G_c(nq) \frac{\sqrt{nG_c(nq)} - nG_c(nq)}{\sqrt{c}\sqrt{nG_c(nq)} + \sqrt{r}G_c(nq)}$$

we need to know the number of extrema, called turning points. For that, we consider the auxiliary map

$$g(q) = \sqrt{c}q \frac{\sqrt{nq} - nq}{\sqrt{c}\sqrt{nq} + \sqrt{r}q}$$

and compute the solutions of the equation

$$g'(q) = 0,$$

which gives us

$$\widetilde{q} = \left(\frac{\sqrt{r} - 3n(1+\sqrt{r}) + \sqrt{(n + (n+1)\sqrt{r})(9n + (1+9n)\sqrt{r})}}{4\sqrt{nr}} \right)^2.$$

Since

$$\varphi'(q) = (g \circ G_c)'(q) = g'(G_c(nq))G_c'(nq) = 0$$

it gives us all the possible turning points. It is easy to see that G_c attains its maximal value when $q = \frac{1}{4nc} = \frac{1}{4n(1+\sqrt{r})^2}$ and $G_c\left(\frac{1}{4n(1+\sqrt{r})^2}\right) = \frac{1}{4(1+\sqrt{r})^2}$. Solving the equation

$$G_c(nq) = \widetilde{q},$$

with solutions

$$q^{\pm} = \frac{1 \pm \sqrt{1 - 4\tilde{q}(1+\sqrt{r})^2} - 2\tilde{q}(1+\sqrt{r})^2}{2n(1+\sqrt{r})^2}$$

gives us that the model has either one maximum point (when $4\tilde{q}(1+\sqrt{r})^2 \geq 1$) or two maximum points and a minimum one (when $4\tilde{q}(1+\sqrt{r})^2 < 1$).

On the other hand, it is easy to see that the equation $\varphi(q) = 0$ has the solutions $q_0 = 0$, $q_1 = \frac{1}{nc}$, and

$$q_{\pm} = \frac{n - 2c \pm \sqrt{n(n-4c)}}{2n^2c}.$$

It is easy to see that when $n > 4c$, then $q_{\pm} < \frac{1}{nc}$. We distinguish 6 cases, called 1A, 1B, 2A, 2B, 3A, and 3B, which correspond with different shapes of the map f, which are shown in Figs. 15.1, 15.2 and 15.3.

Now, the rest of the paper is organized as follows. First, we make a short introduction of basic mathematical background useful to understand the mathematical

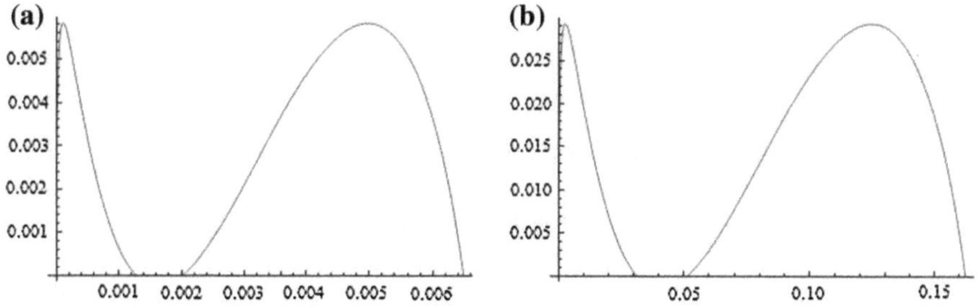

Fig. 15.1 We show the graphs of maps of type 1A (**a**) and 1B (**b**). On the *left*, we fix 26 firms and $r = 2.2$, while on the *right*, the number of firms is 6 and $r = 0.012$. In case 1A, the map fulfills that $f(q^-) \geq q_-$, while in the case 1B, we have the opposite inequality

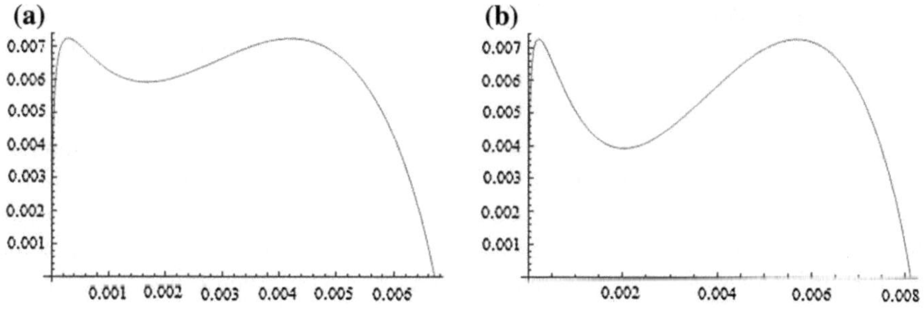

Fig. 15.2 We show the graphs of maps of type 2A (**a**) and 2B (**b**). On the *left*, we fix 21 firms and $r = 3$, while on the *right*, the number of firms is 21 as well and $r = 2.2$. In case 2A, the map fulfills that $f(q^-) \geq \frac{1}{cn}$, while in the case 1B, we have the opposite inequality. In both cases, the maps have two maxima and one minimum

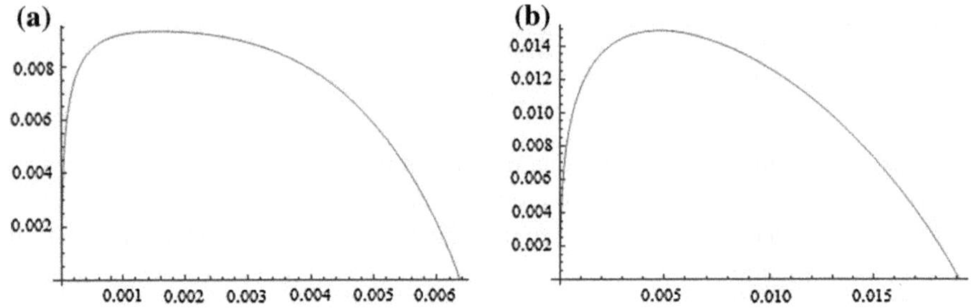

Fig. 15.3 We show the graphs of maps of type 3A (*left*) and 3B (*right*). On the *left*, we fix 16 firms and $r = 5$, while on the *right*, the number of firms is 6 as well and $r = 5$. In case 2A, the map fulfills that $f(\frac{1}{4cn}) \geq \frac{1}{cn}$, while in the case 1B, we have the opposite inequality. In both cases, the maps have one maximum

analysis of the model which will be done in the last section. Some conclusions and open problems can be found at the end of the chapter.

15.3 Mathematical Tools

The models considered in this chapter are given by *difference equations*, which are expressions with the form

$$\begin{cases} x(t+1) = f_t(x(t)), \\ x(0) = x_0, \end{cases}$$

where $f_t : X \to X$, $t \in \mathbb{N}$, is a sequence of maps on a metric space X into itself and $x_0 \in X$. The solution of the above difference equation is called *orbit* or *trajectory* of x_0 under f_t. When the sequence of maps is constant, that is, $f_t = f$, $t \in \mathbb{N}$, we have an autonomous difference equation, which is usually seen as a *discrete dynamical system*, usually denoted by the pair (X, f).[1] Then, the orbit of x_0 under f, denoted $\mathrm{Orb}(x_0, f)$, is given by the sequence $f^t(x_0)$, $t \geq 0$, where $f^t = f \circ f^{t-1}, t > 1, f^1 = f$, and f^0 is the identity on X.

Although one can study topological properties of dynamical systems, in this chapter, we are interested in the case $X = \mathbb{R}^n_{\geq}$, where $\mathbb{R}_{\geq}$ represents the set of nonnegative real numbers. There is a huge literature on discrete dynamical systems either for the one-dimensional case, when $n = 1$ (see, e.g., [2, 8] or [15]) or for more higher dimensions a general spaces (see, e.g., [3, 16]). Here, we introduce some basic results and notation on dynamical systems on general metric spaces which can be easily translated for real maps.

[1] In this section we denote the map defining the dynamical systems with the same letter f that we use in our model. We think that readers will not be confused with this notation.

15.3.1 Periodic Orbits and Topological Dynamics

We consider a metric space (X, d), which is usually compact, and a continuous map $f : X \to X$ and recall that (X,f) denotes a discrete dynamical system. Note that all the definitions below can be expressed either in terms of the map f or the system (X,f).

To understand the dynamics of f, we have to introduce some definitions, which have topological roots, to obtain some knowledge of the system (see, e.g., [8] or [28]). A point $x \in X$ is *periodic* when $f^t(x) = x$ for some $t \geq 1$. The smallest positive integer satisfying this condition is called the *period* of x. Periodic points of period 1 are called *fixed points*. Denote by F(f), P(f), and Per(f) the sets of fixed and periodic points and periods of f, respectively.

Periodic orbits are the simplest orbits that a discrete dynamical system can generate, but there are many other classes of orbit which makes richer the dynamics. For $x \in X$, define its *ω-limit set*, $\omega(x,f)$, as the set of limit points of its orbit $\mathrm{Orb}(x,f)$. If $\omega(x,f)$ is finite, then it is a periodic orbit, but often, the dynamical behavior of a single orbit can be very complicated or unpredictable, and usually, the word chaos is used to refer to dynamical systems which are able to produce such a complicated orbits as we discuss below.

Previously, note that to understand the dynamics, it is enough to do it on small subsets of X called *attractors*, which are non-empty compact sets A that attract all trajectories starting in some neighborhood $\mathcal{U}$ of A, that is, for all $x \in \mathcal{U}$, we have that

$$\lim_{t\to\infty} dist(f^t(x), A) = 0,$$

where $dist(x, A) = \min\{d(x, y) : y \in A\}$. When $\mathcal{U}$ is the whole space X, we have a *global attractor*. The existence of attractors makes easier the understanding of the dynamics, which in principle may be very complex. The existence and approximate location of attractors are usually given by the *absorbing sets*; namely, a subset $B \subset X$ is an *absorbing set* if for any bounded set D of X there is $t_0 = t_0(D)$ such that $f^t(D) \subset B$ for all $n \geq t_0$.

There are many definitions of chaos, but we will focus our interest in the following well-known ones. The map f is *chaotic in the sense of Li and Yorke* (*LY-chaotic*) [21] if there is an uncountable set $S \subset X$ (called *scrambled set* of f) such that for any $x, y \in S$, $x \neq y$, we have that

$$\liminf_{t\to\infty} d(f^t(x), f^t(y)) = 0,$$

$$\limsup_{t\to\infty} d(f^t(x), f^t(y)) > 0.$$

Li and Yorke's definition of chaos became famous because of the famous result *period three implies chaos* which linked periodic orbits and unpredictable dynamical behavior for continuous interval maps. Note that the definition implies the comparison between two orbits or limit points of orbits. Another well-known chaos definition, inspired in the notion of sensitivity with respect to the initial conditions [22], was given by Devaney [16] as follows. The map f is said to be *chaotic in the sense of Devaney (D-chaotic)*; if it is *transitive* (there is $x \in X$ such that $\omega(x,f) = X$), the set of periodic points P(f) is dense in X and it has *sensitive dependence on initial conditions*, that is, there is $\varepsilon > 0$ such that for any $x \in X$ there is an arbitrarily close $y \in X$ and $t \in \mathbb{N}$ such that $d(f^t(x), f^t(y)) > \varepsilon$.[2] Both Li–Yorke chaos and sensitivity to initial conditions are in the dynamical systems folklore.

There is a big field between periodic orbits and chaotic behavior, so it is interesting to explain what is simple dynamics. In fact, sometimes, the chaotic behavior can be also taken as the opposite of simple (or ordered) behavior. We say that f is *strongly simple* (*ST-simple*) if any ω-limit set is a periodic orbit of f. We say that an orbit Orb(x,f), $x \in X$, is approximated by periodic orbits if for any $\varepsilon > 0$, there is $y \in$ P(f) and $t_0 \in \mathbb{N}$ such that $d(f^t(x), f^t(y)) < \varepsilon$ for all $t \geq t_0$. The map f is *LY-simple* [31] if any orbit is approximated by periodic orbits. Finally, f is *Lyapunov stable* (*L-simple*) [19] if it has equicontinuous powers.

The above definitions are quite difficult to verify, and specially when we are working with models which in principle may depend on several parameters, we need some practical methods to try to measure the dynamical complexity of the system. One of them is given by *topological entropy*, which was introduced in the setting of continuous maps on compact topological spaces by Adler, Konheim and McAndrew [1], and Bowen [11].[3] It is remarkable that both definitions agree when the set is metric and compact. It is a conjugacy invariant[4] which is usually taken as a criteria to decide whether the dynamics is complicated or not according to $h(f)$ greater than zero or not. Here, we introduce the equivalent definitions by Bowen [11] when (X, d) is a compact metric space. Given $\varepsilon > 0$, we say that a set $E \subset X$ is *(t, ε, f)-separated* if for any $x, y \in E$, $x \neq y$, there exists $k \in \{0, 1, \ldots, t-1\}$ such that $d(f^k(x), f^k(y)) > \varepsilon$. Denote by $s(t, \varepsilon, f)$ the biggest cardinality of any maximal (t, ε, f)-separated set in X. Then, the topological entropy of f is

$$h(f) = \lim_{\varepsilon \to 0} \limsup_{t \to \infty} \frac{1}{t} \log s(t, \varepsilon, f).$$

There is an equivalent definition using spanning sets as follows. We say that a set $F \subset X$ *(t, ε, f)-spans* X if for any $x \in X$, there exists $y \in F$ such that $d(f^i(x),$

[2]It is proved in [5] that the first two conditions in Devaney's definition implies the third one. The definitions is presented in the original form because of the dynamical meaning of sensitive dependence on initial conditions.

[3]Dinaburg [17] gave simultaneously a Bowen like definition for continuous maps on a compact metric space.

[4]Two continuous maps $f : X \to X$ and $g : Y \to Y$ are said to be topologically conjugate if there is an homeomorphism $\varphi : X \to Y$ such that $g \circ \varphi = \varphi \circ f$. In general, conjugate maps share many dynamical properties.

$f^i(y)) < \varepsilon$ for any $i \in \{0, 1, \ldots, t-1\}$. Denote by $r(t, \varepsilon, f)$ the smallest cardinality of any minimal (t, ε, f)-spanning set in X. Then, topological entropy can be computed as

$$h(f) = \lim_{\varepsilon \to 0} \limsup_{t \to \infty} \frac{1}{t} \log r(t, \varepsilon, f).$$

The above definitions do not depend on the metric d and give us a nice interpretation of topological entropy (see [2, p. 188]) as follows. Imagine that we have a magnifying glass through which we can distinguish two point if and only if they are more than ε-apart. If we know t points of two orbits given by x and y, that is, $(x, f(x), \ldots, f^{t-1}(x))$ and $(y, f(y), \ldots, f^{t-1}(y))$, then we can distinguish between x and y if and only if $\max_{1 \le i \le t} d(f^i(x), f^i(y)) > \varepsilon$. Hence, $s(t, \varepsilon, f)$ gives us how many points of the space X we can see if we know the pieces of orbits of length t. Then, we take the exponential growth rate with t of this quantity and finally the limit of this as we take better and better magnifying glasses. Then, we obtain the topological entropy.

In general, the above chaos definitions are not equivalent and their relations with topological entropy are not homogeneous. For instance, it has been proved that D-chaotic maps are LY-chaotic [20], but the converse is false [31]. On the other hand, positive topological entropy implies LY-chaos [7][5] and the converse is also false [31]. In [4, 23], it studied the relationship between topological entropy and D-chaos. ST-simple maps are LY-simple maps, but the converse is false [31].

More popular than topological entropy are the so-called Lyapunov exponents (see [26]), which make sense when differentiable structures are considered. Namely, assume that X is a smooth finite dimensional manifold and $f : X \to X$ is a $C^{1+\alpha}$ map. Denote, as usual, by T_xX the tangent space at x and the derivative $d_xf : T_xX \to T_{f(x)}X$. The Lyapunov exponent at $x \in X$ in the direction of $\mathbf{v} \in T_xX \backslash \{\mathbf{0}\}$ is defined by

$$\exp(x, \mathbf{v}) = \lim_{t \to \infty} \frac{1}{t} \log ||d_xf^t(\mathbf{v})||$$

if this limit exists. An invariant measure μ is a probability measure on the Borel sets of X such that $\mu(f^{-1}(A)) = \mu(A)$ for any Borel set $A \subseteq X$. This invariant measure μ is ergodic if the equality $f^{-1}(A) = A$ implies that $\mu(A)$ is either 0 or 1. The multiplicative ergodic theorem states that the above limit exists for μ-almost all point in X. We use Lyapunov exponents in some particular cases, to show the existence of physically observable chaos.

Next, we study the particular case of real maps. We will see how the above result is sharpened for continuous interval maps. In addition, we will give some notions on the dynamics of several dimensional real maps.

[5] See also [32] which almost simultaneously states the same result for C^2 diffeomorphisms on compact manifolds of dimension greater than one.

15.3.2 Dynamics of Continuous Interval Maps

In general, for one-dimensional maps, the relevant results are given when $X = [a, b] \subset \mathbb{R}$ is a compact interval, usually $[0, 1]$ due to linear conjugacy. In this setting, Sharkovsky's theorem is a remarkable result which helps to distinguish between simple and complicated dynamics. Recall Sharkovsky's order of natural numbers

$$3 >_s 5 >_s 7 >_s \ldots >_s 2 \cdot 3 >_s 2 \cdot 5 >_s \ldots >_s 2^2 \cdot 3 >_s 2^2 \cdot 5 >_s \ldots$$

$$\ldots >_s 2^k \cdot 3 >_s 2^k \cdot 5 >_s \ldots >_s 2^3 >_s 2^2 >_s 2 >_s 1.$$

Applying Sharkovsky's theorem (see [28] or [2]. Also [18] for an "easy" proof), one can see that for any continuous map $f : \mathbb{R} \to \mathbb{R}$ with one periodic point holds that either $\mathrm{Per}(f) = \mathrm{S}(m) = \{k : m >_s k\} \cup \{m\}$, with $m \in \mathbb{N}$, or $\mathrm{Per}(f) = \mathrm{S}(2^\infty) = \{2^n : n \in \mathbb{N} \cup \{0\}\}$. A map is of *type* $m \in \mathbb{N} \cup \{2^\infty\}$ if $\mathrm{Per}(f) = \mathrm{S}(m)$. A map f is called *S-chaotic* if $\mathrm{Per}(f) = \mathrm{S}(m)$, $m = 2^r q$, $r \geq 0$, and $q > 1$ odd.

On the other hand, for one-dimensional dynamics, the topological entropy is an useful tool to check the dynamical complexity of a map because it is strongly connected with the notion of *horseshoe* (see [2, p. 205]). We say that the map $f : [0, 1] \to [0, 1]$ has a k-horseshoe, $k \in \mathbb{N}$, $k \geq 2$, if there are k disjoint subintervals J_i, $i = 1, \ldots, k$, such that $J_1 \cup \ldots \cup J_k \subseteq f(J_i)$, $i = 1, \ldots, k$.[6] It is well-known that if f has a k-horseshoe, then its topological entropy is greater or equal to $\log k$ ([2, Chap. 4]).

The following result shows some equivalences among the above definitions of chaos and order (see [8, 28, 31]). Note that the situation is simpler than in the general case.

Theorem 15.1 *Let $f : [0, 1] \to [0, 1]$ be a continuous map. Then,*

(a) The map f has positive topological entropy if and only if the map f is S-chaotic.
(b) If f is D-chaotic, then $h(f) > 0$.
(c) If f is either ST-simple or L-simple, then $h(f) = 0$.
(d) If $h(f) > 0$, then f is LY-chaotic, but the converse is false in general. If f is LY-simple, then $h(f) = 0$. The union of LY-chaotic and LY-simple continuous maps is the set of continuous interval maps.

The nature of the above result is topological. If we consider another points of view, we can obtain more information giving rise to apparently strange paradoxes. For instance, there exist maps with positive entropy and therefore chaotic in some sense, such that the orbit of almost all points in $[0, 1]$ (with respect to the Lebesgue measure) converges to a periodic orbit.

Although we will come back to this point later, let us show how to get such example. Consider f a C^3 *unimodal* map such that $f(0) = f(1) = 0$. Recall that a

---

[6] Since Smale's work (see [30]), horseshoes have been in the core of chaotic dynamics, describing what we could call random deterministic systems.

map f is said to be unimodal if there is $c \in [0, 1]$, called *turning point* such that $f|_{[0,c)}$ is strictly increasing and $f|_{[c,1]}$ is strictly decreasing. The *Schwarzian derivative* (see [29] or [34]) is then given by

$$S(f)(x) = \frac{f'''(x)}{f'(x)} - \frac{3}{2}\left(\frac{f''(x)}{f'(x)}\right)^2,$$

at those points whose first derivative does not vanish. Assume that $S(f)(x) < 0$ and that there is a locally attracting periodic orbit, that is, a periodic orbit $P = \{x_1, \dots, x_p\}$ for which there exists a neighborhood V of P such that for any $x \in V$ the distance $d(f^t(x), P) = \min_{1 \le i \le p} d(f^t(x), x_i)$ tends to zero as t tends to infinity. The logistic map $f(x) = 3.83\,x(1-x)$ is a good example of such behavior; almost all trajectory converges to a periodic orbit of period 3, while the topological entropy is positive (see, e.g., [9]). This example and many others in the literature show that it is important to study dynamics from several points of view.

15.3.3 Piecewise Monotone Maps: Entropy and Attractors

Usually, one-dimensional difference equation models in science are given by piecewise monotone maps. A continuous interval map is *piecewise monotone* if there is a finite partition of $[0, 1]$, $0 = x_0 < x_1 < \ldots < x_k = 1$, such that $f|_{[x_i, x_{i+1}]}$ is monotone for $0 \le i < k$. Note that a piecewise monotone map may have constant pieces. The extreme points, intervals included if there exist, of f will be called *turning points* (intervals). For a piecewise monotone map f, let $c(f)$ denote the number of pieces of monotone of f. If g is another piecewise monotone map, it is easy to see that $c(f \circ g) \le c(f)c(g)$. Hence, the sequence $c(f^t)$ gives the number of monotonicity pieces of f^t, and the following result due to Misiurewicz and Szlenk (see [25]) shows that for piecewise monotone maps, topological entropy can be easily understood.

Theorem 15.2 *Let $f : [0, 1] \to [0, 1]$ be a continuous and piecewise monotone map. Then,*

$$h(f) = \lim_{t \to \infty} \frac{1}{t} \log c(f^t).$$

Note that $c(f^t) \le c(f)^t$, and so $h(f) \le \log c(f)$. Hence, a consequence of Misiurewicz–Szlenk theorem is that homeomorphisms on the interval have zero topological entropy. On the other hand, following Theorem 15.2, we can easily see that the logistic map $f(x) = 4x(1-x)$ and the tent map $g(x) = 1 - |2x - 1|$ have topological entropy $\log 2$, since $c(f^t) = c(g^t) = 2^t$ for all $t \in \mathbb{N}$. However, computing topological entropy can be a very complicated task, but we will see in what follows how to make these computations for a suitable class of maps.

The dynamics of smooth enough piecewise monotone maps is well-known in the following sense. Following [24], a metric attractor is a subset $A \subset [0,1]$ such that $f(A) \subseteq A$, $O(A) = \{x : \omega(x,f) \subset A\}$ has positive Lebesgue measure, and there is no proper subset $A' \subsetneq A$ with the same properties. The set $O(A)$ is called the *basin* of the attractor.

By [35], the regularity properties of f imply that there are three possibilities for its metric attractors for a class of piecewise monotone maps, called *multimodal maps*, fulfilling the following assumptions. There are $c_1 < c_2 < \ldots < c_k$, creating a partition on $[0,1]$, such that f is strictly monotone on each element of the partition. f is C^3, and f is non-flat on the turning points $c_1, \ldots, c_k$, that is, for x close to c_i, $i = 1, 2, \ldots, k$,

$$f(x) = \pm|\phi_i(x)|^{\beta_i} + f(c_i),$$

where ϕ_i is C^3, $\phi_i(c_i) = 0$, and $\beta_i > 0$. Then, the metric attractors of such multimodal maps can be of one of the following types:

(A1) A periodic orbit.
(A2) A solenoidal attractor, which is basically a Cantor set in which the dynamics is quasi-periodic. More precisely, the dynamics on the attractor is conjugated to a minimal translation, in which each orbit is dense on the attractor. The dynamics of f restricted to the attractor is simple; neither positive topological entropy nor Li–Yorke chaos can be obtained. Its dynamics is often known as quasi-periodic.
(A3) A union of periodic intervals $J_1, \ldots, J_k$, such that $f^k(J_i) = J_i$ and $f^k(J_i) = J_j$, $1 \le i < j \le k$, and such that f^k is topologically mixing. Topologically, mixing property implies the existence of dense orbits on each periodic interval (under the iteration of f^k).

Moreover, if f has an attractor of type (A2) and (A3), then they must contain the orbit of a turning point, and therefore, its number is bounded by the turning points. In addition, if $Sf(x) < 0$, then the total number of attractors is bounded by k. From a practical point of view, in a computer simulation, we are able to show the existence of attractors of type (A1) and (A3), and only attractors of type (A3) are able to exhibit unpredictable dynamics. As a conclusion of this, if all the turning points of f are attracted by periodic orbits, then the map f will not exhibit physically observable chaos, although it can be topologically chaotic.

The Lyapunov exponents on the turning points can be computed by

$$\exp(c_i) = \lim_{t\to\infty} \frac{1}{t} \log|(f^t)'(c_i)| = \lim_{t\to\infty} \frac{1}{t} \log|f'((f^{t-1})(c_i))|,$$

for $i = 1, 2, \ldots, k$, and all of them are negative when the map f is free of attractors of type (A3). So, positive Lyapunov exponents imply the existence of observable chaos.

15.3.4 Computing Topological Entropy

The above definition of topological entropy is not useful in practice, and counting monotone pieces of an iterated map f^t are not easy. In addition, an exact computation of topological entropy for continuous interval maps cannot be done in general, but there are several papers devoted to compute it approximately for unimodal maps (see [9]) and bimodal maps, that is, with three monotone pieces (see [10]) and four monotone pieces (see [13]). In general, it is possible to make computations for arbitrarily large monotone pieces whenever the number of so-called kneading sequences will not be big enough (see [12]).

Now, we introduce the unimodal case where the topological entropy can be computed by using kneading sequences as follows. Let f be an unimodal map with maximum (turning point) at c. Let $k(f) = (k_1, k_2, k_3, \ldots)$ be its kneading sequence given by the rule

$$k_i = \begin{cases} R & \text{if } f^i(c) > c, \\ C & \text{if } f^i(c) = c, \\ L & \text{if } f^i(c) < c. \end{cases}$$

We fix that $L < C < R$. For two different unimodal maps f_1 and f_2, we fix their kneading sequences $k(f_1) = (k_n^1)$ and $k(f_2) = (k_n^2)$. We say that $k(f_1) \leq k(f_2)$ provided there is $m \in \mathbb{N}$ such that $k_i^1 = k_i^2$ for $i < m$ and either an even number of k_i^1's are equal to R and $k_m^1 < k_m^2$ or an odd number of k_i^1's are equal to R and $k_m^2 < k_m^1$. Then, it is proved in [9] that if $k(f_1) \leq k(f_2)$, then $h(f_1) \leq h(f_2)$. In addition, if $k_m(f)$ denotes the first m symbols of $k(f)$, then if $k_m(f_1) < k_m(f_2)$, then $h(f_1) \leq h(f_2)$.

The algorithm for computing the topological entropy is based on the fact that the tent family

$$g_k(x) = \begin{cases} kx & \textit{if } x \in [0, 1/2], \\ -kx + k & \textit{if } x \in [1/2, 1], \end{cases}$$

with $k \in [1, 2]$ and holds that $h(g_k) = \log k$. The idea of the algorithm is to bound the topological entropy of an unimodal maps between the topological entropies of two tent maps. The algorithm is divided in four steps:

Step 1. Fix $\varepsilon > 0$ (fixed accuracy) and an integer n such that $\delta = 1/n < \varepsilon$.
Step 2. Find the least positive integer m such that $k_m(g_{1+i\delta})$, $0 \leq i \leq n$, are distinct kneading sequences.
Step 3. Compute $k_m(f)$ for a fixed unimodal map f.
Step 4. Find r the largest integer such that $k_m(g_{1+r\delta}) < k_m(f)$. Hence, $\log(1 + r\delta) \leq h(f) \leq \log(1 + (r + 2)\delta)$.

The algorithm is easily programmed. We usually use Mathematica, which has the advantage of computing the kneading invariants of tent maps without round off errors, improving in practice the accuracy of the method.

15.4 Mathematical Analysis of the Model

Below, we analyze our model. Recall that it is given by the difference equation

$$q(t+2) = f(q(t)) = \max\{0, \varphi(q(t))\},$$

where

$$\varphi(q) = \sqrt{c}G_c(nq)\frac{\sqrt{nG_c(nq)} - nG_c(nq)}{\sqrt{c}\sqrt{nG_c(nq)} + \sqrt{r}G_c(nq)}$$

and

$$G_c(nq) = \max\{0, \sqrt{nq/c} - nq\}.$$

Note that we have to make a difference between cases 1A, 2A, and 3A and 1B, 2B, and 3B. The bifurcation diagrams of Figs. 15.4 and 15.5 show us the difference.

Apparently, in cases A, the dynamics is simple because the bifurcation diagrams show us a fixed point, in fact at 0. However, it is easy to see that the topological entropy is positive, and therefore, there is Li–Yorke chaos, although probably contained in a subset of zero Lebesgue measure. Note that in cases 1A, 2A,

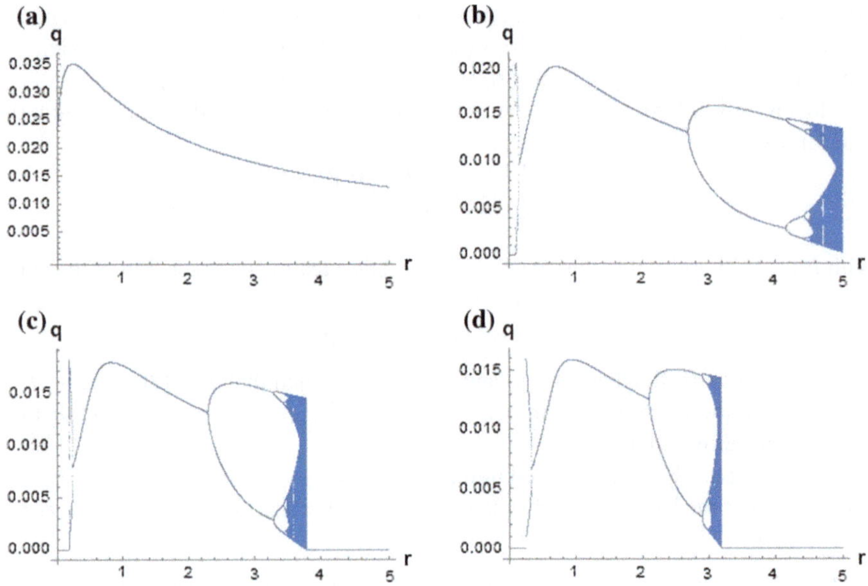

Fig. 15.4 We show the bifurcation diagrams for r ranging the interval [0,5] with step size 0.001 for 5 firms (**a**), 8 (**b**), 9 (**c**) and 10 firms (**d**). It seems that there is just one bifurcation diagram made plotting the last 200 points of orbits of length 1200

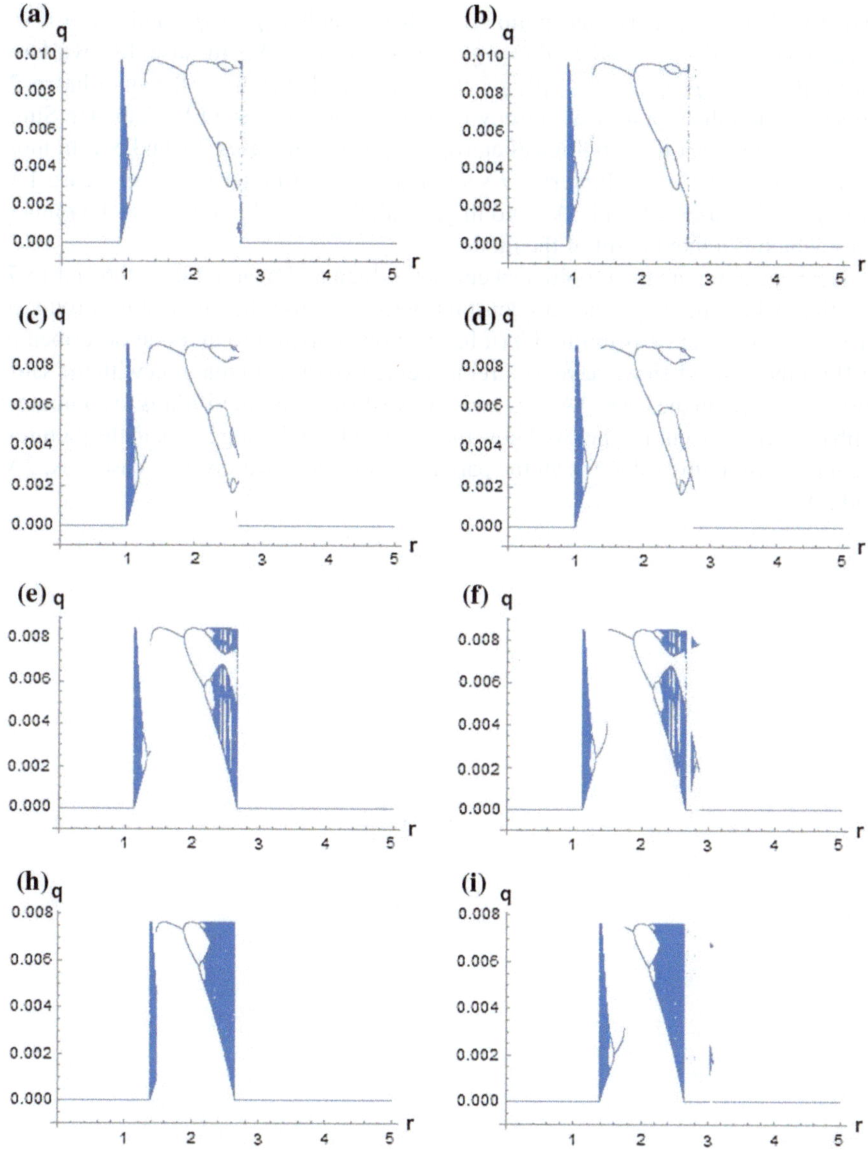

Fig. 15.5 We show the bifurcation diagrams for r ranging the interval [0,5] with step size 0.001 for 16 firms (**a**, **b**) and 17 (**c**, **d**), 18 (**e**, **f**) and 20 firms (*h*, *i*). It seems that there is just two bifurcation diagrams made plotting the last 200 points of orbits of length 1200. The existence of two different diagrams comes from the fact that there are 3 extrema

and 3A, there is at least one point q^* such that either $q^* < q_-$ and $f(q^*) = q_-$ (case 1A), or $q^* < \frac{1}{cn}$ and $f(q^*) = \frac{1}{cn}$ (cases 2A and 3A). In case 1A, we have that $f([0,q^*]) \cap f([q^*,q_-]) \subseteq [0,q^*] \cup [q^*,q_-]$, and therefore, the map has a 2-horseshoe and the topological entropy is greater than log 2 (see [2, Chap. 4]). Similarly, we prove that the topological entropy is positive in cases 2A and 3A. In most of the cases, there is a subinterval J such that for $q \in J$ either $f(q) > q\_$ (case 1A) or $f(q) > \frac{1}{cn}$ (cases 2A and 3A), and in general, the second iterate of such points is zero, which is a fixed point of the model.

Next, we compute the topological entropy, which is shown in Figs. 15.6 and 15.7. For that, when the map f has no constant pieces, we use the algorithm introduced in Sect. 15.3.4 when f is unimodal (it has just one maximum) and that described in [12] for the case when we have 3 extrema, being two of them maxima with the same forward image. In the cases 1A, 2A, and 3A, we do not have algorithms to compute it with prescribed accuracy, but we know that a lower bound is log 2. So, in the pictures, we have chosen this value for all the parameter values which gives us cases 1A, 2A, and 3A.

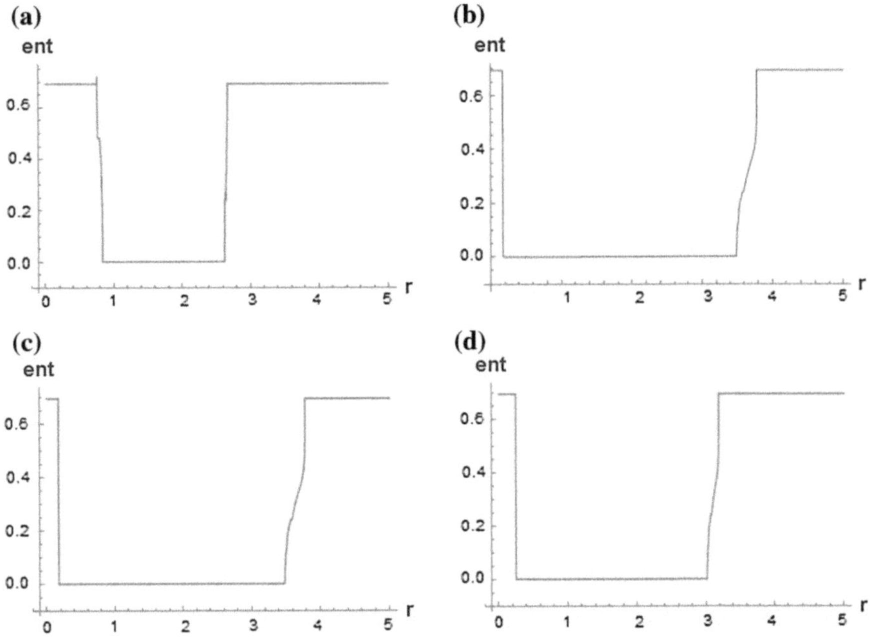

Fig. 15.6 We show the topological entropy with accuracy 10^{-3} when r ranges the interval [0,5] with step size 0.001 for 5 firms (**a**), 8 (**b**), 9 (**c**), and 10 firms (**d**)

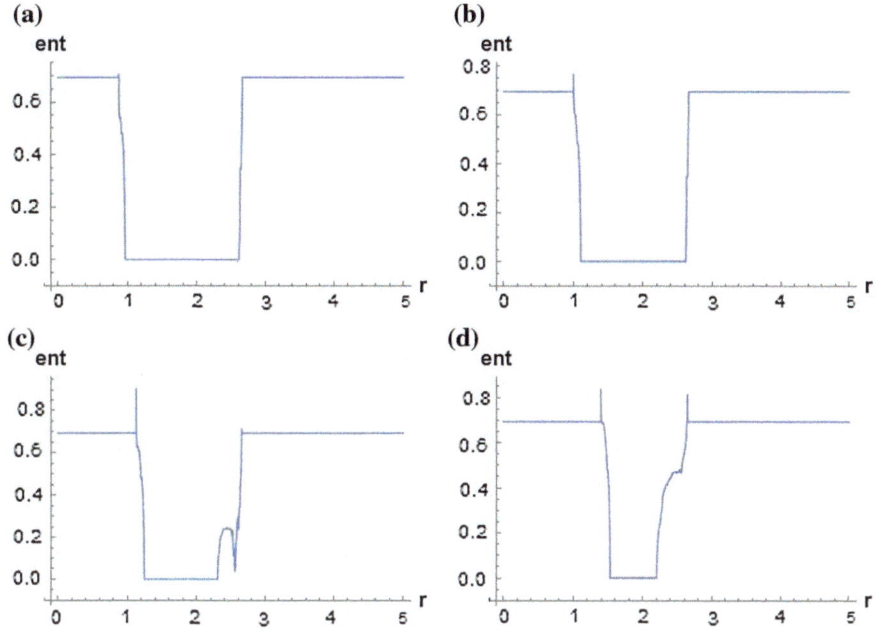

Fig. 15.7 We show the topological entropy with accuracy 10^{-3} when r ranges the interval [0,5] with step size 0.001 for 16 firms (**a**) and 17 (**b**), 18 (**c**), and 20 firms (**d**)

We highlight that the computation of topological entropy is made with algorithms with prescribed accuracy, and therefore, when we obtain a positive value of topological entropy, we are proving mathematically the existence of topological chaos. This goes further than numerical simulations that we made for estimating the Lyapunov exponents below.

When we have zero topological entropy, we can be confident that the dynamics is simple, and when it is positive, it is complicated since it is chaotic in the sense of Li and Yorke. But, as we wrote before, the complexity of cases 1A, 2A, and 3A may remain unobserved. The same can happens when we have positive topological entropy for the cases 1B, 2B, and 3B. Here, we estimate the Lyapunov exponent on the image of turning points and take the maximal obtained value. Figures 15.8 and 15.9 show our computations. When the Lyapunov exponent is negative, we observe a simple dynamics, i.e., convergence to a periodic orbit even if the topological entropy is positive. Recall that for having attractors different from periodic orbits, such attractors must contain the orbit of a turning point.

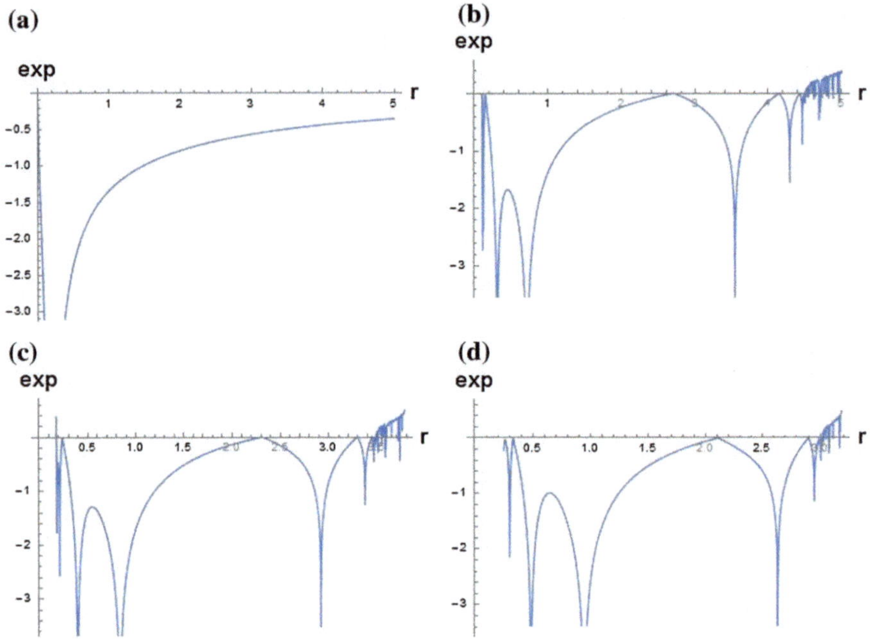

Fig. 15.8 We estimate the Lyapunov exponent for cases B when r ranges the interval [0,5] with step size 0.001 for 5 firms (**a**), 8 (**b**), 9 (**c**), and 10 firms (**d**)

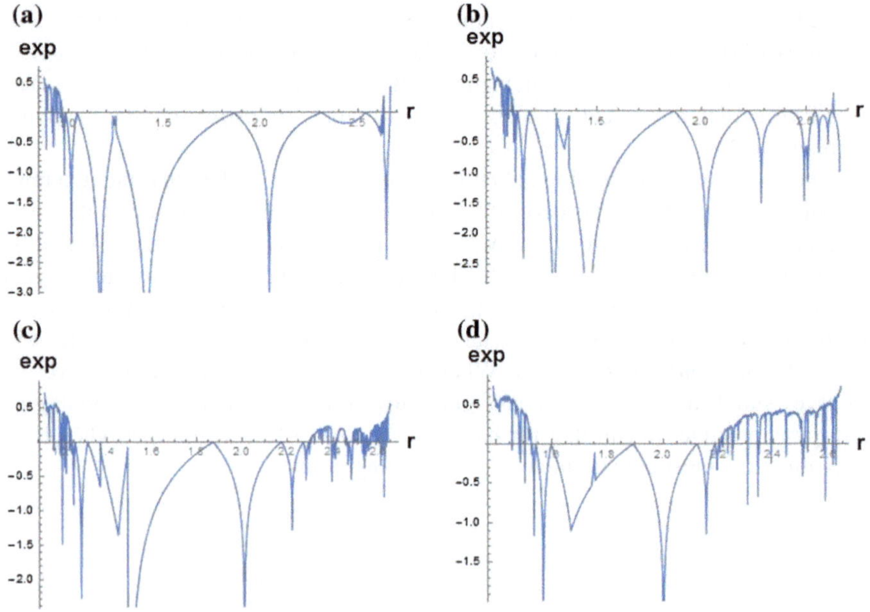

Fig. 15.9 We estimate the Lyapunov exponent for cases B when r ranges the interval [0,5] with step size 0.001 for 16 firms (**a**), 17 (**b**), 18 (**c**), and 20 firms (**d**)

15.5 Conclusions and Final Remark

We have analyzed a restricted oligopoly model which depends on two parameters, the number of firms and the capital rent. We prove that increasing the number of firms and reducing the capital rent may produce a complicated dynamical behavior given by positive topological entropy. We show, however, that this topological complexity may be unobserved on a computer simulation. Even more, increasing the number of firms, we reduce the interval of capital rent where the chaotic behavior can be physically observed, and if the number of firms is big enough (with 29 firms it is not possible to obtain a value of topological entropy smaller than log 2), this complexity remains unobserved for all the values of the capital rent.

We remark that our model is restricted to the diagonal set, where all the firms produce the same. Our simulations show that such set seems to be a global attractor for the model; that is, the dynamics is concentrate on the diagonal, but unfortunately, we do not have an analytical proof of this fact.

Finally, we remark that obtaining the turning points of the model is quite technical. If we want to increase the number of short-run periods, the computation of the turning points is even more complicated and it is not clear if its number increases. This can make the computation of topological entropy more difficult, or even impossible.

Acknowledgments This work has been supported by the grants MTM2014-52920-P from Ministerio de Economía y Competitividad (Spain), and COST Action IS1104 "The EU in the new complex geography of economic systems: models, tools and policy evaluation." Finally, este trabajo es resultado del proyecto de investigación 19294/PI/14 financiado por la Fundación Séneca-Agencia de Ciencia y Tecnología de la Región de Murcia en el marco del PCTIRM 2011–2014.

References

1. Adler, R.L., Konheim, A.G., McAndrew, M.H.: Topological entropy. Trans. Amer. Math. Soc. **114**, 309–319 (1965)
2. Alsedá, L., Llibre, J., Misiurewicz, M.: Combinatorial Dynamics and Entropy in Dimension One. World Scientific Publishing (1993)
3. Aoki, N., Hiraide, K.: Topological Theory of Dynamical Systems: Recent Advances. North–Holland (1994)
4. Balibrea, F., Snoha, L.: Topological entropy of Devaney chaotic maps. Topol. Appl. **133**, 225–239 (2003)
5. Banks, J., Brooks, J., Cairns, G., Davis, G., Stacey, P.: On Devaney's definition of chaos. Amer. Math. Monthly **99**, 332–334 (1992)
6. Bischi, G.I., Chiarella, C., Kopel, M., Szidarovszky, F.: Nonlinear Oligopolies. Springer, Berlin (2010)
7. Blanchard, F., Glasner, E., Kolyada, S., Maass, A.: On Li-Yorke pairs. J. Reine Angew. Math. **547**, 51–68 (2002)
8. Block, L.S., Coppel, W.A.: Dynamics in One Dimension. Lectures Notes in Mathematics, vol. 1513. Springer, Berlin (1992)
9. Block, L.S., Keesling, J., Li, S., Peterson, K.: An improved algorithm for computing topological entropy. J. Statist. Phys. **55**, 929–939 (1989)

10. Block, L., Keesling, J.: Computing the topological entropy of maps of the interval with three monotone pieces. J. Statist. Phys. **66**, 755–774 (1992)
11. Bowen, R.: Entropy for group endomorphism and homogeneous spaces. Trans. Amer. Math. Soc. **153**, 401–414 (1971)
12. Cánovas, J.S., Muñoz-Guillermo, M.: Computing topological entropy for periodic sequences of unimodal maps. Commun. Nonlinear Sci. Numer. Simul. **19**, 3119–3127 (2014)
13. Cánovas, J.S., Muñoz– Guillermo, M.: Computing Topological Entropy for Periodic Sequences of Unimodal Maps, preprint (2014)
14. Cánovas, J.S., Panchuk, A., Puu, T.: Asymptotic dynamics of a piecewise smooth map modelling a competitive market. Math. Comput. Simul. **117**, 20–38 (2015)
15. de Melo, W., van Strien, S.: One-Dimensional Dynamics. Springer, New York (1993)
16. Devaney, R.L.: An Introduction to Chaotic Dynamical Systems. Addison-Wesley, Redwood City (1989)
17. Dinaburg, E.I.: The relation between topological entropy and metric entropy. Soviet Math. **11**, 13–16 (1970)
18. Du, B.S.: A simple proof of Sharkovsky's theorem. Amer. Math. Monthly **111**, 595–599 (2004)
19. Fedorenko, V.V., Sharkovsky, A.N., Smítal, J.: Characterizations of weakly chaotic maps of the interval. Proc. Amer. Math. Soc. **110**, 141–148 (1990)
20. Huang, W., Ye, X.: Devaney's chaos or 2-scattering implies Li-Yorke's chaos. Topology Appl. **117**, 259–272 (2002)
21. Li, T.Y., Yorke, J.A.: Period three implies chaos. Amer. Math. Monthly **82**, 985–992 (1975)
22. Guckhenheimer, J.: Sensitive dependence to initial conditions for one-dimensional maps. Comm. Math. Phys. **70**, 133–160 (1979)
23. Kwietniak, D., Misiurewicz, M.: Exact Devaney chaos and entropy. Qual. Theory Dyn. Syst. **6**, 169–179 (2005)
24. Milnor, J.: On the concept of attractor. Commun. Math. Phys. **99**, 177–195 (1985)
25. Misiurewicz, M., Szlenk, W.: Entropy of piecewise monotone mappings. Studia Math. **67**, 45–63 (1980)
26. Oseledets, V.I.: A multiplicative ergodic theorem. Lyapunov characteristic numbers for dynamical systems. Trans. Moscow Math. Soc. **19**, 197–231, Moscov. Mat. Obsch. **19**, 179–210 (1968)
27. Puu, T.: Oligopoly: Old Ends - New Means. Springer, Berlin (2011)
28. Sharkovsky, A.N., Kolyada, S.F., Sivak, A.G., Fedorenko, V.V.: Dynamics of One–Dimensional Maps. Kluwer Academic Publishers (1997)
29. Singer, D.: Stable orbits and bifurcation of maps of the interval. SIAM J. App. Math. **35**, 260–267 (1978)
30. Smale, S.: Differentiable dynamical systems. Bull. Amer. Math. Soc. **73**, 747–817 (1967)
31. Smítal, J.: Chaotic functions with zero topological entropy. Trans. Amer. Math. Soc. **297**, 269–282 (1986)
32. Sumi, N.: Diffeomorphisms with positive entropy and chaos in the sense of Li-Yorke. Ergod. Th. Dynam. Sys. **23**, 621–635 (2003)
33. Theocharis, R.D.: On the stability of the Cournot solution on the oligopoly problem. Rev. Econ. Stud. **27**, 133–134 (1959)
34. Thunberg, H.: Periodicity versus chaos in one-dimensional dynamics. SIAM Rev. **43**, 3–30 (2001)
35. van Strien, S., Vargas, E.: Real bounds, ergodicity and negative Schwarzian for multimodal maps. J. Amer. Math. Soc. **17**, 749–782 (2004)

Part III
Fundamental and Computational Mechanics

Chapter 16
Two Thermodynamic Laws as the Forth and the Fifth Integral Postulates of Continuum Mechanics

Boris E. Pobedria and Dimitri V. Georgievskii

Abstract A methodological reduction of the known in physics statements of the first and second laws of thermodynamics to general form of integral postulates adopted in classical mechanics of continuous media, is realized. It is shown that the second law should be represented in the Carathéodory form which makes possible to introduce both absolute temperature and entropy as phenomenological values not having recourse to the model of perfect gas and the Carnot cycle. The local equations being consequences of the integral postulates include mass densities of thermodynamic values which must be defined as scalar or vector fields in material.

16.1 The Second Law of Thermodynamics in the Carathéodory Form

In literature, there are a lot of attempts to construct phenomenological thermodynamics by axiomatic, i. e., strictly mathematical way. All these attempts are based on acceptance of the Carathéodory principle. Let us observe some fundamental statements which seem to be necessary for any axiomatic construction of thermodynamics [1–3].

The notion "system **A**" will mean some system being in the state of uniform thermodynamic equilibrium and characterizing by the thermodynamic parameters of state $A_1, A_2, \ldots, A_{n_A}$. Each of the mentioned parameters may be either 0-rank tensor (scalar) or 1-rank tensor (vector) or 2-rank tensor, etc.

Let **B** and **C** be the other systems. By virtue of so-called zero law of thermodynamics, if each of the two systems **A** and **B** is found in heat equilibrium with **C**, then **A** is found in heat equilibrium with **B**. We introduce the symbol "$\sim$" for designation of heat equilibrium for two systems. Let this relation comply with the following properties:

B.E. Pobedria · D.V. Georgievskii (✉)
Moscow State University, Moscow 119991, Russia
e-mail: georgiev@mech.math.msu.su

V.A. Sadovnichiy and M.Z. Zgurovsky (eds.), *Advances in Dynamical Systems and Control*, Studies in Systems, Decision and Control 69,
DOI 10.1007/978-3-319-40673-2_16

1°. $\mathbf{A} \sim \mathbf{A}$ (reflexivity).
2°. $\mathbf{A} \sim \mathbf{B} \Longrightarrow \mathbf{B} \sim \mathbf{A}$ (symmetry).
3°. $\mathbf{A} \sim \mathbf{B} \vee \mathbf{B} \sim \mathbf{C} \Longrightarrow \mathbf{A} \sim \mathbf{C}$ (transitivity).

Then, the relation of heat equilibrium is the equivalence relation. All systems are divided by equivalence classes so that two systems of type **A** will belong to the same class if and only if they are found in mutual heat equilibrium.

Among various kinds of thermodynamic systems, there are such ones that are characterized by only scalar thermodynamic parameter of state T_E. We choose one of these systems and will call it as "the system **E**." So, there exists the functional connection for the systems of each type, for example

$$T_E = \varphi_A(A_1, A_2, \ldots, A_{n_A}), \; T_E = \varphi_B(B_1, B_2, \ldots, B_{n_B}), \; T_E = \varphi_C(C_1, C_2, \ldots, C_{n_C}) \tag{16.1}$$

such that two systems **A** and **B** are found in heat equilibrium if and only if

$$T_E = \varphi_A(A_1, A_2, \ldots, A_{n_A}) = \varphi_B(B_1, B_2, \ldots, B_{n_B}) \tag{16.2}$$

Thus, zero law of thermodynamics leads to definition of the new parameter of state T_E being suitable for all thermodynamic systems. This parameter is said to be empirical temperature; it is convenient to introduce it as an independent parameter of state. Any scalar parameter, for example A_{n_A}, may be expressed as

$$A_{n_A} = \psi_A(A_1, \ldots, A_m, T_E), \quad m = n_A - 1 \tag{16.3}$$

The first law of thermodynamics ensures an introduction of the notions of internal energy E and heat Q. The value δQ is an energy transmitted from one system to another due to a difference of its empirical temperatures. For adiabatic processes,

$$dE + \delta \mathsf{A}^{(\mathrm{int})} = 0 \tag{16.4}$$

where $\delta \mathsf{A}^{(\mathrm{int})}$ is the change of work of internal forces.

The second law of thermodynamics is used for introduction of the notions of absolute temperature scale and entropy. The relation (16.3) demonstrates that every thermodynamic parameter of state, for example the internal energy E, is expressed in terms of thermodynamic parameters of state in the form

$$E = E(A_1, \ldots, A_m, T_E) \tag{16.5}$$

The tensor values $A_1, A_2, \ldots, A_m$ may be considered as generalized displacements; we denote by P_j the corresponding to its generalized forces:

$$P_j = P_j(A_1, \ldots, A_m, T_E), \quad j = 1, \ldots, m \tag{16.6}$$

Therefore

$$\delta \mathsf{A}^{(\mathrm{int})} = \sum_{j=1}^{m} P_j \, dA_j \tag{16.7}$$

The generalized forces (16.6) are connected with the generalized displacements by some constitutive relations (the state equations). If the process is balanced and it is effected so slowly that every generalized force (16.6) corresponds to the state equations in any time moment then the first law of thermodynamics gives the following relation:

$$\delta Q = dE + \delta \mathsf{A}^{(\mathrm{int})} = \sum_{j=1}^{m} \left(\frac{\partial E}{\partial A_j} + P_j \right) dA_j + \frac{\partial E}{\partial T_E} dT_E \tag{16.8}$$

So, according to (16.8) the value δQ for balanced processes is a linear differential form (the Pfaffian form) of independent thermodynamic parameters of state.

C. Carathéodory suggested (1909) the statement of the second law of thermodynamics in the form of following principle.

- *For any state of thermodynamic system, one may produce the state with two properties:*

(a) *it is arbitrarily close to the original state;*
(b) *it is not reached from the original state by means of adiabatic balanced process.*

Because $\delta Q = 0$ for adiabatic balanced process (16.8) becomes an equation in total differentials:

$$\sum_{j=1}^{m} \left(\frac{\partial E}{\partial A_j} + P_j \right) dA_j + \frac{\partial E}{\partial T_E} dT_E = 0 \tag{16.9}$$

According to the Carathéodory principle, there are close states which cannot be joined with the help of the solution of (16.9). Carathéodory established that this means an integrability of the Pfaffian form (16.8) i.e., an existence of the integrating factor $\nu(A_1, \ldots, A_m, T_E)$ and the associated function $M(A_1, \ldots, A_m, T_E)$ such that

$$\frac{\delta Q}{\nu} = dM \tag{16.10}$$

or in detail

$$\frac{1}{\nu}\left[\sum_{j=1}^{m} \left(\frac{\partial E}{\partial A_j} + P_j \right) dA_j + \frac{\partial E}{\partial T_E} dT_E \right] = \sum_{j=1}^{m} \frac{\partial M}{\partial A_j} dA_j + \frac{\partial M}{\partial T_E} dT_E \tag{16.11}$$

It can be shown that among all possible integrating factors ν, there exists unique (to within constant) factor depending on temperature T_E only. It is denoted by $T(T_E)$ and is said to be an absolute temperature. This is an universal function of state applicable to any thermodynamic system. The associated to it function is denoted by $S(A_1, \ldots, A_m, T_E) = S(\mu_1, \ldots, \mu_m, T)$ and is said to be an entropy of the system under consideration. Then, the Eq. (16.10) has the following form

$$dQ = T\,dS = dE + d\mathsf{A}^{(\mathrm{int})} = \sum_{j=1}^{m} \left(\frac{\partial E}{\partial \mu_j} + P_j \right) : d\mu_j + \frac{\partial E}{\partial T}\,dT \qquad (16.12)$$

where symbol ":" means a full contraction of the 2nd rank tensors. It is valid for any balanced process between adjacent states.

We see that the Carathéodory principle allows to introduce both entropy and absolute temperature not having recourse to the model of perfect gas and the Carnot cycle. In the case of gas with the state equation $f(p, V, T) = 0$, we receive from (16.12)

$$dS = \frac{1}{T}\left(\frac{\partial E}{\partial T}\right)_V dT + \frac{1}{T}\left[\left(\frac{\partial E}{\partial T}\right)_T + p\right] dV \qquad (16.13)$$

The number of independent parameters of state equals two, so any Pfaffian form has an integrating factor and

$$\left(\frac{\partial E}{\partial V}\right)_T = T\left(\frac{\partial p}{\partial T}\right)_V - p \qquad (16.14)$$

It is easy to verify a realizability of (16.14) for both perfect gas and the van der Waals gas.

16.2 Legendre Transforms and Thermodynamic Potentials

Side by side with the internal energy $E(\mu_1, \ldots, \mu_m, S)$ let us consider the following thermodynamic potentials:

- *the enthalpy (heat content)* $H(P_1, \ldots, P_m, S)$:

$$H = E + \sum_{j=1}^{m} P_j : \mu_j \qquad (16.15)$$

- *the Helmholtz free energy* $F(\mu_1, \ldots, \mu_m, T)$:

$$F = E - TS \qquad (16.16)$$

- *the Gibbs potential* $G(P_1, \ldots, P_m, T)$:

$$G = H - TS \qquad (16.17)$$

Using both the first and the second laws of thermodynamics, we may write (16.14) in the following way

$$\delta \mathsf{A}^{(\text{int})} = \sum_{j=1}^{m} P_j : d\mu_j \tag{16.18}$$

as well as represent the thermodynamic identity:

$$dE = T\,dS - \sum_{j=1}^{m} P_j : d\mu_j \tag{16.19}$$

In order to pass from one thermodynamic potential to some other, it is efficient to use the Legendre transform of function $\varphi(x_1, x_2, \ldots)$ with the total differential

$$d\varphi = \frac{\partial \varphi}{\partial x_1}\,dx_1 + \frac{\partial \varphi}{\partial x_2}\,dx_2 + \cdots \equiv X_1\,dx_1 + X_2\,dx_2 + \ldots \tag{16.20}$$

The Legendre transform poses the function $\Phi(X_1, X_2, \ldots)$ in correspondence with the function $\varphi(x_1, x_2, \ldots)$ such that

$$\Phi = \varphi - X_1 x_1 - X_2 x_2 - \ldots \tag{16.21}$$

$$d\Phi = d\varphi - X_1\,dx_1 - x_1\,dX_1 - X_2\,dx_2 - x_2\,dX_2 - \ldots \tag{16.22}$$

A transition from the internal energy E to the enthalpy (16.15) is realized by means of the following transform

$$-P_j = \frac{\partial E}{\partial \mu_j}, \quad j = 1, \ldots, m \tag{16.23}$$

Then, repeating (16.19)

$$dE = -\sum_{j=1}^{m} P_j : d\mu_j + T\,dS \tag{16.24}$$

$$dH = dE + \sum_{j=1}^{m} P_j : d\mu_j + \sum_{j=1}^{m} \mu_j : dP_j = T\,dS + \sum_{j=1}^{m} \mu_j : dP_j \tag{16.25}$$

$$\frac{\partial H}{\partial S} = T, \quad \frac{\partial H}{\partial P_j} = \mu_j \tag{16.26}$$

Analogous relations take place in cases of the transition $H \mapsto G$:

$$dG = dH - T\,dS - S\,dT = \sum_{j=1}^{m} \mu_j : dP_j - S\,dT \tag{16.27}$$

$$\frac{\partial G}{\partial T} = -S, \quad \frac{\partial G}{\partial P_j} = \mu_j \tag{16.28}$$

as well as the transition $E \mapsto F$:

$$dF = dE - T\,dS - S\,dT = -\sum_{j=1}^{m} P_j : d\mu_j - S\,dT \tag{16.29}$$

$$\frac{\partial F}{\partial T} = -S, \quad \frac{\partial F}{\partial \mu_j} = -P_j \tag{16.30}$$

It should be noted that both the thermodynamic parameters μ_j and its fluxes P_j must be given by choose of the model.

16.3 Mass Densities of Thermodynamic Potentials

To describe the models in continuum mechanics, it is more convenient to use densities of the thermodynamic functions and potentials under consideration. We remember the statements of the first law of thermodynamics

$$dE + dK = \delta \mathsf{A}^{(\mathrm{ext})} + \delta Q \tag{16.31}$$

or

$$dE = -\delta \mathsf{A}^{(\mathrm{int})} + \delta Q \tag{16.32}$$

and the second law of thermodynamics

$$T\,dS = \delta Q + W^*\,dt \tag{16.33}$$

where W^* is the dispersion function. Excepting δQ from (16.32) and (16.33), we receive the following thermodynamic identity

$$dE = T\,dS - \delta \mathsf{A}^{(\mathrm{int})} - W^*\,dt \tag{16.34}$$

Let us introduce a mass density of internal energy $e(\mathbf{x}, t)$, a mass density of entropy $s(\mathbf{x}, t)$ and a density of dispersion $w^*(\mathbf{x}, t)$ by means of the following relations for arbitrary moving volume V:

$$E = \int\limits_V \rho e \, dV, \quad S = \int\limits_V \rho s \, dV, \quad W^* = \int\limits_V w^* \, dV \tag{16.35}$$

In order to present the expression for the value δQ in (16.31)–(16.33), we consider an arbitrary finite volume V bounded by surface Σ with external unit normal $\mathbf{n}$. Let mass density of heat $q(\mathbf{x}, t)$ is given in any material point of this volume, and normal component $q^{(n)}(\mathbf{y}, t)$ of the heat flux vector $\mathbf{q}$ is given on each square element $d\Sigma$ ($\mathbf{y} \in \Sigma$):

$$q^{(n)} = q_i n_i = \mathbf{q} \cdot \mathbf{n} \tag{16.36}$$

Then a heat influx in the volume V for some time interval dt is equal to

$$\delta Q = -dt \int\limits_\Sigma q^{(n)} \, d\Sigma + dt \int\limits_V \rho q \, dV = dt \int\limits_V (\rho q - \operatorname{div} \mathbf{q}) \, dV \tag{16.37}$$

The sign minus before the surface integral in (16.37) is explained by fact that vector $\mathbf{n}$ is the external normal, whereas positive surface heat influx must be directed from the outside toward the interior the volume V.

Physical dimensions of the introduced mass densities $[e] = \mathrm{L^2T^{-2}}$, $[s] = \mathrm{L^2T^{-2}K^{-1}}$, $[q] = \mathrm{L^2T^{-3}}$, $[q^{(n)}] = \mathrm{MT^{-3}}$, $[w^*] = \mathrm{ML^{-1}T^{-3}}$ demonstrate that the scalar fields $e(\mathbf{x}, t)$, $w^*(\mathbf{x}, t)$, $q(\mathbf{x}, t)$, $q^{(n)}(\mathbf{y}, t)$ have purely mechanical nature (in spite of the word "heat") and may be defined without the notion "temperature."

It follows from the expressions (16.33), (16.35), (16.37) that in any material point of the volume V:

$$\rho T \frac{ds}{dt} = \rho q - q_{i,i} + w^* \tag{16.38}$$

This equation is known as the heat influx equation and it is the local consequence of the second law of thermodynamics.

For a broad class of continuums, the constitutive relations connecting the heat influx vector $\mathbf{q}$ and gradient of temperature grad T are valid. The Fourier law of heat conduction

$$\mathbf{q} = -\Lambda \cdot \operatorname{grad} T \tag{16.39}$$

represents the simplest such relation. Here Λ is a positive definite symmetric tensor of the second rank named the tensor of heat conduction. Using (16.39) the heat influx Eq. (16.38) may be written as

$$\rho T \frac{ds}{dt} = \rho q + (\Lambda_{kl} T_{,l})_{,k} + w^* \tag{16.40}$$

16.4 Two Thermodynamic Laws in the Form of Integral Postulates

The introduced mass densities allow to formulate two thermodynamic laws as the 4th and 5th postulates of continuum mechanics, thus to add them to statements of boundary-value problems. The integral statement of the first law is the following [4, 5].

- *Let $\Omega \in R^3$ be a material volume in actual frame of reference, V be an arbitrary moving volume in Ω and Σ be its boundary with unit external normal* **n**. *Then*

$$\frac{d}{dt}\int\limits_V \rho\Big(e+\frac{|\mathbf{v}|^2}{2}\Big)dV = \int\limits_V \rho(\mathbf{F}\cdot\mathbf{v}+q)\,dV + \int\limits_\Sigma (\mathbf{P}^{(n)}\cdot\mathbf{v}-q^{(n)})\,d\Sigma \tag{16.41}$$

or taking into account the theorem of kinetic energy

$$\frac{d}{dt}\int\limits_V \rho e\,dV = \int\limits_V (\rho q + P : D)\,dV - \int\limits_\Sigma q^{(n)}\,d\Sigma \tag{16.42}$$

Here D is strain rate tensor, **F** is mass force.

Differential consequence of the formulation (16.42) represents the local energy equation

$$\rho\frac{de}{dt} = \rho q - \operatorname{div}\mathbf{q} + P : D \tag{16.43}$$

Integral statement of the second law of thermodynamics may be following [4, 5].

- *Let $\Omega \in R^3$ be a material volume in actual frame of reference, V be an arbitrary moving volume in Ω and Σ be its boundary with unit external normal* **n**. *Then*

$$\frac{d}{dt}\int\limits_V \rho s\,dV = \int\limits_V \frac{\rho q}{T}\,dV - \int\limits_\Sigma \frac{q^{(n)}}{T}\,d\Sigma + \int\limits_V \Big(\frac{w^*}{T} - \frac{\mathbf{q}\cdot\operatorname{grad}T}{T^2}\Big)\,dV \tag{16.44}$$

The last integral in the right hand of (16.44) is said to be the production of entropy. It is always nonnegative by virtue of positive definiteness of the tensor Λ (see (16.39)) as well as the inequalities $w^* \geq 0$, $T > 0$:

$$S^* = \int\limits_V \Big(\frac{w^*}{T} - \frac{\mathbf{q}\cdot\operatorname{grad}T}{T^2}\Big)\,dV = \int\limits_V \Big(\frac{w^*}{T} + \frac{1}{T^2}\operatorname{grad}T\cdot\Lambda\cdot\operatorname{grad}T\Big) \geq 0 \tag{16.45}$$

Substituting the surface integral in (16.44) on volume one:

$$\int_{\Sigma} \frac{q^{(n)}}{T} d\Sigma = \int_{V} \operatorname{div}\left(\frac{\mathbf{q}}{T}\right) dV = \int_{V} \left(\frac{\operatorname{div}\mathbf{q}}{T} - \frac{\mathbf{q}\cdot\operatorname{grad}T}{T^2}\right) dV \tag{16.46}$$

we easily receive the differential consequence of the 5th postulate, namely the equation of heat influx

$$\rho T \frac{ds}{dt} = \rho q - \operatorname{div}\mathbf{q} + w^* \tag{16.47}$$

The models of continuum mechanics for which $w^* = 0$ are said to be reversible ones. The inequality (16.45) demonstrates that the production of entropy may be not equal to zero even for reversible models.

References

1. Germain, P.: Cours de Mécanique des Milieux Continus. T. 1. Théorie Générale. Masson Éditeurs, Paris (1973)
2. Sedov, L.I.: Mechanics of Continuous Media, vols. I, II. World Scientific Publ, Singapore (1997)
3. Ilyushin, A.A.: Mechanics of Continuous Media. Moscow State Univ. Publ, Moscow (1990). [in Russian]
4. Pobedria, B.E., Georgievskii, D.V.: Foundations of Mechanics of Continuous Media. Fizmatlit, Moscow (2006). [in Russian]
5. Pobedria, B.E., Georgievskii, D.V.: Uniform approach to construction of nonisothermal models in the theory of constitutive relations. Continuous and Distributed Systems II. Ser. Studies in Systems, Decision and Control, vol. 30, pp. 341–352 (2015)

Chapter 17
Flow Control Near a Square Prism with the Help of Frontal Flat Plates

Iryna M. Gorban and Olha V. Khomenko

Abstract The case of two symmetrical flat plates fixed in front of a square prism for passive control of a near-body flow pattern is numerically investigated at moderate Reynolds numbers. The plates are used for generation of a pair of the frontal stable vortices which would be able suppress flow separation in the neighbor body edges. The improvement of body loads in this case is achieved by wake constriction and reducing the difference between bottom and frontal pressure. The control scheme presented was found to be sensitive to its geometrical parameters. The dynamic system analysis is attracted for studying the flow topology in the area and deriving optimum parameters of the control device. It was found that the plate length $l \approx 0.2d$ and $r \approx 0.16d$, where d is the prism side and r is the distance between the plate base and the prism edge, is the appropriate choice which permits reduce the prism drag approximately per 20 %. An influence of the Reynolds number on the effectiveness of the control scheme is also investigated.

17.1 Introduction

Square prism, as a circular cylinder, is a fundamental bluff body configuration used in many engineering applications including heat exchangers, architectural structures, and marine equipment. Exploitation of these systems in air and fluid flows is accompanied by vortex shedding from the body that causes large unsteady forces, acoustic noise and resonance, structural vibrations, and other dangerous effects. So, to improve productivity of the equipment and prevent its destruction, different passive and active methods have been proposed for control of vortex dynamics near a bluff body. The

I.M. Gorban
Institute of Hydromechanics, National Academy of Sciences of Ukraine,
Zheliabova St. 8/4, Kyiv 03680, Ukraine
e-mail: ivgorban@gmail.com

O.V. Khomenko (✉)
Institute for Applied System Analysis, National Technical University of Ukraine
"Kyiv Polytechnic Institute", Peremogy Ave. 37, Build 35, Kyiv 03056, Ukraine
e-mail: olgkhomenko@ukr.net

V.A. Sadovnichiy and M.Z. Zgurovsky (eds.), *Advances in Dynamical Systems and Control*, Studies in Systems, Decision and Control 69,
DOI 10.1007/978-3-319-40673-2_17

earliest methods for suppression of vortex shedding and fluid forces of bluff bodies were discussed by Zdravkovich [1] who classified the schemes of flow control with respect to the subject of exposure. He considered surface protrusions affecting separated shear layers, shrouds acting entrainment layers and near wake stabilizers. In most cases, the methods are based on body surface modifications and are passive in the sense that there is no power input. Further development of flow control has lead to creation of the schemes requiring energy supplying from external sources (blowing and suction, injection of micro-bubbles, surface heating or cooling, etc.). In paper [2], those are classified into active open-loop and active closed-loop controls.

Majority of the both passive and active flow control researches deals with a circular cylinder, which is the most popular bluff body configuration. At the same time, the flow control for a square prism may differ from that for a cylinder owing to the fixed separation points. In this case, wake modification does not require any influence on the body boundary layer with the aim to delay the flow separation. Then the control has to be brought to the body wake directly.

Among the most known direct-wake control methods for a square prism, are using a downstream splitter plate [3, 4], installation of the small element, such as a flat plate or rod, upstream of the prism [5, 6], and base blowing/suction [7–9]. Those change vortex wake dynamics; as a result, the fluid forces acting on a square prism are reduced. Notice that optimal control in these researches is derived by systematic changing parameters of the proposed devices, which is a very time-consuming. To overcome this difficulty, the control theory based on the rigorous mathematical apparatus can be applied. Researchers have shown that feedback control algorithms based on mathematical analysis, such as optimal control approach and dynamical systems theory, effectively control strong nonlinear flows generating in bluff body wakes [10–12].

To achieve the desired effects to the flow, not only active but and passive methods are in want of optimization in frames of the chosen control strategy. It is known that one of the successful ways to control flow–body interactions at large Reynolds numbers is connected with modification of near-body flow by creating artificial large-scale vortices there. In passive control, the special surface irregularities, such as cross groves or plates, are usually used for this purpose. This conception known as trapped vortex approach has found its practical application in aviation, marine engineering, and hydraulic systems [13–15].

The principal requirement to the algorithms based on the generation of large-scale stable vortices near a body consists in possibility to forecast and control the behavior of those. Therefore, the control scheme will be effective, if it applies information about critical points, dynamic properties of the vortices created and other topological features of flow field. It was stated in paper [16], a knowledge of critical-point theory is important for interpreting and understanding flow patterns whether they are obtained experimentally or computationally. Modern control algorithms are not only used the above-mentioned information, but also directed to creating the necessary topology in the flow field that includes changing the location and type of flow critical points in accordance with the control goals.

In this work, the trapped vortex approach is used to improve loads of a square prism. Two stable symmetrical vortices are proposed to be generated in front of the prism with the help of special attaching plates. The effects of the control plates on the force coefficients, flow pattern, and vortex shedding frequency of the prism are numerically studied at moderate Reynolds numbers, with *Re* based on the side of prism.

To identify the optimum position and length of the plates, the so-called reduced order model is applied [11]. This concept is based on the nonviscous model of point vortices, in which the vorticity field is represented by a discrete set of isolated circulatory elements whose axes are perpendicular to the flow plane. The flow field in this case is reduced to the finite system of vortices moving along the trajectories of fluid particles. Analyzing the vortex system dynamics in the considered area, one is able to derive the main regularities of the flow pattern there. The model of the vortex dynamics has ensured many important results in regard to the flow control [11, 12, 16–18].

We use here the model with one degree of freedom to study the dynamic behavior of the trapped vortex clamped between the control plate and the prism front side. It is supposed the recirculation zone formed due to flow separation in the plate edge is replaced by a singular vortex. According to the present control strategy, the vortex has to be immovable and prevent the flow separation in the prism leading edge. Then, the problem is reduced to the PDE system relatively coordinates and circulation of the vortex as well as parameters of the control plate.

Numerical modeling of flow patterns around the square prism with two frontal plates is performed in 2D space by the vortex method [19, 20], which belongs to high-resolution Lagrangian-type schemes developed as fast alternative to direct numerical simulations (DNS) [21]. As pointed out by Liu and Kopp [22], the accuracy in the last versions of the vortex method is compared well to nondissipative and high-order finite-difference schemes, especially in large and intermediate scales.

The Reynolds number in the present investigation is changed in the range $Re = 100 \div 500$. The second wake instability connected with 3D transition is known to develop in the square cylinder flow starting from $Re \approx 170$ [23]. But it has been shown from previous researches two-dimensional calculations at higher Reynolds number simulate flow patterns, mean forces, and separation frequency for the square cylinder quite successfully [22].

Position and length of the control plate are set using the information obtained with the help of the simplified model of trapped, or standing, vortices [24]. The objective of this work was to estimate an influence of the control plates on flow patterns and fluid forces of the square prism as well as demonstrate that the reduced order model is able to ensure optimum parameters of the control device. Wake stabilization and a significant decrease of both drag and lateral force are observed in the flow under the control. It follows from the present-study new successful control methods for fluid flows can be developed on base of the dynamic models taking into account the flow topology.

17.2 Problem Statement

The rigid square prism of side d with two symmetrical plates is immersed into the uniform flow of velocity U_∞. The two-dimensional geometrical model of the problem and coordinate system are depicted in Fig. 17.1. The control scheme presented is characterized by two geometrical parameters. Those are the plate length l and space r between the plate and the neighboring square edge. The plates are supposed to be thin and their width is invariable in this investigation. Incompressible flow with constant fluid properties is assumed. The Reynolds number is defined as $Re = U_\infty d/\nu$, where ν is the kinematic viscosity of water. All geometrical lengths are normalized with d, velocities with U_∞, physical times with d/U_∞, and frequencies with U_∞/d. Consequently, the Strouhal number is defined as $St = f_s d/U_\infty$, where f_s is the shedding frequency. Force and pressure coefficients are specified by the dynamic pressure $\rho U_\infty^2/2$, where ρ is the fluid density.

The governing Eqs. (17.1) and (17.2) of continuity and momentum for the flow under consideration are represented as follows:

$$\nabla \mathbf{V} = 0, \tag{17.1}$$

$$\frac{\partial \mathbf{V}}{\partial t} + (\mathbf{V} \cdot \nabla)\mathbf{V} = -\nabla p + \frac{1}{Re}\nabla^2 \mathbf{V}, \tag{17.2}$$

where $\mathbf{V} = (u, v)$ is the velocity vector, p is the pressure, and t is the time.

On the body, the slipping condition must be satisfied

$$\mathbf{V} \cdot \mathbf{n}|_\Sigma = 0, \tag{17.3}$$

$$\mathbf{V} \cdot \tau|_\Sigma = 0, \tag{17.4}$$

where Σ denotes the body contour, and $\mathbf{n}$, τ are the normal and tangential unit vectors to the body.

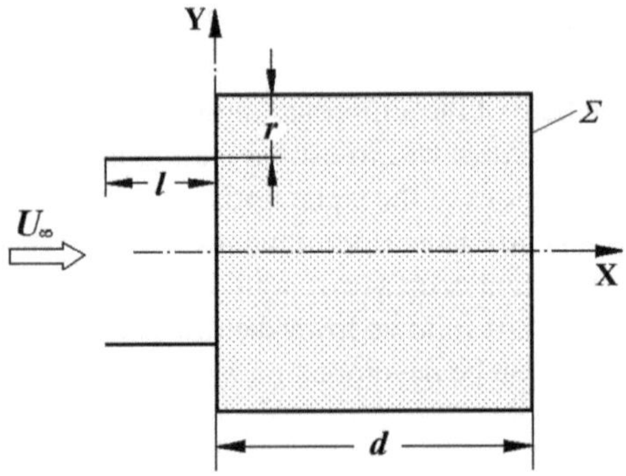

Fig. 17.1 Flow configuration and coordinate system

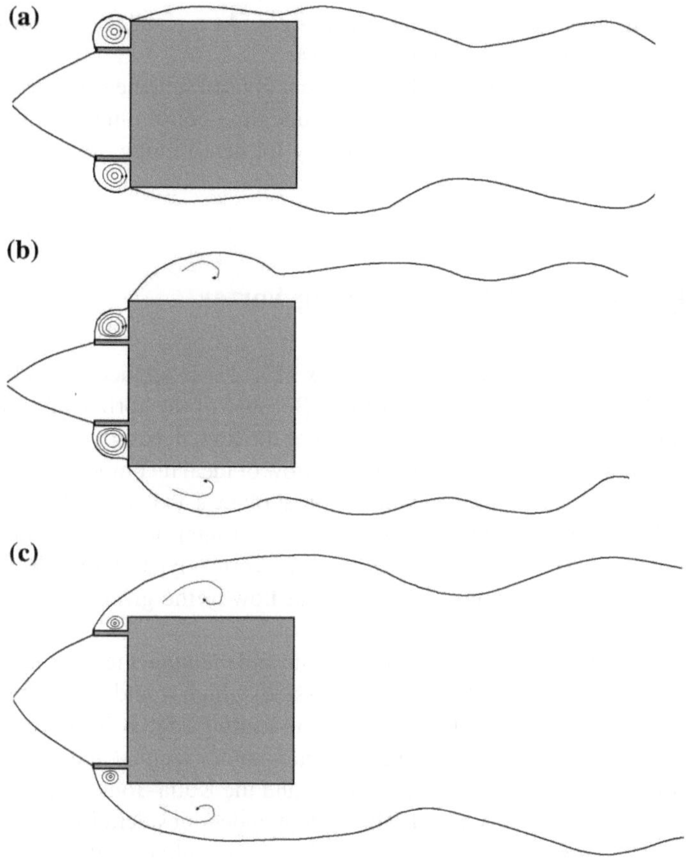

Fig. 17.2 Classification of controlled wake flow patterns. **a** Pattern **A**. **b** Pattern **B**. **c** Pattern **C**

Choice of geometrical parameters of the control plates is seen to be conditioned by the flow patterns generated near the prism with the plates. Note the thickness of the plates is fixed to $0.02d$. Then the square prism flow depends on the relation between the length l and the position r of the control plates. The possible flow patterns around the square prism with the plates are presented in Fig. 17.2. In pattern A, the shear layers that separated from the control plates attach to the leading edges of prism. That leads to depression of the flow separation in these edges and narrowing of the wake behind the prism.

It is well known from Roshko's classical experiments with a circular cylinder [25] that wake narrowing causes reduction of the body drag. So, we expect the essential suppression of fluid forces for this configuration. Pattern B corresponds to the case when the plates are located excessive far off the prism edges. Then, the plate shear layers attach to the frontal side of prism that influences weakly on the flow separation at the leading edges. Pattern C is the most invalid for a flow control because we obtain here the global flow separation in the plate edges. As a result, the prism wake can become even broader than at uncontrolled case.

It follows from the above analysis the present control scheme is in need of correct choice of its geometrical parameters. To exclude time-consuming systematic calculations, we apply here the critical-point theory for determining the optimal sizes of control plates.

17.3 Dynamic Model of a Standing Vortex

In this section, the model of standing vortex [17, 23] is applied to study the flow topology in the domain under consideration. Because of the horizontal symmetry of the problem, it is enough to consider top part of the flow field only. The geometry of interest is presented in Fig. 17.3. The uniform flow of ideal incompressible fluid in the half-plane containing a hemiprism with a frontal plate is analyzed. The recirculation flow in the domain between the plate and the prism frontal side is replaced by a point vortex of circulation Γ_0 and coordinates (x_0, y_0). Vortex dynamics model focuses on equilibrium of the point vortex in the incident flow at the given geometry of flow field.

The practical goal of the control was to create and maintain the stationary circulating flow between the plate and the frontal prim sides which would suppress generation of vorticity in both the plate end **A** and the prism leading edge **B**. Therefore, the theoretical model comes to determining the plate parameters, length l and position r, such that the vortex of Γ_0 is immovable (standing) and the Kutta–Joukowski condition is satisfied in the sharp edges **A** and **B**. The system under consideration has one degree of freedom. Therefore, its evolution is described by a nonlinear differential equation in R^2:

$$\frac{d\mathbf{X}(t)}{dt} = f(\mathbf{X}(t)), \tag{17.5}$$

where $\mathbf{X}(t) \in R^2$ is the vector of vortex coordinates: $\mathbf{X}(t) = (x_0(t), y_0(t))$, the vector function $f: R^2 \to R^2$ sets the vortex velocity: $f(\mathbf{X}(t)) = \mathbf{V}(x_0(t), y_0(t))$.

As the incident flow velocity does not change in time, Eq. (17.5) is autonomous one. Then the vortex moves along the flow lines and the phase space of the dynamical system coincides with the flow domain. It follows that the vortex equilibrium position

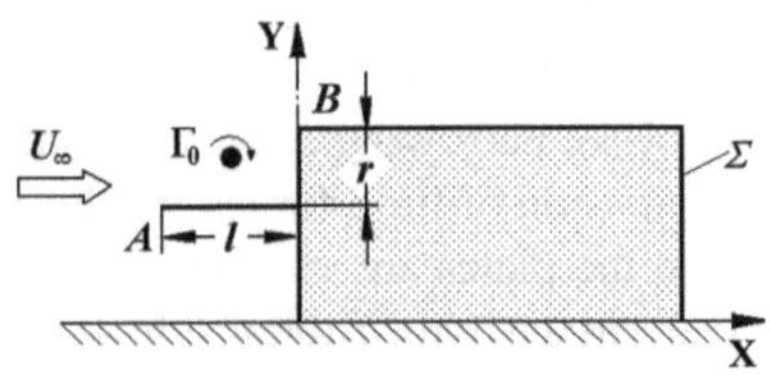

Fig. 17.3 Scheme of the flow with a trapped vortex in the top of the square prism with the attached plate

checks with the flow critical point; that is, its coordinates can be determined from the equation:

$$f(\mathbf{X}) = 0 \tag{17.6}$$

The Kutta–Joukowski theorem states that one will obtain an attached flow in the sharp edges **A** and **B** if the following equalities are satisfied:

$$\mathbf{V}|_A = C_1, \quad \mathbf{V}|_B = C_2. \tag{17.7}$$

Equations (17.6) and (17.7) fully describe the control problem in the sense that they allow to define uniquely the standing vortex characteristics and plate parameters ensuring the nonseparated flow in the sharp edges.

The flow under consideration is potential with the exception of the vortex point (x_0, y_0), so methods of the complex analysis can be employed for the flow analysis. Within this formalism, the position of a point vortex is identified with a point in the complex plane, i.e., $z_0 = x_0 + iy_0 \in C$, where $i = \sqrt{-1}$. The fluid flow at any point $z \in C$ is described by the complex potential that is the Green's function for the Laplace equation. In the flow with a solid boundary, the complex potential is built to satisfy condition (17.3) that the normal velocity component vanishes on all boundaries.

To fulfill the condition on the body surface Σ, the last is modeled by a continuous vortex sheet whose strength γ is induced by the jump in tangential velocity across the sheet. In the computation scheme, the sheet is divided into partitions of equal length and each partition is replaced by a single vortex of circulation $\Gamma_k = \gamma(s)\Delta s$. Here, $\gamma(s)$ is the linear intensity of the sheet in the body point of parametric coordinate s, Δs is the length of the segment between neighboring vortices, $k = 1, 2, \ldots, N$ and N is the number of the vortices in the discrete scheme. On the solid wall coinciding with x axes, the boundary condition is satisfied with applying the "method of mirror images." Then, the complex velocity potential of the problem is given by

$$\Phi(z) = U_\infty z + \frac{\Gamma_0}{2\pi i} \ln \frac{z - z_0}{z - \bar{z}_0} + \frac{1}{2\pi i} \sum_{k=1}^{N} \Gamma_k \ln \frac{z - z_k}{z - \bar{z}_k}. \tag{17.8}$$

In Eq. (17.8) $z_k = x_k + iy_k$ are the complex coordinates of the bound vortices, and the overbar denotes a complex conjugate. Taking into account that the flow complex velocity is

$$\mathbf{V}(z) = (u - iv)(z) = \frac{d\Phi(z)}{dz},$$

one derives the following expression for the free vortex velocity:

$$(u - iv)(z_0) = U_\infty + \frac{\Gamma_0}{4\pi y_0} + \frac{1}{2\pi i} \sum_{k=1}^{N} \Gamma_k \left(\frac{1}{z_0 - z_k} - \frac{1}{z_0 - \bar{z}_k} \right). \tag{17.9}$$

Sharing real and imaginary parts of (17.9), we construct the equations corresponding to condition of vortex equilibrium (17.6):

$$u(z_0) = 0, \tag{17.10}$$

$$v(z_0) = 0. \tag{17.11}$$

Conditions (17.7) are identical to the following equations:

$$\gamma(\mathbf{A}) = 0, \tag{17.12}$$

$$\gamma(\mathbf{B}) = 0. \tag{17.13}$$

So, control problem (17.6) and (17.7) is presented in the discrete scheme by system four transcendental Eqs. (17.10)–(17.13) with respect to the standing vortex parameters (x_0, y_0, Γ_0) and the plate geometrical characteristics. To close the system, the plate length l is supposed to be fixed and other parameters are presented as functions of l.

To determine circulations of bound vortices Γ_k, the method of boundary integral equations is applied [26]. Assuming that the control points which no-through flow boundary condition (17.3) is satisfied in are located in the middle of the segments coupling two neighboring vortices, we obtain the following system of linear algebraic equations with respect to Γ_k:

$$\frac{1}{2\pi}\sum_{k=1}^{N}\Gamma_k(V_n)_{lk} = -U_\infty - \frac{1}{2\pi}\Gamma_0(V_n)_{l0}, \quad l = 1, 2, \ldots N, \tag{17.14}$$

where $(V_n)_{lk}$, $(V_n)_{l0}$ are the normal velocities induced in the l- th control point by the bound vortices and free vortex, respectively. Notice due to the symmetry of the flow field relative to x axis, the condition of constancy of the circulation along a closed contour is satisfied automatically.

The results of calculations show that problem (17.10)–(17.13) has a unique solution, when $0.15 \le l \le 0.65$, i.e., the vortex of circulation Γ_0 located in the point (x_0, y_0) is immovable and it ensures an attached flow in both sharp edges **A** and **B**. Analysis of the eigenvalues of the matrix of the linearized system corresponding to (17.5) in the flow critical point (x_0, y_0) demonstrates that the point is a stable focus. The linear stability makes the present configuration quite interesting from the practical point of view. The picture of streamlines with a pair of standing vortices in frontal part of the square prism is presented in Fig. 17.4. It sustains that optimal choice of the control plate geometrical parameters is able to ensure nonseparated flow in the prism leading edge.

The attributes Γ_0, x_0, y_0 of the standing vortex and the space r between the prism corner and the plate attachment point against the plate length l are shown in Fig. 17.5. It can be observed that the dependencies are of linear type. Among them, the last

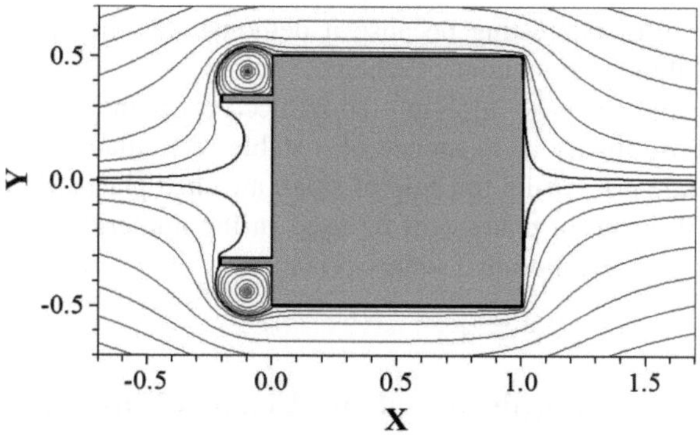

Fig. 17.4 Picture of streamlines with two standing vortices in front of the square prism with control plates at $l = 0.2$, $r = 0.16$

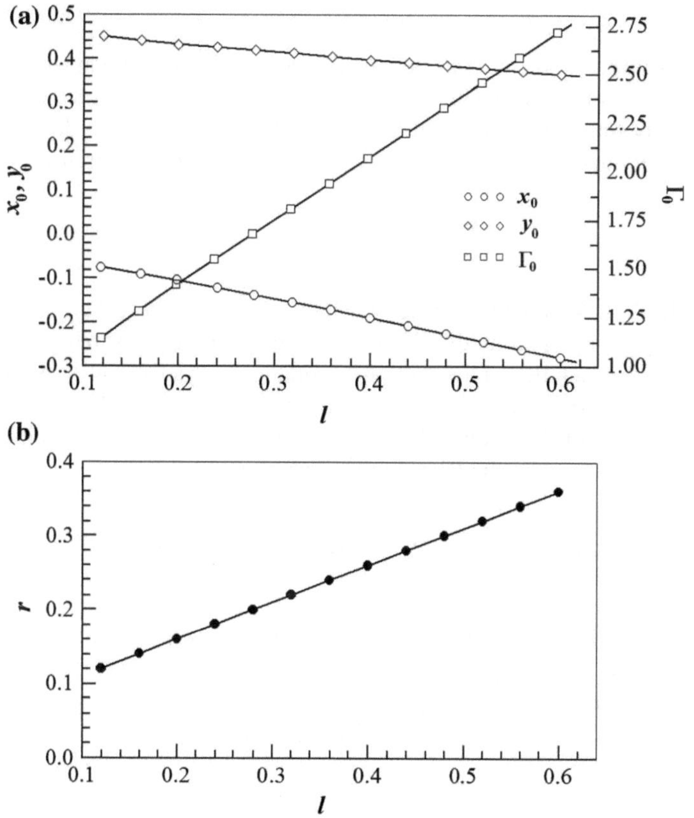

Fig. 17.5 **a** The standing vortex parameters Γ_0, x_0, y_0, **b** the plate position against the plate length l

function is the most interesting because it demonstrates the optimal geometrical characteristics of the present control scheme.

The results obtained with applying the reduced order model are fundamentally important as those justify an existence of a stable recirculation zone generated in front of the square prism with the help of small attached plates. Further the derived parameters of the control plates will be used in the numerical simulations of the viscous flow around the modified square prism.

17.4 Numerical Simulation of the Viscous Flow Past a Square Prism with Attached Frontal Plates

17.4.1 Details of Implementation of the 2D Vortex Method

Numerical simulations of viscous flow field around the square prism with two frontal plates were performed with a high-resolution vortex method, which we have shown can accurately simulate this class of flows [19, 20]. Vortex methods describe translation of vorticity in the flow field. Those are based on the vorticity transport equation:

$$\frac{\partial \omega}{\partial t} + (\mathbf{V} \cdot \nabla)\omega = \frac{1}{Re}\Delta\omega \tag{17.15}$$

with $\omega = \mathbf{k} \cdot \nabla \times \mathbf{V}$ being the vorticity, which is treated as a scalar quantity for $2D$ flows. This approach is preferable due to the absence of the pressure in the equations, automatic implementation of the continuity equation, and adaptability as the domains of concentrated vorticity are only considered when performing the calculations.

The velocity field $\mathbf{V}(\mathbf{r})$ induced by the volume and the surface vorticity, ω and γ, respectively, is defined by the Biot–Savart formula:

$$\mathbf{V}(\mathbf{r}, t) = \int_{\Sigma} \gamma(\mathbf{r}', t)\mathbf{k} \times \nabla G(\mathbf{r}, \mathbf{r}')dl(\mathbf{r}') + \int_{S} \omega(\mathbf{r}', t)\mathbf{k} \times \nabla G(\mathbf{r}, \mathbf{r}')ds(\mathbf{r}'), \tag{17.16}$$

where $\mathbf{r}$ is the radius vector of the point, G is the fundamental solution of Laplace operator for a unbounded domain: $G(\mathbf{r}) = \frac{1}{2\pi} ln|\mathbf{r}|$.

Equation (17.15) is solved by the splitting procedure when we get the equations, which describe separately the vortex convection and diffusion:

$$\frac{\partial \omega}{\partial t} + (\mathbf{V} \cdot \nabla)\omega = 0, \tag{17.17}$$

$$\frac{\partial \omega}{\partial t} = \frac{1}{Re}\Delta\omega. \tag{17.18}$$

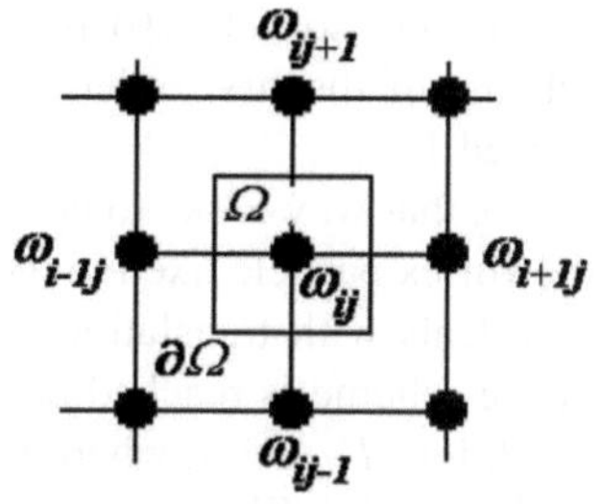

Fig. 17.6 The scheme of discretization of the vorticity field

In the present numerical realization of the vortex method, the finite volume discretization of the vorticity field is performed. The volumes are connected with node points of the orthogonal grid put on the calculation domain (Fig. 17.6). The point vortices located in the middle of each volume are characterized by the vorticity ω_{ij}, where $i = 1, 2, \ldots, N_x$, $j = 1, 2, \ldots, N_y$, N_x, N_y are the numbers of grid cells in x and y directions, respectively. From the divergence theorem, the law of conservation of vorticity in the elementary volume can be described in the form:

$$\frac{\partial}{\partial t}\int\int_{\Omega}\omega_{ij}dq = -\int_{\partial\Omega}\omega_{ij}(\mathbf{V}\cdot\mathbf{n})dl, \tag{17.19}$$

where Ω, $\partial\Omega$ is the discrete volume and its boundary, respectively, $\mathbf{n}$ is the normal to $\partial\Omega$ and $\mathbf{V}$ is the flow velocity on $\partial\Omega$. As Eq. (17.19) defines the vorticity convection across the elementary volume, we obtain the following numerical scheme for Eq. (17.17):

$$\begin{gathered}\frac{\omega_{ij}^{t+\Delta t}-\omega_{ij}^{t}}{\Delta t}\Delta x\Delta y \approx (\omega_{i-1j}^{t}u_{i-1j}^{t}-\omega_{i+1j}^{t}u_{i+1j}^{t})\Delta y+\\ (\omega_{ij-1}^{t}v_{ij-1}^{t}-\omega_{ij+1}^{t}v_{ij+1}^{t})\Delta x-\omega_{ij}^{t}(|u_{ij}^{t}|\Delta y+|v_{ij}^{t}|\Delta x),\end{gathered} \tag{17.20}$$

where Δx and Δy are the steps of space discretization in x and y directions, and Δt is the time step.

It is obvious that scheme (17.20) has the first order in time and the second order in space. Development of this approach on multilayer templates is presented in [27]. Note the scheme is dissipation-free and has improved dispersion properties compared with classical linear schemes.

To simulate the viscous diffusion process, we integrate Eq. (17.18) by the finite-difference method. The scheme of the second order in space written in the nodes of orthogonal grid takes the form:

$$\frac{\omega_{ij}^{t+\Delta t}-\omega_{ij}^{t}}{\Delta t} = \frac{1}{Re}\left(\frac{\omega_{i+1j}^{t}-2\omega_{ij}^{t}+\omega_{i-1j}^{t}}{(\Delta x)^2}+\frac{\omega_{ij+1}^{t}-2\omega_{ij}^{t}+\omega_{ij-1}^{t}}{(\Delta y)^2}\right). \tag{17.21}$$

Discrete Eqs. (17.20) and (17.21) are integrated in time with applying the explicit scheme of the first order. Notice it is stable at the Courant numbers that do not exceed 1.

So, that way looks to changing in time the circulation $\Gamma_{ij}(t) = \omega_{ij}(t)\Delta x \Delta y$ of the vortex particle fixed in the grid node unlike the classical vortex method [21, 22] that deals with translation of free discrete vortices in the flow field. Adaptability of the scheme is reached because of the grid points whose circulation satisfies the condition $|\Gamma_{ij}| < \varepsilon$, where ε is the small value, are only considered.

The Lighthill's mechanism of vorticity creation at a solid wall and linking it to vortex methods are described in detail in [21]. It explains the vorticity generation by changing the circulation γ of the vortex sheet simulating the body surface because of vorticity field modifications. In the numerical schemes of a vortex type, there are different approaches to calculation γ and its incorporation in a boundary condition for vorticity. We determinate the intensity of body sheet from no-through flow boundary condition (17.3), which leads to the following integral equations with respect to γ:

$$\int_{\Sigma} \gamma(\mathbf{r}', t) \frac{\partial G(\mathbf{r}, \mathbf{r}')}{\partial n} dl(\mathbf{r}') + \int_{S} \omega(\mathbf{r}', t) \frac{\partial G(\mathbf{r}, \mathbf{r}')}{\partial n} ds(\mathbf{r}') = 0, \tag{17.22}$$

where $\mathbf{r} \in \Sigma$.

The Kelvin's theorem of circulation conservation in the computational domain must be also satisfied:

$$\int_{\Sigma} \gamma(\mathbf{r}', t) dl(\mathbf{r}') + \int_{S} \omega(\mathbf{r}', t) ds(\mathbf{r}') = 0, \tag{17.23}$$

No-slip condition (17.4) is used to derive a boundary condition for vorticity. Taking into account the velocity jump across the vortex sheet, one has the following relation:

$$(\mathbf{V}_{\tau})_{-} = \mathbf{V}_{\tau}^{0} + \frac{\gamma}{2}, \tag{17.24}$$

where $\mathbf{V}_{\tau}^{0}$ is the tangential velocity of body-surface points calculated from (17.16) and $(\mathbf{V}_{\tau})_{-}$ is the limiting value of tangential velocity at the body, which condition (17.4) has to be satisfied for. Following Wu [28] who divided the strength of the vortex sheet by the distance from the wall to the first mesh point in the computational domain to obtain the vorticity on the body, we get the Dirichlet-type boundary condition for vorticity in the following form:

$$\omega_0 = \frac{2\mathbf{V}_{\tau}^{0}}{\Delta s}, \tag{17.25}$$

where Δs is the grid spacing perpendicularly to the wall.

The vorticity created on smooth walls enters the fluid through a mechanism of viscous diffusion described by formula (17.21). And the sharp edge vorticity is transferred to the flow with applying convection formula (17.20) that is equivalent to implementation the Kutta–Joukowski condition in this point.

17.4.2 Calculation of the Pressure Field and Forces on the Body

The introduction of vorticity and velocity–vorticity formulation of the Navier–Stokes equations allow to decouple purely kinematical problem from the pressure problem. It simplifies significantly numerical modeling of the hydrodynamic fields. But to estimate either the fluid forces acting to a body or sound level in the flow, one is need of calculating the pressure at least on the body. It has to be noted that recovery of the pressure from vorticity and velocity fields is a daunting challenge, which has invited attention of many researchers [29–31]. When direct solving the Poisson equation for the pressure, the problem of the correct choice of boundary condition arises. On the other hand, use of alternative approaches such as variational formulation [29] or Uhlman's integral [30] is difficult due to having sharp edges in the considered geometrical configuration.

We derive the pressure field by direct integrating the Navier–Stokes equations in the Lamb representation [32]:

$$\frac{\partial u}{\partial t} + \frac{1}{2}\frac{\partial}{\partial x}(u^2+v^2) - v\omega = -\frac{1}{\rho}\frac{\partial p}{\partial x} - \frac{1}{Re}\frac{\partial \omega}{\partial y}, \tag{17.26}$$

$$\frac{\partial v}{\partial t} + \frac{1}{2}\frac{\partial}{\partial y}(u^2+v^2) + u\omega = -\frac{1}{\rho}\frac{\partial p}{\partial y} + \frac{1}{Re}\frac{\partial \omega}{\partial x}. \tag{17.27}$$

It is obvious Eqs. (17.26) and (17.27) connect the pressure field with velocity and vorticity fields. Integrating Eq. (17.26) of the variable x and Eq. (17.27) of the variable y, one obtains the following formulae for calculation the dimensionless pressure:

$$\overline{p} = 1 - u^2 - v^2 + 2\int_{-\infty}^{x}\left(v\omega - \frac{\partial u}{\partial t} - \frac{1}{Re}\frac{\partial \omega}{\partial y}\right)dx, \tag{17.28}$$

$$\overline{p} = 1 - u^2 - v^2 + 2\int_{-\infty}^{y}\left(-u\omega - \frac{\partial v}{\partial t} + \frac{1}{Re}\frac{\partial \omega}{\partial x}\right)dy, \tag{17.29}$$

where $\overline{p} = 2(p - p_\infty)/\rho U_\infty^2$.

It depends on the flow field configuration, what equation from (17.28) and (17.29) will be chosen for calculating the pressure. Note that this way allows deriving the total drag including its form and viscous components.

The coefficients of fluid forces on the body are calculated using the pressure distribution:

$$C_x = \int_L \overline{p} n_x \, dx, \quad C_y = \int_L \overline{p} n_y \, dy, \tag{17.30}$$

where C_x, C_y are the coefficients of drag and lift, respectively, and $\mathbf{n} = (n_x, n_y)$ is the internal normal to the body.

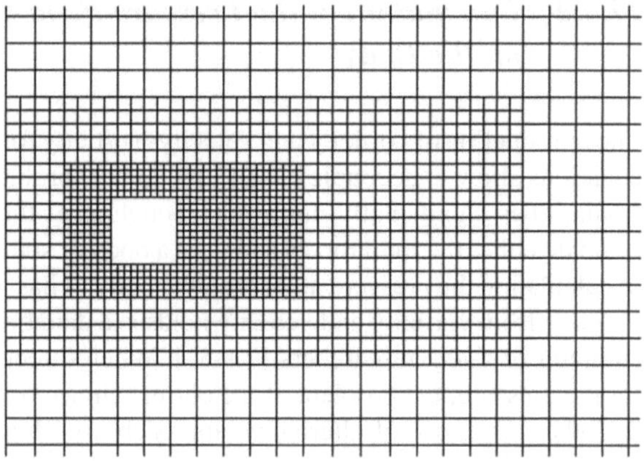

Fig. 17.7 Sketch of the computational grid

In the present numerical scheme, Eqs. (17.28) and (17.29) are integrated with the trapezium method on the base orthogonal grid.

17.4.3 Validation of the Algorithm

With the vortex method described above, the present simulation results for an impulsively started square prism at moderate Reynolds numbers ($Re = 100 \div 600$) are validated against theoretical, experimental, and numerical data available in the literature. In this study, we adopt the three-level rectangular grid with a constant cell size at each level as presented in Fig. 17.7. The grid spacing Δ_1 in the domain adjoining the body coincides with the length of the panels that simulate the bound vortex sheet. And the cell size of each next grid is doubled compared with the previous. The number of the nodes throughout the square side is determined after preliminary tests as $N_s = 50$ that leads to $\Delta_1 = 0.02$. The dimensionless width of the calculation region is 20 and the lengths of upstream and wake regions are 10 and 90, respectively. For all the cases investigated in this paper, the normalized computational time step is equal to $\Delta t = 0.01$.

Figure 17.8 presents the variation of Strouhal number St and mean drag coefficient C_D with Reynolds number for square prism from the present simulations. The shedding frequencies were determined from the power spectra of the nonstationary lift signals, as well as velocity fluctuations in the wake.

Included for comparisons are the known experimental data of Okajima [33, 34] together with the results of $2D$ and $3D$ DNS simulations of Norberg et al. [23]. In spite of the fact that $3D$ effects develop in the square cylinder flow starting from $Re \approx 170$ [23], the present results are seen to be in close agreement with the experimental data and in reasonable agreement with the numerical results. Note, when $Re \geq 150$, the mean drag coefficient obtained matches as experimental as numerical results very good. At the same time, the Strouhal numbers predicted by $3D$ simulations are

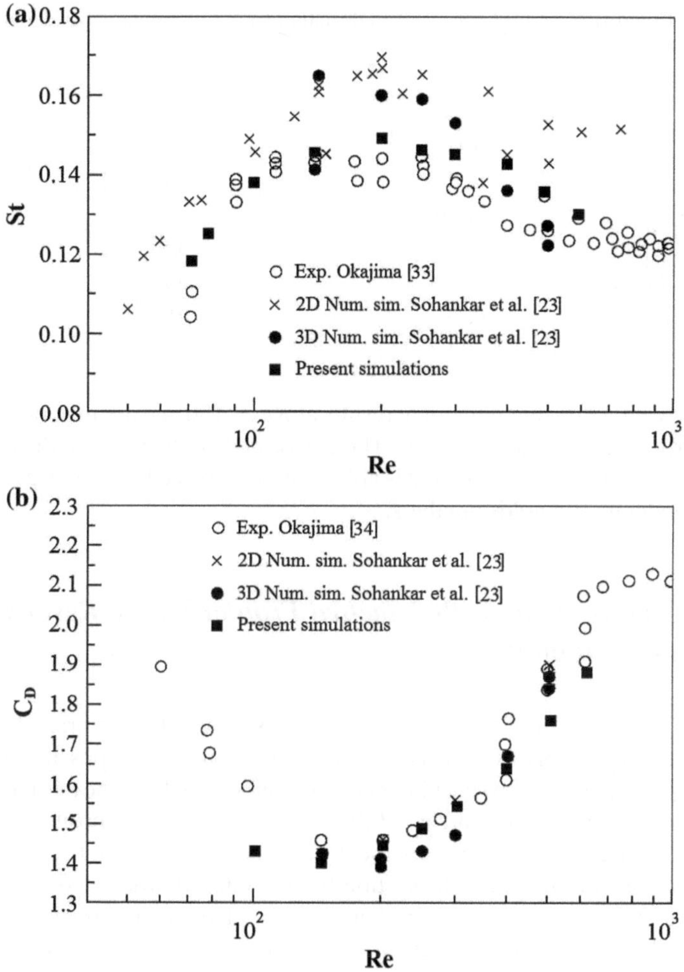

Fig. 17.8 **a** Strouhal number St, **b** mean drag coefficient C_D of a square prism against Re

not necessarily more "accurate" than the present results. Generally, the performed comparisons indicate the good correlation of both the time-mean drag and shedding frequency calculated with known experimental and numerical data.

As for quantitative characteristics of the lift force acting to a square cylinder, those are scarce in the literature. Table 17.1 contains data for the root-mean-square value (i. e., standard deviation) of the lift coefficient $C_{L_{rms}}$ obtained in the present calculations and known from previous researches at $Re = 150$ and $Re = 500$. Among those, data from [27] are only experimental and all other are acquired in numerical simulations. The coefficient $C_{L_{rms}}$ has been shown to be extremely sensitive as to Re variations as to aspect ratio of the computation domain [23, 27] that explains significant discrepancies in the results. Nonetheless, the present $C_{L_{rms}}$ values compare

Table 17.1 Comparison of lift force standard deviation $C_{L_{rms}}$ for a square cylinder at $Re = 150$ and $Re = 500$

Source	$C_{L_{rms}}$	
	$Re = 150$	$Re = 500$
Ali et al. [4]	0.28	–
Sohankar et al. [23]	0.23	1.13–1.22
Doolan [35]	0.296	–
Shimizu et al. [36]	–	0.56–0.72
Hwang and Sue [37]	–	0.9–1.01
Present simulation	0.23	0.9

reasonably well with other numerical results at $Re = 150$ and are in good agreement with experimental data at $Re = 500$. The performed comparisons indicate that the present version of the vortex method is able to predict correctly the flow past a square prism at moderate Reynolds numbers.

17.4.4 Square Prism with Attached Frontal Plates. Results of Simulation

In this section, an effect of two symmetrical plates attached to the prism frontal side on flow structure and prism loads is studied. As we consider the possibility of small control impact upon the flow, the plates are quite short and thin. The normalized plate length and width are $l = 0.2$ and $w = 0.02$, respectively. A plate position toward the adjacent prism edge r is chosen from the dependency presented in Fig. 17.5b, which has been obtained in the previous section by the reduced order model. That has to guarantee a stable recirculation zone between the plate and the prism frontal side. Here, the value of r_{opt} corresponding to the chosen plate length is 0.16.

An effect of the plates is as early as obvious if one compares the flow patterns developed beyond a square prism without control and under the optimal control. In Fig. 17.9, we present the vorticity fields obtained in the uncontrolled flow at $Re = 150$ (Fig. 17.9a) and $Re = 500$ (Fig. 17.9b). In all the figures, solid and dashed lines represent positive and negative vorticity values, respectively. At $Re = 150$, which is still before the onset of $3D$ effects, the wake is seen to be laminar, regular and characterized by the primary instability, the von Karman vortices. The estimated Strouhal number characterizing the vortex shedding frequency is 0.145, which is close to the experimental data of Okajima [33] ($St = 0.148$) and slightly smaller than the computational value of Inoue et al. [38] obtained by high-order direct numerical simulations ($St = 0.151$). As regards calculations at $Re = 500$, those have approximation character because the transition to $3D$ flow behind the prism occurs well before, at $Re \approx 190$ [23]. In particular, significant levels in components of nonspanwise vorticity can be presented near the body at $Re = 500$. However, the flow patters

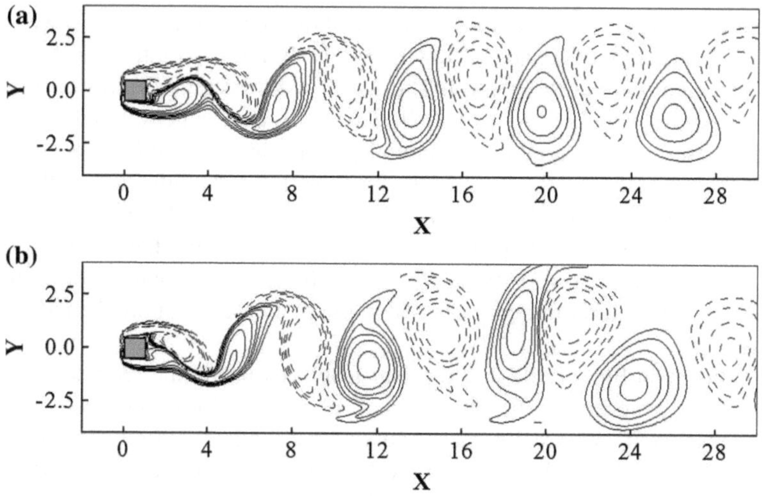

Fig. 17.9 Vorticity contours past a square prism without control at **a** $Re = 150$ and **b** $Re = 500$: *solid line*—positive circulation, *dashed line*—negative circulation

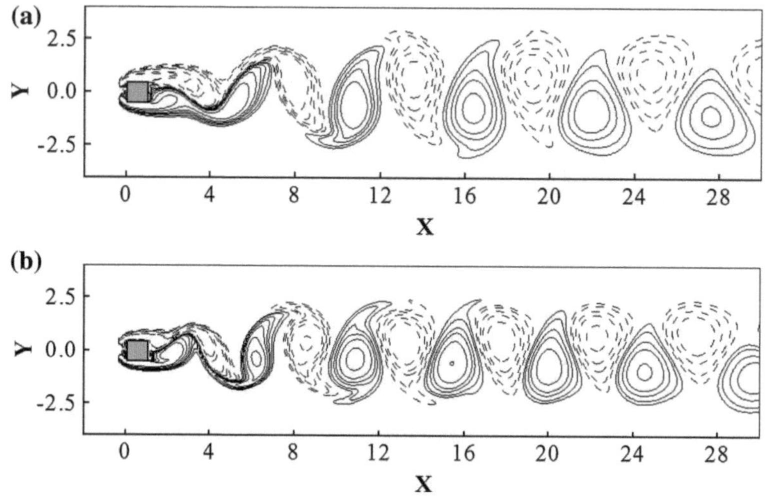

Fig. 17.10 Vorticity contours past a square prism with optimal control at $l = 0.2$, $r = 0.16$, **a** $Re = 150$ and **b** $Re = 500$

and loads obtained in our calculations are close to those observed in nature. So, the Strouhal number is 0.135 that coincides with experimental data of Norberg [39]. Other characteristics are also in good agreement with experimental and numerical data available in the literature that is shown in Fig. 17.8b and Table 17.1.

Figure 17.10 illustrates the wake patterns generated beyond the prism with the attached frontal plates ($l = 0.2$, $r = r_{opt} = 0.16$). Note the vorticity contours in Figs. 17.9 and 17.10 correspond to not only identical Reynolds numbers but also an identical instant when the flow with or without the control plates is well established. The structure of separated flow in Fig. 17.10a, b is seen to be different from those observed without the control. The wake width becomes narrower and much more regular, especially at $Re = 500$. The vortex shedding period in the controlled flow decreases as compared to the natural prism flow that results in the reduction of both the intensity and the scale of the wake vortices. An influence of the control plates on the flow grows significantly when increasing the Reynolds number. The obtained nondimensional frequency of vortex shedding (Strouhal number) is 0.157 at $Re = 150$ against $St = 0.195$ at $Re = 500$. It means the increase of St in comparison with the natural frequency is 8 % in the first case and more than 40 % in the last case.

It is shown in Fig. 17.10a, b the prism front lies inside the recirculation zones generated by plate ends. The phenomenon as well as lowering the recirculation bubble length and wake realignment causes drastic redistribution of pressure over the body. Figure 17.11a, b compares the time-averaged pressure coefficient $C_p = 2(p - p_\infty)/\rho U_\infty^2$ over the prism calculated without control plates and with the plates at $Re = 150$ and $Re = 500$, respectively. The pictures demonstrate equalizing the pressure at the frontal side and increase of the base pressure coefficient C_{pb} in the

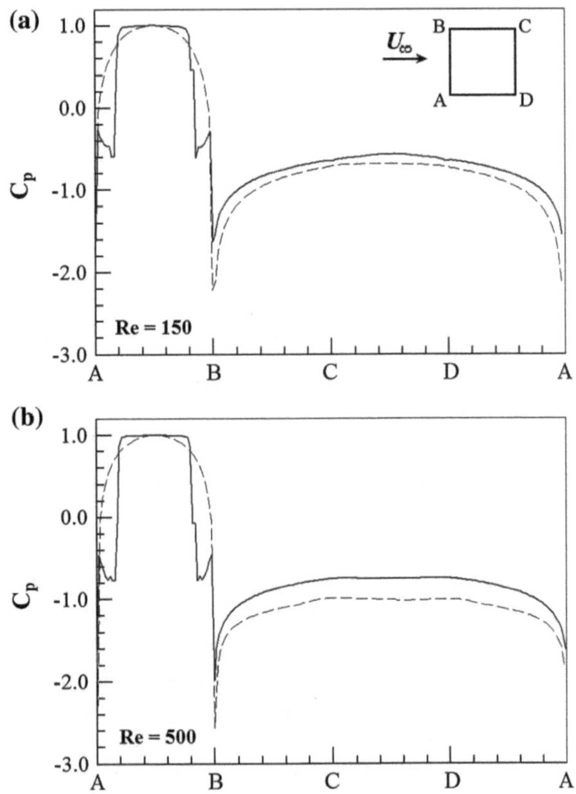

Fig. 17.11 The pressure coefficient over the prism surface without control (*dashed line*) and with optimal control (*solid line*) at **a** $Re = 150$ and **b** $Re = 500$

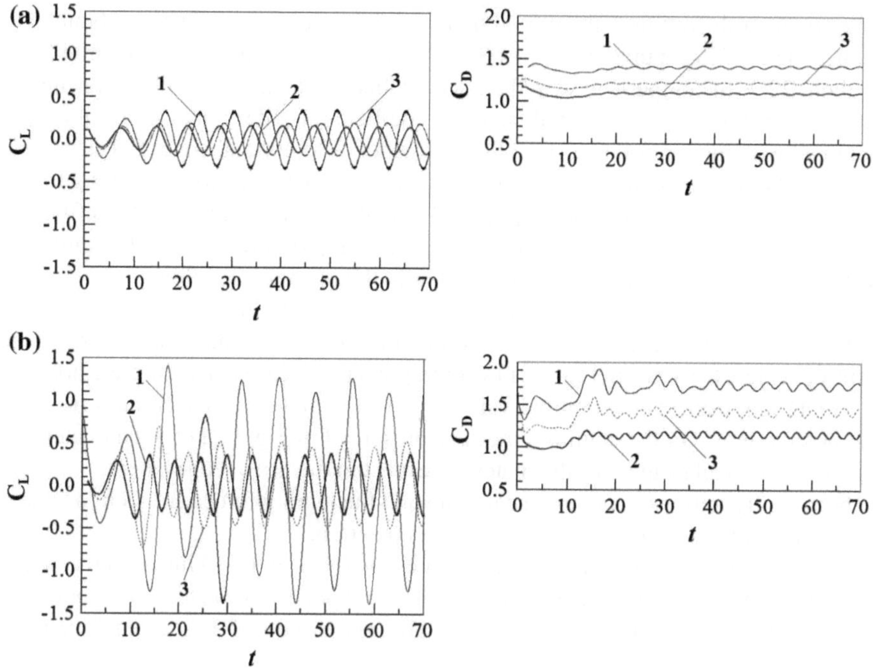

Fig. 17.12 Instantaneous drag coefficient C_D and lift coefficient C_L without control (curves 1), with optimal control (curves 2), with nonoptimal control (curves 3) at **a** $Re = 150$ and **b** $Re = 500$

controlled flow. At $Re = 150$, the base pressure coefficient rises from $C_{pb} = -0.73$ in the natural flow to $C_{pb} = -0.6$ in the controlled flow and at $Re = 500$, the increase is from $C_{pb} = -1.2$ to $C_{pb} = -0.8$. It is obvious the tendency leads to decreasing the prism drag, which is expected to be more significant at $Re = 500$. Note the obtained values of C_{pb} in the natural flow are close to DNS data of Sohankar et al. [23] that is important for the verification of our numerical scheme.

So, the attached frontal plates significantly affect both flow pattern and pressure distribution about the prism and one can thus expect change of the fluid forces as compared to the uncontrolled flow. Figure 17.12a, b shows the temporal traces of the drag (C_D) and lift (C_L) coefficients of square prism for uncontrolled and controlled flows at $Re = 150$ and $Re = 500$, respectively. Here curves labeled 1 correspond to the natural prism flow, curves labeled 2 describe the prism characteristics at optimal control ($l = 0.2, r = 0.16$), and curves as 3 deal with nonoptimal control when plate position r is chosen independently of the results obtained with applying the standing vortex model.

The presented data demonstrate substantial reduction of the hydrodynamic loads of the prism with attached frontal plates. The mean value of the prism drag coefficient at $Re = 150$ derived from its time history in our simulation (curve 1 in Fig. 17.11a) is $\overline{C_D} = 1.4$ that coincides exactly with both experimental [34] and DNS [23, 38] data.

The prism modification with optimal parameters of the control plates is obtained to reduce the coefficient up to $\overline{C_D} = 1.08$, curve 2 in Fig. 17.11a. So, drop in drag force is about 22 % in comparison with the uncontrolled flow. At $Re = 500$, lowering the prism mean drag is more essential; here, the coefficient $\overline{C_D}$ decreases from value 1.76 obtained without control up to 1.12 that corresponds to optimal control. The control effect is about 35 % in this case. Analogous results are achieved for fluctuating forces acting on a square prism. The amplitudes of both the drag and the lift coefficients decrease significantly in the controlled flow. For example, at $Re = 500$, the amplitude of C_L in the controlled flow is a third of that in the uncontrolled flow.

To emphasize an importance of the results derived by the simplified model of a standing vortex, we carried out the simulation with nonoptimal parameters of the control device. Curves 2 in Fig. 17.11a correspond to the case when the plate position $r = 0.22$ that exceeds the optimal value r_{opt}, so the plates are located too far from the prism edges. On the contrary, the case presented in Fig. 17.11b is characterized by too close displacement of the plates in respect to the prism edges, here $r = 0.08$. In both configurations, reducing mean drag and fluctuating forces are seen to be less than at the optimal ratio of plate length to its position.

The plots of instantaneous streamlines presented in Fig. 17.13 interpret the possible flow topology around the prism with the control plates. The results were obtained in the numerical simulations at $Re = 500$. Three snapshots of these streamline plots

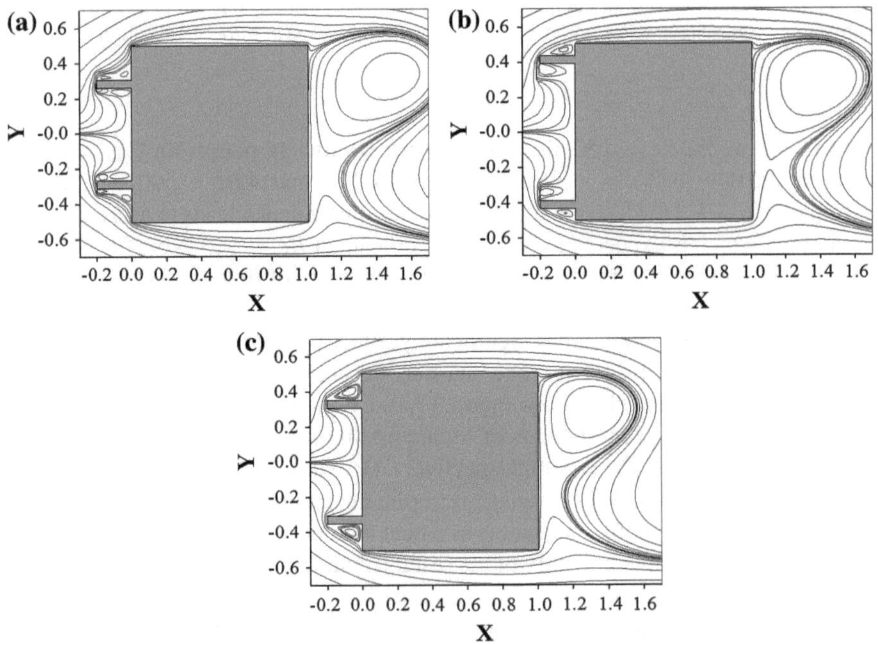

Fig. 17.13 Streamlines around the square prism with control plates at plate length $l = 0.2$, $Re = 500$ and different plate position: **a** $r = 0.22$, **b** $r = 0.08$, **c** $r = 0.16$

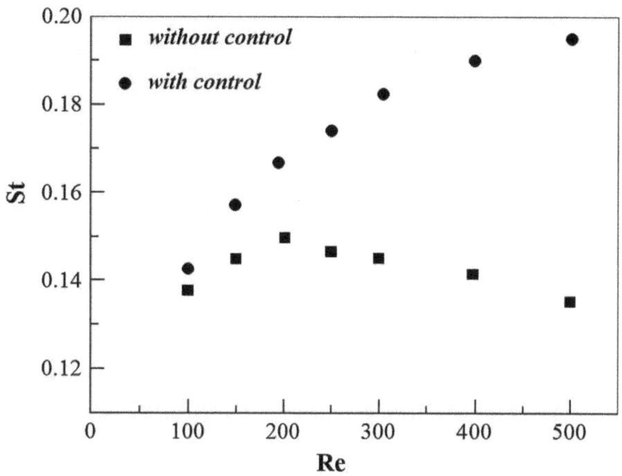

Fig. 17.14 Strouhal number of the square prism without control (*fill squares*) and with control (*fill circles*) against the Reynolds number

correspond to an identical instant inside the period of vortex shedding and differ by the plate location in respect to the prism edge. In the first mode, when $r > r_{opt}$, the streamline get off the plate end meets the prism at the frontal side (Fig. 17.13a). The small vortex restricted by zero streamline weakens but cannot prevent in full the flow separation in the prism leading edge. In the second case, no streamline reattachment to the prism is observed (Fig. 17.13b). Only the last picture derived at $r = r_{opt}$ demonstrates reattachment of zero streamline close to the prism leading edge (Fig. 17.13c). The recirculation zone restricted by the streamline is stable enough and able to suppress the flow separation at the prism edge. As a result, the recirculation bubble in the rear of prism is smaller than in previous modes. The conclusions are identical to those that have been made with applying the standing vortex theory (Fig. 17.4). This fact stresses an importance of development of simple topological models in order to forecast optimal properties of the devices used for a flow control.

Figure 17.14 shows the modification of the Strouhal number St by the frontal plates at its optimal configuration ($l = 0.2$, $r = 0.16$) for different Reynolds numbers. For comparison, values of St in the uncontrolled flow are also represented. Monitoring the vortex shedding frequency is known to be one way to quantify the processes occurring in the body wake. So, an increasing of the prism Strouhal number in the controlled flow indicates the substantial change of the wake pattern as it has been shown in Fig. 17.10. It follows from Fig. 17.14 the changes are typical for all Reynolds numbers from the considered range. The plate effect on the shedding frequency is seen to grow when increasing the Reynolds number.

Figure 17.15 depicts the mean drag coefficient $\overline{C_D}$ against the Reynolds number obtained at the optimal plate control. For the range of Reynolds number considered in the present study, the coefficient $\overline{C_D}$ is seen to reduce greatly in the controlled flow. At the same time, its change with the Reynolds number is rather weak. It points out the fact the relative decrease of the drag force under the control is higher at large Reynolds numbers.

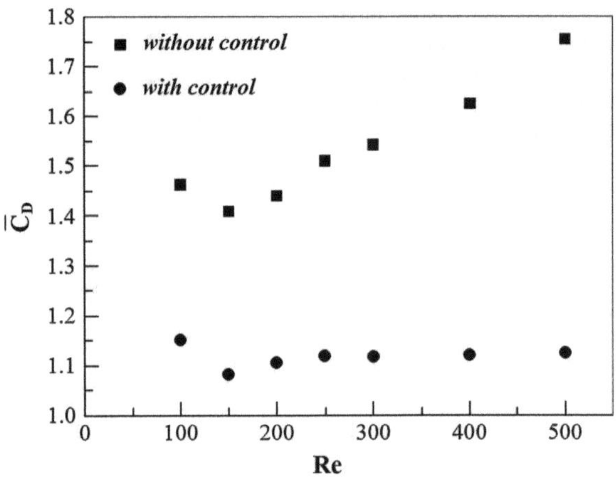

Fig. 17.15 Mean drag coefficient of the square prism without control and with control against the Reynolds number

All the results obtained in the present numerical simulation demonstrate the effectiveness of the frontal plates grows when increasing the Reynolds number of the flow. The conclusion lies in the frames of the chosen control strategy which is directed to creation of a stable recirculation zone, stable vortex, before the body. At small Reynolds numbers from the considered range, the vortex is less pronounced because it is suppressed by viscosity. The results point out the need to take into account the flow topology for the development of successful algorithms of flow control near bluff bodies.

17.5 Conclusion

In this work, passive control of flow around a square prism applying two attached frontal plates is analyzed. The developed control strategy is directed to creating the new flow topology in the region which should include a stable recirculation zone, standing vortex, before the body.

The topological flow pattern is studied with the help of the reduced model of a standing vortex, which allows us to derive the optimal ratio of plate length to its position. In this case, the critical point is a stable focus and it ensures smooth reattachment of the streamline separating in the plate end to the sharp edge of prism.

To estimate an efficiency of the control scheme, numerical simulation of viscous flow around the prism with small frontal plates, when $l = 0.2$, was carried out in the range of moderate Reynolds numbers. The obtained results revealed a considerable regularization of the flow pattern in the prism wake under the control, especially at large Reynolds numbers from the considered range. The positive flow correction in the near-wake results in a shortening of the recirculation bubble and reduction of both the strength and size of shedding vortices. The Strouhal number of prism in the controlled flow exceeds that in the natural flow and the plate effect on the shedding frequency becomes more intense when increasing the Reynolds number.

Due to wake regularization, both the mean drag and fluctuating forces decrease on the prism under the control. At the optimal displacement of the control plate, the reduction of drag coefficient of the square prism changes from 20 % at $Re = 100$ to 35 % at $Re = 500$. Reduction of the amplitude of lift coefficient is even more significant. It lies in the diapason 50–70 % depending on the Reynolds number.

The results of numerical simulations show that the largest decrease of prism loads is observed at the optimal ratio of plate length to its position, which has been derived with the help of the reduced model of a standing vortex. Thus, it is essential to take the flow topology into account when developing new algorithms for flow control near bluff bodies.

References

1. Zdravkovich, M.M.: Review and classification of various aerodynamic and hydrodynamic means for suppressing vortex shedding. J. Wind Eng. Ind. Aerodyn. **7**, 145–189 (1981)
2. Choi, H., Jeon, W.P., Kim, J.: Control of flow over a bluff body. Ann. Rev. Fluid Mech. **40**, 113–139 (2008)
3. Turki, S.: Numerical simulation of passive control of vortex shedding behind square cylinder using splitter plate. J. Eng. Appl. of Comp. Fl. Mech. **2**(4), 514–524 (2008)
4. Ali, M., Doolan, C., Wheatley, V.: Aeolian tones generated by a square cylinder with a detached flat plate. AIAA J. **51**(2), 291–301 (2013)
5. Zhou, C.Y., Wang, C., Islam, S.U., Xiao, Y.Q.: Numerical study of fluid force reduction on a square cylinder using a control plate. In: Proceedings of the Nineteenth International Offshore and Polar Engineering Conference. Osaka, Japan, June 21–26. pp. 1351–1356 (2009)
6. Igarashi, T.: Drag reduction of a square prism by flow control using a small rod. J. Wind Eng. Ind. Aerodyn. **69–71**, 141–153 (2013)
7. Bearman, P.W.: The effect of base bleed on the flow behind a two-dimensional model with a blunt trailing edge. J. Aeronaut. Q. **18**, 207–224 (1967)
8. Akansu, Y.E., Firhat, E.: Control of flow around a square prism by slot injection from the rear surface. J. Exp Ther. Fl. Sci. **34**(7), 906–914 (2010)
9. Firati, E., Akansu Y.E., Hacıaliogullar M.: Active control of flow around a square prism by slot injection. EPJ Web of Conference. **45** (2013)
10. Protas, B., Styszek, A.: Optimal rotary control of the cylinder wake in the laminar regime. J. Phys. Fl. **14**(7), 2073–2087 (2002)
11. Protas, B.: Vortex dynamics models in flow control problems. Nonlinearity **21**(9), 1–54 (2008)
12. Cortelezzi, L.: Nonlinear feedback control of the wake past a plate with a suction point on the downstream wall. J. Fluid Mech. **327**, 303–324 (1996)
13. Wu, J.Z., Vakili, A.D., Wu, J.M.: Review of the physics of enhancing vortex lift by unsteady excitation. Prog. Aerosp. Sci. **28**(2), 73–131 (1991)
14. Mhitaryan, A.M., Lukashchuk, C.A., Trubenok, B.D., Friland B.Ya.: Influence of eddy generators on aerodynamic characteristics of wing and body of rotation. Hydrodynamics of planes [in Russian]. Kiev: Naukova Dumka, pp. 254–263 (1966)
15. Perry, A.E., Chong, M.S.: A description of eddying motions and flow patterns using critical-point concepts. Ann. Rev. Fl. Mech. **19**, 125–155 (1978)
16. Chernyshenko, S.I.: Stabilization of trapped vortices by alternating blowing suction. Phys. Fluids **7**(4), 802–807 (1995)

17. Gorban, O.V., Gorban, I.M.: Dynamics of vortices in near-wall flows: eigenfrequencies, resonant properties, algorithms of control. AGARD Report **827**, 15-11 (1998)
18. Gorban, I.M., Khomenko, O.V.: Dynamics of vortices in near-wall flows with irregular boundaries. In: Zgurovsky, M.Z., Sadovnichiy, V.A. (eds.) Continuous and Distributed Systems: Theory and Applications. Solid Mechanics and Its Applications, vol. 211, pp. 115–128 (2014)
19. Gorban, O.V., Gorban, I.M.: Vortical flow structure near a square prism: numerical model and algorithms of control [in Ukrainian]. J Appl. Hydromech. **7**, 8–26 (2005)
20. Zgurovsky, M.Z, Melnik, V.S., Kasyanov, P.O.: Evolution inclusions and variation inequalities for Earth data processing. Adv. Mech. Math. **25** (2011)
21. Cottet, G.-H., Koumoutsakos, P.: Vortex Methods: Theory and Practice. Cambridge University Press, London (2000)
22. Liu, Z., Kopp, G.A.: High-resolution vortex particle simulations of flows around rectangular cylinders. J. Comp Fluids **40**, 2–21 (2011)
23. Sohankar, A., Norberg, C., Davidson, L.: Simulation of three-dimensional flow around a square cylinder at moderate Reynolds numbers. J. Phys. Fluids **11**(2), 288–306 (1999)
24. Ringleb, F.O.: Two-dimensional flow with standing vortex in ducts and diffusers. J. Fluids Eng. **82**(4), 921–927 (2011)
25. Roshko, A.: On the drag and shedding frequency of two-dimensional bluff bodies. NASA Technical Note **3169** (1954)
26. Lifanov, I.K.: Method of Singular Integral Equations and Numerical Experiment [in Russian]. TOO Janus, Moscow (1995)
27. Danilin, A.V., Goloviznin, V.M.: Scheme KABARE in variables vorticity-velocity for numerical simulation of ideal fluid motions in two-dimensional domains [in Russian]. J. Math. Model. **24**(5), 45–60 (2012)
28. Wu, J.C.: Numerical boundary conditions for viscous flow problems. AIAA J. **24**(5), 1042–1049 (1976)
29. Nowakowski, A., Rocicki, J., Styczek, A.: The pressure problem in the stochastic vortex blob method. ESAIM Proceed. **1**, 125–134 (1996)
30. Uhlman, J.S.: An integral equation formulation of the equations of motion of an incompressible fluid. NUWC-NPT Technical Report. **10086** (1992)
31. Dynnikova, G.Ya.: The integral formula for pressure field in the nonstationary barotropic flows of viscous fluid. J. Math. Fluid Mech. **16**, 145–162 (2014)
32. Lamb, G.: Hydromechanics. Cambridge University Press, London (1916)
33. Okajima, A.: Strouhal number of rectangular cylinders. J. Fluid Mech. **123**, 379–398 (1982)
34. Okajima, A., Sugitani, K.: Strouhal number and base pressure coefficient of a rectangular cylinder. J. Trans. ASME **50**, 2004–2010 (1984)
35. Doolan, C.J.: Flat-plate interaction with the near-wake of a square cylinder. AIAA J. **47**(2), 475–478 (2009)
36. Shimizu, M., Tanida, Y.: On the fluid forces acting on rectangular sectional cylinders. Trans. JSME **44**, 2699–2706 (1978)
37. Hwang, R.R., Sue, J.C.: Numerical simulation of shear effect on vortex shedding behind a square cylinder. Int. J. Numer. Meth. Fl. **25**, 1409–1420 (1977)
38. Inoue, O., Mori, M., Hatakeyama, N.: Aeolian tones radiated from flow past two square cylinders in a side-by-side arrangement. J. Phys. Fluids. **18** (2006)
39. Norberg, C.: Flow around rectangular cylinders: pressure forces and wake frequencies. J. Wind Eng. Ind. Aero. **49**, 187–195 (1993)

Chapter 18
Long-Time Behavior of State Functions for Badyko Models

Nataliia V. Gorban, Mark O. Gluzman, Pavlo O. Kasyanov and Alla M. Tkachuk

Abstract In this note we examine the long-time behavior of state functions for a climate energy balance model (Budyko Model) in the strongest topologies of the phase and the extended phase spaces. Strongest convergence results for all weak solutions are obtained. New structure and regularity properties for global and trajectory attractors are justified.

18.1 Introduction and Setting of the Problem

Let $(\mathscr{M}, \mathbf{g})$ be a C^∞ compact connected oriented two-dimensional Riemannian manifold without boundary (e.g., $\mathscr{M} = S^2$ the unit sphere of $\mathbb{R}^3$). Consider the problem:

$$\frac{\partial u}{\partial t} - \Delta u + R_e(x, u) \in QS(x)\beta(u), \quad (x, t) \in \mathbb{R}_+ \times \mathscr{M}, \tag{18.1}$$

where $\Delta u = \mathrm{div}_{\mathscr{M}}(\nabla_{\mathscr{M}} u)$; $\nabla_{\mathscr{M}}$ is understood in the sense of the Riemannian metric $\mathbf{g}$. Note that (18.1) is the so-called climate energy balance model. It was proposed in Budyko [4] and Sellers [38] and examined also in Díaz et al. [10–13]. The unknown

N.V. Gorban · P.O. Kasyanov
Institute for Applied System Analysis, National Technical University of Ukraine "Kyiv Polytechnic Institute", Peremogy ave., 37, build, 35, Kyiv 03056, Ukraine
e-mail: nata_gorban@i.ua

P.O. Kasyanov
e-mail: kasyanov@i.ua

M.O. Gluzman (✉)
Department of Applied Physics and Applied Mathematics,
Columbia University, New York, NY 10027, USA
e-mail: mark.gluzman@columbia.edu

A.M. Tkachuk
Faculty of Automation and Computer Systems,
National University of Food Technologies, Volodymyrska st., 68, Kyiv 01601, Ukraine
e-mail: tkachukam@ukr.net

V.A. Sadovnichiy and M.Z. Zgurovsky (eds.), *Advances in Dynamical Systems and Control*, Studies in Systems, Decision and Control 69,
DOI 10.1007/978-3-319-40673-2_18

$u(x, t)$ represents the average temperature of the Earth's surface. In Budyko [4] the energy balance is expressed as

$$\text{heat variation} = R_a - R_e + D.$$

Here $R_a = QS(x)\beta(u)$. It represents the solar energy absorbed by the Earth, $Q > 0$ is a solar constant, $S(x)$ is an insolation function (the distribution of solar radiation falling on upper atmosphere), β represents the ratio between absorbed and incident solar energy at the point x of the Earth's surface (so-called the co-albedo function). The term R_e represents the energy emitted by the Earth into space, and as usual, it is assumed to be an increasing function on u. The term D is the heat diffusion, and we assume (for simplicity) that it is constant.

As usual, the term R_e may be chosen according to the Newton cooling law as linear function on u, $R_e = Bu + C$ (here B and C are some positive constants) [4], or according to the Stefan–Boltzmann law, $R_e = \sigma u^4$ [38]. In this note we consider $R_e = Bu$ as in Budyko [4].

Let $S : \mathscr{M} \to \mathbb{R}$ be a function such that $S \in L^\infty(\mathscr{M})$, and there exist $S_0, S_1 > 0$ such that

$$0 < S_0 \le S(x) \le S_1.$$

Suppose also that β is a bounded maximal monotone graph of $\mathbb{R}^2$; that is, there exist $m, M \in \mathbb{R}$, such that for all $s \in \mathbb{R}$ and $z \in \beta(s)$

$$m \le z \le M.$$

Through the note we consider real Hilbert spaces

$$H := L^2(\mathscr{M}), \quad V := \{u \in L^2(\mathscr{M}) \, : \, \nabla_{\mathscr{M}} u \in L^2(T\mathscr{M})\}$$

with respective standard norms $\| \cdot \|_H$, $\| \cdot \|_V$, and inner products $(\cdot, \cdot)_H$, $(\cdot, \cdot)_V$, where $T\mathscr{M}$ represents the tangent bundle and the functional spaces $L^2(\mathscr{M})$ and $L^2(T\mathscr{M})$ are defined in a standard way; see, for example, Aubin [2]. Let V^* be the dual space of the function space V. We remark that

$$V \subset H \subset V^*,$$

and all embeddings are compact and dense; see, for example, Aubin [2, p. 55, Theorem 2.34].

Let $-\infty < \tau < T < +\infty$. A function $u(\cdot) \in L^2(\tau, T; V)$ is called a *weak solution* of Problem (18.1) on $[\tau, T]$, if there exists a measurable function $d : \mathscr{M} \times (\tau, T) \to \mathbb{R}$ such that

$$d(x, t) \in QS(x)\beta(u(x, t)) \text{ for a.e. } (x, t) \in \mathscr{M} \times (\tau, T), \tag{18.2}$$

and

$$\int_{\tau}^{T} \left[\langle -u, \frac{\partial \xi}{\partial t} \rangle - \langle u, \Delta \xi \rangle + \langle R_e(\cdot, t, u), \xi \rangle - \langle d, \xi \rangle \right] dt = 0, \tag{18.3}$$

for all $\xi \in C_0^\infty(\mathscr{M} \times (\tau, T))$, where $\langle \cdot, \cdot \rangle$ denotes the pairing in the space V.

In this manuscript, we examine the long-term dynamics as $t \to +\infty$ of all weak solution for Problem (18.1) in the strongest sense under the assumptions listed above.

We note that the existence of a Lyapunov function for a class of semi-linear parabolic differential reaction-diffusion equations with discontinuous nonlinearities, regularity properties for global and trajectory attractors, and its applications were considered in [16–18]. In [5, 32, 46, 48, 49] authors provided sufficient conditions for the existence of a Lyapunov function for autonomous evolution inclusions of hyperbolic type. The theory of the global and trajectory attractors for parabolic systems in the natural phase and extended phase spaces was considered in [1, 3, 6–9, 14, 19–28, 30, 31, 33, 39–45]. Topological properties of strong and weak solutions were provided in [15, 34–37]. Strong regularity properties of global and trajectory attractors were proved in [10, 26–29].

18.2 Auxiliaries

According to [16], for each $u_0 \in H$ and $T > 0$, there exists at least one weak solution of Problem (18.1) on $[0, T]$. Moreover, each weak solution $u(\cdot)$ of Problem (18.1) on $[0, T]$ is *regular*, that is, $u(\cdot) \in C([\varepsilon, T]; V) \cap L^2(\varepsilon, T; D(A))$ and $u_t(\cdot) \in L^2(\varepsilon, T; H)$, for each $\varepsilon \in (0, T)$; see Gluzman et al. [16, Theorem 14.1], where $D(A) := \{u \in V : Au \in H\}$ and $\langle Au, v \rangle_V = (u, v)_V$ for each $u, v \in V$. Furthermore, each weak solution of Problem (18.1) on $[0, T]$ can be extended to a global one defined on $[0, +\infty)$; see Gluzman et al. [16, p. 235].

Denote by $\mathscr{D}(u_0)$ the set of all weak solutions of Problem (18.1) globally defined on $[0, +\infty)$ with initial data $u(0) = u_0$, $u_0 \in H$. Then, $\mathscr{D}(u_0) \subset L^2_{loc}(0, +\infty; V) \cap C([0, +\infty), H)$ for each $u_0 \in H$. Moreover, $\mathscr{D}(u_0) \subset L^\infty(0, +\infty; H)$ for each $u_0 \in H$.

Consider the family of all weak solutions of Problem (18.1) defined on the semi-infinite time interval $[0, +\infty)$:

$$\mathscr{K} = \cup_{u_0 \in H} \mathscr{D}(u_0).$$

The set $\mathscr{K}_+$ is a *translation invariant* one, that is, $u(\cdot + h) \in \mathscr{K}_+$ for each $u(\cdot) \in \mathscr{K}_+$ and $h \geq 0$.

Let us consider Problem (18.1) on the entire time axis. A function $u \in L^\infty(\mathbb{R}; H)$ is called a *complete trajectory* of Problem (18.1), if $\Pi_+ u(\cdot + h) \in \mathscr{K}_+$ for each $h \geq 0$, where Π_+ is the restriction operator to the interval $[0, +\infty)$. Denote by $\mathscr{K}$ the family of all complete trajectories of Problem (18.1) A complete trajectory

$u(\cdot) \in \mathcal{K}$ is *stationary* if there is $z \in D(A)$ such that $u(t) = z$ for all $t \in \mathbb{R}$. Such z is called a *rest point*. We denote the set of all rest points by Z.

Definition 18.1 The function $E : V \to \mathbb{R}$ is called a *Lyapunov type one* for $\mathcal{K}_+$, if the following conditions hold:
(a) E is continuous on V;
(b) $E(u(t)) \leq E(u(s))$ whenever $u \in \mathcal{K}_+$ and $t \geq s > 0$;
(c) If $E(u(\cdot)) \equiv \text{const}$, for some $u \in \mathcal{K}$, then u is stationary complete trajectory.

Let $\Upsilon(s)$ be a real function such that $\partial\Upsilon(s) = \beta(s)$ for each $s \in \mathbb{R}$ and $\mathbf{1} : \mathcal{M} \to \mathbb{R}$, $\mathbf{1} \equiv 1$. According to Gluzman et al. [16, Theorem 14.2], the following function

$$E(u) = \frac{1}{2}\|u\|_V^2 + \frac{B}{2}\|u\|_H^2 - Q\langle S(\cdot)\Upsilon(u), \mathbf{1}\rangle \quad u \in V, \tag{18.4}$$

is a Lyapunov-type function for $\mathcal{K}_+$. Moreover, the following energy equality holds:

$$E(u(T)) - E(u(\tau)) = -\int_\tau^T \left\| \frac{\partial u}{\partial s}(\cdot, s) \right\|_H^2 ds, \tag{18.5}$$

for each $u \in \mathcal{K}_+$ and $0 < \tau < T < \infty$. The following lemma provides the main convergence result for all weak solutions of Problem (18.1) in the strongest topologies.

Lemma 18.1 (Gluzman et al. [16, Theorem 14.3]) *Let $0 < \tau < T$, $u_\tau \in H$, and $\{u_n(\cdot)\}_{n\geq 1}$ be a sequence of weak solutions for Problem (18.1) on $[\tau, T]$. Furthermore, let $u_n(\tau) \to u_\tau$ weakly in H as $n \to \infty$. Then, there exists a weak solution $u(\cdot)$ for Problem (18.1) on $[\tau, T]$ such that $u(\tau) = u_\tau$, and there exists an increasing sequence of positive integers $\{n_k\}_{k\geq 1}$ such that for each $\varepsilon \in (0, T - \tau)$*

$$\sup_{t\in[\tau+\varepsilon,T]} \|u_{n_k}(t) - u(t)\|_V + \int_{\tau+\varepsilon}^T \left\| \frac{\partial u_{n_k}}{\partial s}(\cdot, s) - \frac{\partial u}{\partial s}(\cdot, s) \right\|_H^2 ds \to 0, \tag{18.6}$$

as $k \to +\infty$.

Definition 18.2 The multivalued map $G : \mathbb{R}_+ \times H \to 2^H \setminus \emptyset$ is called a *strict multivalued semiflow* if:
(a) $G(0, \cdot) = \text{Id}$ (the identity map);
(b) $G(t + s, x) = G(t, G(s, x))\ \forall x \in H, t, s \in \mathbb{R}_+$.

Let us define the multivalued map $G : \mathbb{R}_+ \times H \to 2^H\setminus\{\emptyset\}$ as follows:

$$G(t, u_0) = \{u(t) \mid u(\cdot) \in \mathcal{K}_+,\ u(0) = u_0\}. \tag{18.7}$$

Lemma 18.2 (Zgurovsky et al. [47, Chap. 2]) *The multivalued map $G : \mathbb{R}_+ \times H \to 2^H \setminus \{\emptyset\}$, defined in (18.7), is a strict multivalued semiflow.*

18.3 Main Results

In this section we state that there exist trajectory and global attractors for all weak solutions of Problem (18.1) and provide their structure and regularity properties.

Definition 18.3 A set $\mathscr{A} \subseteq H$ is called *an invariant global attractor* for multi-valued semiflow G if the following conditions hold:
(1) $\mathscr{A}$ is an invariant set, that is $\mathscr{A} = G(t, \mathscr{A})$ for each $t \geq 0$;
(2) $\mathscr{A}$ is an attracting set, that is, for each nonempty bounded subset $B \subset H$,

$$dist_H(G(t, B), \mathscr{A}) \to 0, \quad t \to +\infty,$$

where $dist_H(C, D) = \sup_{c \in C} \inf_{d \in D} \|c - d\|_H$ denote the Hausdorff semidistance between nonempty subsets C and D of space H.
(3) For any closed attracting set $Y \subseteq H$, we have $\mathscr{A} \subseteq Y$.

Theorem 18.1 *The strict multivalued semiflow $G : \mathbb{R}_+ \times H \to 2^H \setminus \emptyset$, defined in (18.7), has a compact invariant global attractor $\mathscr{A}$ in the phase space H.*

Let $\{T(h)\}_{h \geq 0}$ be the translation semigroup acting on $\mathscr{K}_+$, that is, $T(h)u(\cdot) = u(\cdot + h)$, $h \geq 0$, $u(\cdot) \in \mathscr{K}_+$. On $\mathscr{K}_+$, we consider the topology induced from the Fréchet space $C_{loc}(\mathbb{R}_+; H)$. Note that $f_n(\cdot) \to f(\cdot)$ in $C_{loc}(\mathbb{R}_+; H)$ as $n \to \infty$ if and only if $\forall M > 0$ $\Pi_{0,M} f_n(\cdot) \to \Pi_{0,M} f(\cdot)$ in $C([0, M]; H)$ as $n \to \infty$.

Definition 18.4 A set $\mathscr{U} \subset \mathscr{K}_+$ is called a *trajectory attractor* for translation semigroup $\{T(h)\}_{h \geq 0}$ on $\mathscr{K}_+$ in the induced topology of $C_{loc}(\mathbb{R}_+; H)$, if $\mathscr{U} \subset \mathscr{K}_+$ is a global attractor for the translation semigroup $\{T(h)\}_{h \geq 0}$ acting on $\mathscr{K}_+$; see Kasyanov et al. [29, Sect. 3].

Theorem 18.2 *There exists a trajectory attractor $\mathscr{U}$ for $\{T(h)\}_{h \geq 0}$ on $\mathscr{K}_+$ in the induced topology of $C_{loc}(\mathbb{R}_+; H)$. Moreover, the following equalities hold:*

$$\mathscr{U} = \Pi_+ \mathscr{K} = \{u(\cdot) \in \mathscr{K}_+ \mid u(t) \in \mathscr{A} \ \forall t \in \mathbb{R}_+\} = \{u(\cdot) \in \mathscr{K}_+ \mid u(0) \in \mathscr{A}\}; \tag{18.8}$$

The following theorem provides structure and regularity properties for global and trajectory attractors for all weak solutions of Problem (18.1).

Theorem 18.3 *The following statements hold:*

(i) $\mathscr{A}$ *is a compact subset of* V*;*
(ii) $\mathscr{U}$ *is a bounded subset of* $L^\infty(\mathbb{R}_+; V)$ *and* $\Pi_{0,M}\mathscr{U}$ *is a compact subset of* $W(0, M)$ *for each* $M > 0$*, where* $W(0, M) = \{u(\cdot) \in C([0, M]; V) : u_t(\cdot) \in L^2(0, M; H)\}$ *is a real Banach space;*
(iii) $\mathscr{K}$ *is a bounded subset of* $L^\infty(\mathbb{R}; V)$ *and* $\Pi_{0,M}\mathscr{U}$ *a compact subset of* $W(0, M)$ *for each* $M > 0$*;*

(iv) *For each nonempty bounded set* $B \subset H$ $dist_V(G(t, B), \mathscr{A}) \to 0$, $t \to \infty$;

(v) *For any bounded in* $L^\infty(\mathbb{R}_+; H)$ *set* $\mathbf{B} \subset \mathscr{K}_+$ *and any* $M \geq 0$ *the following relation holds:* $dist_{W(0,M)}(\Pi_{0,M}T(t)\mathbf{B}, \Pi_{0,M}\mathscr{U}) \to 0, \quad t \to +\infty$;

(vi) *For each* $u \in \mathscr{K}$ *the limit sets*

$$\alpha(u) = \{z \in V \mid u(t_j) \to z \text{ in } V \text{ for some sequence } t_j \to -\infty\},$$

$$\omega(u) = \{z \in V \mid u(t_j) \to z \text{ in } V \text{ for some sequence } t_j \to +\infty\}$$

are connected subsets of Z *on which* E *is constant. If* Z *is totally disconnected (in particular, if* Z *is countable) the limits in* V

$$z_- = \lim_{t \to -\infty} u(t), \quad z_+ = \lim_{t \to +\infty} u(t) \tag{18.9}$$

exist and z_-, z_+ *are rest points; furthermore,* $u(t)$ *tends in* V *to a rest point as* $t \to +\infty$ *for every* $u \in \mathscr{K}_+$.

18.4 Proof of Theorems 18.1, 18.2 and 18.3

Gluzman et al. [16, Theorem 14.4] yield all the statements of Theorems 18.1, 18.2, and 18.3, because the spaces V, H and operators A, $J_1(\cdot) := \frac{B}{2}\|\cdot\|_H^2$, $J_2(\cdot) := E(\cdot) - \frac{B}{2}\|\cdot\|_H^2 - \frac{1}{2}\|\cdot\|_V^2$ satisfy the assumptions of [16, Theorem 14.4], that is,

(a) $(V; H; V^*)$ is an evolution triple, where V is a real Hilbert space, such that $V \subset H$ with compact imbedding;

(b) $A : V \to V^*$ is a linear symmetric operator such that there exists $c > 0$ such that $\langle Av, v\rangle \geq c\|v\|_V^2$, for each $v \in V$;

(c) $J_i : H \to \mathbb{R}$ is a convex, lower semicontinuous function such that the following assumptions hold: (i) (growth condition) There exists $c_1 > 0$ such that $\|y\|_H \leq c_1(1 + \|u\|_H)$, for each $u \in H$ and $y \in \partial J_i(u)$ and $i = 1, 2$; (ii) (sign condition) there exist $c_2 > 0$, $\lambda \in (0, c)$ such that $(y_1 - y_2, u)_H \geq -\lambda\|u\|_H^2 - c_2$, for each $y_i \in \partial J_i(u)$, $u \in H$, where $\partial J_i(u)$ the subdifferential of $J_i(\cdot)$ at a point u; $i = 1, 2$, $0 < \lambda < \lambda_1$, λ_1 is a first eigenvalue of A. Note that $u^* \in \partial J_i(u)$ if and only if $u^*(v - u) \leq J_i(v) - J_i(u)$ $\forall v \in H$; $i = 1, 2$.

Acknowledgments This work was partially supported by the Ukrainian State Fund for Fundamental Researches under grant GP/F66/14921 and by the National Academy of Sciences of Ukraine under grant 2284.

References

1. Arrieta, J.M., Rodrígues-Bernal, A., Valero, J.: Dynamics of a reaction-diffusion equation with discontinuous nonlinearity. Int. J. Bifurc. Chaos **16**, 2695–2984 (2006)
2. Aubin, T.: Nonlinear Analysis on Manifolds. Monge-Ampére Equations. Springer, Berlin (1980)
3. Balibrea, F., Caraballo, T., Kloeden, P.E., Valero, J.: Recent developments in dynamical systems: three perspectives. Int. J. Bifurc. Chaos (2010). doi:10.1142/S0218127410027246
4. Budyko, M.I.: The effects of solar radiation variations on the climate of the Earth. Tellus **21**, 611–619 (1969)
5. Ball, J.M.: Global attractors for damped semilinear wave equations. DCDS **10**, 31–52 (2004)
6. Barbu, V.: Nonlinear Semigroups and Differential Equations in Banach Spaces. Editura Academiei, Bucuresti (1976)
7. Chepyzhov, V.V., Vishik, M.I.: Trajectory and global attractors of three-dimensional Navier-Stokes systems. Math. Notes **71**, 177–193 (2002). doi:10.1023/A:1014190629738
8. Chepyzhov, V.V., Vishik, M.I.: Trajectory attractor for reaction-diffusion system with diffusion coefficient vanishing in time. Discret. Contin. Dyn. Syst. Ser. A **27**, 1493–1509 (2013)
9. Chepyzhov, V.V., Conti, M., Pata, V.: A minimal approach to the theory of global attractors. Discret. Contin. Dyn. Syst. **32**, 2079–2088 (2012)
10. Díaz, H., Díaz, J.: On a stochastic parabolic PDE arising in climatology. Rev. R. Acad. Cien. Serie A Mat. **96**, 123–128 (2002)
11. Díaz, J., Tello, L.: Infinitely many stationary solutions for a simple climate model via a shooting method. Math. Methods Appl. Sci. **25**, 327–334 (2002)
12. Díaz, J., Hernández, J., Tello, L.: On the multiplicity of equilibrium solutions to a nonlinear diffusion equation on a manifold arising in climatology. J. Math. Anal. Appl. **216**, 593–613 (1997)
13. Díaz, J., Hernández, J., Tello, L.: Some results about multiplicity and bifurcation of stationary solutions of a reaction diffusion climatological model. Rev. R. Acad. Cien. Serie A. Mat. **96**(3), 357–366 (2002)
14. Feireisl, E., Norbury, J.: Some existence and nonuniqueness theorems for solutions of parabolic equations with discontinuous nonlinearities. Proc. R. Soc. Edinb. A. **119**(1–2), 1–17 (1991)
15. Gajewski, H., Gröger, K., Zacharias, K.: Nichtlineare operatorgleichungen und operatordifferentialgleichungen. Akademie-Verlag, Berlin (1974)
16. Gluzman, M.O., Gorban, N.V., Kasyanov, P.O.: Lyapunov functions for differential inclusions and applications in physics, biology, and climatology. Continuous and distributed systems II. Theory and applications. Series studies in systems. Decis. Control **30**, 233–243 (2015). doi:10.1007/978-3-319-19075-4_14
17. Gluzman, M.O., Gorban, N.V., Kasyanov, P.O.: Lyapunov type functions for classes of autonomous parabolic feedback control problems and applications. Appl. Math. Lett. (2015). https://dx.doi.ord/10.1016/j.aml.2014.08.006
18. Gluzman, M.O., Gorban, N.V., Kasyanov, P.O.: Lyapunov functions for weak solutions of reaction-diffusion equations with discontinuous interaction functions and its applications. Nonautonomous Dyn. Syst. (2015). doi:10.1515/msds-2015-0001
19. Goldstein, G.R., Miranville, A.: A Cahn-Hilliard-Gurtin model with dynamic boundary conditions. Discret. Contin. Dyn. Syst. Ser. S **6**, 387–400 (2013)
20. Gorban, N.V., Kasyanov, P.O.: On regularity of all weak solutions and their attractors for reaction-diffusion inclusion in unbounded domai. Solid Mech. Appl. **211**, 205–220 (2014)
21. Gorban, N.V., Kapustyan, O.V., Kasyanov, P.O.: Uniform trajectory attractor for non-autonomous reaction-diffusion equations with Caratheodory's nonlinearity. Nonlinear Anal. Theory Methods Appl. **98**, 13–26 (2014). doi:10.1016/j.na.2013.12.004
22. Gorban, N.V., Kapustyan, O.V., Kasyanov, P.O., Paliichuk, L.S.: On global attractors for autonomous damped wave equation with discontinuous nonlinearity. Solid Mech. Appl. **211**, 221–237 (2014)

23. Efendiev, M., Miranville, A., Zelik, S.: Exponential attractors for a nonlinear reaction-diffusion system in R^3. Comptes Rendus de l'Academie des Sciences-Series I - Mathematics **330**, 713–718 (2000)
24. Kalita, P., Lukaszewicz, G.: Global attractors for multivalued semiflows with weak continuity properties. Nonlinear Anal. Theory Methods Appl. **101**, 124–143 (2014)
25. Kalita, P., Lukaszewicz, G.: Attractors for Navier-Stokes flows with multivalued and non-monotone subdifferential boundary conditions. Nonlinear Anal. Real World Appl. **19**, 75–88 (2014)
26. Kapustyan, O.V., Kasyanov, P.O., Valero, J.: Regular solutions and global attractors for reaction-diffusion systems without uniqueness. Commun. Pure Appl. Anal. **13**, 1891–1906 (2014). doi:10.3934/cpaa.2014.13.1891
27. Kapustyan, O.V., Kasyanov, P.O., Valero, J.: Structure and regularity of the global attractor of a reaction-diffusion equation with non-smooth nonlinear term. Commun. Pure Appl. Anal. **34**, 4155–4182 (2014). doi:10.3934/dcds.2014.34.4155
28. Kapustyan, O.V., Kasyanov, P.O., Valero, J., Zgurovsky, M.Z.: Structure of uniform global attractor for general non-autonomous reaction-diffusion system. Solid Mech. Appl. **211**, 163–180 (2014)
29. Kasyanov, P.O., Toscano, L., Zadoianchuk, N.V.: Regularity of weak solutions and their attractors for a parabolic feedback control problem. Set-Valued Var. Anal. **21**, 271–282 (2013). doi:10.1007/s11228-013-0233-8
30. Kasyanov, P.O.: Multivalued dynamics of solutions of an autonomous differential-operator inclusion with pseudomonotone nonlinearity. Cybern. Syst. Anal. **47**, 800–811 (2011)
31. Kasyanov, P.O.: Multivalued dynamics of solutions of autonomous operator differential equations with pseudomonotone nonlinearity. Math. Notes **92**, 205–218 (2012)
32. Kasyanov, P.O., Toscano, L., Zadoianchuk, N.V.: Long-time behaviour of solutions for autonomous evolution hemivariational inequality with multidimensional "reaction-displacement" law. Abstr. Appl. Anal. **2012**, 21 (2012). doi:10.1155/2012/450984
33. Melnik, V.S., Valero, J.: On attractors of multivalued semiflows and differential inclusions. Set Valued Anal. **6**, 83–111 (1998). doi:10.1023/A:1008608431399
34. Migórski, S.: On the existence of solutions for parabolic hemivariational inequalities. J. Comput. Appl. Math. **129**, 77–87 (2001)
35. Migórski, S., Ochal, A.: Optimal control of parabolic hemivariational inequalities. J. Glob. Optim. **17**, 285–300 (2000)
36. Otani, M., Fujita, H.: On existence of strong solutions for $\frac{du}{dt}(t) + \partial\varphi^1(u(t)) - \partial\varphi^2(u(t)) \ni f(t)$. J. Fac. Sci. The University of Tokyo. Sect. 1 A, Mathematics. **24**(3), 575–605 (1977)
37. Panagiotopoulos, P.D.: Inequality Problems in Mechanics and Applications. Convex and Non-convex Energy Functions. Birkhauser, Basel (1985)
38. Sellers, W.D.: A global climatic model based on the energy balance of the Earth-atmosphere system. J. Appl. Meteorol. **8**, 392–400 (1969)
39. Sell, G.R., You, Y.: Dynamics of Evolutionary Equations. Springer, New York (2002)
40. Temam, R.: Infinite-Dimensional Dynamical Systems in Mechanics and Physics. Springer, New York (1988)
41. Terman, D.: A free boundary problem arising from a bistable reaction diffusion equation. SIAM J. Math. Anal. **14**, 1107–1129 (1983)
42. Terman, D.: A free boundary arising from a model for nerve conduction. J. Differ. Equ. **58**(3), 345–363 (1985)
43. Valero, J.: Attractors of parabolic equations without uniqueness. J. Dyn. Differ. Equ. **13**, 711–744 (2001). doi:10.1023/A:1016642525800
44. Valero, J., Kapustyan, A.V.: On the connectedness and asymptotic behaviour of solutions of reaction-diffusion systems. J. Math. Anal. Appl. (2006). doi:10.1016/j.jmaa.2005.10.042
45. Vishik, M.I., Zelik, S.V., Chepyzhov, V.V.: Strong trajectory attractor for dissipative reaction-diffusion system. Doclady Math. (2010). doi:10.1134/S1064562410060086
46. Zadoianchuk, N.V., Kasyanov, P.O.: Dynamics of solutions of a class of second-order autonomous evolution inclusions. Cybern. Syst. Anal. **48**, 414–428 (2012)

47. Zgurovsky, M.Z., Kasyanov, P.O., Kapustyan, O.V., Valero, J., Zadoianchuk, N.V.: Evolution Inclusions and Variation Inequalities for Earth Data Processing III. Springer, Berlin (2012). doi:10.1007/978-3-642-28512-7
48. Zgurovsky, M.Z., Kasyanov, P.O.: Multivalued dynamics of solutions for autonomous operator differential equations in strongest topologies. Solid Mech. Appl. **211**, 149–162 (2014)
49. Zgurovsky, M.Z., Kasyanov, P.O., Zadoianchuk, N.V.: Long-time behavior of solutions for quasilinear hyperbolic hemivariational inequalities with application to piezoelectricity problem. Appl. Math. Lett. **25**, 1569–1574 (2012). doi:10.1016/j.aml.2012.01.016

Part IV
Optimization, Control and Decision Making

Chapter 19
Adaptive Control of Impulse Processes in Complex Systems Cognitive Maps with Multirate Coordinates Sampling

Mikhail Z. Zgurovsky, Victor D. Romanenko and Yuriy L. Milyavsky

Abstract Cognitive map (CM) is a popular method of complex systems description. The system of first-order difference equations in variables increment, based on weighting coefficients of CM, is used to describe impulse process of the system. If different vertices coordinates of CM are measured with different frequencies multirate sampling impulse process model should be developed. The current paper proposes such a model and adds external control vectors with multirate sampling to allow to affect impulse process dynamics. To stabilize this multirate system's coordinates at predefined levels two optimality criteria are proposed and correspondent control laws are derived. Controls are also multirate, i.e. frequently measured coordinates are affected by controls frequently and infrequent coordinates are affected with longer sampling period. For the case when weighting coefficients of CM are unknown or varying special algorithm of their estimation is developed. The results are verified by simulation performed for CM of a bank.

19.1 Introduction

Cognitive modelling is used when the object of study is high-dimensional complex system with lots of cross couples. Most real-life social, economic, financial, ecological, political systems are among them. The concept of cognitive map (CM) is the basis of cognitive modelling. According to [1, 2], CM is a weighted oriented graph

M.Z. Zgurovsky
National Technical University of Ukraine "Kyiv Polytechnic Institute", Peremogy Ave. 37, Build 35, Kyiv 03056, Ukraine
e-mail: zgurovsm@hotmail.com

V.D. Romanenko · Y.L. Milyavsky (✉)
Institute for Applied System Analysis, National Technical University of Ukraine "Kyiv Polytechnic Institute", Peremogy Ave. 37, Build 35, Kyiv 03056, Ukraine
e-mail: yuriy.milyavsky@gmail.com

V.D. Romanenko
e-mail: ipsa@kpi.ua

V.A. Sadovnichiy and M.Z. Zgurovsky (eds.), *Advances in Dynamical Systems and Control*, Studies in Systems, Decision and Control 69,
DOI 10.1007/978-3-319-40673-2_19

with vertices (nodes) representing components of complex systems (coordinates, concepts) and edges representing relations between these concepts. CM is built by experts. It allows to describe qualitatively and quantitatively cause-effect interrelations between complex systems components by means of weighted graph. Under impulse disturbance of one or several vertices CM switches to dynamic transient process which is called impulse process [1, 2].

Rule of varying CM vertices coordinates under impulse process in free motion is formulated as difference equation in increments [2]:

$$\Delta Y_i(k+1) = \sum_{j=1}^{n} a_{ij} \Delta Y_j(k), \tag{19.1}$$

where $\Delta Y_i(k) = Y_i(k) - Y_i(k-1)$, $i = 1, 2, ..., n$; a_{ij}—weight of orgraph edge connecting jth vertice with ith one; n—number of CM vertices.

In vector form the expression (19.1) is written as

$$\Delta \overline{Y}(k+1) = A \Delta \overline{Y}(k), \tag{19.2}$$

where A—weighted adjacency matrix, $\Delta \overline{Y}$—vector of CM vertices coordinates Y_i increments.

In [3–5] control automation of CM impulse processes by means of closed-loop system implementation is done. Based on automatic control theory methods MIMO discrete controller is designed. It generates control vector directly affecting CM vertices as controlled outputs of complex system. For this purpose forced motion equation under CM impulse process is formulated:

$$\Delta Y_i(k+1) = \sum_{j=1}^{n} a_{ij} \Delta Y_j(k) + b_i \Delta u_i(k), \tag{19.3}$$

where $\Delta u_i(k) = u_i(k) - u_i(k-1)$—controls increments.

In vector form Eq. (19.3) may be written as

$$\Delta \overline{Y}(k+1) = A \Delta \overline{Y}(k) + B \Delta \overline{u}(k), \tag{19.4}$$

where control matrix B usually has diagonal elements equal to ones.

Thus, while forming vector $\Delta \overline{u}(k)$ it is necessary to select CM vertices which can be affected by decision maker by varying available resources.

19.2 Problem Definition

To solve the problem of CM impulse process automated control it is necessary to measure all vertices coordinates in discrete time moments with some sampling period. But complex system coordinates have different response rates. Some CM coordinates are also difficult to measure and it is impossible to detect them with small sampling period T_0 appropriate for other coordinates.

Current paper considers complex systems with some coordinates measured (fixed) with small sampling period T_0 and others measured with $h = mT_0$, where $m > 1$ is integer. To implement automated control of this class of complex systems in dynamic mode it is necessary:

- to develop CM impulse process model of complex system with multirate coordinates sampling;
- based on the developed model to design slow and fast multivariate controllers to stabilise CM coordinates at predefined levels during impulse process.

19.3 Development of Controlled CM Impulse Process Model with Multirate Sampling

Controlled model (19.3) of impulse process with unirate sampling (with sampling period T_0) may be represented as follows:

$$\Delta Y_i\left[\left[\frac{k}{m}\right]h + (l+1)T_0\right] = \sum_{j=1}^{n} a_{ij}\Delta Y_j\left[\left[\frac{k}{m}\right]h + lT_0\right] + b_i\Delta u_i\left[\left[\frac{k}{m}\right]h + lT_0\right], \tag{19.5}$$

where $h = mT_0, l = 0, 1, \ldots, m-1; i = 1, \ldots, n$; $\left[\frac{k}{m}\right]$—integer part of $\frac{k}{m}$.

Suppose that in the CM of n-dimensional complex system there exist p vertices coordinates Y_i that are measured with sampling period T_0 and $n - p$ coordinates that are measured with longer period $h = mT_0$. Then impulse process model (19.5) can be described with multirate coordinates sampling as the following:

$$\Delta Y_i\left[\left[\frac{k}{m}\right]h + (l+1)T_0\right] = \sum_{j=1}^{p} a_{ij}\Delta Y_j\left[\left[\frac{k}{m}\right]h + lT_0\right] + \sum_{j=p+1}^{n} a_{ij}\Delta\tilde{Y}_j\left[\left[\frac{k}{m}\right]h\right] + b_i\Delta u_i\left[\left[\frac{k}{m}\right]h + lT_0\right], \tag{19.6}$$

$$\Delta Y_s\left[\left(\left[\frac{k}{m}\right]+1\right)h\right] = \sum_{j=1}^{p} a_{sj}\Delta\tilde{Y}_j\left[\left[\frac{k}{m}\right]h + lT_0\right] + \sum_{j=p+1}^{n} a_{sj}\Delta Y_j\left[\left[\frac{k}{m}\right]h\right] + b_s\Delta u_s\left[\left[\frac{k}{m}\right]h\right], \tag{19.7}$$

where $i = 1, 2, ..., p$; $s = p + 1, ..., n$; $l = 0, 1, ..., m - 1$;

$$\Delta\tilde{Y}_j\left[\left[\frac{k}{m}\right]h\right] = \begin{cases} Y_j\left[\left[\frac{k}{m}\right]h\right] - Y_j\left[\left(\left[\frac{k}{m}\right]-1\right)h\right], l = 0, \\ 0, l \neq 0 \end{cases}$$

for $j = p+1, ..., n$. Coordinates $\Delta\tilde{Y}_j\left[\left[\frac{k}{m}\right]h + lT_0\right]$ for $j = 1, ..., p$ will be explained below.

Expressions (19.6), (19.7) may be written in the generalised vector-matrix form:

$$\Delta\overline{Y}_1\left[\left[\frac{k}{m}\right]h + (l+1)T_0\right] = A_{11}\Delta\overline{Y}_1\left[\left[\frac{k}{m}\right]h + lT_0\right]$$
$$+A_{12}\Delta\tilde{\overline{Y}}_2\left[\left[\frac{k}{m}\right]h\right] + B_{11}\Delta\overline{u}_1\left[\left[\frac{k}{m}\right]h + lT_0\right], \qquad (19.8)$$

$$\Delta\overline{Y}_2\left[\left(\left[\frac{k}{m}\right]+1\right)h\right] = A_{21}\Delta\tilde{\overline{Y}}_1\left[\left[\frac{k}{m}\right]h + lT_0\right]$$
$$+A_{22}\Delta\overline{Y}_2\left[\left[\frac{k}{m}\right]h\right] + B_{22}\Delta\overline{u}_2\left[\left[\frac{k}{m}\right]h\right], \qquad (19.9)$$

where matrices have dimensions $A_{11}(p \times p)$, $A_{12}(p \times (n-p))$, $A_{21}((n-p) \times p)$, $A_{22}((n-p) \times (n-p))$, B_{11}, B_{22}—diagonal matrices selected by control system designer.

To transform $\Delta Y_i\left[\left[\frac{k}{m}\right]h + (l+1)T_0\right]$ in (19.5) with sampling period T_0 into $\Delta Y_s\left[\left[\frac{k}{m}\right]h + (l+1)T_0\right]$ in (19.7) with period $h = mT_0$ the following Proposition is used.

Proposition 19.1 *If the first differences in* (19.5) *are equal to*

$$\Delta Y_i\left[\left[\frac{k}{m}\right]h + (l+1)T_0\right] = Y_i\left[\left[\frac{k}{m}\right]h + (l+1)T_0\right] - Y_i\left[\left[\frac{k}{m}\right]h + lT_0\right],$$

$l = 0, 1, 2, ..., m-1$, *then it is possible to get these differences with big sampling period* $h = mT_0$ *based on*

$$\Delta Y_i\left[\left(\left[\frac{k}{m}\right]+1\right)h\right] = \sum_{l=1}^{m}\Delta Y_i\left[\left[\frac{k}{m}\right]h + lT_0\right]. \qquad (19.10)$$

Proof Consider a sequence of the first differences $\Delta Y_i\left[\left[\frac{k}{m}\right]h + (l+1)T_0\right]$ for $l = 0, 1, ..., m-1$:

$$\Delta Y_i\left[\left[\frac{k}{m}\right]h+T_0\right]=Y_i\left[\left[\frac{k}{m}\right]h+T_0\right]-Y_i\left[\left[\frac{k}{m}\right]h\right];$$

$$\Delta Y_i\left[\left[\frac{k}{m}\right]h+2T_0\right]=Y_i\left[\left[\frac{k}{m}\right]h+2T_0\right]-Y_i\left[\left[\frac{k}{m}\right]h+T_0\right];$$

$$\cdots\cdots\cdots\cdots\cdots\cdots\cdots\cdots\cdots\cdots$$

$$\Delta Y_i\left[\left[\frac{k}{m}\right]h+(m-1)T_0\right]=Y_i\left[\left[\frac{k}{m}\right]h+(m-1)T_0\right]-Y_i\left[\left[\frac{k}{m}\right]h+(m-2)T_0\right];$$

$$\Delta Y_i\left[\left[\frac{k}{m}\right]h+mT_0\right]=Y_i\left[\left[\frac{k}{m}\right]h+mT_0\right]-Y_i\left[\left[\frac{k}{m}\right]h+(m-1)T_0\right].$$

After adding left and right hand sides of these equalities and combining similar terms we obtain

$$\sum_{l=1}^{m}\Delta Y_i\left[\left[\frac{k}{m}\right]h+lT_0\right]=Y_i\left[\left[\frac{k}{m}\right]h+mT_0\right]-Y_i\left[\left[\frac{k}{m}\right]h\right]$$
$$=\Delta Y_i\left[\left(\left[\frac{k}{m}\right]+1\right)h\right],$$

which proves (19.10). □

In the impulse process model (19.9) vector $\overline{Y}_2$ is measured in discrete moments with sampling period $h=mT_0$ and coordinates $\Delta\overline{Y}_1\left[\left[\frac{k}{m}\right]h+lT_0\right]$ influence the system with small period T_0. To account for this we formulated the following Proposition.

Proposition 19.2 *For calculating vector* $\Delta\overline{Y}_2\left[\left(\left[\frac{k}{m}\right]+1\right)h\right]$ *in the model* (19.9) *discrete coordinates* $\Delta\tilde{\overline{Y}}_1\left[\left[\frac{k}{m}\right]h+lT_0\right]$ *are accounted for with use of the formula:*

$$\Delta\tilde{\overline{Y}}_1\left[\left[\frac{k}{m}\right]h+lT_0\right]=\overline{Y}_1\left[\left[\frac{k}{m}\right]h+(m-1)T_0\right]-\overline{Y}_1\left[\left[\frac{k}{m}\right]h-T_0\right].\tag{19.11}$$

Proof Impact of fast components $\Delta\overline{Y}_1\left[\left[\frac{k}{m}\right]h+lT_0\right]$ for $l=0,1,\ldots,m-1$ on $\Delta\overline{Y}_2\left[\left(\left[\frac{k}{m}\right]+1\right)h\right]$ may be represented as the following sum of increments with sampling period T_0:

$$\Delta\tilde{\overline{Y}}_1\left[\left[\frac{k}{m}\right]h+lT_0\right]=\Delta\overline{Y}_1\left[\left[\frac{k}{m}\right]h\right]+\Delta\overline{Y}_1\left[\left[\frac{k}{m}\right]h+T_0\right]+\ldots$$
$$+\Delta\overline{Y}_1\left[\left[\frac{k}{m}\right]h+(m-2)T_0\right]+\Delta\overline{Y}_1\left[\left[\frac{k}{m}\right]h+(m-1)T_0\right].$$

After expansion of the differences the sum above takes the form:

$$\Delta\tilde{\overline{Y}}_1\left[\left[\frac{k}{m}\right]h + lT_0\right] = \overline{Y}_1\left[\left[\frac{k}{m}\right]h\right] - \overline{Y}_1\left[\left[\frac{k}{m}\right]h - T_0\right]$$
$$+\overline{Y}_1\left[\left[\frac{k}{m}\right]h + T_0\right] - \overline{Y}_1\left[\left[\frac{k}{m}\right]h\right] + \ldots + \overline{Y}_1\left[\left[\frac{k}{m}\right]h + (m-2)T_0\right]$$
$$-\overline{Y}_1\left[\left[\frac{k}{m}\right]h + (m-3)T_0\right] + \overline{Y}_1\left[\left[\frac{k}{m}\right]h + (m-1)T_0\right] - \overline{Y}_1\left[\left[\frac{k}{m}\right]h + (m-2)T_0\right]$$
$$= \overline{Y}_1\left[\left[\frac{k}{m}\right]h + (m-1)T_0\right] - \overline{Y}_1\left[\left[\frac{k}{m}\right]h - T_0\right],$$

that proves Proposition (19.2). □

19.4 Impulse Processes Adaptive Automated Control in CM with Multirate Sampling

To develop an automated control algorithm for CM impulse processes, vectors $\overline{Y}_1(k)$, $\overline{Y}_2(k)$ dynamics are presented in full variables values according to models (19.8), (19.9):

$$\overline{Y}_1\left[\left[\frac{k}{m}\right]h + (l+1)T_0\right] = (I + A_{11} - A_{11}q_1^{-1})\overline{Y}_1\left[\left[\frac{k}{m}\right]h + lT_0\right]$$
$$+B_{11}\Delta\overline{u}_1\left[\left[\frac{k}{m}\right]h + lT_0\right] + A_{12}\Delta\tilde{\overline{Y}}_2\left[\left[\frac{k}{m}\right]h\right]; \quad (19.12)$$

$$\overline{Y}_2\left[\left(\left[\frac{k}{m}\right] + 1\right)h\right] = (I + A_{22} - A_{22}q_2^{-1})\overline{Y}_2\left[\left[\frac{k}{m}\right]h\right]$$
$$+B_{22}\Delta\overline{u}_2\left[\left[\frac{k}{m}\right]h\right] + A_{21}\Delta\tilde{\overline{Y}}_1\left[\left[\frac{k}{m}\right]h + lT_0\right], \quad (19.13)$$

where q_1^{-1}, q_2^{-1}—inverse shift operators with sampling periods T_0 and $h = mT_0$ respectively. Terms $A_{12}\Delta\tilde{\overline{Y}}_2\left[\left[\frac{k}{m}\right]h\right]$ and $A_{21}\Delta\tilde{\overline{Y}}_1\left[\left[\frac{k}{m}\right]h + lT_0\right]$ in Eqs. (19.12), (19.13) respectively are disturbances.

Suppose that the system is stable. To synthesize the first control vector $\Delta\overline{u}_1\left[\left[\frac{k}{m}\right]h + lT_0\right]$ an optimality criterion is formulated:

$$J_1\left[\left[\frac{k}{m}\right]h + (l+1)T_0\right] = E\left\{\left[\overline{Y}_1\left[\left[\frac{k}{m}\right]h + (l+1)T_0\right] - \overline{G}_1\right]^T\right.$$
$$\times\left[\overline{Y}_1\left[\left[\frac{k}{m}\right]h + (l+1)T_0\right] - \overline{G}_1\right]$$
$$\left.+\Delta\overline{u}_1^T\left[\left[\frac{k}{m}\right]h + lT_0\right]R_1\Delta\overline{u}_1\left[\left[\frac{k}{m}\right]h + lT_0\right]\right\}, \quad (19.14)$$

where $\overline{G}_1$—reference-input signal for stabilisation of CM vertices coordinates $\overline{Y}_1$, E—conditional expectation operator, R_1—weighting matrix selected by controller's designer. Based on minimization of given criterion with respect to control vector $\Delta\overline{u}_1$ having considered model (19.12) we obtain the first controller's equation

$$\frac{\partial J_1\left[\left[\frac{k}{m}\right]h+(l+1)T_0\right]}{\partial\Delta\overline{u}_1\left[\left[\frac{k}{m}\right]h+lT_0\right]}=2B_{11}^T\{(I+A_{11}-A_{11}q_1^{-1})$$
$$\times\overline{Y}_1\left[\left[\frac{k}{m}\right]h+lT_0\right]+B_{11}\Delta\overline{u}_1\left[\left[\frac{k}{m}\right]h+lT_0\right]$$
$$+A_{12}\Delta\tilde{\overline{Y}}_2\left[\left[\frac{k}{m}\right]h\right]-\overline{G}_1\right\}+2R_1\Delta\overline{u}_1\left[\left[\frac{k}{m}\right]h+lT_0\right]=0,$$

that results in the first controller's law:

$$\overline{u}_1\left[\left[\frac{k}{m}\right]h+lT_0\right]=\overline{u}_1\left[\left[\frac{k}{m}\right]h+(l-1)T_0\right]$$
$$-(B_{11}^TB_{11}+R_1)^{-1}B_{11}\{(I+A_{11}-A_{11}q_1^{-1})$$
$$\times\overline{Y}_1\left[\left[\frac{k}{m}\right]h+lT_0\right]+A_{12}\Delta\tilde{\overline{Y}}_2\left[\left[\frac{k}{m}\right]h\right]-\overline{G}_1\right\}. \quad (19.15)$$

To synthesize the second control vector $\Delta\overline{u}_2\left[\left[\frac{k}{m}\right]h\right]$ the second optimality criterion is proposed:

$$J_2\left[\left(\left[\frac{k}{m}\right]+1\right)h\right]=E\left\{\left[\overline{Y}_2\left[\left(\left[\frac{k}{m}\right]+1\right)h\right]-\overline{G}_2\right]^T\times\right.$$
$$\left.\times\left[\overline{Y}_2\left[\left(\left[\frac{k}{m}\right]+1\right)h\right]-\overline{G}_2\right]+\Delta\overline{u}_2^T\left[\left[\frac{k}{m}\right]h\right]R_2\Delta\overline{u}_2\left[\left[\frac{k}{m}\right]h\right]\right\}, \quad (19.16)$$

where $\overline{G}_2$—reference-input signal for stabilization of CM vertices coordinates $\overline{Y}_2$, R_2—weighting matrix selected by controller's designer. After minimization of criterion (19.16) with respect to vector $\Delta\overline{u}_2$, taking into consideration (19.15), we obtain the second controller's equation

$$\frac{\partial J_2\left[\left(\left[\frac{k}{m}\right]+1\right)h\right]}{\partial\Delta\overline{u}_2\left[\left[\frac{k}{m}\right]h\right]}=2B_{22}^T\{(I+A_{22}-A_{22}q_2^{-1})$$
$$\times\overline{Y}_2\left[\left[\frac{k}{m}\right]h\right]+B_{22}\Delta u_2\left[\left[\frac{k}{m}\right]h\right]$$
$$+A_{21}\Delta\tilde{\overline{Y}}_1\left[\left[\frac{k}{m}\right]h+lT_0\right]-\overline{G}_2\right\}+2R_2\Delta\overline{u}_2\left[\left[\frac{k}{m}\right]h\right]=0,$$

that results in the second controller's law:

$$\overline{u}_2\left[\left[\frac{k}{m}\right]h\right]=\overline{u}_2\left[\left(\left[\frac{k}{m}\right]-1\right)h\right]$$
$$-(B_{22}^T B_{22}+R_2)^{-1}B_{22}\{(I+A_{22}-A_{22}q_2^{-1})$$
$$\times\overline{Y}_2\left[\left[\frac{k}{m}\right]h\right]+A_{21}\Delta\tilde{\overline{Y}}_1\left[\left[\frac{k}{m}\right]h+lT_0\right]-\overline{G}_2\Big\}. \quad (19.17)$$

Coefficients of adjacency matrices $A_{11}, A_{12}, A_{21}, A_{22}$ of controlled CM impulse process model (19.8), (19.9) with multirate sampling vary with time during complex system functioning. They have to be estimated in real-time. For this purpose we write model (19.6), (19.7) for each CM coordinate $\Delta Y_i (i = 1, 2, ..., p)$ and $\Delta Y_s (s = p+1, p+2, ..., n)$ backward one period T_0 and h respectively:

$$\Delta Y_i\left[\left[\frac{k}{m}\right]h+lT_0\right]=\sum_{j=1}^{p}a_{ij}\Delta Y_j\left[\left[\frac{k}{m}\right]h+(l-1)T_0\right]$$
$$+\sum_{j=p+1}^{n}a_{ij}\Delta\tilde{Y}_j\left[\left[\frac{k}{m}\right]h\right]$$
$$+b_i\Delta u_i\left[\left[\frac{k}{m}\right]h+(l-1)T_0\right]+\xi_i\left[\left[\frac{k}{m}\right]h+lT_0\right], \quad (19.18)$$

$$\Delta Y_s\left[\left[\frac{k}{m}\right]h\right]=\sum_{j=1}^{p}a_{sj}\Delta\tilde{Y}_j\left[\left(\left[\frac{k}{m}\right]-1\right)h+lT_0\right]$$
$$+\sum_{j=p+1}^{n}a_{sj}\Delta Y_j\left[\left(\left[\frac{k}{m}\right]-1\right)h\right]$$
$$+b_s\Delta u_s\left[\left(\left[\frac{k}{m}\right]-1\right)h\right]+\xi_s\left[\left[\frac{k}{m}\right]h\right], \quad (19.19)$$

where $\xi_i\left[\left[\frac{k}{m}\right]h+lT_0\right]$, $\xi_s\left[\left[\frac{k}{m}\right]h\right]$—disturbances which represent uncontrolled dynamic changes in CM impulse processes during control algorithm functioning.

Estimation of coefficients of the models (19.18), (19.19), which are adjacency matrices $A_{11}, A_{12}, A_{21}, A_{22}$ components, is performed based on recursive least squares (RLS) method with multirate sampling of coordinates. To accomplish this model (19.18) with frequently measured coordinates $Y_i, u_i, i = 1, ..., p$ should be written as

$$\Delta Y_i\left[\left[\frac{k}{m}\right]h+lT_0\right]-b_i\Delta u_i\left[\left[\frac{k}{m}\right]h+(l-1)T_0\right]$$
$$=\overline{X}_1^T\left[\left[\frac{k}{m}\right]h+lT_0\right]\overline{\Theta}_{1i}+\xi_i\left[\left[\frac{k}{m}\right]h+lT_0\right], \quad (19.20)$$

where

$$\overline{X}_1^T\left[\left[\frac{k}{m}\right]h + lT_0\right] = \left(\Delta Y_1\left[\left[\frac{k}{m}\right]h + (l-1)T_0\right], \ldots,\right.$$
$$\left.\Delta Y_p\left[\left[\frac{k}{m}\right]h + (l-1)T_0\right], \Delta\tilde{Y}_{p+1}\left[\left[\frac{k}{m}\right]h\right], \ldots, \Delta\tilde{Y}_n\left[\left[\frac{k}{m}\right]h\right]\right); \quad (19.21)$$

$$\overline{\Theta}_{1i} = (a_{i1}, \ldots, a_{ip}, a_{i(p+1)}, \ldots, a_{in})^T. \quad (19.22)$$

Based on Proposition 19.2 model (19.19) with infrequently changing coordinates $Y_s, u_s, s = p+1, \ldots, n$ may be written as

$$\Delta Y_s\left[\left[\frac{k}{m}\right]h\right] - b_s\Delta u_s\left[\left(\left[\frac{k}{m}\right]-1\right)h\right] = \overline{X}_2^T\left[\left[\frac{k}{m}\right]h\right]\overline{\Theta}_{2s} + \xi_s\left[\left[\frac{k}{m}\right]h\right], \quad (19.23)$$

where

$$\overline{X}_2^T\left[\left[\frac{k}{m}\right]h\right] = \left(Y_1\left[\left[\frac{k}{m}\right]h + (m-1)T_0\right] - Y_1\left[\left[\frac{k}{m}\right]h - T_0\right], \ldots,\right.$$
$$Y_p\left[\left[\frac{k}{m}\right]h + (m-1)T_0\right] - Y_p\left[\left[\frac{k}{m}\right]h - T_0\right],$$
$$\left.\Delta Y_{p+1}\left[\left(\left[\frac{k}{m}\right]-1\right)h\right], \ldots, \Delta Y_n\left[\left(\left[\frac{k}{m}\right]-1\right)h\right]\right); \quad (19.24)$$

$$\overline{\Theta}_{2s} = (a_{s1}, \ldots, a_{sp}, a_{s(p+1)}, \ldots, a_{sn})^T. \quad (19.25)$$

RLS algorithm for estimating vector (19.22) $\hat{\overline{\Theta}}_{1i}\left[\left[\frac{k}{m}\right]h + lT_0\right]$ is performed according to [6] each sampling period T_0 for the model (19.20) for each $i = 1, \ldots, p$ with common vector of measured coordinates (19.21) $\overline{X}_1\left[\left[\frac{k}{m}\right]h + lT_0\right]$. RLS for estimating vector (19.25) $\hat{\overline{\Theta}}_{2s}\left[\left[\frac{k}{m}\right]h\right]$ is performed each sampling period h for the model (19.23) for each $s = p+1, \ldots, n$ with common vector (19.24) $\overline{X}_2\left[\left[\frac{k}{m}\right]h\right]$.

19.5 Practical Example

Consider CM which represents stable operating of a bank [5] (Fig. 19.1). This CM has vertices with the following meaning:

1 Regional network.
2 Capital.
3 Loans.
4 Deposits.

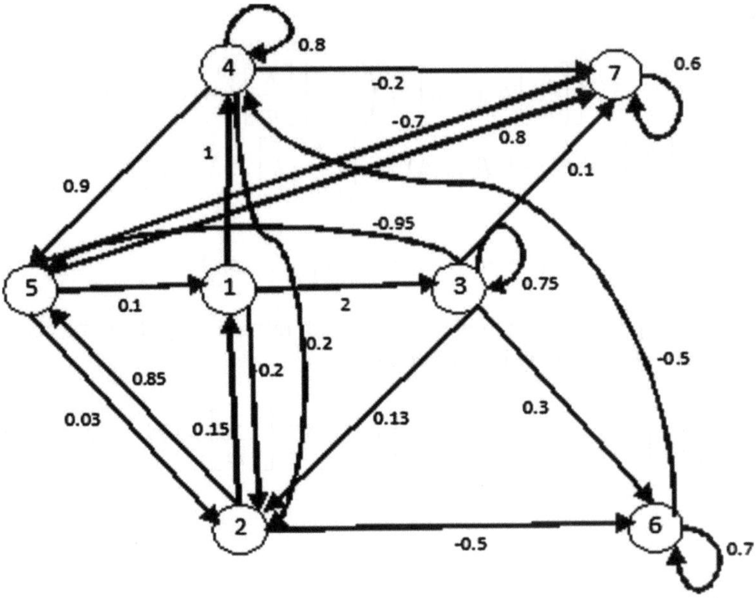

Fig. 19.1 Cognitive map of a bank

5 Liquid assets.
6 Stability risk measure.
7 Liquidity risk measure.

Control inputs Δu_i in (19.3) are formed from outside by changing resources of CM vertices. Vertices 1, 2, 3, 4 and 5 are measured daily with sample period $T_0 = 1$ and vertices 6, 7 (stability and liquidity risk measures) are measured monthly with period $h = 4T_0$.

Adjacency matrices in models (19.8), (19.9) composed of CM's weighting coefficients are equal to:

$$A_{11} = \begin{pmatrix} 0 & 0.15 & 0 & 0 & 0.1 \\ -0.2 & 0 & 0.13 & -0.2 & 0.03 \\ 2 & 0 & 0.75 & 0 & 0 \\ 1 & 0 & 0 & 0.8 & 0 \\ 0 & 0.85 & -0.95 & 0.9 & 0 \end{pmatrix}, \; A_{12} = \begin{pmatrix} 0 & 0 \\ 0 & 0 \\ 0 & 0 \\ -0.5 & 0 \\ 0 & -0.7 \end{pmatrix},$$

$$A_{21} = \begin{pmatrix} 0 & -0.5 & 0.3 & 0 & 0 \\ 0 & 0 & 0.1 & -0.2 & 0.8 \end{pmatrix}, \; A_{22} = \begin{pmatrix} 0.7 & 0 \\ 0 & 0.6 \end{pmatrix}.$$

The problem is to move all CM vertices coordinates of the bank to other reference levels in the impulse process mode with multirate coordinates sampling (19.12), (19.13) based on control inputs $\overline{u}_1\left[\left[\frac{k}{m}\right]h + lT_0\right]$ and $\overline{u}_2\left[\left[\frac{k}{m}\right]h\right]$ synthesis according to the proposed algorithms (19.15), (19.17). Let initial values of the vertices

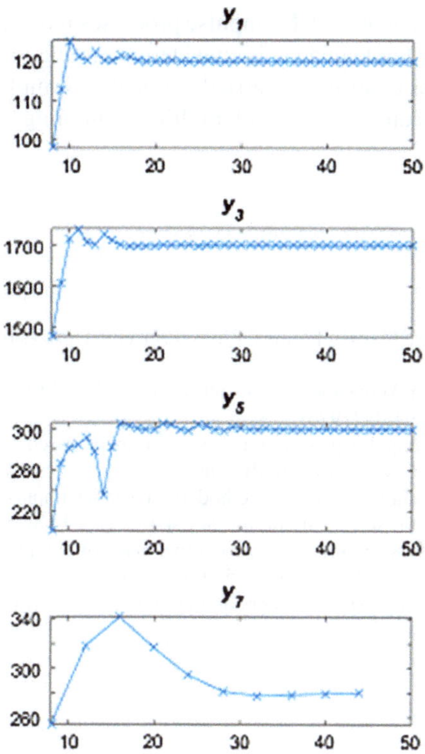

Fig. 19.2 Simulation results

coordinates be equal to $\left(100\ 500\ 1500\ 1000\ 200\ 150\ 250\right)$ and new desired values be $\left(120\ 500\ 1700\ 1200\ 300\ 180\ 280\right)$. Digital simulation results are presented in Fig. 19.2.

19.6 Summary

The paper presents the adaptive control method for complex system described by the model of impulse processes in CM [2]. CM vertices coordinates are measured in discrete time with multirate sampling (with periods T_0 and $h = mT_0, m > 1$). External controls vectors which affect CM vertices directly in closed-loop system are synthesized with multirate sampling. Control inputs are implemented by means of varying available resources in CM vertices. Propositions about transformation of coordinates differences for controlled impulse processes models with multirate sampling are formulated and proven in the paper.

Estimation algorithm for weighting coefficients of CM impulse processes models with multirate sampling, based on recursive least squares, is considered. Simulation of the method developed for impulse process control was carried out on the example of the CM of a bank to stabilize CM coordinates vertices with multirate sampling at predefined levels.

References

1. Axelrod, R.: The Structure of Decision: Cognitive Maps of Political Elites. Princeton University Press, Princeton (1976)
2. Roberts, F.: Discrete Mathematical Models with Applications to Social, Biological, and Environmental Problems. Prentice-Hall, Englewood Cliffs (1976)
3. Romanenko, V., Milyavsky, Y.: Stabilizing of impulse processes in cognitive maps based on state-space models. Syst. Res. Inf. Technol. **1**, 26–42 (2014). (in Russian)
4. Romanenko, V., Milyavsky, Y., Reutov, A.: Adaptive control method for unstable impulse processes in cognitive maps based on reference model. J. Autom. Inf. Sci. **47**(3), 11–23 (2015)
5. Romanenko, V., Milyavsky, Y.: Impulse processes stabilization in cognitive maps of complex systems based on modal regulators. Cybern. Comput. Eng. **179**, 43–55 (2015). (in Russian)
6. Astrom, K., Wittenmark, B.: Computed Controlled Systems. Theory and Design. Prentice-Hall, Englewood Cliffs (1984)

Chapter 20
Estimation of Consistency of Fuzzy Pairwise Comparison Matrices using a Defuzzification Method

Nataliya D. Pankratova and Nadezhda I. Nedashkovskaya

Abstract A definition of consistency of a fuzzy pairwise comparison matrix (FPCM) is developed in the paper. It is supposed that FPCM elements are fuzzy sets with membership functions of any shape. Such FPCMs may be a result of evaluation of decision alternatives by a group of experts when aggregating individual expert judgments made in traditional crisp scales. A comparative analysis of suggested definition with other known definitions of consistent FPCM is done. Usage of suggested definition makes it possible to evaluate the admissibility of inconsistency of expert judgments when calculating weights of decision alternatives and to reveal intransitive expert judgments.

20.1 Introduction

The method of qualitative pairwise comparisons is a powerful instrument to evaluate coefficients of relative importance (weights, priorities) of a set $X = \{x_1, x_2, ..., x_n\}$ of decision-making elements, such as decision criteria and alternatives. Pairwise comparisons may be modeled using preference relation $D : (x_i, x_j) \in X \times X$ defined on X, where $d_{ij} = D(x_i, x_j) \in R$ is a quantitative representation of an intensity of preference of element x_i over element x_j given by an expert. The relation D is formalized as a pairwise comparison matrix (PCM) if cardinality of the set X is small. Many methods are known for calculation weights on basis of a PCM [1–7].

The notion of consistency [1] is used to evaluate quality of pairwise expert judgments. Coefficients of consistency CR, GCI, HCR, CI^{tr} and k_y and several consistency criteria are known for evaluation of admissibility of inconsistency of *crisp*

N.D. Pankratova (✉) · N.I. Nedashkovskaya
Institute for Applied System Analysis, National Technical University of Ukraine "Kyiv Polytechnic Institute", Peremogy ave., 37, build, 35, 03056 Kyiv, Ukraine
e-mail: natalidmp@gmail.com

N.I. Nedashkovskaya
e-mail: n.nedashkivska@gmail.com

V.A. Sadovnichiy and M.Z. Zgurovsky (eds.), *Advances in Dynamical Systems and Control*, Studies in Systems, Decision and Control 69,
DOI 10.1007/978-3-319-40673-2_20

PCMs when making a decision (analysis and comparative study of these coefficients and criteria are made, for example, in [8, 9]).

Fuzzy PCMs [10–13] and interval PCMs [14–18] are used in some decision-making problems. Elements of the fuzzy PCMs are fuzzy sets mainly with triangle or trapezoid membership functions. When calculating weights of decision alternatives on basis of an interval PCM (IPCM) $I = \{([l_{ij}, u_{ij}])|0 < l_{ij} \leq u_{ij}, i, j = 1, ..., n\}$ the LUAM method [14], the TLGP method [15], two-stage methods [3] and other are known. The Chang's method [19] is widely used to derive weights on basis of a fuzzy PCM with triangle fuzzy elements $Trmf = \{((l_{ij}, m_{ij}, u_{ij}))|0 < l_{ij} \leq m_{ij} \leq u_{ij}, i, j = 1, ..., n\}$. However, not enough attention is given to problems of evaluation of consistency of fuzzy and interval PCMs. Thus, the Chang's method [19] does not contain any stage for consistency evaluation. The two-stage model [3] on basis of an IPCM has first stage devoted to consistency evaluation. But this model as well as other known methods and models [14–18] on basis of an IPCM do not evaluate admissibility of inconsistency of an IPCM in a process of decision making. Also not enough attention is given to problems of consistency increasing of interval and fuzzy PCMs and cycles (intransitive elements) elimination in these matrices.

Purpose of the paper is an analysis of different approaches for definition of a fuzzy PCM consistency, and also an investigation of their usage to evaluate an admissible inconsistency of a fuzzy PCM and consistency increasing of a fuzzy PCM while providing a decision making process.

20.2 A Problem Statement

Definition: *A fuzzy pairwise comparison matrix (FPCM)* of n decision alternatives (DAs) is a matrix

$$\tilde{D} = \begin{pmatrix} 1 & \tilde{d}_{12} & \dots & \tilde{d}_{1n} \\ \tilde{d}_{21} & 1 & \dots & \tilde{d}_{2n} \\ \vdots & \vdots & \vdots & \vdots \\ \tilde{d}_{n1} & \tilde{d}_{n2} & \dots & 1 \end{pmatrix}, \qquad (20.1)$$

where fuzzy set $\tilde{d}_{ij} = (x, \mu_{ij}(x))$ represents intensity of preference of DA a_i over DA a_j, $x \in R$, R—a set of real numbers, $\mu_{ij}(x)$—value of membership function for a fuzzy preference relation of DA a_i over DA a_j and $\tilde{d}_{ii} = 1$.

Let us suppose that an expert, performing pairwise comparisons of n DAs, gives fuzzy estimate for preference degree of one DA over some other DA. For example, such estimates as "preference degree is between weak and very weak intensity" or "preference degree is near strong intensity" are specified using membership functions.

Also a fuzzy PCM may be a logical result of evaluation of DAs by a group of experts. Suppose m experts use the traditional Saaty scale [1] when providing pairwise comparisons of DAs. Let us denote $D(k) = \{(d_{ij}(k))\}$ a PCM, given by a kth expert. Then a fuzzy PCM $\tilde{D} = \{(\tilde{d}_{ij})\}$ may be a result of an aggregation of these

PCMs, where $\tilde{d}_{ij}$ is a discrete fuzzy set that represents distribution among elements $d_{ij}(k)$ of PCMs $D(k)$, $k = 1, ..., m$.

It should be noted that in the known problem statements [7, 10–19], that deals with FPCMs, it is supposed that an expert provides pairwise comparisons in the Saaty fundamental scale [1], as in the traditional Analytic Hierarchy Process [1], and after that his/her estimates are formalized using fuzzy fundamental scales with elements in a form of symmetrical triangle or trapezoidal normalized convex fuzzy sets. As a rule, heights of these fuzzy sets correspond to expert judgments in the Saaty fundamental scale and the fuzzy sets may have different width. In the given paper, as opposed to the mentioned problem statement, elements of a FPCM (20.1) have an arbitrary form, and also may be discrete, built on basis of statistical results of group assessment.

Let $\tilde{D} = \{(\tilde{d}_{ij})\}$ be a FPCM (20.1) of n DAs. The problem is to evaluate a FPCM $\tilde{D}$ consistency; to improve (increase) $\tilde{D}$ consistency up to admissible level that is acceptable for a calculation of coefficients of relative importance (weights, priorities) of DAs.

20.3 Definitions of Consistency of a FPCM

In the paper we propose intuitive definitions of strong and weak consistency of a FPCM (20.1) that use notion of consistency of some crisp PCM, built on basis of the FPCM. Then to evaluate and increase a FPCM consistency an extensive knowledge [1, 8, 9, 20] about evaluation and increasing of a traditional (crisp) PCM consistency may be used. Similar approaches are used in [10, 21, 22].

Let us consider a PCM D which elements are positive real numbers and are results of defuzzification of fuzzy sets $\tilde{d}_{ij}$—elements of a FPCM (20.1):

$$D = \{(d_{ij})\} \in R^{+}_{n \times n} \tag{20.2}$$

(1) $d_{ij} = Deffuz(\tilde{d}_{ij})$ if $\tilde{d}_{ij} \geq 1$,
(2) $d_{ij} = 1/d_{ji}$ otherwise.

It should be noted that the second condition ensures a necessary property of inverse symmetry of a PCM D.

Definition 20.1 A FPCM $\tilde{D}$ (20.1) is *consistent*, if corresponding defuzzified PCM $D = \{(d_{ij})\}$ (20.2) is consistent, namely $d_{ij} = d_{ik}d_{kj}$ for $\forall i, j, k = 1, ..., n$.

Definition 20.2 A FPCM $\tilde{D}$ (20.1) is admissibly inconsistent, if corresponding defuzzified PCM $D = \{(d_{ij})\}$ (20.2) is admissibly inconsistent, namely $CR(D) \leq CR^*$ or $GCI(D) \leq GCI^*$, or $HCR(D) \leq HCR^*$, or $CI^{tr}(D) \leq CI^{tr*}$ (depending on coefficient of consistency which is used), where CR, GCI, HCR, CI^{tr} are consistency ratio, geometric consistency index, harmonic consistency ratio and consistency index of transitivities, and CR^*, GCI^*, HCR^*, CI^{tr*} are threshold values of corresponding coefficients.

Definition 20.3 A FPCM $\tilde{D}$ (20.1) is weak consistent, if corresponding defuzzified PCM $D = \{(d_{ij})\}$ (20.2) is weak consistent, namely the following ordinal transitivities are hold: $(d_{ij} > 1) \wedge (d_{jk} > 1) \Rightarrow (d_{ik} > 1)$, $(d_{ij} = 1) \wedge (d_{jk} > 1) \Rightarrow (d_{ik} > 1)$, $(d_{ki} > 1) \wedge (d_{ij} = 1) \Rightarrow (d_{kj} > 1)$ and $(d_{ij} = 1) \wedge (d_{jk} = 1) \Rightarrow (d_{ik} = 1)$.

The following example illustrates usage of the most famous defuzzification methods, such as the centroid method, the median method and the centre of maxima method [23]. It is worth noted that the mentioned defuzzification methods give the same results for a symmetrical unimodal fuzzy number.

Example 20.1 Let us consider a FPCM with elements $\tilde{d}_{ij}$ in a form of trapezoid fuzzy numbers with membership functions

$$\tilde{d}_{ij} = Trap(x, a_{ij}, b_{ij}, c_{ij}, d_{ij}) = \max\left(\min\left(\frac{x - a_{ij}}{b_{ij} - a_{ij}}, 1, \frac{d_{ij} - x}{d_{ij} - c_{ij}}\right), 0\right), \tag{20.3}$$

where $a_{ij}, b_{ij}, c_{ij}, d_{ij}$ are parameters.

For example, a symmetrical fuzzy number $\tilde{d}_{ij}$ with parameters $a_{ij} = 1$, $b_{ij} = 2$, $c_{ij} = 4$, $d_{ij} = 5$ may be used to model such expert judgment as "preference degree is near to weak intensity". The $\tilde{d}_{ij}$ defuzzification using the centriod method, the median method and the centre of maxima method results in the same crisp value $d_{ij} = 3$.

In the case of unsymmetrical fuzzy number $\tilde{d}_{ij}$ (20.3), for example, with parameters $a_{ij} = 1$, $b_{ij} = 2$, $c_{ij} = 4$, $d_{ij} = 6$, the centroid method gives the same result $d_{ij} = 3$, the median method—the value $d_{ij} = 3.25$ and the centre of maxima method—the value $d_{ij} = 3.29$.

20.4 A Comparative Study of Definitions of a FPCM Consistency

Let us consider several known definitions of a FPCM consistency and their drawbacks, and a comparative study of proposed definitions of a FPCM consistency with the known definitions.

Definition 20.4 ([10]) A fuzzy positive inverse symmetrical matrix A is consistent if condition $a_{ij} \otimes a_{jk} = a_{ik}$ holds for all $i, j = 1, ..., n$, where $\otimes$ is an extended binary operation of multiplication.

Definition 20.4 allows to estimate a consistency of general type FPCMs (20.1). It is a direct extension of a traditional well-known Definition [1] of consistency of a crisp PCM with real-valued elements on a fuzzy PCM. Later several Definitions [3, 11, 15, 21, 22] where proposed, which are particular cases of Definition 20.4 for special typed FPCMs.

Definition 20.5 ([21]) An IPCM $\tilde{D} = \{\tilde{d}_{ij} = [l_{ij}, u_{ij}]\}$ is consistent if the following PCMs D^L and D^U (20.4) are consistent:

$$D^L = \begin{pmatrix} 1 & l_{12} & \dots & l_{1n} \\ u_{21} & 1 & \dots & l_{2n} \\ \vdots & \vdots & \vdots & \vdots \\ u_{n1} & u_{n2} & \dots & 1 \end{pmatrix}, D^U = \begin{pmatrix} 1 & u_{12} & \dots & u_{1n} \\ l_{21} & 1 & \dots & u_{2n} \\ \vdots & \vdots & \vdots & \vdots \\ l_{n1} & l_{n2} & \dots & 1 \end{pmatrix}, \tag{20.4}$$

Definition 20.6 ([22]) A FPCM $\tilde{D} = \{\tilde{d}_{ij} = (l_{ij}, m_{ij}, u_{ij})\}$ with triangle fuzzy elements is consistent, if PCMs D^L, D^U (20.4) and $D^M = \{(m_{ij})\}$ are consistent.

Let us show that usage of Definitions 20.4, 20.5 and 20.6, described above, may lead to contradictory results. Thus, if a task is to compare only two objects, then a result of such pairwise comparison is always consistent [1], an inconsistency by nature may appear only when three objects should be compared. Therefore a PCM of dimension 2×2 must be consistent regardless of crispness or fuzziness of its elements. Let us consider a FPCM of dimension 2×2, for example, with triangle elements

$$\tilde{D}_{2\times 2} = (\tilde{d}_{ij})_{2\times 2} = \begin{pmatrix} 1 & (l_{12}, m_{12}, u_{12}) \\ \left(\frac{1}{u_{12}}, \frac{1}{m_{12}}, \frac{1}{l_{12}}\right) & 1 \end{pmatrix}.$$

The Definition 20.4 results in an inconsistency of this FPCM, since $\tilde{d}_{12} \otimes \tilde{d}_{21} \neq \tilde{d}_{11}$ in general case.

The proposed Definition 20.1, in turn, does not have such contradiction, since a result of defuzzification (20.2) for a FPCM (20.1) is always a consistent PCM when $n = 2$.

Now let us consider a FPCM of higher dimension $n \geq 3$, for example, with triangle elements

$$\tilde{A} = (\tilde{a}_{ij})_{n\times n} = \begin{pmatrix} 1 & (l_{12}, m_{12}, u_{12}) & \dots & (l_{1n}, m_{1n}, u_{1n}) \\ (l_{21}, m_{21}, u_{21}) & 1 & \dots & (l_{2n}, m_{2n}, u_{2n}) \\ \vdots & \vdots & \vdots & \vdots \\ (l_{n1}, m_{n1}, u_{n1}) & (l_{n2}, m_{n2}, u_{n2}) & \dots & 1 \end{pmatrix}.$$

where $u_{ij} \geq m_{ij} \geq l_{ij} > 0$, $l_{ij} = \frac{1}{u_{ji}}$, $m_{ij} = \frac{1}{m_{ji}}$, $u_{ij} = \frac{1}{l_{ji}}$ for each $i, j = 1, ..., n$. Then Definitions 20.4 and 20.6 require satisfaction of the following equations:

$$l_{ij} \cdot l_{jk} = l_{ik}, \tag{20.5}$$
$$u_{ij} \cdot u_{jk} = u_{ik} \text{ and } m_{ij} \cdot m_{jk} = m_{ik}.$$

$$\forall i, j, k = 1, ..., n, \ i < j < k.$$

The simultaneous satisfaction of the equations (20.5) is quite strict requirement, and only FPCMs of special types meet the equations (20.5) in practical decision problems. To illustrate this statement the following Example 20.2 is given.

Example 20.2 Suppose an expert gives a multiplicative preference relation on a set $A = \{a_1, a_2, a_3, a_4\}$ of four DAs, such that an intensity of preference of DA a_1 over DAs a_2 and a_4 is near weak intensity, DA a_2 is nearly equivalent to DA a_4 and so on. Suppose triangle fuzzy numbers (l_{ij}, m_{ij}, u_{ij}), $0 < l_{ij} \leq m_{ij} \leq u_{ij}$, $l_{ij} = \frac{1}{u_{ji}}$, $m_{ij} = \frac{1}{m_{ji}}$, $u_{ij} = \frac{1}{l_{ji}}$, $i, j = 1, ..., n$ are used to formalize these expert pairwise comparison judgments, and as a result the following FPCM is obtained

$$\tilde{D} = \begin{pmatrix} 1 & (1,3,5) & (4,6,8) & (1,3,5) \\ (1,3,5)^{-1} & 1 & (1,2,4) & (1,1,3) \\ (4,6,8)^{-1} & (1,2,4)^{-1} & 1 & (1,2,4)^{-1} \\ (1,3,5)^{-1} & (1,1,3)^{-1} & (1,2,4) & 1 \end{pmatrix}. \qquad (20.6)$$

The FPCM (20.6) is obviously inconsistent in terms of Definitions 20.4 and 20.6, since equalities $l_{ij} \cdot l_{jk} = l_{ik}$ and $u_{ij} \cdot u_{jk} = u_{ik}$ in (20.5) are not satisfied. But equations $m_{ij} \cdot m_{jk} = m_{ik}$ are satisfied for $\forall i, j, k = 1, ..., n$. Therefore there may be a contradiction of results about consistency of expert judgments on basis of crisp and fuzzified PCMs. Namely, in this example the formalization of expert judgments using crisp PCM results in consistency of the judgments (since the PCM $D^M = \{(m_{ij})\}$ is consistent), and the formalization of the judgments using fuzzy triangle numbers (FPCM) results in inconsistency which is moreover not admissible. This result follows from the fact that PCMs D^L and D^U (20.4), built on basis of the FPCM (20.6), which have to be consistent according to Definitions 20.4 and 20.6, have quite high level of inconsistency: $CR(D^L) = 0.93 > CR^* = 0.08$, $CR(D^U) = 0.06 < CR^* = 0.08$. Besides the inconsistency is inadmissible for the PCM D^L. Such contradiction could be decreased using triangle numbers of smaller width. However, we can't assure conditions $l_{ij} \cdot l_{jk} = l_{ik}$ and $u_{ij} \cdot u_{jk} = u_{ik}$ (20.5) for the given FPCM (20.6) , and only admissible inconsistency of the PCMs D^L and D^U could be achieved.

According to Definition 20.1 proposed in the Sect. 20.3, the FPCM (20.6) is consistent, since the result of its defuzzification is the consistent PCM $D^M = \{(m_{ij})\}$. Contradiction results on basis of crisp and fuzzyfied expert judgments (i.e. the PCM D^M) do not occur when using the proposed Definition 20.1.

The following Definition 20.7 represents another approach to consistency evaluation and is used in the methods [3, 11, 15] for calculation weights of DAs on basis of an interval PCM (IPCM).

Definition 20.7 ([3, 11, 15]) An IPCM $\tilde{D} = \{\tilde{d}_{ij} = [l_{ij}, u_{ij}]\}$ is called consistent if a vector of weights w exists such that $w_i \in R$, $w_i > 0$, $\sum_{i=1}^{n} w_i = 1$ and $l_{ij} \leq w_i/w_j \leq u_{ij}$, $i = 1, 2, ..., n-1, j = 2, 3, ..., n$.

Statement 1 *An IPCM* $\tilde{D} = \{\tilde{d}_{ij} = [l_{ij}, u_{ij}]\}$ *is consistent in terms of Definition 20.7 if and only if an inequality* $\max_k(l_{ik}l_{kj}) \leq \min_k(u_{ik}u_{kj})$ *holds for* $\forall i < j$.

Necessary and sufficient condition to satisfy Definition 20.7 is an existence of some consistent PCM, such that all its elements are in corresponding intervals of the initial IPCM. This definition, obviously, is weaker in comparison with Definitions 20.1 and 20.4 and more IPCMs become consistent.

A comparative study of the known Definitions 20.4–20.7 with the proposed in the Sect. 20.3 ones results in the following conclusions:

(1) to evaluate inconsistency level of a FPCM the proposed Definitions 20.1–20.3 enable to apply all known results about evaluation of inconsistency of traditional crisp PCMs and opposite to the known described above definitions enable to identify weak consistency of a FPCM and evaluate admissibility of inconsistency of a FPCM when calculating weights of DAs;
(2) the proposed Definitions 20.1–20.3, opposite to all known described above definitions, enable to increase consistency of a FPCM rather easy, in particular, enable to find the most inconsistent and intransitive elements of a FPCM, using methods developed for crisp PCMs;
(3) the proposed Definitions 20.1–20.3 may be used to evaluate inconsistency of a FPCM with fuzzy elements of any shape (triangle, trapezoidal, gaussian and other, and also discrete fuzzy sets);
(4) contradiction results does not appear when the proposed Definition 20.1 is applied to evaluate consistency of crisp and fuzzyfied PCM, opposite to the known Definitions 20.4–20.6, which use extended binary arithmetic operations;
(5) the proposed Definition 20.1 does not lead to contradiction results about consistency of a FPCM of dimension $n = 2$;
(6) in special case when evaluating inconsistency of interval PCM, the proposed Definition 20.1 is more strong in comparison with the known and widely used Definition 20.7. More precisely, IPCMs which are consistent in terms of Definition 20.7 are not always consistent in terms of the proposed Definition 20.1. However, these IPCM, are in general admissibly inconsistent (see Definition 20.2) and therefore may be used for calculation of weights.

20.5 Illustrative Examples

Let us illustrate the proposed Definitions 20.1–20.3 on FPCMs of special case, namely, on interval PCMs (IPCMs). It makes possible to compare results on basis of Definitions 20.1–20.3 with results on basis of Definition 20.7. IPCMs of different inconsistency level are considered in the following Examples 20.3–20.5.

Example 20.3 The following IPCM $\tilde{D}$ is consistent in terms of Definition 20.7: the possible vector of weights is $w = (0.45\ 0.22\ 0.11\ 0.22)$. However this IPCM is inconsistent in terms of Definition 20.4, since the following PCMs D^L and D^U, calculated according to (20.4), are inconsistent:

$$\tilde{D} = \begin{pmatrix} 1 & [2,5] & [2,4] & [1,3] \\ \left[\frac{1}{5},\frac{1}{2}\right] & 1 & [1,3] & [1,2] \\ \left[\frac{1}{4},\frac{1}{2}\right] & \left[\frac{1}{3},1\right] & 1 & \left[\frac{1}{2},1\right] \\ \left[\frac{1}{3},1\right] & \left[\frac{1}{2},1\right] & [1,2] & 1 \end{pmatrix},$$

$$D^L = \begin{pmatrix} 1 & 2 & 2 & 1 \\ 1/2 & 1 & 1 & 1 \\ 1/2 & 1 & 1 & 1/2 \\ 1 & 1 & 2 & 1 \end{pmatrix} \text{ and } D^U = \begin{pmatrix} 1 & 5 & 4 & 3 \\ 1/5 & 1 & 3 & 2 \\ 1/4 & 1/3 & 1 & 1 \\ 1/3 & 1/2 & 1 & 1 \end{pmatrix}.$$

The IPCM $\tilde{D}$ is also inconsistent according to the proposed Definition 20.1. Definition 20.2 makes it possible to evaluate inconsistency level and admissibility of inconsistency of the IPCM $\tilde{D}$. In the example the IPCM $\tilde{D}$ is admissibly inconsistent since the consistency ratio does not exceed its threshold value $CR(Deffuz(\tilde{D})) = 0.04 < CR^{threshold} = 0.08$.

Example 20.4 The following IPCM $\tilde{D}$ is weak consistent in terms of Definition 20.3 and has no ordinal intransitive elements (cycles):

$$\tilde{D} = \begin{pmatrix} 1 & \left[\frac{1}{3},1\right] & \left[\frac{1}{4},\frac{1}{2}\right] & \left[\frac{1}{7},\frac{1}{5}\right] & \left[\frac{1}{3},1\right] \\ [1,3] & 1 & [3,5] & [4,6] & [1,3] \\ [2,4] & \left[\frac{1}{5},\frac{1}{3}\right] & 1 & \left[\frac{1}{3},1\right] & \left[\frac{1}{5},\frac{1}{3}\right] \\ [5,7] & \left[\frac{1}{6},\frac{1}{4}\right] & [1,3] & 1 & \left[\frac{1}{4},\frac{1}{2}\right] \\ [1,3] & \left[\frac{1}{3},1\right] & [3,5] & [2,4] & 1 \end{pmatrix}, D = \begin{pmatrix} 1 & \frac{1}{2} & \frac{1}{3} & \frac{1}{6} & \frac{1}{2} \\ 2 & 1 & 4 & 5 & 2 \\ 3 & \frac{1}{4} & 1 & \frac{1}{2} & \frac{1}{4} \\ 6 & \frac{1}{5} & 2 & 1 & \frac{1}{3} \\ 2 & \frac{1}{2} & 4 & 3 & 1 \end{pmatrix}.$$

Inconsistency level of the defuzzified PCM D is inadmissible since the consistency ratio $CR(D) = 0.202$ exceeds the threshold value 0.1. Therefore the IPCM $\tilde{D}$ in the example is inadmissibly inconsistent and should not be used for calculation of weights, it requires a correction to increase its consistency.

To compare results, Definition 20.7 also defines this IPCM $\tilde{D}$ as inconsistent. Weights of DAs calculated on basis of the IPCM $\tilde{D}$ using several known methods [3, 14–16] are shown in Table 20.1. It is worth noted that in the two-stage methods TLGP [15] and 2SLGP [3] first stage deals with evaluation of IPCM consistency. Weights in Table 20.1 show that different methods lead to the same ranking of DAs. It seems to be a result of the IPCM $\tilde{D}$ weak consistency. Index J^* in the TLGP model [15] has nonzero value and therefore indicate an inconsistency of the IPCM $\tilde{D}$.

Example 20.5 The following IPCM $\tilde{D}$ is not weak consistent in terms of Definition 20.3 since the condition of ordinal transitivity of the defuzzified PCM D is violated:

Table 20.1 Weights of DAs on basis of the IPCM $\tilde{D}$ calculated using different methods (Example 20.4)

Weights	TLGP [15]	TLGP [15] (defuzz)	LGPPM [16]	2SLGP [3]	LUAM [14]
	$J^* = 1,610$				
w_1	[0,268;0,749]	0,080	0,04	1	0,109
w_2	[1,958;3,107]	0,396	0,424	5,576	0,273
w_3	[0,514;0,720]	0,097	0,094	1,28	0,119
w_4	[0,621;1,000]	0,127	0,148	2,021	0,125
w_5	[1,599;2,245]	0,301	0,281	3,655	0,177

$$\tilde{D} = \begin{pmatrix} 1 & [1;5] & [4;8] & [3;7] & [2;6] \\ \left[\frac{1}{5};1\right] & 1 & \left[\frac{1}{6};\frac{1}{2}\right] & \left[\frac{1}{7};\frac{1}{3}\right] & [3;7] \\ \left[\frac{1}{8};\frac{1}{4}\right] & [2;6] & 1 & \left[\frac{1}{4};1\right] & \left[\frac{1}{7};\frac{1}{3}\right] \\ \left[\frac{1}{7};\frac{1}{3}\right] & [3;7] & [1;4] & 1 & [1;5] \\ \left[\frac{1}{6};\frac{1}{2}\right] & \left[\frac{1}{7};\frac{1}{3}\right] & [3;7] & \left[\frac{1}{5};1\right] & 1 \end{pmatrix}, D = \begin{pmatrix} 1 & 3 & 6 & 5 & 4 \\ \frac{1}{3} & 1 & \frac{1}{4} & \frac{1}{5} & 5 \\ \frac{1}{6} & 4 & 1 & \frac{1}{3} & \frac{1}{5} \\ \frac{1}{5} & 5 & 3 & 1 & 3 \\ \frac{1}{4} & \frac{1}{5} & 5 & \frac{1}{3} & 1 \end{pmatrix}.$$

Namely, a triple of elements exists in the PCM D, such that $(d_{53} > 1) \wedge (d_{32} > 1) \wedge (d_{52} < 1)$ and therefore a cycle in D exists.

A consistency ratio of the PCM D equals to $CR(D) = 0.514$ and considerably increases the threshold value 0.1, therefore an inconsistency level of the PCM D is quite high. As a result the IPCM $\tilde{D}$ should not be used for calculation of weights and requires a correction to increase its consistency. The most inconsistent element of the IPCM $\tilde{D}$ that leads to a cycle should be found and corrected (changed) (see an Example 20.6).

The IPCM $\tilde{D}$ is inconsistent also in terms of Definition 20.7. Weights of DAs calculated on basis of the IPCM $\tilde{D}$ using several known methods [3, 14–16] are shown in Table 20.2. It is worth noted that in the two-stage methods TLGP [15] and 2SLGP [3] first stage deals with evaluation of IPCM consistency. Results in Table 20.5 indicate a contradiction in rankings of DAs when using different methods of weights calculation.

Index J^* in the TLGP model [15] has quite large nonzero value and therefore an inconsistency level of the IPCM $\tilde{D}$ is high. However, neither Definition 20.7 nor index J^* does not allow to evaluate an admissibility of a FPCM inconsistency level, as opposed to the proposed Definition 20.2.

It is known that a vector of weights does not exist that satisfies a weak inconsistent PCM with a cycle (cycles). A method for finding the most inconsistent element in a FPCM which leads to a cycle in a FPCM is shown in next Sect. 20.6.

Table 20.2 Weights of DAs on basis of the IPCM $\tilde{D}$ calculated using different methods (Example 20.5)

Weights	TLGP [15]	TLGP (defuzz)	LGPPM [16]	2SLGP [3]	LUAM [14]
	$J^* = 2,036$				
w_1	[2,442; 4,091]	0,483	0,481	1	0,25
w_2	[0,324; 0,813]	0,084	0,128	0,359	0,133
w_3	[0,422; 0,804]	0,091	0,059	0,206	0,133
w_4	[1,000; 1,888]	0,213	0,226	0,467	0,136
w_5	[0,556; 1,185]	0,129	0,063	0,262	0,133

20.6 Finding of the Most Inconsistent Element in a FPCM

A modified M_Outflow method [24] is proposed for finding the most inconsistent element and a cycle (a triple of intransitive elements) elimination in a FPCM (20.1). Let us consider a PCM D as a result of a FPCM (20.1) defuzzification according to (20.2). The M_Outflow method has several stages:

(1) To calculate the inflow Φ_i^- and outflow Φ_i^+ values for each DA a_i, $i = 1, 2, ..., n$. Let Φ_i^+ be a number of DAs a_j, such that a_i outperforms a_j, namely $d_{ij} > 1$. Let Φ_i^- be a number of DAs a_j, such that a_j outperforms a_i, namely $d_{ji} > 1$.
(2) To find the maximum of differences $\Phi_j^+ - \Phi_i^+$ and $\Phi_i^- - \Phi_j^-$. Then element $d_{i^*j^*}$ is the most inconsistent one:

$$d_{i^*j^*} : \max_{ij}(\max(\Phi_j^+ - \Phi_i^+, \Phi_i^- - \Phi_j^-)), \text{ if } i \neq j, d_{ij} > 1. \qquad (20.7)$$

Suppose that several elements $d_{i^*j^*}$ satisfy the condition (20.7). Then an element among them is found which lead to more inconsistency, namely an element which results in maximum value of the expression

$$\gamma_{ij} = \frac{1}{n-2}\sum_{k=1}^{n}(\ln d_{ij} - \ln(d_{ik}d_{kj})), \text{ where } k \neq i \neq j. \qquad (20.8)$$

Hence, the most inconsistent element in an initial FPCM (20.1) is the element $d_{i^*j^*}$, which corresponds to the maximum value of the expression (20.8).

Example 20.6 Let us consider defuzzyfied PCMs shown in the Examples 20.4 and 20.5, and let us find the most inconsistent elements of these matrices using the M_OutFlow method:

(a) the inflow Φ^- and outflow Φ^+ vectors for the PCM D from the Example 20.4 are $\Phi^- = (4, 0, 3, 2, 1)$ and $\Phi^+ = (0, 4, 1, 2, 3)$. Elements d_{25}, d_{31}, d_{43} and d_{54}

satisfy the condition (20.7). Element d_{31} is the most inconsistent, since it has the maximum value among values γ_{ij} (20.8). After correction of this element (it is assigned a new value $d_{31} = 1$) the consistency level of the PCM D is increased up to the value $CR(D) = 0.166$;

(b) the inflow Φ^- and outflow Φ^+ vectors for the PCM D from the Example 20.5 are $\Phi^- = (0, 3, 3, 1, 3)$ and $\Phi^+ = (4, 1, 1, 3, 1)$. Elements d_{25}, d_{32} and d_{53} satisfy the condition (20.7). Element d_{25} is the most inconsistent, since it has the maximum value among values $\gamma_{25} = 2.682$, $\gamma_{32} = 2.52$ and $\gamma_{53} = 2.473$ calculated according to (20.8). After correction of d_{25} (it is assigned a new value $d_{25} = 1/5$) the cycle in the PCM D is eliminated (D becomes weak consistent) and the consistency level of the PCM D is increased up to the value $CR(D) = 0.224$.

As a sequence, the elements d_{31} and d_{25} are the most inconsistent elements of the initial IPCMs from the Examples 20.4 and 20.5, respectively.

20.7 Conclusions

A new definition of consistency of a fuzzy pairwise comparison matrix (FPCM) is developed in the paper under a suggestion that FPCM elements are fuzzy sets with membership functions of any shape. Such FPCMs may be a result of evaluation of decision alternatives by a group of experts when aggregating individual expert judgments made in traditional crisp scales. A comparative analysis of proposed definition with other known definitions of consistent FPCM is done. New definitions are suggested to evaluate the admissibility of inconsistency of expert judgments when calculating weights of decision alternatives, to reveal weak inconsistent and intransitive expert judgments. Usage of these definitions allows to increase a FPCM inconsistency in a quite easy way. The most inconsistent and intransitive FPCM elements (expert judgments) are defined using a modified M_Outflow method.

References

1. Saaty, T.L., Vargas, L.G.: Decision Making with the Analytic Network Process: Economic, Political, Social and Technological Applications with Benefits, Opportunities, Costs and Risks. Springer, New York (2006)
2. Ramanathan, R., Ramanathan, U.: A qualitative perspective to deriving weights from pairwise comparison matrices. Omega **38**(3–4), 228–232 (2010)
3. Tsyganok, V.: Investigation of the aggregation effectiveness of expert estimates obtained by the pairwise comparison method. Math. Comput. Model. **52**(3), 538–544 (2010)
4. Jalao, E.R., Wu, T., Shunk, D.: An intelligent decomposition of pairwise comparison matrices for large-scale decisions. Eur. J. Op. Res. **238**(1), 270–280 (2014)
5. Durbach, I., Lahdelma, R., Salminen, P.: The analytic hierarchy process with stochastic judgements. Eur. J. Op. Res. **238**(2), 552–559 (2014)
6. Koczkodaj, W.W., Szybowski, J.: Pairwise comparisons simplified. Appl. Math. Comput. **253**, 387–394 (2015)

7. Pankratova, N.D., Nedashkovskaya, N.I.: Models and methods of analysis of hierarchies. Theory Appl. Kiev, p. 371 (2010). (in ukrainian)
8. Pankratova, N., Nedashkovskaya, N.: The method of estimating the consistency of paired comparisons. Int. J. Inf. Technol. Knowl. **7**(4), 347–361 (2013)
9. Nedashkovskaya, N.I.: Method of consistent pairwise comparisons when estimating decision alternatives in terms of qualitative criterion. Syst. Res. Inf. Technol. **4**, 67–79 (2013). Access mode: http://journal.iasa.kpi.ua/article/view/33943 (in ukrainian)
10. Buckley, J.J.: Fuzzy hierarchical analysis. Fuzzy Sets Syst. **17**(3), 233–247 (1985)
11. Mikhailov, L.: Deriving priorities from fuzzy pairwise comparison judgements. Fuzzy Sets Syst. **134**(3), 365–385 (2003)
12. Wang, Y.M., Chin, K.S.: Fuzzy analytic hierarchy process: a logarithmic fuzzy preference programming methodology. Int. J. Approx. Reason. **52**(4), 541–553 (2011)
13. Wang, J., Lan, J., Ren, P., Luo, Y.: Some programming models to derive priority weights from additive interval fuzzy preference relation. Knowl.-Based Syst. **27**, 69–77 (2012)
14. Sugihara, K., Ishii, H., Tanaka, H.: Interval priorities in AHP by interval regression analysis. Eur. J. Op. Res. **158**, 745–754 (2004)
15. Wang, Y.M., Yang, J.B., Xu, D.L.: A two-stage logarithmic goal programming method for generating weights from interval comparison matrices. Fuzzy Sets Syst. **152**, 475–498 (2005)
16. Wang, Y.M., Elhag, T.M.S.: A goal programming method for obtaining interval weights from an interval comparison matrix. Eur. J. Op. Res. **177**, 458–471 (2007)
17. Z.S, Xu, Chen, J.: Some models for deriving the priority weights from interval fuzzy preference relations. Eur. J. Op. Res. **184**(1), 266–280 (2008)
18. Liu, F., Zhang, W.G., Fu, J.H.: A new method of obtaining the priority weights from an interval fuzzy preference relation. Inf. Sci. **185**(1), 32–42 (2012)
19. Chang, D.Y.: Applications of the extent analysis method on fuzzy AHP. Eur. J. Op. Res. **95**(3), 649–655 (1996)
20. Pankratova, N., Nedashkovskaya, N.: Methods of evaluation and improvement of consistency of expert pairwise comparison judgements. Int. J. Inf. Theor. Appl. **22**(3), 203–223 (2015)
21. Liu, F.: Acceptable consistency analysis of interval reciprocal comparison. Matrices Fuzzy Sets Syst. **160**, 2686–2700 (2009)
22. Liu, F., et al.: Consistency analysis of triangular fuzzy reciprocal preference relations. Eur. J. Op. Res. **235**, 718–726 (2014)
23. Ross, T.J.: Fuzzy Logic with Engineering Applications, 3rd edn, p. 606. Wiley, New York (2010)
24. Nedashkovskaya, N.I.: The M_Outflow method for finding the most inconsistent elements of a pairwise comparison matrix. System analysis and information technologies: materials of international scientific and technical conference SAIT 2015 (June 22–25, Kyiv). – 95 p. Access mode: http://sait.kpi.ua/books/

Chapter 21
Approximate Optimal Control for Parabolic–Hyperbolic Equations with Nonlocal Boundary Conditions and General Quadratic Quality Criterion

Volodymyr O. Kapustyan and Ivan O. Pyshnograiev

Abstract Control theory recaptures an increasingly prominent place in modern science. We constructed an approximate optimal control for parabolic–hyperbolic equations with nonlocal boundary conditions and general quadratic quality criterion in special norm. We considered the problem for distributed and divided control. Also, we proved the convergence of approximate control and provided the numerical experiments that characterized its properties.

21.1 Introduction

Control theory recaptures an increasingly prominent place in modern science. Research results in this direction can be seen in many areas such as economics [1] and physics [2]. Mathematical models with investigated control problems are becoming more complex (starting from simple models with ordinary differential equations of the first order [3] to boundary problems of parabolic equations [4]).

In this paper, we construct an approximate optimal control for parabolic–hyperbolic equations with nonlocal boundary conditions and general quadratic quality criterion in special norm. We consider the problem for distributed and divided control.

V.O. Kapustyan (✉) · I.O. Pyshnograiev
National Technical University of Ukraine "Kyiv Polytechnic Institute",
Prosp. Peremohy, 37, Kyiv 03056, Ukraine
e-mail: kapustyanv@ukr.net

I.O. Pyshnograiev
e-mail: pyshnograiev@gmail.com

V.A. Sadovnichiy and M.Z. Zgurovsky (eds.), *Advances in Dynamical Systems and Control*, Studies in Systems, Decision and Control 69,
DOI 10.1007/978-3-319-40673-2_21

21.2 The Problem with Distributed Control

Let the controlled process $y(x,t) \in C^1(\bar{D}) \cap C^2(D_-) \cap C^{2,1}(D_+)$ in D satisfy the equation

$$Ly(x,t) = \hat{u}(x,t) \tag{21.1}$$

with initial

$$y(x,-\alpha) = \varphi(x) \tag{21.2}$$

and boundary conditions

$$y(0,t) = 0, y'(0,t) = y'(1,t), -\alpha \le t \le T, \tag{21.3}$$

where $D = \{(x,t) : 0 < x < 1, -\alpha < t \le T, \alpha, T > 0\}$, $D_- = \{(x,t) : 0 < x < 1, -\alpha < t \le 0\}$, $D_+ = \{(x,t) : 0 < x < 1, 0 < t \le T\}$,

$$Ly = \begin{cases} y_t - y_{xx}, \text{ t} > 0, \\ y_{tt} - y_{xx}, \text{ t} < 0. \end{cases}$$

and

$$\hat{u}(x,t) = \begin{cases} u(x,t), \text{ t} \ge 0, \\ v(x,t), \text{ t} < 0. \end{cases}$$

This boundary value problem was solved in [5].

It is required to find the control $\hat{u}^*(x,t) \in \mathscr{K}$, which minimizes the functional

$$I(\hat{u}) = 0.5(\hat{\alpha}||y(.,T) - \psi(.)||_D^2$$
$$+\hat{\beta}_1 \int_{-\alpha}^{0} ||y(.,t)||_D^2 dt + \hat{\beta}_2 \int_0^T ||y(.,t)||_D^2 dt$$
$$+\hat{\gamma}_1 \int_{-\alpha}^{0} ||v(.,t)||_D^2 dt + \hat{\gamma}_2(||u(.,0)||_D^2 + \int_0^T ||u_t(.,t)||_D^2 dt))$$
$$= 0.5 \sum_{i=0}^{\infty} (\hat{\alpha}(y_i(T) - \psi_i)^2 + \hat{\beta}_1 \int_{-\alpha}^{0} y_i^2(t)dt + \hat{\beta}_2 \int_0^T y_i^2(t)dt$$
$$+\hat{\gamma}_1 \int_{-\alpha}^{0} v_i^2(t)dt + \hat{\gamma}_2(u_i^2(0) + \int_0^T \dot{u}_i^2(t)dt)), \tag{21.4}$$

where $\psi(x)$ is fixed function, $\hat{\alpha}$, $\hat{\beta}_i \ge 0, \hat{\gamma}_i > 0, i = \overline{1,2}$; $\hat{\alpha} + \hat{\beta}_1 + \hat{\beta}_2 > 0$; the function class $\mathscr{K}$ and expansion of the functions by the Riss basis are shown in [5].

In [6], it is shown that the optimal control can be found from equation system

$$\hat{\gamma}_1 v_0(t) + \int_{-\alpha}^{0} \mathscr{K}_{0,1}^{(1)}(t,\tau) v_0(\tau) d\tau + \mathscr{K}_{0,2}^{(1)}(t) u_0(0) + \int_{0}^{T} \mathscr{K}_{0,3}^{(1)}(t,\tau)\xi_0(\tau) d\tau$$
$$= \mathscr{M}_{0,1}^{(1)}(t)\varphi_0 + \mathscr{M}_{0,2}^{(1)}(t)\psi_0, \ t \in [-\alpha, 0),$$

$$\hat{\gamma}_2 u_0(0) + \int_{-\alpha}^{0} \mathscr{K}_{0,1}^{(2)}(\tau) v_0(\tau) d\tau + \mathscr{K}_{0,2}^{(2)} u_0(0) + \int_{0}^{T} \mathscr{K}_{0,3}^{(2)}(\tau)\xi_0(\tau) d\tau$$
$$= \mathscr{M}_{0,1}^{(2)}\varphi_0 + \mathscr{M}_{0,2}^{(2)}\ \psi_0,$$

$$\hat{\gamma}_2 \xi_0(t) + \int_{-\alpha}^{0} \mathscr{K}_{0,1}^{(3)}(t,\tau) v_0(\tau) d\tau + \mathscr{K}_{0,2}^{(3)}(t) u_0(0) + \int_{0}^{T} \mathscr{K}_{0,3}^{(3)}(t,\tau)\xi_0(\tau) d\tau$$
$$= \mathscr{M}_{0,1}^{(3)}(t)\varphi_0 + \mathscr{M}_{0,2}^{(3)}(t)\psi_0, \ t \in (0, T],$$

$$\hat{\gamma}_1 v_i(t) + \sum_{j=2k-1}^{2k} \Big(\int_{-\alpha}^{0} \mathscr{K}_{j,1}^{(1,i)}(t,\tau)\, v_j(\tau) d\tau + \mathscr{K}_{j,2}^{(1,i)}(t) u_j(0)$$
$$+ \int_{0}^{T} \mathscr{K}_{j,3}^{(1,i)}(t,\tau)\, \xi_j(\tau) d\tau \Big)$$
$$= \sum_{j=2k-1}^{2k} \big(\mathscr{M}_{j,1}^{(1,i)}(t)\varphi_j + \mathscr{M}_{j,2}^{(1,i)}(t)\ \psi_j\big), \ t \in [-\alpha, 0),$$

$$\hat{\gamma}_2 u_i(0) + \sum_{j=2k-1}^{2k} \Big(\int_{-\alpha}^{0} \mathscr{K}_{j,1}^{(2,i)}(\tau) v_j(\tau) d\tau + \mathscr{K}_{j,2}^{(2,i)} u_j(0)$$
$$+ \int_{0}^{T} \mathscr{K}_{j,3}^{(2,i)}(\tau)\xi_j(\tau) d\tau \Big)$$
$$= \sum_{j=2k-1}^{2k} \big(\mathscr{M}_{j,1}^{(2,i)}\ \varphi_j + \mathscr{M}_{j,2}^{(2,i)}\psi_j\big),$$

$$\hat{\gamma}_2 \xi_i(t) + \sum_{j=2k-1}^{2k} \Big(\int_{-\alpha}^{0} \mathscr{K}_{j,1}^{(3,i)}(t,\tau)\, v_j(\tau) d\tau + \mathscr{K}_{j,2}^{(3,i)}(t) u_j(0)$$
$$+ \int_{0}^{T} \mathscr{K}_{j,3}^{(3,i)}(t,\tau)\, \xi_j(\tau) d\tau \Big) = \sum_{j=2k-1}^{2k} \big(\mathscr{M}_{j,1}^{(3,i)}(t)\varphi_j + \mathscr{M}_{j,2}^{(3,i)}(t)\ \psi_j\big),$$
$$t \in (0, T],\ i = \overline{2k-1, 2k}, \quad (21.5)$$

where $\mathscr{K}$ and $\mathscr{M}$ are notations, which can be found from solution of the boundary value problem.

21.2.1 Approximate Optimal Control

Let us consider the approximate optimal control and take the finite number of its elements. So we get

$$v^{(N)}(x,t) = v_0(t)X_0(x) + \sum_{k=1}^{N} (v_{2k-1}(t)X_{2k-1}(x) + v_{2k}(t)X_{2k}(x)),$$

$$u^{(N)}(x,0) = u_0(0)X_0(x) + \sum_{k=1}^{N} (u_{2k-1}(0)X_{2k-1}(x) + u_{2k}(0)X_{2k}(x)),$$

$$\xi^{(N)}(x,t) = \xi_0(t)X_0(x) + \sum_{k=1}^{N} (\xi_{2k-1}(t)X_{2k-1}(x) + \xi_{2k}(t)X_{2k}(x)).$$

Then, the following theorem is obvious.

Theorem 21.1 *Let the functions $\varphi(x)$, $\psi(x)$ of the problem (21.1)–(21.3), (21.4) satisfy the conditions from [6]. Then the functions $v^{(N)}(x,t)$, $u^{(N)}(x,t)$ are the approximate solution of the optimal control problem. And the next equality is correct.*

$$\lim_{N\to\infty} ||v^* - v^{(N)}||_{C(0,1)\times C(-\alpha,0)} = 0,$$

$$\lim_{N\to\infty} ||u^* - u^{(N)}||_{C(0,1)\times C(0,T)} = 0,$$

$$\lim_{N\to\infty} |I(\hat{u}^*) - I^{(N)}(\hat{u}^{(N)})| = 0.$$

21.2.2 Example of Calculations

Let $\varphi(x) = x, \psi(x) = 10x^2$, and $\alpha = 2, T = 5, \hat{\gamma} = 10, \hat{\alpha} = \hat{\beta}_1 = \hat{\beta}_2 = \hat{\gamma}_1 = \hat{\gamma}_2 = 1$. Then, we use numerical algorithms for the solution of (21.5).

On the Fig. 21.1, changing criterion values are presented depending on the number N of component series. Obviously, the criterion was changed a little for $N > 1$.

Thus, Fig. 21.2 shows the solution of the problem. Value criterion in this case is $I = 17,87$.

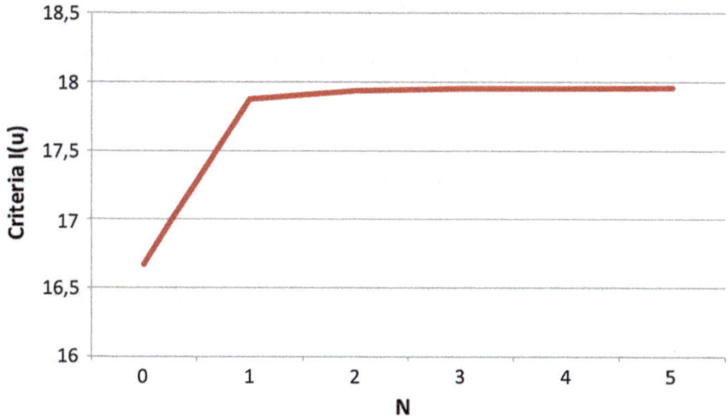

Fig. 21.1 Criterion value

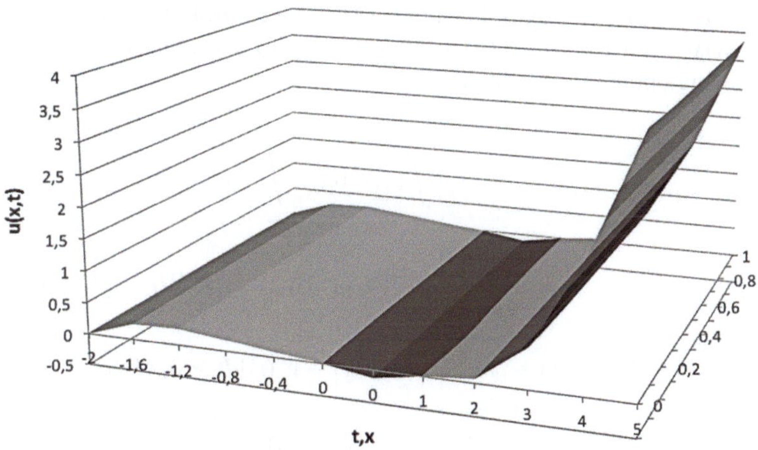

Fig. 21.2 Optimal control

21.3 The Problem with Divided Control

Let the controlled process $y(x, t)$ is described by the boundary value problem

$$Ly(x, t) = g(t)\hat{u}(x) \tag{21.6}$$

with conditions (21.2)–(21.3). It is needed to find the control $v^*(t) \in C[-\alpha, 0)$:$|v^*(t)| \leq 1$; $|u^*(0)| \leq l_0$; $\xi^*(t) \in L_2[0, T] : |\xi^*(t)| \leq l_1$ almost everywhere on $[0, T]$, which minimizes the functional

$$I(\hat{u}) = 0.5(\hat{\alpha}||y(.,T) - \psi(.)||_D^2 + \hat{\beta}_1 \int_{-\alpha}^{0} ||y(.,t)||_D^2 dt + \hat{\beta}_2 \int_{0}^{T} ||y(.,t)||_D^2 dt$$
$$+\hat{\gamma}_1 \int_{-\alpha}^{0} v^2(t)dt + \hat{\gamma}_2(u^2(0) + \int_{0}^{T} \dot{u}^2(t)dt))$$
$$= 0.5(\sum_{i=0}^{\infty}(\hat{\alpha}\,(y_i(T) - \psi_i)^2 + \hat{\beta}_1 \int_{-\alpha}^{0} y_i^2(t)dt + \hat{\beta}_2 \int_{0}^{T} y_i^2(t)dt$$
$$+\hat{\gamma}_1 \int_{-\alpha}^{0} v^2(t)dt + \hat{\gamma}_2(u^2(0) + \int_{0}^{T} \dot{u}^2(t)dt)). \quad (21.7)$$

Because of strict convexity functional (21.7) by control, it has a single point of minimum $(v^*(t), u^*(0), \xi^*(t))\ C[-\alpha, 0) \times R^1 \times L_2(0, T)$, which is characterized by the following optimality conditions

$$\int_{-\alpha}^{0} [\hat{\gamma}_1 v^*(t) + \int_{-\alpha}^{0} \mathcal{K}_1^{(1)}(t,\tau)v^*(\tau)d\tau +$$
$$\mathcal{K}_2^{(1)}(t)u^*(0) + \int_{0}^{T} \mathcal{K}_3^{(1)}(t,\tau)\xi^*(\tau)d\tau$$
$$-\mathcal{M}_1^{(1)}(t,\varphi) - \mathcal{M}_2^{(1)}(t,\psi)][v(t) - v^*(t)]dt \geq 0, \forall |v(t)| \leq 1,$$
$$(\hat{\gamma}_2 u^*(0) + \int_{-\alpha}^{0} \mathcal{K}_1^{(2)}(\tau)v^*(\tau)d\tau + \mathcal{K}_2^{(2)}u^*(0) + \int_{0}^{T} \mathcal{K}_3^{(2)}(\tau)\xi^*(\tau)d\tau$$
$$-\mathcal{M}_1^{(2)}(\varphi) + \mathcal{M}_2^{(2)}(\psi))[u(0) - u^*(0)] \geq 0, \forall |u(0)| \leq l_0,$$
$$\int_{0}^{T} [\hat{\gamma}_2 \xi^*(t) + \int_{-\alpha}^{0} \mathcal{K}_1^{(3)}(t,\tau)v^*(\tau)d\tau$$
$$+\mathcal{K}_2^{(3)}(t)u^*(0) + \int_{0}^{T} \mathcal{K}_3^{(3)}(t,\tau)\xi^*(\tau)d\tau$$
$$-\mathcal{M}_1^{(3)}(t,\varphi) - \mathcal{M}_2^{(3)}(t,\psi)][\xi(t) - \xi^*(t)]dt \geq 0, |\xi(t)| \leq l_1, \quad (21.8)$$

where $\mathcal{K}$ and $\mathcal{M}$ are notations, which can be found from solution of the boundary value problem.

This problem is investigated in [6].

21.3.1 Approximate Control

Let us consider two cases of this problem. There are problem with approximate unbounded control and approximate bounded control.

Unbounded Control

In this case, the system of variational inequalities (21.8) takes the form

$$\hat{\gamma}_1 v^*(t) + \int_{-\alpha}^{0} \mathscr{K}_1^{(1)}(t,\tau)v^*(\tau)d\tau + \mathscr{K}_2^{(1)}(t)u^*(0) + \int_{0}^{T} \mathscr{K}_3^{(1)}(t,\tau)\xi^*(\tau)d\tau$$
$$= \mathscr{M}_1^{(1)}(t,\varphi) + \mathscr{M}_2^{(1)}(t,\psi), t \in [-\alpha, 0),$$
$$\hat{\gamma}_2 u^*(0) + \int_{-\alpha}^{0} \mathscr{K}_1^{(2)}(\tau)v^*(\tau)d\tau + \mathscr{K}_2^{(2)}u^*(0) + \int_{0}^{T} \mathscr{K}_3^{(2)}(\tau)\xi^*(\tau)d\tau$$
$$= \mathscr{M}_1^{(2)}(\varphi) + \mathscr{M}_2^{(2)}(\psi),$$
$$\hat{\gamma}_2 \xi^*(t) + \int_{-\alpha}^{0} \mathscr{K}_1^{(3)}(t,\tau)v^*(\tau)d\tau + \mathscr{K}_2^{(3)}(t)u^*(0) + \int_{0}^{T} \mathscr{K}_3^{(3)}(t,\tau)\xi^*(\tau)d\tau$$
$$= \mathscr{M}_1^{(3)}(t,\varphi) + \mathscr{M}_2^{(3)}(t,\psi), t \in (0,T]. \quad (21.9)$$

Since the core and right parts of the system (21.9) are the sequences, we consider the approximate control, which can be found from the system of equations

$$\hat{\gamma}_1 v^{(N)}(t) + \int_{-\alpha}^{0} \mathscr{K}_1^{(1,N)}(t,\tau)v^{(N)}(\tau)d\tau + \mathscr{K}_2^{(1,N)}(t)u^{(N)}(0)$$
$$+ \int_{0}^{T} \mathscr{K}_3^{(1,N)}(t,\tau)\,\xi^{(N)}(\tau)d\tau$$
$$= \mathscr{M}_1^{(1,N)}(t,\varphi) + \mathscr{M}_2^{(1,N)}(t,\psi), t \in [-\alpha, 0),$$
$$\hat{\gamma}_2 u^{(N)}(0) + \int_{-\alpha}^{0} \mathscr{K}_1^{(2,N)}(\tau)v^{(N)}(\tau)d\tau + \mathscr{K}_2^{(2,N)}u^{(N)}(0)$$
$$+ \int_{0}^{T} \mathscr{K}_3^{(2,N)}(\tau)\xi^{(N)}(\tau)d\tau = \mathscr{M}_1^{(2,N)}(\varphi) + \mathscr{M}_2^{(2,N)}(\psi),$$
$$\hat{\gamma}_2 \xi^{(N)}(t) + \int_{-\alpha}^{0} \mathscr{K}_1^{(3,N)}(t,\tau)v^{(N)}(\tau)d\tau + \mathscr{K}_2^{(3,N)}(t)u^{(N)}(0)$$

$$+\int_0^T \mathcal{K}_3^{(3,N)}(t,\tau)\xi^{(N)}(\tau)d\tau = \mathcal{M}_1^{(3,N)}(t,\varphi) + \mathcal{M}_2^{(3,N)}(t,\psi), t \in (0,T], \quad (21.10)$$

where $\mathcal{K}_i^{(j,N)}(), \mathcal{M}_l^{(j,N)}(), i,j = \overline{1,3}, andl = \overline{1,2}$ are the finite sums.

Clearly, the system of equations (21.10) represents optimality conditions for optimal control with criteria

$$I^{(N)}(\hat{u}^{(N)}) = 0.5(\sum_{i=0}^{N}(\hat{\alpha}(y_i^{(N)}(T) - \psi_i)^2 + \hat{\beta}_1 \int_{-\alpha}^{0} (y_i^{(N)}(t))^2 dt$$

$$+\hat{\beta}_2 \int_0^T (y_i^{(N)}(t))^2\, dt + \hat{\gamma}_1 \int_{-\alpha}^{0} (v^{(N)}(t))^2 dt$$

$$+\hat{\gamma}_2((u^{(N)}(0))^2 + \int_0^T (\xi^{(N)}(t))^2 dt)), \quad (21.11)$$

where $y_i^{(N)}(t) andi = \overline{0, 2N}$ are solutions of Fourier coefficients of the boundary problem with the control $\hat{u}^{(N)}(t)$.

Theorem 21.2 *Let the functions $\varphi(x)$, $\psi(x)$, $g(x)$ in optimal control problem (21.6), (21.2), (21.3) and (21.7) ($\hat{u}(x,t) = g(x)\hat{u}(t)$) satisfy the lemmas conditions in [6]. Then continuous functions $v^{(N)}(t)$, $u^{(N)}(t)$ are the approximate control of original optimal control problem, in other words*

$$\lim_{N\to\infty} ||v^* - v^{(N)}||_{L_2(-\alpha,0)} = 0,$$

$$\lim_{N\to\infty} ||u^* - u^{(N)}||_{C(0,T)} = 0,$$

$$\lim_{N\to\infty} ||y^* - y^{(N)}||_{C(0,1)\times C(-\alpha,T)} = 0,$$

$$\lim_{N\to\infty} |I(\hat{u}^*) - I^{(N)}(\hat{u}^{(N)})| = 0.$$

Justification. Let us define the operator

$$\hat{\mathcal{A}}^{(N)}\theta^{(N)}(.) = \Gamma_{3\times3}\theta^{(N)}(t) + \mathcal{A}^{(N)}\theta^{(N)}(.),$$

where $(\theta^{(N)}(t))' = (v^{(N)}(t), u^{(N)}(0), \xi^{(N)}(t)) \in L_2(-\alpha,0) \times R^1 \times L_2(0,T)$, *and* operator $\mathcal{A}^{(N)}$ is determined by the remaining members of equations left parts (21.10).

Since the operator $\hat{\mathcal{A}}^{(N)}$ is positively identified, the system (21.10) has a unique solution in space $C(-\alpha,0) \times R^1 \times L_2(0,T)$.

Let us denote the vectors of right-hand sides of (21.9) and (21.10) by $F(t,\varphi,\psi)$, $F^{(N)}(t,\varphi,\psi)$. Then, these systems can be written as operator equations

$$\begin{aligned}\Gamma_{3\times 3}\theta(t)+\mathscr{A}\theta(.)&=F(t,\varphi,\psi),\\ \Gamma_{3\times 3}\theta^{(N)}(t)+\mathscr{A}^{(N)}\theta^{(N)}(.)&=F^{(N)}(t,\varphi,\psi).\end{aligned} \tag{21.12}$$

Let us denote $\Delta^{(N)}\Theta(t)=\Theta(t)-\ \Theta^{(N)}(t)$, $\Delta^{(N)}\mathscr{A}(.)=\mathscr{A}(.)-\ \mathscr{A}^{(N)}(.)$, $\Delta^{(N)}F(t,\varphi,\psi)=\ F(t,\varphi,\psi)-F^{(N)}(t,\varphi,\psi)$. Then, discrepancy $\Delta^{(N)}\Theta(t)$ satisfies the operator equation

$$\Gamma_{3\times 3}\Delta^{(N)}\Theta(t)+\mathscr{A}\Delta^{(N)}\Theta(.)=\Delta^{(N)}F(t,\varphi,\psi)-\Delta^{(N)}\mathscr{A}\Theta^{(N)}(.). \tag{21.13}$$

The operator from the left-hand side of the Eq. (21.13) is positive defined. Then for solutions of this equation correct estimation

$$||\Delta^{(N)}\Theta(.)||_3\le C\,(||\Delta^{(N)}F(.,\varphi,\psi)||_3+|||\Delta^{(N)}\mathscr{A}|||_3||\Theta^{(N)}(.)||_3), \tag{21.14}$$

where $|||\mathscr{A}|||_3$ is the norm of operator $\mathscr{A}$.

From (21.12) and lemma in [6], it follows that numerical sequence $\{||\Theta^{(N)}(.)||_3\}$ coincides. Moreover,

$$\lim_{N\to\infty}||\Delta^{(N)}F(.,\varphi,\psi)||_3=0.$$

Indeed, the estimates imply inequality

$$\begin{aligned}||\Delta^{(N)}F(.,\varphi,\psi)||_3\le C\sum_{k=N+1}^{\infty}&(|g_{2k-1}|+|g_{2k}|)\,((|\varphi_{2k-1}|+|\varphi_{2k}|)\\ &\times(\frac{\hat{\alpha}}{\lambda_k\exp(\lambda_k^2T)}+\frac{\hat{\beta}_1}{\lambda_k}+\frac{\hat{\beta}_2}{\lambda_k^3})+(|\psi_{2k-1}|\,+|\psi_{2k}|)\frac{\hat{\alpha}}{\lambda_k}).\end{aligned}$$

The convergence of this sequence provides the required behavior of $\Delta^{(N)}F(t,\varphi,\psi)$.

Let us find the upper estimate for the norm of the operator $\mathscr{A}$. In accordance with [7],

$$|||\mathscr{A}|||_3=\sup_{\theta\ne 0}\frac{\langle\mathscr{A}\theta(.),\theta(.)\rangle_3}{||\Theta(.)||_3^2}.$$

Using (21.9), we find

$$\begin{aligned}\langle\mathscr{A}\theta(.),\theta(.)\rangle_3\le{}&||\mathscr{K}_1^{(1)}||_{L_2(-\alpha,0)\times L_2(-\alpha,0)}\,||v||^2_{L_2(-\alpha,0)}+2||\mathscr{K}_2^{(1)}||_{L_2(-\alpha,0)}|u(0)|\\ &\times||v||_{L_2(-\alpha,0)}+2||\mathscr{K}_3^{(1)}||_{L_2(-\alpha,0)\times L_2(0,T)}\,||\xi||_{L_2(0,T)}||v||_{L_2(-\alpha,0)}+|\mathscr{K}_2^{(2)}|u^2(0)\\ &+2||\mathscr{K}_3^{(2)}||_{L_2(0,T)}||\xi||_{L_2(0,T)}\,|u(0)|+||\mathscr{K}_3^{(3)}||_{L_2(0,T)\times L_2(0,T)}||\xi||^2_{L_2(0,T)}\end{aligned}$$

$$< (||\mathscr{K}_1^{(1)}||_{L_2(-\alpha,0)\times L_2(-\alpha,0)} + ||\mathscr{K}_2^{(1)}||_{L_2(-\alpha,0)} + ||\mathscr{K}_3^{(1)}||_{L_2(-\alpha,0)\times L_2(0,T)} + |\mathscr{K}_2^{(2)}| + ||\mathscr{K}_3^{(2)}||_{L_2(0,T)} + ||\mathscr{K}_3^{(3)}||_{L_2(0,T)\times L_2(0,T)})\ ||\Theta(.)||_3^2.$$

Hence, we find the required estimate

$$|||\mathscr{A}|||_3 < ||\mathscr{K}_1^{(1)}||_{L_2(-\alpha,0)\times L_2(-\alpha,0)} + ||\mathscr{K}_2^{(1)}||_{L_2(-\alpha,0)} + ||\mathscr{K}_3^{(1)}||_{L_2(-\alpha,0)\times L_2(0,T)} + |\mathscr{K}_2^{(2)}| + ||\mathscr{K}_3^{(2)}||_{L_2(0,T)} + ||\mathscr{K}_3^{(3)}||_{L_2(0,T)\times L_2(0,T)}. \tag{21.15}$$

From (21.15), we establish estimate

$$|||\Delta^{(N)}\mathscr{A}|||_3 < C(\hat{\alpha} + \hat{\beta}_1 + \hat{\beta}_2) \sum_{k=N+1}^{\infty} \frac{1}{\lambda_k^2},$$

that is,

$$\lim_{N\to\infty} |||\Delta^{(N)}\mathscr{A}|||_3 = 0.$$

Using the obtained boundary equality from (21.14), we get

$$\lim_{N\to\infty} ||\Delta^{(N)}\Theta(.)||_3 = 0,$$

which provides the first two equalities of theorem. Substituting the solutions of (21.10) in solutions of the boundary value problem, we find $y^{(N)}(x,t)$. Then, we find the following estimates

$$||y^*(x,t) - y^{(N)}(x,t)||_{C(0,1)\times C(-\alpha,T)} \le C[\sum_{k=N+1}^{\infty} (|\varphi_{2k-1}| + |\varphi_{2k}| + \frac{|g_{2k-1}| + |g_{2k}|}{\lambda_k} (||v^*||_{L_2(-\alpha,0)} + ||u^*||_{C(0,T)})) + (||v^* - v^{(N)}||_{L_2(-\alpha,0)} + ||u^* - u^{(N)}||_{C(0,T)}) \sum_{k=0}^{N} \frac{|g_{2k-1}| + |g_{2k}|}{\lambda_k})],$$

from which the equality of third theorems follows. Equality of last theorems is a consequence of the previous equalities.

Bounded Control

Let the optimal control can be found from

$$
\begin{aligned}
v^{(N)}(t) = -1, -1 + \int_{\bar{\xi}_1^{(N)}}^{0} \mathscr{K}_1^{(1,N)}(t,\tau)\, v^{(N)}(\tau)d\tau > \mathscr{M}_1^{(1,N)}(t,\varphi) \\
+ \mathscr{M}_2^{(1,N)}(t,\psi) + \int_{-\alpha}^{\bar{\xi}_1^{(N)}} \mathscr{K}_1^{(1,N)}(t,\tau)d\tau, t \in [-\alpha, \bar{\xi}_1^{(N)}); \\
v^{(N)}(t) + \int_{\bar{\xi}_1^{(N)}}^{0} \mathscr{K}_1^{(1,N)}(t,\tau)\, v^{(N)}(\tau)d\tau = \mathscr{M}_1^{(1,N)}(t,\varphi) + \mathscr{M}_2^{(1,N)}(t,\psi) \\
+ \int_{-\alpha}^{\bar{\xi}_1^{(N)}} \mathscr{K}_1^{(1,N)}(t,\tau)d\tau, |v^{(N)}(t)| < 1, t \in [\bar{\xi}_1^{(N)}, 0).
\end{aligned}
\tag{21.16}
$$

The number $\bar{\xi}_1^{(N)}$ defines as the solution of equation

$$
v^{(N)}(\bar{\xi}_1^{(N)}) = -1, \tag{21.17}
$$

where $v^{(N)}(t)$ is the solution of the equation from (21.16).

Suppose that an approximate control satisfies the conditions

$$
\begin{aligned}
\lim_{N\to\infty} v^{(N)}(t) = -1,\ t \in [-\alpha, \lim_{N\to\infty} \bar{\xi}_1^{(N)}); \\
|v^{(N)}(t)| < 1, t \in [\lim_{N\to\infty} \bar{\xi}_1^{(N)} + 0, 0).
\end{aligned}
\tag{21.18}
$$

Then, we prove the theorem.

Theorem 21.3 *Suppose that for optimal control problem (21.6), (21.2), (21.3) and (21.7) the conditions from [6] is performed. Then formulas (21.16) is approximate control for problem (21.6), (21.2), (21.3) and (21.7), that is,*

$$
\begin{aligned}
\lim_{N\to\infty} |\bar{\xi}_1 - \bar{\xi}_1^{(N)}| = 0, \\
\lim_{N\to\infty} |v^*(t) - v^{(N)}(t)| = 0, t \in [-\alpha, 0), \\
\lim_{N\to\infty} |y^*(x,t) - y^{(N)}(x,t)| = 0, x \in [0,1], t \in [-\alpha, T), \\
\lim_{N\to\infty} |I(v^*) - I(v^{(N)})| = 0.
\end{aligned}
\tag{21.19}
$$

Justification. With fixed $\bar{\xi}_1$, the equation from (21.16) has a unique solution in the space $C(\bar{\xi}_1, 0)$ in full and approximate cases.

From the uniqueness of solutions of equations (21.17) in full and approximate cases, it follows that

$$\bar{\xi}_1 = \lim_{N\to\infty} \bar{\xi}_1^{(N)}.$$

Suppose that $\bar{\xi}_1^{(N)} < \bar{\xi}_1$. Let us consider the difference $\Delta v^{(N)}(t) = v^*(t) - v^{(N)}(t)$. Then

$$\begin{aligned} \Delta v^{(N)}(t) &= 0, t \in [-\alpha, \bar{\xi}_1^{(N)}); \\ \Delta v^{(N)}(t) &= -1 - v^{(N)}(t), t \in [\bar{\xi}_1^{(N)}, \bar{\xi}_1); \\ \Delta v^{(N)}(t) &+ \int_{\bar{\xi}_1}^{0} \mathscr{K}_1^{(1)}(t,\tau)\, \Delta v^{(N)}(\tau) d\tau \\ &= \Delta \mathscr{M}_1^{(1,N)}(t,\varphi) + \Delta \mathscr{M}_2^{(1,N)}(t,\psi) \\ - \int_{\bar{\xi}_1}^{0} \mathscr{K}_1^{(1)}(t,\tau)\, v^{(N)}(\tau) d\tau &+ \int_{\bar{\xi}_1^{(N)}}^{0} \mathscr{K}_1^{(1,N)}(t,\tau)\, v^{(N)}(\tau) d\tau \\ + \int_{-\alpha}^{\bar{\xi}_1} \mathscr{K}_1^{(1)}(t,\tau) d\tau &- \int_{-\alpha}^{\bar{\xi}_1^{(N)}} \mathscr{K}_1^{(1,N)}(t,\tau) d\tau, t \in [\bar{\xi}_1, 0), \end{aligned} \tag{21.20}$$

where

$$\begin{aligned} \Delta \mathscr{M}_1^{(1,N)}(t,\varphi) &= \mathscr{M}_1^{(1)}(t,\varphi) - \mathscr{M}_1^{(1,N)}(t,\varphi), \\ \Delta \mathscr{M}_2^{(1,N)}(t,\varphi) &= \mathscr{M}_2^{(1)}(t,\psi) - \mathscr{M}_2^{(1,N)}(t,\psi). \end{aligned}$$

From (21.20), it follows that function $\Delta v^{(N)}(t)$ is continuous on $t \in [-\alpha, 0)$ and following inequality is correct.

$$\lim_{N\to\infty} \Delta v^{(N)}(t) = 0, t \in [-\alpha, 0). \tag{21.21}$$

Indeed, from (21.20) with $t \in (\bar{\xi}_1^{(N)}, \bar{\xi}_1)$, we get

$$\Delta v^{(N)}(t) = -\frac{dv^{(N)}(\Theta_1^{(N)})}{dt}(t - \bar{\xi}_1^{(N)}), \Theta_1^{(N)} \in (\bar{\xi}_1^{(N)}, \bar{\xi}_1). \tag{21.22}$$

From (21.16), it follows that the sequence $\{dv^{(N)}(\Theta_1^{(N)})/dt\}$ is limited when $N \to \infty$. Indeed, firstly, the formula (21.22) occurs when a function $dv^{(N)}(t)/dt$ is continuous. Let us find conditions on the functions, to ensure its continuity. For this purpose using the Eq. (21.16), we obtain the estimate

$$||\frac{d^2 v^{(N)}(.)}{dt^2}||_{L_2(\bar{\xi}_1^{(N)},0)} \le C\,(||\frac{\partial^2 \mathscr{K}_1^{(1,N)}(.,.)}{\partial t^2}||_{L_2(-\alpha,0)\times L_2(-\alpha,0)}$$

$$+\,||\frac{d^2\mathcal{M}_1^{(1,N)}(.,\varphi)}{dt^2}||_{L_2(-\alpha,0)} + ||\frac{d^2\mathcal{M}_2^{(1,N)}(.,\psi)}{dt^2}||_{L_2(-\alpha,0)}). \qquad (21.23)$$

We find the estimates of second derivatives from the right-hand side of inequality (21.23) using the definition of the relevant functions. From some N, we have:

$$||\frac{\partial^2\mathcal{K}_1^{(1,N)}(.,.)}{\partial t^2}||_{L_2(-\alpha,0)\times L_2(-\alpha,0)} \le \sum_{k=1}^{N} g_{2k-1}^2(\frac{C_1\hat{\alpha}}{\lambda_k} + C_2\hat{\beta}_1 + \frac{C_3\hat{\beta}_2}{\lambda_k})$$

$$+2\,|g_{2k-1}||g_{2k}|(\frac{C_4\hat{\alpha}}{\lambda_k} + C_5\hat{\beta}_1 + \frac{C_6\hat{\beta}_2}{\lambda_k}) + g_{2k}^2(\frac{C_7\hat{\alpha}}{\lambda_k^2} + C_8\hat{\beta}_1 + \frac{C_9\hat{\beta}_2}{\lambda_k^2})$$

$$\le \sum_{k=1}^{N}(g_{2k-1}^2 + |g_{2k-1}||g_{2k}| + g_{2k}^2),$$

$$||\frac{d^2\mathcal{M}_1^{(1,N)}(.,\varphi)}{dt^2}||_{L_2(-\alpha,0)} \le \sum_{k=1}^{N}|g_{2k-1}|(|\varphi_{2k-1}|(\frac{C_1\hat{\alpha}}{\lambda_k} + C_2\hat{\beta}_1\lambda_k + \frac{C_3\hat{\alpha}}{\lambda_k})$$

$$+|\varphi_{2k}|(\frac{C_4\hat{\alpha}}{\lambda_k} + C_5\hat{\beta}_1\lambda_k + \frac{C_6\hat{\alpha}}{\lambda_k})) + |g_{2k}|(|\varphi_{2k-1}|(\frac{C_7\hat{\alpha}}{\lambda_k^2} + C_8\hat{\beta}_1\lambda_k + \frac{C_9\hat{\alpha}}{\lambda_k^2})$$

$$+|\varphi_{2k}|(\frac{C_{10}\hat{\alpha}}{\lambda_k^2} + C_{11}\hat{\beta}_1\lambda_k + \frac{C_{12}\hat{\alpha}}{\lambda_k^2}))) \le C\sum_{k=1}^{N}\lambda_k(|g_{2k-1}| + |g_{2k}|)(|\varphi_{2k-1}| + |\varphi_{2k}|)$$

$$||\frac{d^2\mathcal{M}_2^{(1,N)}(.,\psi)}{dt^2}||_{L_2(-\alpha,0)} \le C\sum_{k=1}^{N}(\frac{|g_{2k-1}|}{\lambda_k}(|\psi_{2k-1}| + |\psi_{2k}|) + \frac{|g_{2k}||\psi_{2k}|}{\lambda_k^2}).$$

If the conditions from [6] is performed, then the following equality is correct.

$$\lim_{N\to\infty}\,\Delta v^{(N)}(t) \;=\; 0.$$

The equation from (21.20) is uniquely solvable in the space $C(\bar{\xi}_1, 0)$. Other equalities of theorem are proved analogous to Theorem 21.2.

21.3.2 Example of Calculations

Let $\varphi(x) = 2x^3 - 3x^2 + x$, $\psi(x) = 4.5 * x^2$, $g(x) = x$, and $\alpha = 0.5$, $T = 3$, $\gamma = 10$. Then numerically solving a system of integral equation (21.9), we find for different N the values of criteria I (Fig. 21.3).

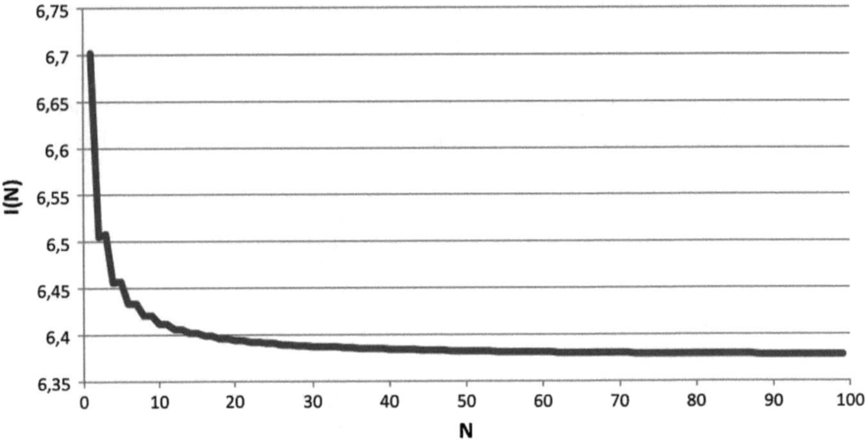

Fig. 21.3 Criterion values

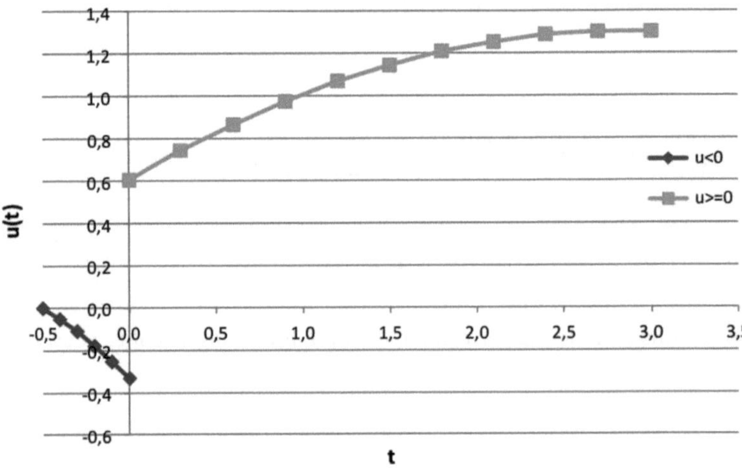

Fig. 21.4 Optimal control

For $N > 7$, criterion value does not change significantly, so we choose $N = 7$. Then, the found control is shown in Fig. 21.4, and criterion value is $I = 6, 42$.

References

1. Romanovskij, M.Ju., Romanovskij, Ju.M.: Introduction to Econophysics. The statistical and dynamic models. IKI, Moscow (2012) (Russian)
2. Egorov, A.I.: Fundamentals of Control Theory. Fizmatlit, Moscow (2004) (Russian)
3. Lotov, A.V.: Introduction to the Economic and Mathematical Modeling. Nauka, Moscow (1984)

4. Kapustyan, V.O., Lazarenko, I.S.: Problems with the minimum energy for parabolic equations with nonlocal boundary conditions. Bull. Dnipropetr. Univ. **17**(8), 47–60 (2009) (Russian)
5. Kapustyan, V.O., Pyshnograiev, I.O.: The conditions of existence and uniqueness of the solution of a parabolic-hyperbolic equation with nonlocal boundary conditions. Sci. News NTUU "KPI" **4**, 72–86 (2012) (Ukrainian)
6. Kapustyan, V.O., Pyshnograiev, I.O.: Distributed control with the general quadratic criterion in a special norm for systems described by parabolic-hyperbolic equations with nonlocal boundary conditions. Cybern. Syst. Anal. **51**(3), 438–447 (2015)
7. Berezanskij, JuM, Us, G.V., Sheftel, Z.G.: Functional Analysis. Vysha Shkola, Kyiv (1990)

Chapter 22
On Approximate Regulator in Linear-Quadratic Problem with Distributed Control and Rapidly Oscillating Parameters

Oleksiy V. Kapustyan and Alina V. Rusina

Abstract In this chapter, we have substantiated the approximate feedback formula for distributed optimal control in linear-quadratic problem with fast-oscillating coefficients.

22.1 Introduction

In optimal control theory of infinite-dimensional systems [7], one of the important problems is to obtain the optimal control in the feedback form (regulator). In the case of linear-quadratic model, this problem can be reduced to a set of finite-dimensional problems by expressing functions as a Fourier series and the exact formula of regulator can be found [1, 2]. If the original problem describes a process in micro-inhomogeneous media, its coefficients and the formula of optimal regulator are typically expressed via rapidly oscillating parameters. Since the numerical implementation of the optimal control problem with oscillatory coefficients is problematic, our goal was to justify the use of approximate optimal regulator, obtained by replacing rapidly oscillating coefficients with their averaged values.

The aforementioned problem was addressed in [6] for the controls of the form $g(x)u(t)$ under certain assumptions on the input data. In this chapter, we focus on

O.V. Kapustyan (✉) · A.V. Rusina
Taras Shevchenko National University of Kyiv,
Volodymyrska str., 64, Kyiv 01601, Ukraine
e-mail: kapustyanav@gmail.com

A.V. Rusina
e-mail: rusina.alina@gmail.com

O.V. Kapustyan
Institute for Applied System Analysis, National Technical University of Ukraine
"Kyiv Polytechnic Institute", Kyiv, Ukraine

V.A. Sadovnichiy and M.Z. Zgurovsky (eds.), *Advances in Dynamical Systems and Control*, Studies in Systems, Decision and Control 69,
DOI 10.1007/978-3-319-40673-2_22

the optimal stabilization of solutions of a parabolic equation with rapidly oscillating coefficients and bounded distributed control $u(t,x)$. The case of finite time interval was considered in [4]. Using the exact formula of optimal control in the feedback form, we justify the formula of approximate optimal regulator, in which rapidly oscillating coefficients are replaced with their homogenized values and infinite sums are replaced with finite.

22.2 Statement of the Problem

Let $\Omega \subset R^n$ be a bounded domain, $\varepsilon \in (0,1)$ be a small parameter, $\xi = (\xi_i)_{i=1}^{\infty} \in l_2$ be a fixed vector. In $Q = (0,+\infty) \times \Omega$, we consider the following optimal control problem:

$$\begin{cases} \frac{dy}{dt} = A^{\varepsilon} y(t) + u(t,x), & (t,x) \in Q, \\ y|_{\partial\Omega} = 0, \\ y|_{t=0} = y_0^{\varepsilon}, \end{cases} \tag{22.1}$$

$$J(y,u) = \int_0^{+\infty}\int_{\Omega} \left(y^2(t,x) + u^2(t,x)\right) dt dx \to \inf, \tag{22.2}$$

$$u \in U_{\varepsilon} = \left\{ v \in L^2(Q): \ \forall i \ge 1 \ \left| \int_{\Omega} v(t,x) X_i^{\varepsilon}(x) dx \right| \le \xi_i \ \text{ for a.e. } t > 0 \right\}, \tag{22.3}$$

where $A^{\varepsilon} = div(a^{\varepsilon}\nabla)$, $a^{\varepsilon}(x) = a(\frac{x}{\varepsilon})$, $a = ((a_{ij}))$ is a measurable symmetric periodic matrix which satisfies the condition of uniform ellipticity: $\exists\, \nu_1 > 0,\ \nu_2 > 0\ \forall \eta, x \in R^n$

$$\nu_1 \sum_{i=1}^n \eta_i^2 \le \sum_{i,j=1}^n a_{ij}(x)\eta_i\eta_j \le \nu_2 \sum_{i=1}^n \eta_i^2, \tag{22.4}$$

$\{X_i^{\varepsilon}\}$, $\{\lambda_i^{\varepsilon}\}$ are solutions of the following spectrum problem

$$\begin{cases} A^{\varepsilon} X_i^{\varepsilon} = -\lambda_i^{\varepsilon} X_i^{\varepsilon}, \\ X_i^{\varepsilon}|_{\partial\Omega} = 0, \end{cases} \tag{22.5}$$

$\{X_i^{\varepsilon}\} \subset H_0^1(\Omega)$ is an orthonormal basis in $L^2(\Omega)$, $0 < \lambda_1^{\varepsilon} \le \lambda_2^{\varepsilon} \le \cdots$, $\lambda_i^{\varepsilon} \to \infty$, $i \to \infty$.

It is known [7] that the optimal control problem (22.1)–(22.3) has the unique solution $\{y^{\varepsilon}, u^{\varepsilon}\}$ in $W \times L^2(Q)$, where

$$W = \{y \in L^2(0,+\infty; H_0^1(\Omega)) \mid \frac{dy}{dt} \in L^2(0,+\infty; H^{-1}(\Omega))\}.$$

The aim of this chapter was to justify the form of approximate optimal regulator $u = u[y]$ of the problem (22.1)–(22.3).

22.3 Main Results

Let $\|\cdot\|$ and $(\cdot,\cdot)$ denote a norm and a scalar product in $L^2(\Omega)$ respectively.

The problem (22.1)–(22.3) is equivalent to the set of one-dimensional optimal stabilization problems in $W \times L^2(Q)$

$$\begin{cases} \frac{d}{dt} y_i^\varepsilon(t) = -\lambda_i^\varepsilon y_i^\varepsilon + u_i^\varepsilon(t), \\ y_i^\varepsilon(0) = (y_0^\varepsilon, X_i^\varepsilon), \end{cases} \tag{22.6}$$

$$J(y_i^\varepsilon, u_i^\varepsilon) = \int\limits_0^{+\infty} \left(y_i^{\varepsilon 2}(t) + u_i^{\varepsilon 2}(t)\right) dt \to \inf, \tag{22.7}$$

$$|u_i^\varepsilon(t)| \le \xi_i \text{ a.e.}, \tag{22.8}$$

where $y^\varepsilon(t,x) = \sum\limits_{i=1}^{\infty} y_i^\varepsilon(t) X_i^\varepsilon(x)$, $u^\varepsilon(t,x) = \sum\limits_{i=1}^{\infty} u_i^\varepsilon(t) X_i^\varepsilon(x)$.

For each $i \ge 1$, the optimal control for the problem (22.6)–(22.8) in the feedback form can be defined from Bellman equation as follows [2]:

$$u_i^\varepsilon[y_i^\varepsilon(t)] = \begin{cases} \xi_i, \text{ if } y_i^\varepsilon(t) < -\frac{\xi_i}{R_i^\varepsilon}; \\ -R_i^\varepsilon y_i^\varepsilon(t), \text{ if } y_i^\varepsilon(t) \in [-\frac{\xi_i}{R_i^\varepsilon}, \frac{\xi_i}{R_i^\varepsilon}]; \\ -\xi_i, \text{ if } y_i^\varepsilon(t) > \frac{\xi_i}{R_i^\varepsilon}; \end{cases} \tag{22.9}$$

where $R_i^\varepsilon = -\lambda_i^\varepsilon + \sqrt{\lambda_i^{\varepsilon 2} + 1}$.

Then,

$$u^\varepsilon[y^\varepsilon(t)] = \sum_{i=1}^{\infty} u_i^\varepsilon\left[(y^\varepsilon(t), X_i^\varepsilon)\right] X_i^\varepsilon(x) \tag{22.10}$$

is the optimal control in the feedback form for the problem (22.1)–(22.3).

Let us analyze the formula (22.9).

If

$$\left|y_i^\varepsilon(0)\right| \le \frac{\xi_l}{R_i^\varepsilon}, \tag{22.11}$$

from (22.6), then the open-loop optimal control can be expressed as

$$u_i^\varepsilon(t) = -R_i^\varepsilon y_i^\varepsilon(0) e^{-\sqrt{\lambda_i^{\varepsilon 2}+1}t}; \tag{22.12}$$

thus, in this case

$$\forall t \geq 0 \;\; u_i^\varepsilon[y_i^\varepsilon(t)] = -R_i^\varepsilon y_i^\varepsilon(t). \tag{22.13}$$

If

$$\left|y_i^\varepsilon(0)\right| > \frac{\xi_i}{R_i^\varepsilon}, \tag{22.14}$$

it follows from (22.6) the optimal control is given by

$$u_i^\varepsilon[y_i^\varepsilon(t)] = \begin{cases} -\xi_i sign(y_i^\varepsilon(t)), \text{ if } t \in [0, t_i^\varepsilon]; \\ -R_i^\varepsilon y_i^\varepsilon(t), \text{ if } t > t_i^\varepsilon; \end{cases} \tag{22.15}$$

where $t_i^\varepsilon > 0$ is the unique solution of equation

$$y_i^\varepsilon(t) sign(y_i^\varepsilon(t)) = \frac{\xi_i}{R_i^\varepsilon}, \tag{22.16}$$

and y_i^ε is the solution of (22.6) with the control (22.15).

Let y_0^ε be a function such that (22.14) holds $\forall i \geq 1$. Then t_i^ε is determined by

$$t_i^\varepsilon = \frac{1}{\lambda_i^\varepsilon} \ln \left(\frac{R_i^\varepsilon}{\sqrt{\lambda_i^{\varepsilon 2} + 1}} \left(1 + \frac{\lambda_i^\varepsilon}{\xi_i} sign(y_i^\varepsilon(t)) y_i^\varepsilon(t)\right) e^{\lambda_i^\varepsilon t} \right), \tag{22.17}$$

where $y_i^\varepsilon(t), \; t \in [0, t_i^\varepsilon]$ is the solution of (22.6) with the control (22.15).

By the above assumptions, (22.10) can be expressed as

$$u^\varepsilon[y^\varepsilon(t)] = \sum_{i=1}^{\infty} \left(\alpha_i^\varepsilon(y^\varepsilon(t), X_i^\varepsilon) + \beta_i^\varepsilon\right) X_i^\varepsilon(x), \tag{22.18}$$

where

$$\alpha_i^\varepsilon = \alpha_i^\varepsilon(y^\varepsilon(t)) = \begin{cases} 0, \;\; t \in [0, t_i^\varepsilon], \\ -R_i^\varepsilon, \; t > t_i^\varepsilon, \end{cases}$$

$$\beta_i^\varepsilon = \beta_i^\varepsilon(y^\varepsilon(t)) = \begin{cases} -\xi_i sign(y_i^\varepsilon(t)), \;\; t \in [0, t_i^\varepsilon], \\ 0, \;\; t > t_i^\varepsilon, \end{cases}$$

$$t_i^\varepsilon = t_i^\varepsilon(y^\varepsilon(t)) = \frac{1}{\lambda_i^\varepsilon} \ln \left(\frac{R_i^\varepsilon}{\sqrt{\lambda_i^{\varepsilon 2} + 1}} \left(1 + \frac{\lambda_i^\varepsilon}{\xi_i} sign((y^\varepsilon(t), X_i^\varepsilon))(y^\varepsilon(t), X_i^\varepsilon)\right) e^{\lambda_i^\varepsilon t} \right), \quad t \in [0, t_i^\varepsilon],$$

where y^ε is the solution of (22.1) with the control (22.18).

Let us construct the law of approximated averaged synthesis which is based on formula (22.18).

Let a^0 be a constant homogenized matrix for $a\left(\frac{x}{\varepsilon}\right)$ [3], $A^0 = div(a^0\nabla)$, $\{X_i^0\}$, $\{\lambda_i^0\}$ are solutions of following spectrum problem:

$$\begin{cases} A^0X_i^0 = -\lambda_i^0 X_i^0, \\ X_i^0|_{\partial\Omega} = 0, \end{cases}$$

and besides, a spectrum of A^0 is simple, i.e.,

$$0 < \lambda_1^0 < \lambda_2^0 < \cdots < \lambda_k^0 <, \cdots, \lambda_i^0 \to \infty, i \to \infty. \tag{22.19}$$

With condition (22.19) we can claim [3]

$$\forall i \geq 1 \;\; \lambda_i^\varepsilon \to \lambda_i^0, \; X_i^\varepsilon \to X_i^0 \text{ in } L^2(\Omega) \text{ as } \varepsilon \to 0. \tag{22.20}$$

Assume

$$a^\varepsilon \xrightarrow{G} a^0, \; y_0^\varepsilon \to y_0 \;\text{ weekly in } L^2(\Omega) \text{ as } \varepsilon \to 0. \tag{22.21}$$

We note that a class of symmetric matrices which satisfies (22.4) is compact in the sense of G-convergence [3].

Denote $\forall i \geq 1$

$$t_i^0 = \frac{1}{\lambda_i^0} \ln\left(\frac{R_i^0}{\sqrt{\lambda_i^{02}+1}} \left(1 + \frac{\lambda_i^0}{\xi_i} sign(y_i^0) y_i^0 \right) \right), \tag{22.22}$$

$$\alpha_i^0 = \alpha_i^0(t) = \begin{cases} 0, & t \in [0, t_i^0], \\ -R_i^0, & t > t_i^0, \end{cases}$$

$$\beta_i^0 = \beta_i^0(t) = \begin{cases} -\xi_i sign(y_i^0), & t \in [0, t_i^0], \\ 0, & t > t_i^0, \end{cases}$$

where $R_i^0 = -\lambda_i^0 + \sqrt{\lambda_i^{02}+1}$, $y_i^0 = (y_0, X_i^0)$.

Let us define an approximate (parametric) regulator as

$$u_N^0[t, x, y_N^\varepsilon(t,x)] = \sum_{i=1}^{N} \left(\alpha_i^0 (y_N^\varepsilon(t), X_i^0) + \beta_i^0 \right) X_i^0(x), \tag{22.23}$$

where $y_N^\varepsilon(t,x)$ is the solution of the problem (22.1) with the control (22.23).

If for some $i \geq 1$ inequality (22.11) holds for y_0^ε, it also holds for y_i^0; thus, by setting $t_i^\varepsilon = t_i^0 = 0$, $\beta_i^\varepsilon \equiv \beta_i^0 \equiv 0$, $\alpha_i^\varepsilon \equiv -R_i^\varepsilon$, $\alpha_i^0 \equiv -R_i^0$, we can use formula (22.23) in this case as well.

Theorem 22.1 *Suppose that the assumptions (22.4), (22.19)–(22.21) are satisfied and*

$$\exists \gamma > 0 : \ \overline{\lim}_{i\to\infty} \xi_i e^{\gamma \lambda_i^0} > 0.$$

Then formula (22.23) determines the approximate optimal regulator of the problem (22.1)–(22.3), i.e., $\forall\ \eta > 0 \ \ \forall\ 0 < \tau < T \ \ \exists\ \overline{\varepsilon} \in (0, 1) \ \ \exists \overline{N} \geq 1$ *such that* $\forall \varepsilon \in (0, \overline{\varepsilon}) \ \ \forall N \geq \overline{N}$

$$\left\| u^{\varepsilon}[y^{\varepsilon}] - u_N^0[y_N^{\varepsilon}] \right\|_{L_2(Q)} < \eta, \tag{22.24}$$

$$\max_{t\in[\tau,T]} \left\| y^{\varepsilon}(t) - y_N^{\varepsilon}(t) \right\| < \eta, \tag{22.25}$$

$$\left| J(y^{\varepsilon}, u^{\varepsilon}) - J(y_N^{\varepsilon}, u_N^0[y_N^{\varepsilon}]) \right| < \eta, \tag{22.26}$$

where $\{y^{\varepsilon}, u^{\varepsilon}\}$ *is the optimal process of the problem (22.1)–(22.3), and* y_N^{ε} *is the solution of (22.1) with the control (22.23).*

Proof Consider

$$\begin{cases} \frac{\partial z}{\partial t} = A^{\varepsilon} z + u^0[t, x, z], \\ z|_{\partial\Omega} = 0, \\ z|_{t=0} = y_0^{\varepsilon}, \end{cases} \tag{22.27}$$

where

$$u^0[t, x, z] = \sum_{i=1}^{\infty} \left(\alpha_i^0(t) \left(z(t), X_i^0 \right) + \beta_i^0(t) \right) X_i^0(x).$$

Since $\left\| u^0[t, z] \right\|^2 \leq 2 \left(\|z\|^2 + \|\xi\|^2 \right)$, $\left\| u^0[t, z_1] - u^0[t, z_2] \right\| \leq \|z_1 - z_2\|$, the problem (22.27) has the unique solution $\forall T > 0$ [8] $z^{\varepsilon} = z^{\varepsilon}(t, x)$ in the class

$$W(0, T) = \left\{ y \in L^2(0, T; H_0^1(\Omega)) \mid \frac{dy}{dt} \in L^2(0, T; H^{-1}(\Omega)) \right\}$$

which is defined on $[0, +\infty)$. For a.e. $t > 0$ for z^{ε} the following estimate holds

$$\frac{1}{2}\frac{d}{dt} \|z^{\varepsilon}(t)\|^2 + \nu_1 \|z^{\varepsilon}(t)\|_{H_0^1}^2 \leq \sum_{i=1}^{\infty} \left(\alpha_i^0 (z^{\varepsilon}(t), X_i^0)^2 + \beta_i^0 (z^{\varepsilon}(t), X_i^0) \right). \tag{22.28}$$

Since $\alpha_i^0(t) \leq 0$, $\left| \beta_i^0(t) \right| \leq \xi_i$, with (22.28) for a.e. $t > 0$ we have:

$$\frac{1}{2}\frac{d}{dt} \|z^{\varepsilon}(t)\|^2 + \nu_1 \|z^{\varepsilon}(t)\|_{H_0^1}^2 \leq \|\xi\| \|z^{\varepsilon}(t)\|. \tag{22.29}$$

Thus, $\forall t > 0$

$$\|z^\varepsilon(t)\|^2 \le \|y_0^\varepsilon\|^2 e^{-\nu_1 \lambda t} + \frac{\|\xi\|^2}{\nu_1^2 \lambda^2}, \tag{22.30}$$

where a constant $\lambda > 0$ is from the Poincare inequality.

With (22.29) and (22.30) and the compactness lemma [8], we can derive the existence of $z \in W(0, T)$ such that subsequently

$$\begin{array}{l} z^\varepsilon \to z \text{ weekly in } L^2(0, T; H_0^1(\Omega)), \\ \frac{\partial z^\varepsilon}{\partial t} \to \frac{\partial z}{\partial t} \text{ weekly in } L^2(0, T; H^{-1}(\Omega)), \\ z^\varepsilon \to z \text{ in } L^2((0, T) \times \Omega) \text{ and a.e. in } (0, T) \times \Omega. \end{array} \tag{22.31}$$

Since

$$u^0[z^\varepsilon] \to u^0[z] \text{ in } L^2((0, T) \times \Omega), \tag{22.32}$$

$A^\varepsilon \xrightarrow{G} A^0$, from [5] we can claim z is the unique solution of (22.27) when $\varepsilon = 0$; moreover,

$$z^\varepsilon \to z \text{ in } C([\delta, T]; L^2(\Omega))\ \forall\, \delta > 0. \tag{22.33}$$

Thus, we can claim that the function z, which is defined on $[0, +\infty)$, is the solution of (22.27) when $\varepsilon = 0$. Besides (22.31)–(22.33) hold $\forall T > 0$.

Furthermore, denoting

$$J_T(y, u) = \int_0^T \int_\Omega \left(y^2(t, x) + u^2(t, x)\right) dt dx,$$

we have

$$J_T(z^\varepsilon, u^0[z^\varepsilon]) \to J_T(z, u^0[z]),\ \varepsilon \to 0. \tag{22.34}$$

Let us show (22.34) on an infinite interval.

If $\exists \gamma > 0 : \ \overline{\lim}_{i \to \infty} \xi_i e^{\gamma \lambda_i^0} > 0$, from (22.22), it follows that

$$\exists T > 0\ \ \forall i \ge 1\ t_i^0 \le T. \tag{22.35}$$

Then from (22.28) for all $s > 2T$, we obtain

$$2\nu_1 \int_{2T}^{s} \|z^\varepsilon(t)\|_{H_0^1}^2 d\tau \le \|z^\varepsilon(2T)\|^2 \le \|z^\varepsilon(T)\|^2 e^{-2\nu_1 \lambda T} \le$$

$$\le e^{-2\nu_1 \lambda T} \left(\|y_0^\varepsilon\|^2 e^{-\nu_1 \lambda t} + \frac{\|\xi\|^2}{\nu_1^2 \lambda^2} \right). \tag{22.36}$$

This gives a constant $C_1 > 0$ (which does not depend on $\varepsilon,\ T$) such that

$$\int_{2T}^{\infty} \|z^{\varepsilon}(t)\|^2 dt \le C_1 e^{-2\nu_1 \lambda T}. \tag{22.37}$$

Then from inequality

$$\forall s \ge T \int_{s}^{\infty} \|u^0[z^{\varepsilon}(t)]\|^2 dt \le \int_{s}^{\infty} \|z^{\varepsilon}(t)\|^2 dt$$

and with (22.31), we conclude

$$z^{\varepsilon} \to z \text{ in } L^2(Q), \tag{22.38}$$

$$u^0[z^{\varepsilon}] \to u^0[z] \text{ in } L^2(Q), \tag{22.39}$$

$$J(z^{\varepsilon}, u^0[z^{\varepsilon}]) \to J(z, u^0[z]). \tag{22.40}$$

Let us compare solutions y_N^{ε} and z^{ε}. For $\omega_N^{\varepsilon} = y_N^{\varepsilon} - z^{\varepsilon}$, we have

$$\begin{cases} \frac{\partial \omega_N^{\varepsilon}}{\partial t} = A^{\varepsilon} \omega_N^{\varepsilon} + \sum_{i=1}^{N} \alpha_i^0(t)(\omega_N^{\varepsilon}(t), X_i^0) X_i^0(x) + f_N^{\varepsilon}(t, x), \\ \omega_N^{\varepsilon}|_{\partial \Omega} = 0, \\ \omega_N^{\varepsilon}|_{t=0} = 0, \end{cases} \tag{22.41}$$

where

$$f_N^{\varepsilon}(t, x) = -\sum_{i=N+1}^{\infty} (\alpha_i^0(z^{\varepsilon}(t), X_i^0) + \beta_i^0) X_i^0(x).$$

For $\forall T > 0$ let us show that $\forall \eta > 0\ \exists N_1 \ge 1\ \exists \varepsilon_1 \in (0, 1)\ \forall N \ge N_1\ \forall \varepsilon \in (0, \varepsilon_1)$

$$\begin{aligned} &\sup_{t \in [0,T]} \|\omega_N^{\varepsilon}(t)\|^2 + \int_0^T \|u_N^0[y_N^{\varepsilon}] - u^0[z^{\varepsilon}]\|^2 dt < \eta, \\ &\left| J_T(y_N^{\varepsilon}, u_N^0[y_N^{\varepsilon}]) - J_T(z^{\varepsilon}, u^0[z^{\varepsilon}]) \right| < \eta. \end{aligned} \tag{22.42}$$

Applying Parseval's identity, we obtain for $t \in [0, T]$

$$\begin{aligned} \|f_N^{\varepsilon}(t)\|^2 &= \sum_{i=N+1}^{\infty} \left(\alpha_i^0(t)(z^{\varepsilon}(t), X_i^0) + \beta_i^0(t)\right)^2 \le 2 \sum_{i=N+1}^{\infty} (\alpha_i^0(t))^2 (z^{\varepsilon}(t), X_i^0)^2 + \\ &+ 2 \sum_{i=N+1}^{\infty} (\beta_i^0(t))^2 \le 2 \sum_{i=N+1}^{\infty} (z^{\varepsilon}(t), X_i^0)^2 + 2 \sum_{i=N+1}^{\infty} \xi_i^2 \le \\ &\le 4 \sum_{i=N+1}^{\infty} (z(t), X_i^0)^2 + 4\|z^{\varepsilon}(t) - z(t)\|^2 + 2 \sum_{i=N+1}^{\infty} \xi_i^2, \end{aligned} \tag{22.43}$$

where z is the solution of (22.27) when $\varepsilon = 0$.

From (22.41), we have the following estimates for a.e. $t \in (0, T)$:

$$\frac{1}{2}\frac{d}{dt}\|\omega_N^\varepsilon(t)\|^2 + \nu_1\|\omega_N^\varepsilon(t)\|_{H_0^1}^2 \le \|\omega_N^\varepsilon(t)\|^2 + \left(f_N^\varepsilon(t), \omega_N^\varepsilon(t)\right),$$

$$\frac{d}{dt}\|\omega_N^\varepsilon(t)\|^2 \le 3\|\omega_N^\varepsilon(t)\|^2 + \|f_N^\varepsilon(t)\|^2.$$

Then by Gronwall's lemma $\forall t \in [0, T]$

$$\|\omega_N^\varepsilon(t)\|^2 \le \int_0^T \|f_N^\varepsilon(t)\|^2 dt \cdot e^{3T}. \tag{22.44}$$

With (22.43) and (22.44), we conclude $\exists\, C_2 > 0\ \ \forall t \in [0, T]$

$$\|\omega_N^\varepsilon(t)\|^2 \le C_2\left(\int_0^T \sum_{i=N+1}^{\infty}(z(t), X_i^0)^2 dt + \int_0^T \|z^\varepsilon(t) - z(t)\|^2 dt + \sum_{i=N+1}^{\infty}\xi_i^2\right). \tag{22.45}$$

Since $\forall t \in [0, T]$ by Bessel's inequality

$$\sum_{i=N+1}^{\infty}(z(t), X_i^0)^2 \to 0 \text{ as } N \to \infty,$$

and taking into account $\left|\alpha_i^0(t)\right| \le 1$, $\left|\beta_i^0(t)\right| \le \xi_i$, with (22.28) and Gronwall's lemma we have

$$\left|\sum_{i=N+1}^{\infty}(z(t), X_i^0)^2\right| \le (\|y_0\|^2 + \|\xi\|^2 T)e^{3T}.$$

Thus, it follows from the Lebesgue theorem that the first term in (22.45) converges to 0 as $N \to \infty$. Then with (22.38)–(22.40) $\forall \eta_1 > 0$, $\exists N_1 \ge 1\ \exists \varepsilon_1 \in (0, 1)$ such that $\forall N \ge N_1\ \ \forall \varepsilon \in (0, \varepsilon_1)$

$$\sup_{t\in[0,T]}\|\omega_N^\varepsilon(t)\|^2 + \int_0^T \|u_N^0[t, y_N^\varepsilon] - u^0[t, z^\varepsilon]\|^2 dt < \eta. \tag{22.46}$$

By (22.46) and the following estimate

$$\left|J_T(y_N^\varepsilon, u_N^0[y_N^\varepsilon]) - J_T(z^\varepsilon, u^0[z^\varepsilon])\right| \le$$
$$\le C_3\left(\|y_N^\varepsilon(T) - z^\varepsilon(T)\| + \left(\int_0^T \|u_N^0[t, y_N^\varepsilon] - u^0[t, z^\varepsilon]\|^2 dt\right)^{\frac{1}{2}}\right) \tag{22.47}$$

we obtain (22.42).

For all $t \geq 0$, we have

$$\frac{1}{2}\frac{d}{dt}\|\omega_N^\varepsilon(t)\|^2 + \nu_1\|\omega_N^\varepsilon(t)\|^2_{H_0^1} \leq \left\|f_N^\varepsilon(t)\right\| \left\|\omega_N^\varepsilon(t)\right\|,$$

$$\|f_N^\varepsilon(t)\|^2 = \sum_{i=N+1}^{\infty}(\alpha_i^0(z^\varepsilon(t),X_i^0)+\beta_i^0)^2 \leq 2R_{N+1}^{02}\|z^\varepsilon(t)\|^2 + 2\sum_{i=N+1}^{\infty}\xi_i^2 \leq$$
$$(\text{by}(22.30)) \leq r_{N+1}^2 \to 0,\ N \to \infty.$$

Hence,

$$\frac{1}{2}\frac{d}{dt}\|\omega_N^\varepsilon(t)\|^2 + \nu_1\|\omega_N^\varepsilon(t)\|^2_{H_0^1} \leq r_{N+1}\|\omega_N^\varepsilon(t)\|.$$

It follows that $\forall t \geq 0$

$$\|\omega_N^\varepsilon(t)\|^2 \leq \|\omega_N^\varepsilon(0)\|^2 e^{-\lambda\nu_1 t} + e^{-\lambda\nu_1 t}\int_0^t \frac{r_{N+1}^2}{\lambda\nu_1} e^{\lambda\nu_1 s}ds \leq \frac{r_{N+1}^2}{\lambda^2\nu_1^2}.$$

For $t \geq T$

$$\|f_N^\varepsilon(t)\|^2 = \sum_{i=N+1}^{\infty} R_i^{0\,2}(z^\varepsilon(t),X_i^0)^2 \leq R_{N+1}^{0\,2}\|z^\varepsilon(t)\|^2.$$

Thus,

$$\lambda\nu_1\int_{2T}^{\infty}\|\omega_N^\varepsilon(t)\|^2dt \leq \|\omega_N^\varepsilon(2T)\|^2 + R_{N+1}^{0\,2}\int_{2T}^{\infty}\|z^\varepsilon(t)\|^2dt \leq \frac{r_{N+1}^2}{\lambda^2\nu_1^2} + R_{N+1}^{0\,2}C_4e^{-2\lambda\nu_1 T}. \tag{22.48}$$

Combining the above estimate with (22.42), we have $\forall\eta > 0\ \exists\bar{\varepsilon}\ \exists\bar{N}\ \forall\varepsilon \in (0,\bar{\varepsilon})\ \forall N \geq \bar{N}$

$$\|\omega_N^\varepsilon\|_{L^2(Q)} < \eta. \tag{22.49}$$

Since

$$\begin{gathered}\int_{2T}^{\infty}\|u_N^0[y_N^\varepsilon] - u^0[z^\varepsilon]\|^2dt = \int_{2T}^{\infty}\|\sum_{i=1}^{N}\alpha_i^0(\omega_N^\varepsilon(t),X_i^0)X_i^0 + f_N^\varepsilon\|^2dt \leq \\ R_{N+1}^{0\,2}\int_{2T}^{\infty}\|\omega_N^\varepsilon(t)\|^2dt + R_{N+1}^{0\,2}\int_{2T}^{\infty}\|z^\varepsilon(t)\|^2dt \leq C_5R_{N+1}^{0\,2},\end{gathered} \tag{22.50}$$

we obtain the required estimates on $[0,+\infty)$

It remains to prove that $\{y^\varepsilon,\ u^\varepsilon[y^\varepsilon]\}$ converges to $\{z,\ u^0[z]\}$ in the sense of (22.24)–(22.26). For y^ε the estimates (22.28)–(22.30) hold; hence, there exists $y \in W(0,T)$ such that $y^\varepsilon \to y$ in the sense of (22.31).

Let us prove that

$$u^\varepsilon[y^\varepsilon] \to u^0[y] \text{ in } L^2(0,T;L^2(\Omega)). \tag{22.51}$$

By (22.20) and (22.21) $\forall i \geq 1$ $t_i^\varepsilon \to t_i^0$ hence, for a.e. $t > 0$

$$\alpha_i^\varepsilon(t) \to \alpha_i^0(t),\ \beta_i^\varepsilon(t) \to \beta_i^0(t),\ \varepsilon \to 0. \tag{22.52}$$

Therefore, for a.e. (t,x)

$$\left(\alpha_i^\varepsilon(t)(y^\varepsilon(t),X_i^\varepsilon) + \beta_i^\varepsilon(t)\right)X_i^\varepsilon(x) \to \left(\alpha_i^0(t)(y(t),X_i^0) + \beta_i^0\right)X_i^0(x). \tag{22.53}$$

Then, from (22.30) applying the Lebesgue theorem $\forall M \geq 1$, we obtain

$$\sum_{i=1}^{M}\left(\alpha_i^\varepsilon(y^\varepsilon,X_i^\varepsilon) + \beta_i^\varepsilon\right)X_i^\varepsilon \to \sum_{i=1}^{M}\left(\alpha_i^0(y,X_i^0) + \beta_i^0\right)X_i^0 \text{ in } L^2(0,T;L^2(\Omega)),$$

$$\int_0^T\left\|\sum_{i=M+1}^{\infty}\left(\alpha_i^0(y,X_i^0) + \beta_i^0\right)X_i^0\right\|^2 dt = \int_0^T\sum_{i=M+1}^{\infty}\left(\alpha_i^0(y,X_i^0) + \beta_i^0\right)^2 dt \leq$$
$$2R_{M+1}^{0\,2}\int_0^T\|y(t)\|^2dt + 2T\sum_{i=M+1}^{\infty}\xi_i^2 \to 0,\ M \to \infty.$$

Since $\alpha_i^\varepsilon(y^\varepsilon,X_i^\varepsilon) + \beta_i^\varepsilon = u_i^\varepsilon[(y^\varepsilon,X_i^\varepsilon)] \in [-\xi_i,\xi_i]$ for a.e. t, it follows that

$$\left\|\sum_{i=M+1}^{\infty}\left(\alpha_i^\varepsilon(y^\varepsilon(t),X_i^\varepsilon) + \beta_i^\varepsilon\right)X_i^\varepsilon\right\|^2 \leq \sum_{i=M+1}^{\infty}\xi_i^2 \to 0,\ M \to \infty.$$

Thus, (22.51) holds; hence, $y \equiv z$ is the solution of (22.27) when $\varepsilon = 0$,

$$\begin{gathered} y^\varepsilon \to z \text{ in } C([\delta,T];L^2(\Omega)), \\ J_T(y^\varepsilon,u^\varepsilon[y^\varepsilon]) \to J_T(y,u^0[y]). \end{gathered} \tag{22.54}$$

Let us show that $u^\varepsilon[y^\varepsilon] \to u^0[y]$ in $L^2(Q)$. For this purpose, we use the fact that the process $\{y^\varepsilon, u^\varepsilon\}$ is optimal.

By Bellman's principle of optimality, we obtain inequality

$$\int_{2T}^{\infty}\left(\|y^\varepsilon(t)\|^2 + \|u^\varepsilon(t)\|^2\right)dt \leq \int_{2T}^{\infty}\|\tilde{y}^\varepsilon(t)\|^2dt, \tag{22.55}$$

where $\tilde{y}^{\varepsilon}$ is the solution of (22.1) on $[2T, +\infty)$ with control $u \equiv 0$ and initial condition $\tilde{y}^{\varepsilon}(2T) = y^{\varepsilon}(2T)$.

Then

$$\int_{2T}^{\infty} \|\tilde{y}^{\varepsilon}(t)\|^2 dt \leq \sum_{i=1}^{\infty} \frac{1}{2\lambda_i^{\varepsilon}} \left(y_i^{\varepsilon}(2T)\right)^2. \tag{22.56}$$

Hence, $\forall \eta > 0 \; \exists i_0 \geq 1 \; \forall \varepsilon \in (0, 1)$

$$\sum_{i=i_0+1}^{\infty} \frac{1}{2\lambda_i^{\varepsilon}} \left(y_i^{\varepsilon}(2T)\right)^2 < \frac{\eta}{2}. \tag{22.57}$$

On the other hand, $\exists \varepsilon_0 = \varepsilon_0(i_0) \in (0, 1)$ such that $\forall \varepsilon \in (0, \varepsilon_0) \; \forall i \in \overline{1, i_0} \; t_i^{\varepsilon} < T$. Then for $i \in \overline{1, i_0}$ on $[T, 2T]$

$$\frac{d}{dt} y_i^{\varepsilon}(t) = -(\lambda_i^{\varepsilon} + R_i^{\varepsilon}) y_i^{\varepsilon}(t).$$

Thus,

$$y_i^{\varepsilon}(2T) \leq y_i^{\varepsilon}(T) e^{-\lambda_i^{\varepsilon} T} \leq y_i^{\varepsilon}(T) e^{-\nu_1 \lambda T}.$$

It follows that for some constant $C_5 > 0$

$$\sum_{i=1}^{i_0} \frac{1}{2\lambda_i^{\varepsilon}} \left(y_i^{\varepsilon}(2T)\right)^2 \leq \frac{C_5}{2\lambda_1^0} e^{-2\nu_1 \lambda T} \|y^{\varepsilon}(T)\|^2. \tag{22.58}$$

With (22.57), (22.58), and (22.30), we derive

$$y^{\varepsilon} \to y \text{ in } L^2(Q).$$

In the same manner, when $\varepsilon = 0$ we can see that

$$u^{\varepsilon}[y^{\varepsilon}] \to u^0[y] \text{ in } L^2(Q)$$

and the proof is complete. □

References

1. Egorov, A.I.: Optimal Control by Linear Systems. Vyscha Shkola, Kyiv (1988)
2. Egorov, A.I., Mihailova, T.F.: Optimal control synthesis of heat process with bounded control. Part 1. Avtomatika **3**, 57–61 (1990)
3. Jikov, V.V., Kozlov, S.M., Oleynik, O.A.: Homogenization of Differential Operators and Integral Functions. Springer, Berlin (1994)

4. Kapustyan, O.V., Rusina, A.V.: Approximate synthesis of distributed bounded control for a parabolic problem with rapidly oscillating coefficients. Ukr. Math. J. **67**, 355–365 (2015)
5. Kapustyan, O.V., Shklyar, T.B.: Global attractor of a parabolic inclusion with nonautonomous main part. J. Math. Sci. **187**, 458–470 (2012)
6. Kapustyan, O.V., Kapustian, O.A., Sukretna, A.V.: Approximate Bounded Synthesis for Distributed Systems. LAP LAMBERT Academic Publishing, Saarbrucken (2013)
7. Lions, J.L.: Optimal Control of Systems Governed by Partial Differential Equations. Springer, New York (1971)
8. Sell, G.R., You, Y.: Dynamics of Evolutionary Equations. Springer, New York (2002)

Chapter 23
The Optimal Control Problem with Minimum Energy for One Nonlocal Distributed System

Olena A. Kapustian and Oleg K. Mazur

Abstract We obtain sufficient conditions for resolvability of optimal control problem with minimum energy on the solutions of parabolic equation with nonlocal boundary conditions in a circular sector.

23.1 Introduction

It is known, unlike the well-developed finite-dimensional case [1], the theory of optimal control problems with minimum energy for partial derivative equations is far from final construction. In particular, it is connected with infinite dimension of the moment problem in applying the Fourier method [2]. Significant progress in solving this problem has been achieved by applying the variational methods [2–4]. However, these methods are not effective for finding classic solution.

In this chapter we investigate the classic solvability of the energy minimization problem for parabolic equation in sectorial domain with non-local boundary conditions [5]. The elliptic case of such a problem has been considered in [6]. By using the biorthonormal basis systems of functions and the Fourier–Bessel series, we reduce the initial optimal control problem to the infinite-dimensional moment problem. In a class of stationary controls the structure of such moment problem allows an explicit

O.A. Kapustian (✉)
Taras Shevchenko National University of Kyiv, Volodymyrska str., 64, Kyiv 01601, Ukraine
e-mail: olena_kap@gmail.com

O.A. Kapustian
Institute for Applied System Analysis, National Technical University of Ukraine "Kyiv Polytechnic Institute", Kyiv, Ukraine

O.K. Mazur
National University of Food Technologies, Volodymyrska str., 68, Kyiv 01601, Ukraine
e-mail: okmazur@ukr.net

V.A. Sadovnichiy and M.Z. Zgurovsky (eds.), *Advances in Dynamical Systems and Control*, Studies in Systems, Decision and Control 69,
DOI 10.1007/978-3-319-40673-2_23

solving. This allows to substantiate classic solvability of the initial problem for a wide class of initial data.

23.2 Setting of the Problem

In domain $Q = (0, T) \times \Omega$, $\Omega = \{(r, \theta)|r \in (0, 1),\ \theta \in (0, \pi)\}$ we consider the problem: to find a state function $y = y(t, r, \theta)$ and control $u = u(r, \theta)$ such that

$$\begin{cases} \frac{\partial y}{\partial t} = \Delta y + q(t)u(r, \theta),\ (t, r, \theta) \in Q, \\ y(t, 1, \theta) = 0,\ t \in (0, T),\ \theta \in (0, \pi), \\ y(t, r, 0) = 0,\ t \in (0, T),\ r \in (0, 1), \\ \frac{\partial y}{\partial \theta}(t, r, 0) = \frac{\partial y}{\partial \theta}(t, r, \pi),\ t \in (0, T),\ r \in (0, 1), \end{cases} \tag{23.1}$$

$$y(0, r, \theta) = h(r, \theta), \tag{23.2}$$

$$y(T, r, \theta) = z(r, \theta), \tag{23.3}$$

$$J(u) = \int_0^1 r\|u(r)\|^2 dr \to \inf, \tag{23.4}$$

where $\Delta y := \frac{1}{r}\frac{\partial}{\partial r}(r\frac{\partial y}{\partial r}) + \frac{1}{r^2}\frac{\partial^2 y}{\partial \theta^2}$ be Laplace operator in polar coordinates, $q \in C([0, T])$, $h, z \in C(\overline{\Omega})$ are given functions, $\|\cdot\|$ is a norm in $L^2(0, \pi)$. Because of boundary conditions of the problem (23.1), we will use biorthonormal and complete in $L^2(0, \pi)$ systems of functions [5]

$$\Psi = \{\psi_0 = \frac{2}{\pi^2},\ \psi_{2n} = \frac{4}{\pi^2}(\pi - \theta)\sin 2n\theta,\ \psi_{2n-1} = \frac{4}{\pi^2}\cos 2n\theta\},$$

$$\Phi = \{\varphi_0 = \theta,\ \varphi_{2n} = \sin 2n\theta,\ \varphi_{2n-1} = \theta\cos 2n\theta\}.$$

The norm in $L^2(0, \pi)$ is given by the equality

$$\forall v \in L^2(0, \pi)\ \|v\| = \left(\sum_{n=0}^{\infty}(\int_0^{\pi} v(\theta)\psi_n(\theta)d\theta)^2\right)^{1/2}.$$

By solving the problem (23.1)–(23.4), the main difficulty is a nonlocality of the boundary conditions. Using Fourier method it does not allow to obtain a sequence of independent one-dimensional problems. In this paper for a fairly wide class of input data we were able to get the solution of mentioned infinite-dimensional moment problem and, thereby, to substantiate the classical solvability of the problem (23.1)–(23.4).

23.3 The Classical Solvability of the Problem (23.1)–(23.4)

For fixed control

$$u(r,\theta)=\sum_{n=0}^{\infty}u_n(r)\varphi_n(\theta) \tag{23.5}$$

we will find the solution of the problem (23.1), (23.2) in the form

$$y(t,r,\theta)=y_0(t,r)\varphi_0(\theta)+\sum_{n=1}^{\infty}\left(y_{2n-1}(t,r)\varphi_{2n-1}(\theta)+y_{2n}(t,r)\varphi_{2n}(\theta)\right), \tag{23.6}$$

where the functions $\{y_n(t,r)\}_{n=0}^{\infty}$ are defined from the following initial-boundary value problems in domain $\Pi=(0,T)\times(0,1)$:

$$\begin{cases}\frac{\partial y_0}{\partial t}=\frac{1}{r}\frac{\partial}{\partial r}(r\frac{\partial y_0}{\partial r})+q(t)u_0(r),\ (t,r)\in\Pi,\\ y_0(t,1)=0,\ t\in(0,T),\\ y_0(0,r)=h_0(r),\ r\in(0,1),\end{cases} \tag{23.7}$$

$$\begin{cases}\frac{\partial y_{2n-1}}{\partial t}=\frac{1}{r}\frac{\partial}{\partial r}(r\frac{\partial y_{2n-1}}{\partial r})-(\frac{2n}{r})^2y_{2n-1}+q(t)u_{2n-1}(r),\ (t,r)\in\Pi,\\ y_{2n-1}(t,1)=0,\ t\in(0,T),\\ y_{2n-1}(0,r)=h_{2n-1}(r),\ r\in(0,1),\end{cases} \tag{23.8}$$

$$\begin{cases}\frac{\partial y_{2n}}{\partial t}=\frac{1}{r}\frac{\partial}{\partial r}(r\frac{\partial y_{2n}}{\partial r})-(\frac{2n}{r})^2y_{2n}-\frac{4n}{r^2}y_{2n-1}+q(t)u_{2n}(r),\ (t,r)\in\Pi,\\ y_{2n}(t,1)=0,\ t\in(0,T),\\ y_{2n}(0,r)=h_{2n}(r),\ r\in(0,1),\end{cases} \tag{23.9}$$

where $\forall n\geq 0\ h_n=\int_0^{\pi}h(r,\theta)\cdot\psi_n(\theta)d\theta$.

Thus, the original optimal control problem (23.1)–(23.4) is reduced to the following one: among admissible pairs $\{u_n(r),y_n(t,r)\}_{n=0}^{\infty}$ of the problem (23.7)–(23.9) one should find such pairs, which minimize the cost functional

$$J=\int_0^1 ru_0^2(r)dr+\sum_{n=1}^{\infty}\int_0^1 r(u_{2n-1}^2(r)+u_{2n}^2(r))dr \tag{23.10}$$

and satisfy conditions

$$y_0(T,r)=z_0(r)=\frac{2}{\pi^2}\int_0^{\pi}z(r,\theta)d\theta, \tag{23.11}$$

$$\forall n\geq 1\ y_{2n-1}(T,r)=z_{2n-1}(r)=\frac{4}{\pi^2}\int_0^{\pi}z(r,\theta)\cos 2n\theta d\theta, \tag{23.12}$$

$$\forall n \geq 1 \; y_{2n}(T, r) = z_{2n}(r) = \frac{4}{\pi^2} \int_0^{\pi} z(r, \theta)(\pi - \theta) \sin 2n\theta d\theta. \qquad (23.13)$$

Herewith the optimal process $\{\tilde{u}_n, \tilde{y}_n\}_{n=0}^{\infty}$ should be such that the formula (23.5) defines a function $\tilde{u} \in C(\overline{\Omega})$ and the formula (23.6) defines a function $\tilde{y} \in C(\bar{Q})$ for which

$$\frac{\partial \tilde{y}}{\partial \theta} \in C(\bar{Q}), \; \frac{\partial \tilde{y}}{\partial t}, \frac{\partial^2 \tilde{y}}{\partial r^2}, \frac{\partial^2 \tilde{y}}{\partial r \partial \theta}, \frac{\partial^2 \tilde{y}}{\partial \theta^2} \in C(Q). \qquad (23.14)$$

To solve such a task we make additional assumptions on the initial data: let for some $N \geq 0$

$$\forall r \in [0, 1] \; h(r, \cdot), z(r, \cdot) \in L_N := \text{span}\{\varphi_0, \varphi_1, ..., \varphi_{2N}\}, \qquad (23.15)$$

i.e.

$$h(r, \theta) = \sum_{n=0}^{2N} h_n(r)\varphi_n(\theta), \; z(r, \theta) = \sum_{n=0}^{2N} z_n(r)\varphi_n(\theta).$$

From the condition (23.15) it follows that for $n > N$ $h_{2n-1} = h_{2n} = 0, \; z_{2n-1} = z_{2n} = 0$. So, the minimum of the cost functional

$$J_n := \int_0^1 r(u_{2n-1}^2(r) + u_{2n}^2(r))dr$$

is equal to zero and it is achieved at admissible in (23.8), (23.9), (23.12) and (23.13) pairs $\{u_{2n-1} = 0, \; y_{2n-1} = 0\}$, $\{u_{2n} = 0, \; y_{2n} = 0\}$.

In this manner, under the condition (23.15), on the optimal process the series (23.5), (23.6) contain only a finite number of nonzero members. This provides the condition (23.14) as soon as we find the solution of the problem (23.7)–(23.13) for $n = \overline{1, N}$.

To solve the problems (23.7)–(23.9), we will use Fourier–Bessel series [7, 8]

$$\sum_{m=1}^{\infty} A_m^{(n)}(f) J_n(\lambda_m^{(n)} r), \qquad (23.16)$$

where

$$A_m^{(n)}(f) = \frac{\int_0^1 rf(r)J_n(\lambda_m^{(n)} r)dr}{\int_0^1 rJ_n^2(\lambda_m^{(n)} r)dr},$$

J_n is Bessel function of order n, $n \geq 0$, $\{\lambda_m^{(n)}\}_{m=1}^{\infty}$ is a positive monotonically increasing sequence of the solutions of the equation $J_n(\lambda) = 0$. For these solutions the following asymptotical formula holds

$$\forall m \geq 1\ \lambda_m^{(n)} = \pi \cdot m + q + \frac{L \cdot \theta}{m}, \tag{23.17}$$

where $q = q(n) \in Z,\ L = L(n) > 0,\ \theta = \theta(m, n) \in [0, 1]$.

It is known [8] that if the function $f \in C^{p+1}([0, 1]),\ f^{(k)}(0) = f^{(k)}(1) = 0,\ k = \overline{0, p-1}$, then the series (23.16) is absolutely and uniformly convergent on [0, 1], and the following estimate holds

$$\exists C = C(n)\ \forall m \geq 1\ |A_m^{(n)}(f)| \leq \frac{C}{(\lambda_m^{(n)})^{p+\frac{1}{2}}} \cdot \max_{x \in [0,1]} \sum_{i=0}^{p+1} |f^{(i)}(x)|. \tag{23.18}$$

Taking into account the condition (23.15) and the fact that we do not need the exact values of the constants in the estimates like (23.18), further we will denote by the letter C the constants that depend only on $n \in \overline{0, N}$.

For $n \geq 0$ let us consider the problem

$$\begin{cases} \frac{\partial y}{\partial t} = \frac{1}{r}\frac{\partial}{\partial r}(r\frac{\partial y}{\partial r}) - (\frac{2n}{r})^2 y + \sum\limits_{m=1}^{\infty} C_m(t) J_{2n}(\lambda_m^{(2n)} r),\ (t, r) \in \Pi, \\ y(t, 1) = 0,\ y(0, r) = 0,\ t \in (0, T),\ r \in (0, 1), \end{cases} \tag{23.19}$$

where $\{C_m\}_{m=1}^{\infty} \subset C([0, T])$ be given functions, which satisfy an estimate

$$\forall m \geq 1\ \max_{t \in [0,T]} |C_m(t)| \leq \frac{C}{\lambda_m^{(2n)}}. \tag{23.20}$$

Taking into account estimates [7]

$$\forall n \geq 0\ \forall r \geq 0\ |J_n(r)| \leq 1, \tag{23.21}$$

$$\forall n \geq 0\ \forall r \leq 0\ |J_n(r)| \leq \frac{C}{\sqrt{r}} \tag{23.22}$$

and such recurrent formulas

$$J_0'(r) = -J_1(r),\ 2J_n'(r) = J_{n-1}(r) - -J_{n+1}(r),\ n \geq 1, \tag{23.23}$$

from the conditions (23.17) and (23.20), by the Weierstrass M-test, we get that the formula

$$y(t, r) = \sum_{m=1}^{\infty} \left(\int_0^t C_m(s) e^{-(\lambda_m^{(2n)})^2 (t-s)} ds \right) J_{2n}(\lambda_m^{(2n)} r) \tag{23.24}$$

defines the classical solution of the problem (23.19).

Then, by using conditions

$$h_0 \in C^2([0,1]),\ h_0(0) = h_0(1) = 0, \tag{23.25}$$

$$\forall n \geq 1\ h_{2n-1} \in C^2([0,1]),\ h_{2n-1}(0) = h_{2n-1}(1) = 0, \tag{23.26}$$

from the estimate (23.18) with fixed controls

$$u_0(r) = \sum_{m=1}^{\infty} u_m^{(0)} J_0(\lambda_m^{(0)} r),\ u_{2n-1}(r) = \sum_{m=1}^{\infty} u_m^{(2n-1)} \cdot J_{2n}(\lambda_m^{(2n)} r), n \geq 1,$$

where

$$|u_m^{(0)}| \leq \frac{C}{\lambda_m^{(0)}},\ |u_m^{(2n-1)}| \leq \frac{C}{\lambda_m^{(2n)}},\ m \geq 1, \tag{23.27}$$

the classical solutions of the problems (23.7) and (23.8) have the following form

$$\begin{aligned} y_0(t,r) = &\sum_{m=1}^{\infty} A_m^{(0)}(h_0) J_0(\lambda_m^{(0)} r) e^{-(\lambda_m^{(0)})^2 t} + \\ &+ \sum_{m=1}^{\infty} u_m^{(0)} \int_0^t q(s) e^{-(\lambda_m^{(0)})^2 (t-s)} ds \cdot J_0(\lambda_m^{(0)} r), \end{aligned} \tag{23.28}$$

$$\begin{aligned} y_{2n-1}(t,r) = &\sum_{m=1}^{\infty} A_m^{(2n)}(h_{2n-1}) J_{2n}(\lambda_m^{(2n)} r) e^{-(\lambda_m^{(2n)})^2 t} + \\ &+ \sum_{m=1}^{\infty} u_m^{(2n-1)} \int_0^t q(s) e^{-(\lambda_m^{(2n)})^2 (t-s)} ds \cdot J_{2n}(\lambda_m^{(2n)} r). \end{aligned} \tag{23.29}$$

Then, from the equalities (23.11), (23.12) we obtain the following relations

$$u_m^{(0)} \cdot \int_0^T q(t) e^{-(\lambda_m^{(0)})^2 (T-t)} dt = \alpha_m^{(0)} := A_m^{(0)}(z_0) - A_m^{(0)}(h_0) e^{-(\lambda_m^{(0)})^2 T}, \tag{23.30}$$

$$\begin{aligned} &u_m^{(2n-1)} \cdot \int_0^T q(t) e^{-(\lambda_m^{(2n)})^2 (T-t)} dt = \alpha_m^{(2n-1)} \\ &:= A_m^{(2n)}(z_{2n-1}) - A_m^{(2n)}(h_{2n-1}) e^{-(\lambda_m^{(2n)})^2 T}. \end{aligned} \tag{23.31}$$

From these relations the values of controls $u_m^{(0)}$, $u_m^{(2n-1)}$ are defined uniquely. Suppose the following conditions hold

$$\exists q_0 > 0\ \forall t \in [0,T]\ q(t) \geq q_0, \tag{23.32}$$

$$z_0 \in C^4([0,1]),\ z_0^{(k)}(0) = z_0^{(k)}(1) = 0,\ k = \overline{0,2}, \tag{23.33}$$

$$z_{2n-1} \in C^4([0,1])\ z_{2n-1}^{(k)}(0) = z_{2n-1}^{(k)}(1) = 0,\ k = \overline{0,2}. \tag{23.34}$$

Then, from (23.18), (23.30), (23.31), we get that for the controls

$$u_m^{(0)} = \alpha_m^{(0)} \left(\int_0^T q(t) e^{-(\lambda_m^{(0)})^2 (T-t)} dt \right)^{-1}, \tag{23.35}$$

$$u_m^{(2n-1)} = \alpha_m^{(2n-1)} \left(\int_0^T q(t) e^{-(\lambda_m^{(2n)})^2 (T-t)} dt \right)^{-1} \tag{23.36}$$

the following estimates hold

$$\forall m \geq 1 \ \ |u_m^{(0)}| \leq \frac{C}{(\lambda_m^{(0)})^{\frac{3}{2}}}, \ \ |u_m^{(2n-1)}| \leq \frac{C}{(\lambda_m^{(2n)})^{\frac{3}{2}}}, \tag{23.37}$$

and, in particular, from (23.37) it follows (23.27).

Therefore, if the function h satisfies the conditions (23.15), (23.25), (23.26) and the function z satisfies the conditions (23.15), (23.33), (23.34), the function q satisfies the condition (23.32), then formulas

$$u_0(r) = \sum_{m=1}^{\infty} u_m^{(0)} J_0(\lambda_m^{(0)} r), \ \ u_{2n-1}(r) = \sum_{m=1}^{\infty} u_m^{(2n-1)} J_{2n}(\lambda_m^{(2n)} r), \tag{23.38}$$

define the unique admissible controls in the problems (23.7), (23.11) and (23.8), (23.12), moreover, the estimate (23.37) holds for them.

To solve the problem (23.9), we introduce the functions

$$f_k^{(2n)}(r) := \frac{1}{r^2} J_{2n}(\lambda_k^{(2n)} r), \ \ k \geq 1, \ \ n \geq 0. \tag{23.39}$$

By using the recurrent formula [7]

$$2n \frac{J_n(r)}{r} = J_{n-1}(r) + J_{n+1}(r),$$

we can write each function from (23.39) in the form

$$f_k^{(2n)}(r) = (\lambda_k^{(2n)})^2 \left(a_n J_{2n+2}(\lambda_k^{(2n)} r) + b_n J_{2n}(\lambda_k^{(2n)} r) + c_n J_{2n-2}(\lambda_k^{(2n)} r) \right), \tag{23.40}$$

where the positive constants $a_n, \ b_n, c_n$ depend only on n.

At $n > 1$ the functions $f_k^{(2n)} \in C^2([0, 1])$, $f_k^{(2n)}(0) = f_k^{(2n)}(1) = 0$. So, from (23.18) and (23.40) we obtain

$$\forall m \geq 1 \ \ \forall k \geq 1 \ \ |A_m^{(2n)}(f_k^{(2n)})| \leq \frac{C(\lambda_k^{(2n)})^4}{(\lambda_m^{(2n)})^{\frac{3}{2}}}. \tag{23.41}$$

At $n = 1$ $f_k^{(2)}(1) = 0$, $f_k^{(2)}(0) = C_2 \cdot (\lambda_k^{(2)})^2$, so, from [8] we obtain

$$\forall m \geq 1 \ \forall k \geq 1 \ |A_m^{(2)}(f_k^{(2)})| \leq \frac{C(\lambda_k^{(2)})^4}{\lambda_m^{(2)}}. \tag{23.42}$$

Formally, the solution of the problem (23.9) is defined by the formula

$$\begin{aligned} y_{2n}(t,r) = &\sum_{m=1}^{\infty} A_m^{(2n)}(h_{2n}) J_{2n}(\lambda_m^{(2n)} r) e^{-(\lambda_m^{(2n)})^2 t} + \\ &+\sum_{m=1}^{\infty} u_m^{(2n)} \cdot \left(\int_0^t q(s) e^{-(\lambda_m^{(2n)})^2 (t-s)} ds\right) J_{2n}(\lambda_m^{(2n)} r) + \overline{y}(t,r), \end{aligned} \tag{23.43}$$

where $\overline{y}(t,r)$ is the solution of the problem (23.19) with

$$\begin{aligned} C_m(t) = &-4n \sum_{k=1}^{\infty} A_m^{(2n)}(f_k^{(2n)}) \cdot A_k^{(2n)}(h_{2n-1}) e^{-(\lambda_k^{(2n)})^2 t} - \\ &-4n \sum_{k=1}^{\infty} A_m^{(2n)}(f_k^{(2n)}) \cdot u_k^{(2n-1)} \cdot \int_0^t q(s) e^{-(\lambda_k^{(2n)})^2 (t-s)} ds. \end{aligned} \tag{23.44}$$

Let us strengthen the conditions on the functions h and z to the following ones:

$$h_{2n-1},\ z_{2n-1} \in C^6([0,1]),\ h_{2n-1}^{(k)}(0) = z_{2n-1}^{(k)}(0) = h_{2n-1}^{(k)}(1) = z_{2n-1}^{(k)}(1) = 0,\ k = \overline{0,5}. \tag{23.45}$$

Then, from (23.18)

$$\forall m \geq 1 \ |A_k^{(2n)}(h_{2n-1})| \leq \frac{C}{(\lambda_k^{(2n)})^{5+\frac{1}{2}}}, \tag{23.46}$$

and for the set $\{u_m^{(2n-1)}\}_{m=1}^{\infty}$, defined by the formula (23.36), the following formula holds

$$\forall m \geq 1 \ |u_m^{(2n-1)}| \leq \frac{C}{(\lambda_m^{(2n)})^{3+\frac{1}{2}}}. \tag{23.47}$$

The estimates (23.41), (23.42), (23.46), (23.47) guarantee the condition (23.20) for the functions $\{C_m(t)\}_{m=1}^{\infty}$. So, if h_{2n} satisfies (23.26), $u_m^{(2n)}$ satisfies (23.27), then the formula (23.43) defines the classical solution (23.9).

From the equality (23.13) we obtain a relation

$$\begin{aligned} &u_m^{(2n)} \cdot \int_0^T q(t) e^{-(\lambda_m^{(2n)})^2 (T-t)} dt = \alpha_m^{(2n)} := A_m^{(2n)}(z_{2n}) - A_m^{(2n)}(h_{2n}) e^{-(\lambda_m^{(2n)})^2 T} + \\ &+4n \sum_{k=1}^{\infty} A_m^{(2n)}(f_k^{(2n)}) \cdot A_k^{(2n)}(h_{2n-1}) e^{-(\lambda_m^{(2n)})^2 T} \cdot \int_0^T e^{((\lambda_m^{(2n)})^2 - (\lambda_k^{(2n)})^2) t} dt + \\ &+4n \sum_{k=1}^{\infty} A_m^{(2n)}(f_k^{(2n)}) \cdot u_k^{(2n-1)} \cdot e^{-(\lambda_m^{(2n)})^2 T} \cdot \int_0^T \int_0^t e^{(\lambda_k^{(2n)})^2 s} \cdot e^{((\lambda_m^{(2n)})^2 - (\lambda_k^{(2n)})^2) t} ds dt. \end{aligned} \tag{23.48}$$

We need to show that the set of functions $\{u_m^{(2n)}\}_{m=1}^{\infty}$, defined by (23.48), satisfies the estimate (23.27), i.e.

$$\forall m \geq 1 \ \ |u_m^{(2n)}| \leq \frac{C}{\lambda_m^{(2n)}}. \tag{23.49}$$

Due to (23.32) it's sufficient to prove that

$$\forall m \geq 1 \ \ |\alpha_m^{(2n)}| \leq \frac{C}{(\lambda_m^{(2n)})^3}. \tag{23.50}$$

Let the following condition holds

$$z_{2n} \in C^4([0, 1]), \ \ z_{2n}^{(k)}(0) = z_{2n}^{(k)}(1) = 0, \ \ k = \overline{0, 2}. \tag{23.51}$$

Hence $A_m^{(2n)}(z_{2n})$ satisfies the inequality (23.50). Since h_{2n} satisfies (23.26), i.e.

$$h_{2n} \in C^2([0, 1]), \ \ h_{2n}(0) = h_{2n}(1) = 0, \tag{23.52}$$

then $A_m^{(2n)}(h_{2n}) \cdot e^{-(\lambda_m^{(2n)})^2 T}$ satisfies (23.50) also.
Because of inequalities (23.42), (23.46) we get

$$\begin{aligned} &|4n \sum_{k=1}^{\infty} A_m^{(2n)}(f_k^{(2n)}) \cdot A_k^{(2n)}(h_{2n-1}) \cdot e^{-(\lambda_m^{(2n)})^2 T} \cdot \int_0^T e^{((\lambda_m^{(2n)})^2 - (\lambda_k^{(2n)})^2)t} dt| \leq \\ &\leq C \cdot \sum_{k=1}^{\infty} \frac{1}{\lambda_m^{(2n)}} \cdot \frac{1}{(\lambda_k^{(2n)})^{\frac{3}{2}}} \cdot e^{-(\lambda_m^{(2n)})^2 T} \cdot \frac{1}{(\lambda_m^{(2n)})^2} e^{(\lambda_m^{(2n)})^2 T} \leq \frac{C}{(\lambda_m^{(2n)})^3}. \end{aligned} \tag{23.53}$$

From (23.42), (23.47) we obtain

$$\begin{aligned} &|4n \sum_{k=1}^{\infty} A_m^{(2n)}(f_k^{(2n)}) \cdot u_k^{(2n-1)} \cdot e^{-(\lambda_m^{(2n)})^2 T} \cdot \int_0^T e^{(\lambda_m^{(2n)})^2 t} \cdot \left(e^{-(\lambda_k^{(2n)})^2 t} \cdot \int_0^t e^{(\lambda_k^{(2n)})^2 s} ds \right) dt| \leq \\ &\leq C \cdot \sum_{k=1}^{\infty} \frac{(\lambda_k^{(2n)})^{\frac{1}{2}}}{\lambda_m^{(2n)}} \cdot e^{-(\lambda_m^{(2n)})^2 T} \cdot \frac{1}{(\lambda_m^{(2n)})^2} \cdot \frac{1}{(\lambda_k^{(2n)})^2} \cdot e^{(\lambda_m^{(2n)})^2 T} \leq \frac{C}{(\lambda_m^{(2n)})^3}. \end{aligned} \tag{23.54}$$

Thus, the above mentioned arguments have proved the following theorem.

23.4 The Main Result

The main result of this paper has the following formulation:

Theorem 1 *Let in the problem (23.1)–(23.4) the following conditions hold*

$$q \in C([0, 1]), \ \ \exists q_0 > 0 \ \ \forall t \in [0, T] \ \ q(t) \geq q_0, \tag{23.55}$$

for some $N \geq 0$

$$h(r,\theta)=\sum_{n=0}^{2N} h_n(r)\varphi_n(\theta),\ z(r,\theta)=\sum_{n=0}^{2N} z_n(r)\varphi_n(\theta), \tag{23.56}$$

$$h_0 \in C^2([0,1]),\ \ h_0(0)=h_0(1)=0,\ k=\overline{0,2}, \tag{23.57}$$

$$z_0 \in C^4([0,1]),\ \ z_0^{(k)}(0)=z_0^{(k)}(1)=0,\ k=\overline{0,2}, \tag{23.58}$$

$$\forall n=\overline{1,N}\ \ h_{2n} \in C^2([0,1]),\ \ h_{2n}(0)=h_{2n}(1)=0,\ k=\overline{0,2}, \tag{23.59}$$

$$\forall n=\overline{1,N}\ \ z_{2n} \in C^4([0,1]),\ z_{2n}^{(k)}(0)=z_{2n}^{(k)}(1)=0,\ k=\overline{0,2}, \tag{23.60}$$

$$h_{2n-1} \in C^6([0,1]),\ \ h_{2n-1}^{(k)}(0)=h_{2n-1}^{(k)}(1)=0,\ k=\overline{0,4}, \tag{23.61}$$

$$z_{2n-1} \in C^6([0,1]),\ \ z_{2n-1}^{(k)}(0)=z_{2n-1}^{(k)}(1)=0,\ k=\overline{0,4}. \tag{23.62}$$

Then the optimal control problem (23.1)–(23.4) has the unique classical solution

$$\overline{u}(r,\theta)=\sum_{n=0}^{2N} \overline{u}_n(r)\varphi_n(\theta),\ \overline{y}(t,r,\theta)=\sum_{n=0}^{2N} \overline{y}_n(t,r)\varphi_n(\theta),$$

where $\{\overline{u}_n, \overline{y}_n\}_{n=0}^{2N}$ *can be found from the the problems (23.7)–(23.9) and the conditions (23.11)–(23.13).*

23.5 Conclusion

In this chapter, in the class of stationary controls we prove the classical solvability of the optimal control problem with minimum energy for parabolic equation with nonlocal boundary conditions in the sectorial two-dimensional domain.

References

1. Krasovskiy, N.N.: Theory of Control of Motion. Nauka, Moscow (1968)
2. Egorov, A.I.: Optimal control in heat and diffusion processes. Nauka, Moscow (1978)
3. Ivanenko, V.I., Melnik, V.S.: Variational Methods in Control Problems for Systems with Distributed Parameters. Naukova Dumka, Kyiv (1988)
4. Nakonechnyi, O.G.: Optimal Control and Estimation for Partial Differential Equations. Vydavnytstvo Kyivskogo universitetu, Kyiv (2004)

5. Moiseev, E.I., Ambarzumyan, V.E.: About resolvability of non-local boundary-value problem with equality of fluxes. Differential equations. **46**(5), 718–725 (2010)
6. Kapustyan, V.O., Kapustian, O.A., Mazur, O.K.: Problem of optimal control for the Poisson equation with nonlocal boundary conditions. Journal of Mathematical Sciences **201**(3), 325–334 (2014)
7. Watson, G.N.: Theory of Bessel Functions. University Press, Cambridge (1945)
8. Scherberg, M.G.: The degree of convergence of a series of Bessel functions. Trans. Amer. Math. Soc. **35**, 172–183 (1933)

Chapter 24
Optimality Conditions for L^1-Control in Coefficients of a Degenerate Nonlinear Elliptic Equation

Peter I. Kogut and Olha P. Kupenko

Abstract In this article, we study an optimal control problem for a nonlinear elliptic equation of p-Laplace type with a coefficient in the leading order of differentiation taken as control in $L^1(\Omega)$. We allow such controls to vanish on zero Lebesgue sets. As a result, we deal with degenerate elliptic Dirichlet problems that can exhibit the Lavrentiev phenomenon and non-uniqueness of weak solutions. Moreover, the non-differentiability of the term $|\nabla y|^{p-2}\nabla y$ at 0 implying the non-differentiability of the state $y(u)$ with respect to the control necessitates refined concepts in order to derive optimality conditions.

24.1 Introduction

The main goal of this paper is to derive optimality conditions to the following optimal control problem

$$\text{Minimize } \left\{ I(u,y) = \int_{\Omega} \left[|y - y_d|^p + |\nabla y|^p u(x) \right] dx \right\} \tag{24.1}$$

P.I. Kogut
Department of Differential Equations, Dnipropetrovsk National University, Gagarin Av., 72, Dnipropetrovsk 49010, Ukraine
e-mail: p.kogut@i.ua

O.P. Kupenko (✉)
Department of System Analysis and Control, National Mining University, Yavornitskyi Av., 19, Dnipro 49005, Ukraine
e-mail: kogut_olga@bk.ru

O.P. Kupenko
Institute for Applied and System Analysis of National Technical University of Ukraine "Kyiv Polytechnic Institute", Peremogy Av., 37, Building 35, Kyiv 03056, Ukraine

V.A. Sadovnichiy and M.Z. Zgurovsky (eds.), *Advances in Dynamical Systems and Control*, Studies in Systems, Decision and Control 69,
DOI 10.1007/978-3-319-40673-2_24

subject to the constraints

$$u \in \mathfrak{A}_{ad} \subset BV(\Omega), \quad y \in W_0^{1,1}(\Omega), \tag{24.2}$$

$$-\operatorname{div}\left(u(x)|\nabla y|^{p-2}\nabla y\right) = f \quad \text{in } \Omega, \tag{24.3}$$

$$y = 0 \quad \text{on } \partial\Omega, \tag{24.4}$$

where $\mathfrak{A}_{ad}$ is a class of admissible controls, $\Omega \subset \mathbb{R}^N$ is a bounded open domain with a Lipschitz boundary, $f \in L^q(\Omega)$, $y_d \in L^p(\Omega)$, $p \geq 2$, $q = p/(p-1)$, and $N \geq 2$. We consider below some special description of the class of admissible controls $\mathfrak{A}_{ad}$. Namely, we suppose that the controls are L^1-integrable functions such that $u(x) > 0$ for a. e. $x \in \Omega$ and $u + u^{-1/(p-1)} \in L^1(\Omega)$. The set $\mathfrak{A}_{ad}$ has an empty L^1-topological interior. We give the precise definition of such controls in Sect. 24.2 and show in Sect. 24.4 that the optimal control problem (24.1)–(24.4) admits a solution. We point out that this problem can exhibit the Lavrentiev phenomenon and non-uniqueness of weak solutions (see, for instance, [7, 33, 34]). At present, there are many different concepts of solutions for degenerate partial differential equations: weak solutions, H-solutions, W-solutions, variational solutions, T-solutions, shift T-solutions, and others. In general, the mapping

$$\text{admissible control} \quad \mapsto \quad \text{the corresponding solution of BVP}$$

is multivalued. As a result, the corresponding optimal control problem can be stated in different formats according to the space-setting. We study the optimal control problem mentioned above in the class of weak solutions in the sense of Minty. Using the direct method in the calculus of variations, we discuss existence of solutions to this optimal control problem.

Optimal control in coefficients for partial differential equations is a classical subject initiated by Lurie [26]. Murat [28] showed examples of non-existence for such problems. Since the range of OCPs in coefficients is very wide, including optimal shape design problems, problems in structural mechanics, and many others, this topic has been widely studied by many authors. In particular, it leads to the possibility to optimize material properties what are extremely important for material sciences. However, most of the results and methods rely on linear PDEs, while only very few articles deal with nonlinear problems (see O. Kogut [8] and P. Kogut and Leugering [17]). Another point of interest is degeneration in the coefficients which is typically avoided by assuming lower bounds on the coefficients. Even though numerous articles are devoted to variational and non-variational approaches to problems related to (24.3) and (24.4), only few deal with optimal control problems for degenerate partial differential equations (see, for example, [3, 4, 17, 19]). In Kogut and Leugering [18] and in Kupenko and Manzo [23, 24], this problem has been considered in the context of linear problems. The nonlinear case was considered in [9, 20, 21].

The paper is organized as follows. In Sect. 24.2, we give some preliminaries and prescribe the class of admissible controls to problem (24.1)–(24.4). In Sects. 24.3 and 24.4, we give a precise statement of optimal control problem (24.1)–(24.4) and show existence of a solution. The aim of Sect. 24.5 is to discuss the so-called directional stability properties of weighted Sobolev spaces and their application to the study of asymptotic behavior of minimizing sequences. In Sect. 24.6, we provide results concerning the differentiability properties of the Lagrange functional associated with problem (24.1)–(24.4) and show that it admits a Gâteaux derivative with respect to so-called non-degenerate directions $h \in W_0^{1,p}(\Omega, u\,dx)$ in weighted Sobolev spaces. In Sect. 24.7, we discuss the formal approach in deriving first-order optimality conditions for optimal control problem (24.1)–(24.4). In order to derive an optimality system, we apply the Lagrange principle. It is well known that the proof of this principle is different for different classes of optimal control problems (see, for instance, [5, 6, 20]). The complexity of this procedure significantly depends on the form of the extremal problem under consideration. In this article, we deal with the case when we cannot apply the well-known classical results (see, for instance, [13, 16]), because for a given distribution $f \in L^q(\Omega)$, the mapping $u \mapsto y(u)$ is not Fréchet differentiable on the class of admissible controls, in general, and the class $\mathfrak{A}_{ad}$ has an empty topological interior. With that in mind, we apply an indirect approach to derive optimality conditions which is based on the notion of a quasi-adjoint state ψ_ε to an optimal solution $y_0 \in W_0^{1,p}(\Omega, u_0\,dx)$, first proposed for non-degenerate linear problems by Serovajskiy [31]). In order to derive optimality conditions in the framework of more appropriate assumptions, we provide in Sect. 24.8 the analysis for the corresponding variational problem for ψ_θ and describe the asymptotic behavior of its solutions as the parameter θ tends to zero. As a result, we give sufficient conditions in order to show that the optimality system for the original problem can be recovered in an explicit form. In Sect. 24.9, following the well-known Hardy–Poincaré inequality, we study the well-posedness of variational problem for the adjoint system and show that the adjoint state, in spite of possible degeneration, can be defined in a unique way.

24.2 Notation and Preliminaries

Let Ω be a bounded open subset of $\mathbb{R}^N$ ($N \geq 1$) with a Lipschitz boundary. Let $\Gamma \subset \Omega$ be a manifold of positive $(N-1)$-dimensional measure. Let $C_0^\infty(\mathbb{R}^N; \Gamma) = \{\varphi \in C_0^\infty(\mathbb{R}^N) \,:\, \varphi = 0 \text{ on } \Gamma\}$. We define the Banach space $H_0^1(\Omega; \Gamma)$ as the closure of $C_0^\infty(\mathbb{R}^N; \Gamma)$ with respect to the norm $\|y\| = \left(\int_\Omega \|\nabla y\|^2_{\mathbb{R}^N}\,dx\right)^{1/2}$. Let $H^{-1}(\Omega; \Gamma)$ be the dual space to $H_0^1(\Omega; \Gamma)$. For any subset $E \subset \Omega$, we denote by $|E|$ its N-dimensional Lebesgue measure $\mathscr{L}^N(E)$. Let $u : \Omega \to \mathbb{R}$ be an integrable function on Ω such that $u(x) \geq 0$ for a. a. $x \in \Omega$. Then, u gives rise to a measure on the measurable subsets of Ω through integration: $u(E) = \int_E u\,dx$ for measurable sets $E \subset \Omega$. Let p be a real number such that $2 \leq p < \infty$. We will use the standard notation $L^p(\Omega, u\,dx)$ for the set of measurable functions f on Ω such that $\|f\|_{L^p(\Omega, u\,dx)} = \left(\int_\Omega |f|^p u\,dx\right)^{1/p} < +\infty$. We say that a function $u : \Omega \to \mathbb{R}_+$ is a weight on Ω if

$$u(x) > 0 \quad \text{a.e. in } \Omega \quad \text{and} \quad u + u^{-1/(p-1)} \in L^1(\Omega), \tag{24.5}$$

Note that in this case the elements of $L^p(\Omega, u\,dx)$ are Lebesgue integrable on Ω. To each weight function u, we may associate two weighted Sobolev spaces $W_u = W_0^{1,p}(\Omega, u\,dx)$ and $H_u = H_0^{1,p}(\Omega, u\,dx)$, where W_u is the set of functions $y \in W_0^{1,1}(\Omega)$ for which the norm

$$\|y\|_{1,p,u} = \Big(\int_\Omega \big(y^p + u\,|\nabla y|^p\big)\,dx\Big)^{1/p} \tag{24.6}$$

is finite, and H_u is the closure of $C_0^\infty(\Omega)$ in W_u. Note that due to the estimates

$$\int_\Omega |y|\,dx \le \Big(\int_\Omega |y|^p\,dx\Big)^{1/p} |\Omega|^{p/(p-1)} \le C\|y\|_{1,p,u}, \tag{24.7}$$

$$\int_\Omega \sum_{i=1}^N \left|\frac{\partial y}{\partial x_i}\right| dx \le C_1 \Big(\int_\Omega u|\nabla y|^p\,dx\Big)^{1/p} \Big(\int_\Omega u^{-1/(p-1)}\,dx\Big)^{p/p-1} \le C\|y\|_{1,p,u}, \tag{24.8}$$

where the space W_u is complete with respect to the norm $\|\cdot\|_{1,p,u}$. It is clear that $H_u \subset W_u$, $\big(W_u, \|\cdot\|_{1,p,u}\big)$, $\big(H_u, \|\cdot\|_{1,p,u}\big)$ are reflexive separable Banach spaces and that the embedding $W_u \hookrightarrow L^1(\Omega)$ is compact. If u is bounded between two positive constants, or u belongs to the Muckenhoupt class A_p (see below), then it is easy to verify that $W_u = H_u = W_0^{1,p}(\Omega)$. However, for a "typical" weight u, the space of smooth functions $C_0^\infty(\Omega)$ is not dense in W_u. Hence, the identity $W_u = H_u$ is not always valid (for the corresponding examples, we refer to [7, 33]).

Remark 24.1 We recall that the dual space of the weighted Sobolev space H_u is equivalent to $H_u^* = W^{-1,q}(\Omega, u^{1-q}\,dx)$, where q is the conjugate of p, i.e., $q = \frac{p}{p-1}$ (for more details, see [10]). Moreover, if there exists a value $\nu \in \left(\frac{N}{p}, +\infty\right) \cap \left[\frac{1}{p-1}, +\infty\right)$ such that $u^{-\nu} \in L^1(\Omega)$, then the expression (see [10, pp. 46]):

$$||y||_{H_u} = \left[\int_\Omega |\nabla y|^p\,dx\right]^{1/p} \tag{24.9}$$

can be considered as a norm on H_u and it is equivalent to the norm (24.6). Moreover, in this case, the embedding $H_u \hookrightarrow L^p(\Omega)$ is compact.

Let $\{a_\varepsilon\}_{\varepsilon>0}$ be a bounded sequence in $L^1(\Omega)$. We recall that $\{a_\varepsilon\}_{\varepsilon>0}$ is called equi-integrable if for any $\delta > 0$, there is $\tau = \tau(\delta)$ such that $\int_S |a_\varepsilon|\,dx < \delta$ for every measurable subset $S \subset \Omega$ of Lebesgue measure $|S| < \tau$. Then, the following assertions are equivalent:

(i) a sequence $\{a_\varepsilon\}_{\varepsilon>0}$ is weakly compact in $L^1(\Omega)$;
(ii) the sequence $\{a_\varepsilon\}_{\varepsilon>0}$ is equi-integrable.

Theorem 24.1 (Scheffe's Theorem) *If $a_\varepsilon \geq 0$ for all $\varepsilon > 0$, $a_\varepsilon \to a$ almost everywhere in Ω, and $\int_\Omega a_\varepsilon\, dx \to \int_\Omega a\, dx$, then $a_\varepsilon \to a$ in $L^1(\Omega)$.*

Theorem 24.2 (Lebesgue's Theorem) *If a sequence $\{a_\varepsilon\}_{\varepsilon>0} \subset L^1(\Omega)$ is equi-integrable and $a_\varepsilon \to a$ almost everywhere in Ω, then $a_\varepsilon \to a$ in $L^1(\Omega; \mathbb{S}^N)$.*

The space of all nonnegative Radon measures on Ω will be denoted by $M_+(\Omega)$. If μ is a nonnegative Radon measure on Ω, we will use $L^r(\Omega, d\mu)$, $1 \leq r \leq \infty$ to denote the usual Lebesgue space with respect to the measure μ with the corresponding norm $\|f\|_{L^r(\Omega,d\mu)} = \left(\int_\Omega |f(x)|^r\, d\mu\right)^{1/r}$. By $BV(\Omega)$, we denote the space of all functions in $L^1(\Omega)$ with bounded variation. Under the norm $\|f\|_{BV(\Omega)} = \|f\|_{L^1(\Omega)} + \int_\Omega |Df|$, $BV(\Omega)$ is a Banach space (see [15]). It is well known that uniformly bounded set in the BV-norm is relatively compact in $L^1(\Omega)$. Moreover, a sequence $\{f_k\}_{k=1}^\infty \subset BV(\Omega)$ weakly* converges to some $f \in BV(\Omega)$, and we write $f_k \stackrel{*}{\rightharpoonup} f$ if and only if the following conditions hold: $f_k \to f$ strongly in $L^1(\Omega)$, and $Df_k \stackrel{*}{\rightharpoonup} Df$ in $M(\Omega; \mathbb{R}^N)$. In the proposition below, we provide a compactness result related to this convergence, together with the lower semicontinuity property (see [15]):

Proposition 24.1 *Let $\{f_k\}_{k=1}^\infty$ be a sequence in $BV(\Omega)$ strongly converging to some f in $L^1(\Omega)$ and satisfying $\sup_{k\in\mathbb{N}} \int_\Omega |Df_k| < +\infty$. Then,*
(i) $f \in BV(\Omega)$ and $\int_\Omega |Df| \leq \liminf_{k\to\infty} \int_\Omega |Df_k|$; (ii) $f_k \stackrel{*}{\rightharpoonup} f$ in $BV(\Omega)$.

Let $\{\mu_k\}_{k\in\mathbb{N}}$, μ be Radon measures such that $\mu_k \stackrel{*}{\rightharpoonup} \mu$ in $M_+(\Omega)$, i.e.,

$$\lim_{k\to\infty} \int_\Omega \varphi\, d\mu_k = \int_\Omega \varphi\, d\mu \qquad \forall \varphi \in C_0(\mathbb{R}^N). \tag{24.10}$$

Let us recall the definition and main properties of convergence in the variable L^p-space (see [34]).

1. A sequence $\{v_k \in L^p(\Omega, d\mu_k)\}$ is called bounded if $\limsup\limits_{k\to\infty} \int_\Omega |v_k|^p\, d\mu_k < +\infty$.
2. A bounded sequence $\{v_k \in L^p(\Omega, d\mu_k)\}$ converges weakly to $v \in L^p(\Omega, d\mu)$, if

$$\lim_{k\to\infty} \int_\Omega v_k \varphi\, d\mu_k = \int_\Omega v\varphi\, d\mu \qquad \forall \varphi \in C_0^\infty(\Omega),$$

   and it is written as $v_k \rightharpoonup v$ in $L^p(\Omega, d\mu_k)$.
3. Strong convergence $v_k \to v$ in $L^p(\Omega, d\mu_k)$ means that $v \in L^p(\Omega, d\mu)$ and

$$\lim_{k\to\infty} \int_\Omega v_k z_k\, d\mu_k = \int_\Omega v z\, d\mu \qquad \text{as } z_k \rightharpoonup z \text{ in } L^q(\Omega, d\mu_k),\ q = p/(p-1). \tag{24.11}$$

In particular, if $d\mu_k = u_k\,dx$, $0 \le u_k \rightharpoonup u$ in $L^1(\Omega)$, and $v_k \rightharpoonup v$ in $L^p(\Omega, u_k\,dx)$, then $v_k u_k \rightharpoonup vu$ in $L^1(\Omega)$, i.e.,

$$\lim_{k\to\infty} \int_\Omega \varphi v_k u_k\,dx = \int_\Omega \varphi v u\,dx, \quad \forall \varphi \in C_0^\infty(\Omega), \tag{24.12}$$

and if $a, b \in L^\infty(\Omega)$, $a(x) \ge \alpha > 0$ a.e. in Ω, $d\mu_k = u_k\,dx$, and $0 \le u_k \rightharpoonup u$ in $L^1(\Omega)$, then

$$v_k \rightharpoonup v \text{ in } L^p(\Omega, u_k\,dx) \quad \text{implies} \quad bv_k \rightharpoonup bv \text{ in } L^p(\Omega, u_k\,dx), \tag{24.13}$$

$$v_k \rightharpoonup v \text{ in } L^p(\Omega, u_k\,dx) \quad \text{if and only if} \quad av_k \rightharpoonup av \text{ in } L^p(\Omega, u_k\,dx). \tag{24.14}$$

In spite of the fact that the following property of convergence in a variable-weighted L^p-space is rather obvious, it plays an important role in nonlinear problems and allows us to extend the class of admissible functions in relation (24.10). Let $d\mu_k = u_k\,dx$, where $0 \le u_k \rightharpoonup u$ in $L^1(\Omega)$, and let $\varphi \in C_0^\infty(\Omega; \mathbb{R}^N)$ be an arbitrary vector-valued function. Then,

$$|\varphi|^{p-2}\varphi \to |\varphi|^{p-2}\varphi \text{ in } L^p(\Omega, d\mu_k)^N; \text{ in particular } \lim_{k\to\infty} \int_\Omega |\varphi|^p\,d\mu_k = \int_\Omega |\varphi|^p\,d\mu. \tag{24.15}$$

Let $m, \alpha, \delta, \gamma \in \mathbb{R}_+$ be some positive values, and let ξ_1, ξ_2 be the given elements of $L^1(\Omega)$ satisfying the conditions

$$0 \le \xi_1(x) \le \xi_2(x) \text{ a.e. in } \Omega, \tag{24.16}$$

$$\text{and there exists a value } \nu \in \left(\frac{N}{p}, +\infty\right) \cap \left[\frac{1}{p-1}, +\infty\right) \text{ such that } \xi_1^{-\nu} \in L^1(\Omega). \tag{24.17}$$

We assume that there exists a closed subdomain $Q \subset \Omega$ with nonzero Lebesgue measure such that the Hausdorff–Pompeiu distance between the closed sets Q and $\Omega^c := \mathbb{R}^N \setminus \Omega$ satisfies condition

$$\text{dist } (\Omega^c, Q) \ge \delta, \tag{24.18}$$

$$\text{the "volume" of the set } \Omega\setminus Q \text{ is small enough, i.e., } |\Omega\setminus Q| \le \delta, \tag{24.19}$$

$$\xi_1, \xi_2 \in L^\infty(\Omega\setminus Q), \text{ and } \xi_1(x) \ge \alpha > 0 \text{ a.e. in } \Omega\setminus Q. \tag{24.20}$$

We define the class of admissible BV-controls $\mathfrak{A}_{ad}$ as follows:

$$\mathfrak{A}_{ad} = \left\{ u \in BV(\Omega) \,\middle|\, \int_\Omega u\,dx = m, \int_\Omega |Du| \le \gamma, \ \xi_1(x) \le u(x) \le \xi_2(x) \text{ a.e. in } \Omega \right\}. \tag{24.21}$$

It is clear that $\mathfrak{A}_{ad}$ is a convex, relatively compact subset of $L^1(\Omega)$ with an empty topological interior. Hereinafter, we assume that the set of admissible controls $\mathfrak{A}_{ad}$ is always non-empty.

Remark 24.2 As a pathological example of function ξ_1 satisfying the properties (24.16) and (24.17), we consider:

$$\xi_1(x) = (\varepsilon + |f_K(x)|)^{-1/\nu} \text{ in } Q, \quad \text{and } \xi_1(x) = \alpha \text{ in } \Omega\setminus Q, \tag{24.22}$$

where $\varepsilon > 0$ is a positive value, and $f_K : Q \to \mathbb{R}$ is an L^1-function that is essentially unbounded on every non-empty open subset of Q (for the details, we refer to Kovalevsky [22]). Since f_K does not have a pointwise majorant, it follows that ξ_1, given by (24.22), is nonnegative, $\xi_1 \in L^\infty(\Omega)$, $\|\xi_1\|_{L^\infty(\Omega)} \le \max\{\varepsilon^{-1/\nu}, \alpha\}$, and $\xi_1 : \Omega \to \mathbb{R}$ vanishes almost everywhere in Q.

Let $u \in \mathfrak{A}_{ad}$ be an admissible control such that $H_u \neq W_u$, and let V_u be an intermediate space $H_u \subseteq V_u \subseteq W_u$. Let V_u^* be the dual space. We say that the nonlinear operator $\Delta_p(u,\cdot) : V_u \to V_u^*$ is the generalized p-Laplacian if it has a representation

$$\Delta_p(u,y) = -\operatorname{div}\left(u(x)|\nabla y|^{p-2}\nabla y\right), \text{ where } |\nabla y|^{p-2} := |\nabla y|^{p-2}_{\mathbb{R}^N} = \left(\sum_{i=1}^N \left|\frac{\partial y}{\partial x_j}\right|^2\right)^{\frac{p-2}{2}},$$

or via the pairing $\langle \Delta_p(u,y), v\rangle_{V_u^*;V_u} = \int_\Omega u(x)|\nabla y|^{p-2}(\nabla y, \nabla v)_{\mathbb{R}^N}\,dx$, $\forall v \in V_u$. It is easy to see that if the condition (24.17) holds true, then for every admissible control $u \in \mathfrak{A}_{ad}$ and for $V_u = H_u$, the operator $\Delta_p(u,\cdot) : H_u \to H_u^* = W^{-1,q}(\Omega, u^{1-q}\,dx)$ turns out to be coercive, i.e.,

$$\langle \Delta_p(u,y), y\rangle_{H_u^*;H_u} = \int_\Omega |\nabla y|^p u(x)\,dx \overset{\text{by Remark 24.1}}{=} \|y\|^p_{H_u},$$

and semicontinuous, where by the semicontinuity property we mean the continuity of the scalar function $t \to \langle \Delta_p(u, y+tw), v\rangle_{H_u^*;H_u}$ for all $y, v, w \in H_u$. Indeed, in order to obtain the required relation

$$\lim_{t\to 0}\langle \Delta_p(u, y+tw), v\rangle_{H_u^*;H_u} = \langle \Delta_p(u,y), v\rangle_{H_u^*;H_u},$$

it is enough to observe that $|\nabla y + t\nabla w|^{p-2}(\nabla y + t\nabla w) \to |\nabla y|^{p-2}\nabla y$ almost everywhere in Ω and to recall Lebesgue's dominated convergence theorem. Moreover, in this case, the p-Laplacian $\Delta_p(u,y)$ is a strictly monotone operator on H_u for each $u \in \mathfrak{A}_{ad}$. Indeed, having applied the estimate

$$\int_{\Omega} |\nabla y|^{p-1}|\nabla v|u(x)\,dx \leq \left(\int_{\Omega} |\nabla y|^p u\,dx\right)^{(p-1)/p} \left(\int_{\Omega} |\nabla v|^p u\,dx\right)^{1/p} =: \|y\|_{H_u}^{p-1}\|v\|_{H_u},$$

it is easy to verify

$$\left\langle \Delta_p(u,y) - \Delta_p(u,v), y-v\right\rangle_{H_u^*;H_u} \geq 2^{2-p}\left|\|y\|_{H_u} - \|v\|_{H_u}\right|^p > 0 \quad \forall y,v \in H_u,\ y \neq v.$$

Remark 24.3 By well-known existence results for nonlinear elliptic equations with strictly monotone semicontinuous coercive operators (see [14, 25]), one can conclude that for every $u \in \mathfrak{A}_{ad}$ and $f \in H_u^* = W^{-1,q}(\Omega, u^{1-q}\,dx)$, the nonlinear Dirichlet boundary value problem

$$\Delta_p(u,y) = f \quad \text{in} \quad \Omega, \qquad y \in H_u, \tag{24.23}$$

admits a unique weak solution in H_u, or shortly, H_u-solution. However, if $H_u \neq W_u$ the generalized p-Laplacian $\Delta_p(u,\cdot) : V_u \to V_u^*$ may not in general admit a weak solution for an intermediate space V_u with $H_u \subset V_u \subseteq W_u$.

Let us recall that for a given control $u \in \mathfrak{A}_{ad}$, a function y is the V_u-solution of (24.23) if

$$y \in V_u, \tag{24.24}$$

$$\int_{\Omega} u(x)|\nabla y|^{p-2}(\nabla y, \nabla v)_{\mathbb{R}^N}\,dx = \langle f, v\rangle_{V_u^*;V_u}, \quad \forall v \in V_u. \tag{24.25}$$

24.3 Setting of the Optimal Control Problem

Let $y_d \in L^p(\Omega)$ and $f \in L^q(\Omega)$ be the given distributions, where $q = p/(p-1)$ is the conjugate of $p \geq 2$. The optimal control problem we consider in this paper is to minimize the discrepancy between distribution $y_d \in L^p(\Omega)$ and the solution of the following boundary value problem

$$\Delta_p(u,y) = f \quad \text{in} \quad \Omega, \qquad y = 0 \quad \text{on} \ \partial\Omega, \tag{24.26}$$

by choosing an appropriate weight function $u \in \mathfrak{A}_{ad}$, i.e.,

$$u \in BV(\Omega),\ \int_{\Omega} u\,dx = m,\ \int_{\Omega} |Du| \leq \gamma,\ \xi_1(x) \leq u(x) \leq \xi_2(x) \text{ a.e. in } \Omega, \tag{24.27}$$

where functions ξ_1 and ξ_2 satisfy the conditions (24.16)–(24.20). More precisely, we are concerned with the following optimal control problem

$$\text{Minimize } \left\{ I(u,y) = \int_\Omega \Big[|y-y_d|^p + |\nabla y|^p u(x) \Big] dx \right\} \tag{24.28}$$

subject to the constraints (24.26) and (24.27) and (24.16)–(24.20).

Definition 24.1 We say that for a given admissible control $u \in \mathfrak{A}_{ad}$, $y \in W_u := W_0^{1,p}(\Omega, u\,dx)$ is a weak solution (in the sense of Minty) to problem (24.26) if

$$\int_\Omega u(x)|\nabla\varphi|^{p-2}\,(\nabla\varphi, \nabla\varphi - \nabla y)_{\mathbb{R}^N}\, dx \geq \int_\Omega f(\varphi - y)\,dx, \quad \forall \varphi \in C_0^\infty(\Omega). \tag{24.29}$$

Remark 24.4 As follows from this definition, the set of weak solutions to the problem (24.26) is convex and closed. Moreover, a function $y \in V_u$, $H_u \subseteq V_u \subseteq W_u$, is a V_u-solution to (24.26) (see relations (24.24) and (24.25)) if and only if y satisfies the inequality (24.29) for each $\varphi \in V_u$ (for the details, we refer to Propositions 3.2 in [30]). Hence, taking $\varphi = y$ in (24.25), we arrive at the energy equality

$$\int_\Omega |\nabla y|^p u\,dx = \int_\Omega fy\,dx. \tag{24.30}$$

In particular, if $V_u = H_u$, then the energy equality (24.30) immediately leads us to the following a priori estimate:

$$\|y\|_{H_u} \leq C\|f\|_{L^q(\Omega)}^{1/(p-1)}, \quad \forall u \in \mathfrak{A}_{ad}, \tag{24.31}$$

where constant C comes from Friedrichs inequality. It is worth to note that this estimate is valid for H_u-solutions only, and we have no a priori estimate for other types of solutions like weak solutions (see Definition 24.1) or V_u-solutions.

It is clear that the question of uniqueness of a weak solution leads us to the problem of density of the subspace of smooth functions $C_0^\infty(\Omega)$ in W_u for $u \in \mathfrak{A}_{ad}$. However, as was indicated in [36], for a "typical" weight function $u \in \mathfrak{A}_{ad}$, the subspace $C_0^\infty(\Omega)$ is not dense in W_u, and hence, no uniqueness of weak solutions can be expected (for more details and other types of solutions, we refer to [2, 34, 36]). Taking this fact into account, we introduce the set of admissible pairs to the problem (24.26)–(24.28)

$$\Xi = \{(u,y) \mid u \in \mathfrak{A}_{ad},\ y \in W_u,\ (u,y) \text{ are related by inequality (24.29)}\}. \tag{24.32}$$

Note that due to the assumption (24.17), the set Ξ is always non-empty (see Remark 24.3). Therefore, in this case, the minimization problem

$$\inf_{(u,y)\in\Xi} I(u,y) \tag{24.33}$$

is regular. Moreover, as we show in Theorem 24.5 under some additional assumptions, the structure of the set Ξ can be fundamentally simplified. In view of this, we adopt the following concept.

Definition 24.2 We say that a pair $(u^0, y^0) \in BV(\Omega) \times W_0^{1,p}(\Omega, u^0\,dx)$ is a weak optimal solution to the problem (24.26)–(24.28) if (u^0, y^0) is a minimizer for (24.33).

24.4 Existence of Weak Optimal Solutions

Our prime interest in this section is the solution existence in the class of weak solutions to the optimal control problem (24.26)–(24.28). To this end, we provide some auxiliary results.

Lemma 24.1 *Let $\{u_k\}_{k\in\mathbb{N}}$ be a sequence of controls in $\mathfrak{A}_{ad}$ such that $u_k \to u$ in $L^1(\Omega)$. Then, $u \in \mathfrak{A}_{ad}$ and*

$$u_k^{-1/(p-1)} \to u^{-1/(p-1)} \text{ strongly in } L^1(\Omega), \tag{24.34}$$

$$(u_k)^{-1} \to u^{-1} \text{ in the variable space } L^q(\Omega, u_k\,dx) \text{ with } q = p/(p-1). \tag{24.35}$$

Proof By the properties of the set of admissible controls $\mathfrak{A}_{ad}$, we have $u_k^{-\nu} \leq \xi_1^{-\nu}$ for every $k \in \mathbb{N}$, where ν is defined by (24.17). Hence, the sequence $\{u_k^{-\nu}\}_{k\in\mathbb{N}}$ is equi-integrable on Ω. Then, up to a subsequence, we have $u_k \to u$ a.e. in Ω, and therefore, Lebesgue's theorem (see Theorem 24.2) implies

$$u_k^{-\nu} \to u^{-\nu} \text{ in } L^1(\Omega). \tag{24.36}$$

In view of the estimate (here, $(p-1)\nu \geq 1$ by (24.17))

$$\int_\Omega u_k^{-1/(p-1)}\,dx \leq \|u_k^{-\nu}\|_{L^1(\Omega)}^{\frac{1}{(p-1)\nu}} |\Omega|^{\frac{(p-1)\nu-1}{(p-1)\nu}}, \quad \forall k \in \mathbb{N},$$

condition (24.36) guarantees that the sequence $\left\{u_k^{-1/(p-1)}\right\}_{k\in\mathbb{N}}$ is bounded in $L^1(\Omega)$ and equi-integrable. Since $u_k^{-1/(p-1)} \to u^{-1/(p-1)}$ a.e. in Ω, by Lebesgue's theorem, we arrive at the required assertion (24.34).

Let $\varphi \in C_0^\infty(\Omega)$ be a fixed function. Since by initial suppositions

$$\sup_{k\in\mathbb{N}} \int_\Omega \left(u_k^{-1}\right)^q u_k\,dx = \sup_{k\in\mathbb{N}} \int_\Omega u_k^{-1/(p-1)}\,dx < +\infty,$$

it follows that the sequence $\{u_k^{-1}\}_{k\in\mathbb{N}}$ is bounded in $L^q(\Omega, u_k dx)$. Then, the equality

$$\int_\Omega u_k^{-1}\varphi\, u_k\,dx = \int_\Omega \varphi\,dx = \int_\Omega u^{-1}\varphi\, u\,dx \quad \forall k \in \mathbb{N}$$

leads us to the weak convergence $u_k^{-1} \rightharpoonup u^{-1}$ in $L^q(\Omega, u_k dx)$. Taking into account the strong convergence $u_k^{-1/(p-1)} \to u^{-1/(p-1)}$ in $L^1(\Omega)$ and the fact that Ω is a bounded domain, we get

$$\lim_{k\to\infty} \int_\Omega \left(u_k^{-1}\right)^q u_k\, dx = \lim_{k\to\infty} \int_\Omega u_k^{-1/(p-1)}\, dx = \int_\Omega u^{-1/(p-1)}\, dx = \int_\Omega \left(u^{-1}\right)^q u\, dx.$$

Hence, by the strong convergence criterion in the variable space $L^q(\Omega, u_k dx)$ (see [35]), we arrive at the property (24.35). It remains to note that by Proposition 24.1, the set $\mathfrak{A}_{ad}$ is sequentially closed with respect to the strong convergence in $L^1(\Omega)$. Hence, $u \in \mathfrak{A}_{ad}$ and this concludes the proof.

As a direct consequence of this result, we can prove the following assertions.

Lemma 24.2 *Let $\{u_k\}_{k\in\mathbb{N}}$ and $\{w_k\}_{k\in\mathbb{N}}$ be sequences such that*

$$u_k \in \mathfrak{A}_{ad},\ \ \forall k \in \mathbb{N},\ \ u_k \to u \ \text{ in } L^1(\Omega), \tag{24.37}$$

$$w_k \in L^p(\Omega, u_k\, dx),\ \ \forall k \in \mathbb{N},\ \ w_k \rightharpoonup w \ \text{ in } L^p(\Omega, u_k\, dx). \tag{24.38}$$

Then, $w_k, w \in L^1(\Omega)$ and $w_k \rightharpoonup w$ in $L^1(\Omega)$ in the following sense

$$\lim_{k\to\infty} \int_\Omega w_k\varphi\, dx = \int_\Omega w\varphi\, dx,\ \ \forall \varphi \in C_0^\infty(\Omega). \tag{24.39}$$

Proof Due to the estimate

$$\begin{aligned}\int_\Omega w_k\, dx &= \int_\Omega u_k^{-1/p} w_k u_k^{1/p}\, dx \le \|w_k\|_{L^p(\Omega,u_k\,dx)} \|u_k^{-1/(p-1)}\|_{L^1(\Omega)}^{(p-1)/p}\\ &\le \|w_k\|_{L^p(\Omega,u_k\,dx)} \|u_k^{-\nu}\|_{L^1(\Omega)}^{\frac{1}{p\nu}} |\Omega|^{\frac{(p-1)\nu-1}{p\nu}}\\ &\le \left(\sup_{k\in\mathbb{N}} \|w_k\|_{L^p(\Omega,u_k\,dx)}\right) \|\xi_1^{-\nu}\|_{L^1(\Omega)}^{\frac{1}{p\nu}} |\Omega|^{\frac{(p-1)\nu-1}{p\nu}} < +\infty,\end{aligned}$$

the inclusions $w_k, w \in L^1(\Omega)$ are obvious. It remains to prove the convergence (24.39). By Lemma 24.1, we have: $(u_k)^{-1} \to u^{-1}$ in variable space $L^q(\Omega, u_k dx)$. Therefore,

$$\int_\Omega w_k\varphi\, dx = \int_\Omega u_k^{-1} w_k \varphi u_k\, dx \overset{\text{by (24.11)}}{\to} \int_\Omega u^{-1} w\varphi u\, dx = \int_\Omega w\varphi\, dx$$

for each $\varphi \in C_0^\infty(\Omega)$. The proof is complete.

Lemma 24.3 *Let $\{u_k\}_{k\in\mathbb{N}} \subset \mathfrak{A}_{ad}$ and $\{y_k \in W_{u_k}\}_{k\in\mathbb{N}}$ be sequences such that*

$$u_k \to u \ \text{ in } L^1(\Omega),\ \ y_k \rightharpoonup y \ \text{ in } L^p(\Omega),\ \ \nabla y_k \rightharpoonup \mathbf{v} \ \text{ in } L^p(\Omega, u_k\, dx)^N. \tag{24.40}$$

Then, $y \in W_u$ and $\nabla y = \mathbf{v}$.

Proof By estimates (24.7) and (24.8) and Lemma 24.2, we conclude: $y_k \in W_0^{1,1}(\Omega)$ for all $k \in \mathbb{N}$, and

$$\int_\Omega y_k \varphi\, dx \to \int_\Omega y\varphi\, dx, \quad \forall \varphi \in C_0^\infty(\Omega), \text{ and}$$

$$\int_\Omega (\nabla y_k, \psi)_{\mathbb{R}^N}\, dx \to \int_\Omega (\mathbf{v}, \psi)_{\mathbb{R}^N}\, dx, \quad \forall \psi \in C_0^\infty(\Omega)^N, \quad \text{where } \mathbf{v} \in L^1(\Omega)^N.$$

Hence, $y \in W_0^{1,1}(\Omega)$ and $\nabla y = \mathbf{v}$ by completeness of the Sobolev space $W_0^{1,1}(\Omega)$. It remains to note that $\|y\|_{1,p,u} < +\infty$ by (24.40). The proof is complete.

We now concentrate on the topological properties of the set of admissible solutions Ξ to the problem (24.26)–(24.28).

Definition 24.3 A sequence $\{(u_k, y_k) \in \Xi\}_{k\in\mathbb{N}}$ is called bounded if

$$\sup_{k\in\mathbb{N}} \left[\|u_k\|_{BV(\Omega)} + \|y_k\|_{L^p(\Omega)} + \|\nabla y_k\|_{L^p(\Omega, u_k dx)^N} \right] < +\infty.$$

Definition 24.4 We say that a bounded sequence $\{(u_k, y_k) \in \Xi\}_{k\in\mathbb{N}}$ of admissible solutions τ-converges to a pair $(u, y) \in BV(\Omega) \times W_0^{1,1}(\Omega)$ if

(a) $u_k \overset{*}{\rightharpoonup} u$ in $BV(\Omega)$;
(d) $y_k \rightharpoonup y$ in $L^p(\Omega)$;
(e) $\nabla y_k \rightharpoonup \nabla y$ in the variable space $L^p(\Omega, u_k dx)^N$.

Note that due to assumptions (24.5) and estimates like (24.7) and (24.8), the inclusion $y \in W_0^{1,1}(\Omega)$ is obvious.

The following result is crucial for our further analysis.

Theorem 24.3 *Let $\{(u_k, y_k)\}_{k\in\mathbb{N}} \subset \Xi$ be a bounded sequence. Then, there is a pair $(u, y) \in BV(\Omega) \times W_0^{1,1}(\Omega)$ such that up to a subsequence, $(u_k, y_k) \overset{\tau}{\longrightarrow} (u, y)$ and $(u, y) \in \Xi$.*

Proof By compactness criterion for the weak convergence in variable spaces, there exists a subsequence of $\{(u_k, y_k) \in \Xi\}_{k\in\mathbb{N}}$, still denoted by the same indices, and functions $u \in BV(\Omega)$, $y \in L^p(\Omega)$, and $v \in L^p(\Omega, u\, dx)^N$ such that

$$u_k \to u \text{ in } L^1(\Omega), \tag{24.41}$$

$$y_k \rightharpoonup y \text{ in } L^p(\Omega), \quad \nabla y_k \rightharpoonup v \text{ in the variable space } L^p(\Omega, u_k\, dx)^N. \tag{24.42}$$

Then, by Lemmas 24.1 and 24.3, we have: $u \in \mathfrak{A}_{ad}$, $y \in W_u$, and $\nabla y = \mathbf{v}$. It remains to show that the τ-limit pair (u, y) is related by inequality (24.29). With that in mind, we write down the Minty relation for (u_k, y_k):

$$\int_\Omega |\nabla \varphi|^{p-2} (\nabla\varphi, \nabla\varphi - \nabla y_k)_{\mathbb{R}^N}\, u_k\, dx \geq \int_\Omega f(\varphi - y_k)\, dx, \quad \forall \varphi \in C_0^\infty(\Omega). \tag{24.43}$$

Taking into account properties (24.41) and (24.42) and the fact that

$$\lim_{k\to\infty}\int_\Omega |\nabla\varphi|^{p-2}\,(\nabla\varphi,\nabla\varphi)_{\mathbb{R}^N}\,u_k\,dx \overset{\text{by }(24.15)_2}{=} \int_\Omega |\nabla\varphi|^{p-2}\,(\nabla\varphi,\nabla\varphi)_{\mathbb{R}^N}\,u\,dx,$$

$$\lim_{k\to\infty}\int_\Omega |\nabla\varphi|^{p-2}\,(\nabla\varphi,\nabla y_k)_{\mathbb{R}^N}\,u_k\,dx \overset{\text{by }(24.15)_1}{=} \int_\Omega |\nabla\varphi|^{p-2}\,(\nabla\varphi,\nabla y)_{\mathbb{R}^N}\,u\,dx$$

as a product of strongly and weakly convergent sequences, we can pass to the limit in relation (24.43) as $k\to\infty$. As a result, we arrive at the inequality (24.29), which means that $y\in W_u$ is a weak solution to the boundary value problem (24.26) in the sense of Minty. This fact together with $u\in\mathfrak{A}_{ad}$ leads us to the conclusion: $(u,y)\in\Xi$, i.e., the limit pair (u,y) is admissible to optimal control problem (24.26)–(24.28). The proof is complete.

Remark 24.5 If the τ-limit pair (u,y) in the statement of Theorem 24.3 is such that $y\in H_u$, then it is equivalent to state that this pair satisfies the energy equality (24.30). Indeed, if $y\in H_u$, then by density of $C_0^\infty(\Omega)$ in H_u, we can take any $\varphi\in H_u$ for a test function in (24.29). Therefore, after taking $\varphi=y\pm tw$, $w\in H_u$, $t>0$ and passing to the limit in this relation as $t\to 0$, we obtain

$$\pm\int_\Omega u(x)|\nabla y\pm t\nabla w|^{p-2}\,(\nabla y\pm t\nabla w,\nabla w)_{\mathbb{R}^N}\,dx \ge \pm\int_\Omega fw\,dx,\quad \forall\, w\in H_u$$

which obviously yields

$$\int_\Omega u(x)|\nabla y|^{p-2}\,(\nabla y,\nabla w)_{\mathbb{R}^N}\,dx = \int_\Omega fw\,dx,\quad \forall\, w\in H_u.$$

So, in this case, the energy equality (24.30) holds true.

In conclusion of this section, we give the existence result for weak optimal pairs to the problem (24.26)–(24.28).

Theorem 24.4 *Let $y_d\in L^p(\Omega)$ and $f\in L^q(\Omega)$ be the given functions. Then, optimal control problem (24.26)–(24.28) admits at least one weak solution $(u^{opt},y^{opt})\in\Xi\subset BV(\Omega)\times W_0^{1,1}(\Omega)$, $y^{opt}\in W_{u^{opt}}$.*

Proof Since the set of admissible pairs Ξ is non-empty and the cost functional is bounded below on Ξ, it follows that there exists a minimizing sequence $\{(u_k,y_k)\in\Xi\}_{k\in\mathbb{N}}$ to the problem (24.33). Then, the inequality

$$\inf_{(u,y)\in\Xi} I(u,y) = \lim_{k\to\infty}\int_\Omega\Big[|y_k(x)-y_d(x)|^p + |\nabla y_k(x)|^p u_k\Big]\,dx < +\infty,$$

implies that there is a constant $C>0$ such that

$$\sup_{k\in\mathbb{N}}\|y_k\|_{L^p(\Omega)}\le C,\quad \sup_{k\in\mathbb{N}}\|\nabla y_k\|_{L^p(\Omega;u_k dx)^N}\le C.$$

Hence, in view of the definition of the class $\mathfrak{A}_{ad}$, the sequence $\{(u_k, y_k) \in \Xi\}_{k\in\mathbb{N}}$ is bounded in the sense of Definition 24.3. Hence, by Theorem 24.3, there exist elements $u^* \in \mathfrak{A}_{ad}$ and $y^* \in W_{u^*}$ such that up to a subsequence, $(u_k, y_k) \xrightarrow{\tau} (u^*, y^*)$ and $(u^*, y^*) \in \Xi$. To conclude the proof, it is enough to observe that by properties of weak convergence in variable spaces, the cost functional I is sequentially lower τ-semicontinuous. Thus,

$$I(u^*, y^*) \leq \liminf_{k\to\infty} I(u_k, y_k) = \inf_{(u,y)\in\Xi} I(u, y).$$

Hence, (u^*, y^*) is an optimal pair, and we arrive at the required conclusion.

24.5 "Directional Stability" of Weighted Sobolev Spaces

In this section, we proceed to discuss the limit properties of sequences of admissible solutions $\{(u_k, y_k)\}_{k\in\mathbb{N}} \subset \Xi$. As Theorem 24.3 indicates, if $(u, y) \in \Xi$ is a τ-limit of the sequence $\{(u_k, y_k)\}_{k\in\mathbb{N}} \subset \Xi$ as $k \to \infty$, then it is not known in general, whether this pair is related by $y \in V_u$ for some intermediate space V_u, $H_u \subseteq V_u \subseteq W_u$. In other words, what kind of conditions guarantee that the τ-limit pair $(u, y) \in \Xi$ is such that the function y is V_u-solution to the boundary problem (24.26)? Since for a given u the set of weak solutions to the problem (24.26) is not a singleton, in general (see Remark 24.4), it is clear that the fulfillment of the condition $y \in V_u$ depends on the choice of the sequence $\{(u_k, y_k)\}_{k\in\mathbb{N}} \subset \Xi$ (for the details and counterexamples we refer to [34]). The following result can be considered as a specification of Theorem 24.3.

Lemma 24.4 *Let $u \in \mathfrak{A}_{ad}$ and $\{u_\varepsilon\}_{\varepsilon>0} \subset L^1(\Omega)$ be such that $\xi_1(x) \leq u(x) \leq u_\varepsilon(x) \leq \xi_2(x)$ almost everywhere in Ω for all $\varepsilon > 0$, and $u_\varepsilon \to u$ in $L^1(\Omega)$ as $\varepsilon \to 0$. For each $\varepsilon > 0$, let $y_\varepsilon = y(u_\varepsilon)$ be the corresponding H_{u_ε}-solutions to the boundary value problem (24.26). Then, up to a subsequence, we have $(u_\varepsilon, y_\varepsilon) \xrightarrow{\tau} (u, y)$ as $\varepsilon \to 0$, where $y \in H_u$ and y is an H_u-solution to boundary value problem (24.26) for the given control u.*

Proof In view of Remark 24.3, the sequence $\{(u_\varepsilon, y_\varepsilon)\}_{\varepsilon>0}$ is defined in a unique way. Since $y_\varepsilon \in H_{u_\varepsilon}$ for all $\varepsilon > 0$, it follows from a priori estimates that the sequence $\{(u_\varepsilon, y_\varepsilon)\}_{\varepsilon>0}$ is bounded. Hence, by Theorem 24.3, this sequence is relatively τ-compact and each of its τ-cluster pairs (u, y) belongs to the set Ξ. It remains to show that $y \in H_u$ and y is an H_u-solution to the problem (24.26). Then, the τ-limit pair (u, y) is unique by Remark 24.3.

By Theorem 24.3, we have: $y \in W_u$ is a weak solution of (24.26) in the sense of Minty and, within a subsequence,

$$y_\varepsilon \rightharpoonup y \text{ in } L^p(\Omega), \quad \nabla y_\varepsilon \rightharpoonup \nabla y \text{ in the variable space } L^p(\Omega, u_\varepsilon\, dx)^N. \qquad (24.44)$$

Therefore, in order to prove the inclusion $y \in H_u$, it is enough to show that

$$\nabla y_\varepsilon \in L^p(\Omega, u\,dx)^N \text{ for all } \varepsilon > 0 \text{ and } \nabla y_\varepsilon \rightharpoonup \nabla y \text{ in } L^p(\Omega, u\,dx)^N. \quad (24.45)$$

The first assertion in (24.45) is obvious because $\nabla y_\varepsilon \in L^p(\Omega, u_\varepsilon\,dx)^N$ and $u_\varepsilon \geq u$ almost everywhere in Ω for all $\varepsilon > 0$. As for the weak convergence property $(24.45)_2$, we note that

(1) $u_\varepsilon - u \to 0$ almost everywhere in Ω (by the initial assumptions);
(2) $\nabla y_\varepsilon \rightharpoonup \nabla y$ in $L^1(\Omega)^N$ by Lemma 24.2.

Hence, $\nabla y_\varepsilon(u_\varepsilon - u) \to 0$ almost everywhere in Ω and in view of estimate

$$\int_\Omega |\nabla y_\varepsilon| u\,dx \leq \int_\Omega |\nabla y_\varepsilon| u_\varepsilon\,dx \leq \|\nabla y_\varepsilon\|_{L^p(\Omega, u_\varepsilon\,dx)^N} \|u_\varepsilon\|^{(p-1)/p}_{L^1(\Omega)},$$

the sequence $\{\nabla y_\varepsilon(u_\varepsilon - u)\}_{\varepsilon>0}$ is equi-integrable. Therefore, Lebesgue theorem implies that

$$\nabla y_\varepsilon(u_\varepsilon - u) \to 0 \quad \text{in } L^1(\Omega)^N \text{ as } \varepsilon \to 0. \quad (24.46)$$

Taking (24.46) into account the fact that the smooth compactly supported functions are dense in $L^p(\Omega, u\,dx)^N$, for every $\varphi \in C_0^\infty(\Omega)$, we get

$$\begin{aligned}\left| \int_\Omega (\nabla y_\varepsilon, \nabla\varphi)_{\mathbb{R}^N} u\,dx - \int_\Omega (\nabla y, \nabla\varphi)_{\mathbb{R}^N} u\,dx \right| &\leq \int_\Omega \left|(\nabla y_\varepsilon(u_\varepsilon - u), \nabla\varphi)_{\mathbb{R}^N}\right| dx \\ &+ \left| \int_\Omega (\nabla y_\varepsilon, \nabla\varphi)_{\mathbb{R}^N} u_\varepsilon\,dx - \int_\Omega (\nabla y, \nabla\varphi)_{\mathbb{R}^N} u\,dx \right| = I_1 + I_2,\end{aligned}$$

where I_1 tends to zero as $\varepsilon \to 0$ by (24.46) and $I_2 \to 0$ by (24.44). Thus, $\nabla y_\varepsilon \rightharpoonup \nabla y$ weakly in $L^p(\Omega, u\,dx)^N$, and hence, $y \in H_u$.

Lemma 24.5 *Under the assumptions of Lemma 24.4, we have the strong convergence property: $y_\varepsilon \to y$ in $L^p(\Omega)$, $\nabla y_\varepsilon \to \nabla y$ in the variable space $L^p(\Omega, u_\varepsilon\,dx)^N$.*

Proof Taking into account the result of Lemma 24.4 and the following arguments of Remarks 24.4 and 24.5, for each $\varepsilon > 0$, we have the energy equalities

$$\int_\Omega u_\varepsilon(x)|\nabla y_\varepsilon|^p\,dx = \int_\Omega f y_\varepsilon\,dx, \quad \int_\Omega u(x)|\nabla y|^p\,dx = \int_\Omega f y\,dx. \quad (24.47)$$

Since $\nabla y_\varepsilon \rightharpoonup \nabla y$ in the variable space $L^p(\Omega, u_\varepsilon\,dx)^N$, we derive from (24.47) the following relation:

$$\begin{aligned}\int_\Omega u(x)|\nabla y|^p\,dx &\leq \liminf_{\varepsilon\to 0} \int_\Omega u_\varepsilon(x)|\nabla y_\varepsilon|^p\,dx \\ &= \liminf_{\varepsilon\to 0} \int_\Omega f y_\varepsilon\,dx = \int_\Omega f y\,dx = \int_\Omega u(x)|\nabla y|^p\,dx.\end{aligned}$$

Hence, $\lim_{\varepsilon\to 0}\int_\Omega u_\varepsilon(x)|\nabla y_\varepsilon|^p\,dx = \int_\Omega u(x)|\nabla y|^p\,dx$ and this implies the strong convergence $\nabla y_\varepsilon \rightharpoonup \nabla y$ in the variable space $L^p(\Omega, u_\varepsilon\,dx)^N$. Using the fact that $H_{u_\varepsilon} \subset H_u$ (because $u_\varepsilon \ge u$ in Ω) and the embedding $H_u \hookrightarrow L^p(\Omega)$ is compact, we finally conclude the strong convergence $y_\varepsilon \to y$ in $L^p(\Omega)$. The proof is complete.

Lemma 24.6 *Let $u \in \mathfrak{A}_{ad}$ and $\{u_\varepsilon\}_{\varepsilon>0} \subset L^1(\Omega)$ be such that $\xi_1(x) \le u_\varepsilon(x) \le u(x) \le \xi_2(x)$ almost everywhere in Ω for all $\varepsilon > 0$, and $u_\varepsilon \to u$ in $L^1(\Omega)$ as $\varepsilon \to 0$. For each $\varepsilon > 0$, let $y_\varepsilon = y(u_\varepsilon)$ be W_{u_ε}-solutions to the boundary value problem (24.26). Then, up to a subsequence, we have $(u_\varepsilon, y_\varepsilon) \xrightarrow{\tau} (u, y)$ as $\varepsilon \to 0$, where $y \in W_u$ and y is a W_u-solution to the boundary value problem (24.26) for the given control u. Moreover, in this case, we have*

$$\nabla y_\varepsilon \to \nabla y \text{ in the variable space } L^p(\Omega, u_\varepsilon\,dx)^N.$$

Proof By the arguments of the proof of Lemma 24.4, we conclude that up to a subsequence, $(u_\varepsilon, y_\varepsilon) \xrightarrow{\tau} (u, y)$ as $\varepsilon \to 0$, where $y \in W_u$ is a weak solution of (24.26) in the sense of Minty. Taking into account the definition of W_{u_ε}-solution, for each $\varepsilon > 0$, we have

$$\int_\Omega u_\varepsilon(x)|\nabla\varphi|^{p-2}(\nabla\varphi, \nabla\varphi - \nabla y_\varepsilon)_{\mathbb{R}^N}\,dx \ge \int_\Omega f(\varphi - y_\varepsilon)\,dx, \quad \forall\varphi \in W_{u_\varepsilon}. \tag{24.48}$$

However, for an arbitrary function $\varphi \in W_u$, because of inequality $u_\varepsilon \le u$, we have: $\varphi \in W_{u_\varepsilon}$. Moreover, the strong convergence $u_\varepsilon \to u$ in $L^1(\Omega)$ and Lebesgue theorem imply

$$\lim_{\varepsilon\to 0}\int_\Omega |\nabla\varphi|^p u_\varepsilon\,dx = \int_\Omega |\nabla\varphi|^p u\,dx,$$

i.e., $\nabla\varphi \to \nabla\varphi$ strongly in the variable space $L^p(\Omega, u_\varepsilon\,dx)^N$. Therefore, passing to the limit in (24.48) with an arbitrary $\varphi \in W_u$, we arrive at the relation

$$\int_\Omega u(x)|\nabla\varphi|^{p-2}(\nabla\varphi, \nabla\varphi - \nabla y)_{\mathbb{R}^N}\,dx \ge \int_\Omega f(\varphi - y)\,dx, \quad \forall\varphi \in W_u, \tag{24.49}$$

i.e., y is the W_u-solution to the boundary value problem (24.26). The strong convergence properties for the sequence $\left\{\nabla y_\varepsilon \in L^p(\Omega;\, u_\varepsilon\,dx)^N\right\}_{\varepsilon>0}$ can be established by analogy with Lemma 24.5.

Remark 24.6 It is easy to show that Lemma 24.6 remains true if we replace the conditions $\xi_1(x) \le u_\varepsilon(x) \le u(x) \le \xi_2(x)$ almost everywhere in Ω for all $\varepsilon > 0$, and $u_\varepsilon \to u$ in $L^1(\Omega)$ as $\varepsilon \to 0$ by the following stability property of the Sobolev space W_u: if $u_\varepsilon \to u$ in $L^1(\Omega)$, then $W_{u_\varepsilon} \supseteq W_u$ for $\varepsilon > 0$ small enough. It should be emphasized that from an optimal control theory point of view, an L^1-approximation of element $u \in \mathfrak{A}_{ad}$ by the sequence $\{u_\varepsilon\}_{\varepsilon>0}$ is meaningful if only $\{u_\varepsilon\}_{\varepsilon>0}$ are admissible controls. However, the set $\mathfrak{A}_{ad}$ has an empty topological interior. Therefore, the existence of L^1-convergent sequences of admissible controls $\{u_\varepsilon\}_{\varepsilon>0} \subset \mathfrak{A}_{ad}$, with monotone property $\xi_1(x) \le u_\varepsilon(x) \le u(x) \le \xi_2(x)$ or $\xi_1(x) \le u(x) \le u_\varepsilon(x) \le \xi_2(x)$ in Ω, is an unrealistic assumption.

Taking this observation into account, it is reasonable to introduce the following concept.

Definition 24.5 Let $u, \widehat{u} \in \mathfrak{A}_{ad}$ be a given pair of admissible controls. Let $u_\varepsilon := u + \varepsilon(\widehat{u} - u)$ for each $\varepsilon \in [0, 1]$. We say that the weighted Sobolev space H_u is stable along the direction $\widehat{u} - u$ if $H_u = \lim_{\varepsilon\to 0} H_{u_\varepsilon}$ in the following sense:

(K_1) for every $y \in H_u$, there exists a sequence $\left\{y_\varepsilon \in H_{u_\varepsilon}\right\}_{\varepsilon>0}$ such that $(u_\varepsilon, y_\varepsilon) \xrightarrow{\tau} (u, y)$ as $\varepsilon \to 0$;

(K_2) if $\{\varepsilon_k\}_{k\in\mathbb{N}}$ is a sequence converging to 0, and $\{y_k\}_{k\in\mathbb{N}}$ is a sequence such that $y_k \in H_{u_{\varepsilon_k}}$ for every $k \in \mathbb{N}$ and $(u_{\varepsilon_k}, y_k) \xrightarrow{\tau} (u, y)$, then $y_k \to y$ strongly in $L^p(\Omega)$ and $y \in H_u$.

The definition of the limit $W_u = \lim_{\varepsilon\to 0} W_{u_\varepsilon}$ can be done in a similar manner. As a result, Lemmas 24.4–24.6 can be easily generalized to the following assertion.

Lemma 24.7 *Assume that for a given $u, \widehat{u} \in \mathfrak{A}_{ad}$, the weighted Sobolev space H_u (respectively, W_u) is stable along the direction $\widehat{u} - u$. For each $\varepsilon > 0$, let $u_\varepsilon := u + \varepsilon(\widehat{u} - u)$ and let $y_\varepsilon = y(u_\varepsilon)$ be H_{u_ε}-solutions (resp., W_{u_ε}-solutions) to the boundary value problem (24.26). Then, up to a subsequence, we have*

$$u_\varepsilon \to u \ \text{in} \ L^1(\Omega), \tag{24.50}$$

$$y_\varepsilon \to y \ \text{in} \ L^p(\Omega), \quad \nabla y_\varepsilon \to \nabla y \ \text{in variable space} \ L^p(\Omega, u_\varepsilon\, dx)^N, \tag{24.51}$$

where $y \in H_u$ is the H_u-solution (resp., (24.50) and (24.51) take place and $y \in W_u$ is the W_u-solution) to the boundary value problem (24.26) for the given control u.

As for the proof, it is enough to apply Definition 24.5 and repeat the main arguments of the proofs of Lemmas 24.4–24.6.

Our next observation deals with some specification of the set of admissible controls $\mathfrak{A}_{ad}$. Having supposed that the functions ξ_1 and ξ_2 are extended to the whole space of $\mathbb{R}^N$ such that

$$\xi_1, \xi_2 \in L^1_{loc}(\mathbb{R}^N), \ 0 \le \xi_1(x) \le \xi_2(x) \text{ a.e. in } \Omega, \ \text{ and } \ \xi_1^{-\nu} \in L^1_{loc}(\mathbb{R}^N),$$

we assume that there exists a constant $C > 0$ such that

$$\sup_{B\in\mathbb{R}^N} \left(\frac{1}{|B|}\int_B \xi_2\, dx\right)\left(\frac{1}{|B|}\int_B \xi_1^{-1/(p-1)}\, dx\right)^{p-1} \le C, \tag{24.52}$$

where B is a ball in $\mathbb{R}^N$. In this case, we have the following result.

Theorem 24.5 *Assume the condition (24.52) holds true for some constant $C > 0$. Then, boundary value problem (24.26) has a unique weak solution for each $u \in \mathfrak{A}_{ad}$.*

Proof The main idea of this proof is to show that $W_u = H_u$ for each $u \in \mathfrak{A}_{ad}$ and this relies on the fact that such u belongs to the class of Muckenhoupt weights A_p. We omit the details.

Corollary 24.1 *Let $u, \widehat{u} \in \mathfrak{A}_{ad}$ be a given pair of admissible controls. Let $u_\varepsilon := u + \varepsilon(\widehat{u} - u)$ for each $\varepsilon \in [0, 1]$. Assume there exists a constant $C > 0$ such that the estimate (24.52) is valid. For each $\varepsilon > 0$, let $y_\varepsilon = y(u_\varepsilon)$ be the corresponding weak solutions to the boundary value problem (24.26). Then, up to a subsequence, the properties (24.50) and (24.51) hold true, where $y \in H_u$ is the weak solutions to the boundary value problem (24.26) for the given control u.*

Proof As follows from Theorem 24.5, the weak solutions $\left\{y_\varepsilon = y(u_\varepsilon) \in W_{u_\varepsilon}\right\}_{\varepsilon>0}$ can be defined in a unique way. Moreover, by a priori estimate (24.31), we see that the sequence $\{(u_\varepsilon, y_\varepsilon)\}_{\varepsilon>0}$ is bounded. Hence, by Theorem 24.3, this sequence is relatively τ-compact and each of its τ-cluster pairs (u, y) belongs to the set Ξ. Since, for each admissible control $u \in \mathfrak{A}_{ad}$, the boundary value problem (24.26) admits a unique weak solution, it follows that the τ-cluster pair (u, y) is uniquely defined. In order to show the strong convergence property (24.51), it remains to repeat the trick coming from the proof of Lemma 24.5.

24.6 On Differentiability of Lagrange Functional

In this section, we discuss the differentiable properties of the Lagrange functional associated with optimal control problem (24.26)–(24.28). Since the relations (24.26) can be seen as constraints, we define the Lagrangian as follows:

$$\begin{aligned}\Lambda(u, y, \mu) &= I(u, y) + a_u(y, \mu) - \int_\Omega f\mu\, dx \\ &= \|y - y_d\|^p_{L^p(\Omega)} + \|\nabla y\|^p_{L^p(\Omega, u\,dx)^N} + a_u(y, \mu) - \int_\Omega f\mu\, dx, \qquad (24.53)\end{aligned}$$

where $\mu \in W_u := W_0^{1,p}(\Omega, u\,dx)$ is a Lagrange multiplier and

$$a_u(y, \mu) = \langle -\Delta_p(u, y), \mu\rangle_{W_u^*; W_u} = \int_\Omega u(x)\left(|\nabla y|^{p-2}\nabla y, \nabla\mu\right)_{\mathbb{R}^N} dx.$$

In what follows, to each distribution $y \in W_u$, where $u \in \mathfrak{A}_{ad}$, we associate the following sets:

$$S_0(y) = \{x \in \Omega\ :\ |\nabla y(x)| = 0\}, \quad S_1(y) = \operatorname{int} S_0(y). \qquad (24.54)$$

For our further analysis, we adopt the following concept.

Definition 24.6 Let $u \in \mathfrak{A}_{ad}$ be a given control. We say that an element $y \in W_u$ is a regular point for the Lagrangian (24.53) if

$$\Omega\setminus\overline{S}_1(y) \text{ is a connected set with Lipschitz boundary,} \tag{24.55}$$

$$\text{and } S_0(y)\setminus S_1(y) \text{ has zero } u\text{-measure, i.e. } u(S_0) := \int_{S_0} u\,dx = 0, \tag{24.56}$$

and we say that an element $y \in W_u$ is a strongly regular point for the Lagrangian (24.53) if y is its regular point and $S_1(y) = \emptyset$.

Remark 24.7 In the non-degenerate case, i.e., when $u + u^{-1} \in L^\infty(\Omega)$, due to the results of Manfredi (see [27]), the notion of regularity is not too restrictive. In this case, the set $S_0 := \{x \in \Omega \,:\, |\nabla y| = 0\}$ for non-constant solutions of the p-Laplace equation (a p-harmonic function) has zero Lebesgue measure (see [27]). However, in the case of degenerate p-Laplacian $\Delta_p(u, y)$ with $u \in \mathfrak{A}_{ad}$, the regularity assumption is not obvious. Therefore, we put forward a hypothesis that if $y \in W_u$ is a regular point of the functional $\Lambda(u, y, \mu)$, then for every $v \in W_u$ there exists a positive number $\alpha \in \mathbb{R}$ $(\alpha \neq 0)$ such that each point of the segment $[y, \alpha v] = \{y + t(\alpha v - y) \,:\, \forall t \in [0, 1]\} \subset W_u$ is also regular for $\Lambda(u, y, \mu)$.

We are now ready to study the differentiability properties of the Lagrangian $\Lambda(u, y, \lambda)$. We begin with the following result.

Lemma 24.8 *Let $u \in \mathfrak{A}_{ad}$ be the given control, and let $y \in W_u$ be a regular point of the Lagrangian (24.53). Then, the mapping*

$$W_u \ni y \mapsto \Delta_p(u, y) = -\operatorname{div}\left(u(x)|\nabla y|^{p-2}\nabla y\right) \in W_u^*$$

is Gâteaux differentiable at y and its Gâteaux derivative $\left(-\Delta_p(u, y)\right)'_G \in \mathscr{L}\left(W_u, W_u^*\right)$ exists and takes the form:

$$\left(\Delta_p(u, y)\right)'_G[h] = \begin{cases} -\operatorname{div}\left(u(x)|\nabla y|^{p-2}\nabla h\right) & \\ -(p-2)\operatorname{div}\left(u(x)|\nabla y|^{p-4}\left(\nabla y, \nabla h\right)_{\mathbb{R}^N}\nabla y\right), & \text{in } \Omega\setminus S_1(y), \\ 0, & \text{in } S_1(y); \end{cases} \tag{24.57}$$

for $p > 2$, and

$$\left(\Delta_p(u, y)\right)'_G[h] = -\operatorname{div}\left(u(x)\nabla h\right) \quad \text{in } \Omega, \quad \text{if } p = 2. \tag{24.58}$$

Proof Let $y \in W_u$ be a regular point for the Lagrangian (24.53), and let $h \in W_u$ be an arbitrary distribution. Following the definition of Gâteaux derivative, we have to deduce the following equality:

$$\lim_{\lambda\to+0}\left\|\frac{\Delta_p(u, y + \lambda h) - \Delta_p(u, y)}{\lambda} - \left(\Delta_p(u, y)\right)'_G[h]\right\|_{W_u^*} = 0,$$

where $\left(\Delta_p(u,y)\right)'_G[h]$ is defined by (24.57). With that in mind, let us consider the vector-valued function $g(\lambda) := |\nabla y + \lambda\nabla h|^{p-2}(\nabla y + \lambda\nabla h)$ for which the Taylor's expansion with the remainder term in the Lagrange form leads to the relation

$$|g(\lambda) - g(0)| \le |g'(\theta)|\lambda, \quad \theta \in (0,\lambda),$$

where $g(0) = |\nabla y|^{p-2}\nabla y$ and

$$\begin{aligned} g'(\theta) &= |\nabla y + \theta\nabla h|^{p-2}\nabla h \\ &\quad + (p-2)|\nabla y + \theta\nabla h|^{p-2}\left(\theta|\nabla h|^2 + (\nabla y, \nabla h)_{\mathbb{R}^N}\right)\left(\nabla y + \theta\nabla h\right)\frac{1}{|\nabla y + \theta\nabla h|^2} \\ &= |\nabla y + \theta\nabla h|^{p-2}\nabla h + (p-2)|\nabla y + \theta\nabla h|^{p-2}\frac{(\nabla y + \theta\nabla h, \nabla h)_{\mathbb{R}^N}}{|\nabla y + \theta\nabla h|}\frac{\nabla y + \theta\nabla h}{|\nabla y + \theta\nabla h|} \end{aligned}$$

Let $\delta > 0$ be an arbitrary value. Let us consider the following decomposition:

$$\Omega = S_1(y) \cup \left(S_0(y)\setminus S_1(y)\right) \cup \Omega'_\delta \cup \Omega''_\delta,$$

where the sets $S_0(y)$ and $S_1(y)$ are defined in (24.54), and Ω'_δ and Ω''_δ are u-measurable subsets of Ω such that

$$\Omega'_\delta = \{x \in \Omega\setminus S_1(y) \;:\; |\nabla y(x)| \ge \delta\}, \quad \Omega''_\delta = \{x \in \Omega\setminus S_1(y) \;:\; 0 < |\nabla y(x)| < \delta\}.$$

Closely following [1, p.598] (see also [11]), it can be shown that for every $\varepsilon > 0$, there exists a positive value $\delta_0 > 0$ such that

$$\left\|g'(\theta) - (p-2)|\nabla y|^{p-4}\left(\nabla y, \nabla h\right)_{\mathbb{R}^N}\nabla y - |\nabla y|^{p-2}\nabla h\right\|_{L^q(\Omega'_\delta;u\,dx)^N} < \frac{\varepsilon}{2}, \tag{24.59}$$

$$\left\|g'(\theta) - (p-2)|\nabla y|^{p-4}\left(\nabla y, \nabla h\right)_{\mathbb{R}^N}\nabla y - |\nabla y|^{p-2}\nabla h\right\|_{L^q(\Omega''_\delta;u\,dx)^N} < \frac{\varepsilon}{2} \tag{24.60}$$

for all $\delta \in (0,\delta_0)$, $\theta \in (0,\lambda)$, and $\lambda > 0$ small enough. Moreover, as immediately follows from (24.54), we have the following relations:

$$\begin{aligned} \left\|g'(\theta)\right\|_{L^q(S_1(y);u\,dx)^N} &= (p-1)\theta^{p-2}\|\nabla h\|^{p-1}_{L^p(S_1(y);u\,dx)^N} \to 0 \text{ as } \theta \to +0 \text{ if } p > 2, \\ \left\|g'(\theta) - \nabla h\right\|_{L^q(S_1(y);u\,dx)^N} &= \|\nabla h - \nabla h\|_{L^q(S_1(y);u\,dx)^N} = 0 \text{ if } p = 2. \end{aligned}$$

Since $u\big(S_0(y)\setminus S_1(y)\big) = 0$ by the initial assumptions, it follows from (24.59) and (24.60) that the vector-valued function $|\nabla y|^{p-2}\nabla y$ is Gâteaux differentiable. Hence, the operator $\Delta_p(u,y) = -\operatorname{div}\left(u(x)|\nabla y|^{p-2}\nabla y\right)$ is Gâteaux differentiable at each regular point $y \in W_u$ and its Gâteaux derivative takes the form (24.57).

As an obvious consequence of this result and the fact that Gâteaux differentiability of operator $y \mapsto \Delta_p(u, y)$ implies existence of Gâteaux derivative for the functional $\varphi : W_u \to \mathbb{R}$, where

$$\varphi(y) = \langle -\Delta_p(u, y), \mu \rangle_{W^*;W_u} = \int_\Omega u(x) \left(|\nabla y|^{p-2} \nabla y, \nabla \mu \right)_{\mathbb{R}^N} dx$$

and $\langle \varphi'_G(y), h \rangle_{W_u^*;W_u} = \langle \left(-\Delta_p(u, y) \right)'_G [h], \mu \rangle_{W_u^*;W_u}$, $\forall \mu \in W_u$, we arrive at the following obvious assertion.

Corollary 24.2 *Let $u \in \mathfrak{A}_{ad}$ be a given element, and let $y \in W_u$ be a regular point of the Lagrangian (24.53). Then, the mapping*

$$W_u \ni y \mapsto \Lambda(u, y, \mu) = I(u, y) + a_u(y, \mu) - \langle f, \mu \rangle_{W_u^*;W_u} \in \mathbb{R}$$

is Gâteaux differentiable at y and its Gâteaux derivative $\Lambda'_G(u, y, \mu) \in W_u^*$ exists and takes the form:

$$\begin{aligned}
\langle \Lambda'_G(u, y, \mu), h \rangle_{W_u^*;W_u} &= p \int_{\Omega \setminus \overline{S}_1(y)} |\nabla y|^{p-2} (\nabla y, \nabla h)_{\mathbb{R}^N} u \, dx \\
+ p \int_\Omega |y - y_d|^{p-2} (y - y_d) h \, dx &+ \int_{\Omega \setminus \overline{S}_1(y)} \left(u(x) |\nabla y|^{p-2} \nabla \mu, \nabla h \right)_{\mathbb{R}^N} dx \\
&+ (p-2) \int_{\Omega \setminus \overline{S}_1(y)} u(x) |\nabla y|^{p-4} (\nabla y, \nabla \mu)_{\mathbb{R}^N} \left(\nabla y, \nabla h \right)_{\mathbb{R}^N} dx \\
&= p \int_\Omega |\nabla y|^{p-2} (\nabla y, \nabla h)_{\mathbb{R}^N} u \, dx \\
+ p \int_\Omega |y - y_d|^{p-2} (y - y_d) h \, dx &+ \int_\Omega \left(u(x) |\nabla y|^{p-2} \nabla \mu, \nabla h \right)_{\mathbb{R}^N} dx \\
&+ (p-2) \int_\Omega u(x) |\nabla y|^{p-4} (\nabla y, \nabla \mu)_{\mathbb{R}^N} \left(\nabla y, \nabla h \right)_{\mathbb{R}^N} dx.
\end{aligned} \tag{24.61}$$

Remark 24.8 Taking into account the equality $(\nabla y, \nabla \mu)_{\mathbb{R}^N} \nabla y = \left[\nabla y \otimes \nabla y \right] \nabla \mu$, the last term in (24.61) can be rewritten as follows:

$$(p-2) \int_\Omega u(x) |\nabla y|^{p-4} \left(\left[\nabla y \otimes \nabla y \right] \nabla \mu, \nabla h \right)_{\mathbb{R}^N} dx.$$

Before deriving the optimality conditions, we need the following auxiliary result.

Lemma 24.9 *Let $u \in \mathfrak{A}_{ad}$, $y \in W_u$, and $v \in W_u$ be the given distributions. Assume that each point of the segment $[y, v] = \{ y + \alpha(v - y) \ : \ \forall \alpha \in [0, 1] \} \subset W_u$ is regular for the mapping $v \to \Lambda(u, v, \mu)$. Then, there exists a positive value $\varepsilon \in (0, 1)$ such that*

$$
\begin{aligned}
\Lambda(u, v, \mu) - \Lambda(u, y, \mu) &= \langle \Lambda'_G(u, y + \varepsilon h, \mu), h\rangle_{W_u^*; W_u} \\
&= \int_\Omega |\nabla y + \varepsilon \nabla h|^{p-2} \Big(\nabla \mu, \nabla h\Big)_{\mathbb{R}^N} u\, dx \\
&+ p \int_\Omega |\nabla y + \varepsilon \nabla h|^{p-2} \Big(\nabla y + \varepsilon \nabla h, \nabla h\Big)_{\mathbb{R}^N} u\, dx \\
&+ p \int_\Omega |y + \varepsilon h - y_d|^{p-2} \Big(y + \varepsilon h - y_d\Big) h\, dx \\
&+ (p-2) \int_\Omega |\nabla y + \varepsilon \nabla h|^{p-4} \Big(\big[(\nabla y + \varepsilon \nabla h) \otimes (\nabla y + \varepsilon \nabla h)\big] \nabla \mu, \nabla h \Big)_{\mathbb{R}^N} u\, dx
\end{aligned}
\tag{24.62}
$$

with $h = v - y$.

Proof For given u, μ, y_d, y, and v, let us consider the scalar function $\varphi(t) = \Lambda(u, y + t(v - y), \mu)$. Since by Corollary 24.2, the functional $\Lambda(u, \cdot, \mu)$ is Gâteaux differentiable at each point of the segment $[y, v]$, it follows that the function $\varphi = \varphi(t)$ is differentiable on $(0, 1)$ and

$$
\varphi'(t) = \langle \Lambda'_G(u, y + t(v - y), \mu), v - y\rangle_{W_u^*; W_u}, \quad \forall t \in (0, 1).
$$

To conclude the proof, it remains to take into account (24.61) and apply the mean value theorem: $\varphi(1) - \varphi(0) = \varphi'(\varepsilon)$ for some $\varepsilon \in (0, 1)$.

24.7 Formalism of the Quasi-adjoint Technique

Let $(u_0, y_0) \in \Xi$ be an optimal pair for problem (24.26)–(24.28). Let $\mathfrak{A}_{ad}^{stab}$ be a subset of $L^1(\Omega)$ such that $\mathfrak{A}_{ad}^{stab} \subset \mathfrak{A}_{ad}$,

$$
\|\nabla y_0\|^p_{L^p(\Omega, \widehat{u}\, dx)^N} := \int_\Omega |\nabla y_0|^p \widehat{u}\, dx < +\infty \ \text{ and } \ \widehat{u}/u_0 \in L^\infty(\Omega), \ \text{ for all } \widehat{u} \in \mathfrak{A}_{ad}^{stab},
\tag{24.63}
$$

and the weighted Sobolev space H_{u_0} is stable along the direction $\widehat{u} - u_0$ for each $\widehat{u} \in \mathfrak{A}_{ad}^{stab}$. It is clear that $\mathfrak{A}_{ad}^{stab}$ is always non-empty because $u_0 \in \mathfrak{A}_{ad}^{stab}$ by definition of this set.

We begin with the following assumption:

(H0) y_0 is an H_{u_0}-solution to the boundary value problem (24.26).
(H1) For a given distribution $f \in L^q(\Omega)$, the optimal state $y_0 \in H_{u_0}$ is a regular point of the mapping $y \mapsto \Lambda(u, y, \lambda)$ in the sense of Definition 24.6.
(H2) The set $\mathfrak{A}_{ad}^{stab}$ is not a singleton.

Then,

$$\Delta\Lambda = \Lambda(u, y, \lambda) - \Lambda(u_0, y_0, \lambda) \geq 0, \forall\,(u, y) \in \Xi, \quad \forall\,\lambda \in C_0^\infty(\Omega). \qquad (24.64)$$

Since the set of admissible controls $\mathfrak{A}_{ad} \subset BV(\Omega)$ has an empty topological interior, we justify the choice of perturbation for an optimal control as follows: $u_\theta := u_0 + \theta(\widehat{u} - u_0)$, where $\widehat{u} \in \mathfrak{A}_{ad}^{stab}$ and $\theta \in [0, 1]$. As was indicated in Remark 24.3, for each $\theta \in [0, 1]$, there exists a unique H_{u_θ}-solution $y_\theta := y\,(u_\theta) = y\,(u_0 + \theta(\widehat{u} - u_0))$ to boundary value problem (24.34) and (24.35). Then, due to Hypotheses (H0)–(H2), we can suppose that the segment $[y_0, y_\theta]$ belongs to H_{u_0} for θ small enough (by the directional stability property). We also assume that

(H3) for each $\widehat{u} \in \mathfrak{A}_{ad}^{stab}$, there exists a numerical sequence $\{\theta_k\}_{k\in\mathbb{N}} \subset (0, 1]$ such that $\theta_k \to 0$ as $k \to \infty$, and $\left\{y_{\theta_k} := y\left(u_{\theta_k}\right)\right\}_{k\in\mathbb{N}}$ are strongly regular points for the mappind $v \to \Lambda(u, v, \lambda)$.

We note that if $y_0 \in H_{u_0}$ is a strongly regular point of the mapping $y \mapsto \Lambda(u, y, \lambda)$, then fulfillment of Hypothesis (H3) is obvious. As a result, we obtain

$$\begin{aligned} \Delta\Lambda =& \Lambda(u_\theta, y_\theta, \lambda) - \Lambda(u_0, y_0, \lambda) = \Lambda(u_\theta, y_\theta, \lambda) - \Lambda(u_\theta, y_0, \lambda) \\ &+ \Lambda(u_\theta, y_0, \lambda) - \Lambda(u_0, y_0, \lambda) = \Lambda(u_\theta, y_\theta, \lambda) - \Lambda(u_\theta, y_0, \lambda) \\ &+ \Lambda(u_\theta - u_0, y_0, \lambda) = \Delta_{y_\theta}\Lambda(u_\theta, y_0, \lambda) + \theta\Lambda(\hat{u} - u_0, y_0, \lambda) \geq 0. \quad (24.65) \end{aligned}$$

Hence, by Lemma 24.9, there exists a value $\varepsilon_\theta \in (0, 1)$ such that condition (24.65) can be represented as follows:

$$\begin{aligned} &\Delta\Lambda = \Lambda(u_\theta, y_\theta, \lambda) - \Lambda(u_0, y_0, \lambda) \\ &= \langle \Lambda'_G(u_\theta, y_0 + \varepsilon_\theta(y_\theta - y_0), \lambda), y_\theta - y_0\rangle_{H^*_{u_0};H_{u_0}} + \theta\Lambda(\widehat{u} - u_0, y_0, \lambda) \geq 0. \end{aligned} \qquad (24.66)$$

Using (24.61), we obtain

$$\begin{aligned} \Delta\Lambda =& p\int_\Omega |\nabla y_{\varepsilon_\theta,\theta}|^{p-2}\Big(\nabla y_{\varepsilon_\theta,\theta}, \nabla y_\theta - \nabla y_0\Big)_{\mathbb{R}^N} u_\theta\,dx + \theta\int_\Omega (\widehat{u} - u_0)\,|\nabla y_0|^p\,dx \\ &+ p\int_\Omega |y_{\varepsilon_\theta,\theta} - y_d|^{p-2}\Big(y_{\varepsilon_\theta,\theta} - y_d\Big)\,(y_\theta - y_0)\,dx \\ &+ \int_\Omega |\nabla y_{\varepsilon_\theta,\theta}|^{p-2}\Big(\nabla\lambda, \nabla y_\theta - \nabla y_0\Big)_{\mathbb{R}^N} u_\theta\,dx \\ &+ (p-2)\int_\Omega |\nabla y_{\varepsilon_\theta,\theta}|^{p-4}\Big(\big[\nabla y_{\varepsilon_\theta,\theta} \otimes \nabla y_{\varepsilon_\theta,\theta}\big]\nabla\lambda, \nabla y_\theta - \nabla y_0\Big)_{\mathbb{R}^N} u_\theta\,dx \\ &+ \theta\int_\Omega (\widehat{u} - u_0)\,\big(|\nabla y_0|^{p-2}\nabla y_0, \nabla\lambda\big)_{\mathbb{R}^N}\,dx \geq 0, \quad \forall\,\widehat{u} \in \mathfrak{A}_{ad}^{stab}, \end{aligned} \qquad (24.67)$$

where $y_{\varepsilon_\theta,\theta} = y_0 + \varepsilon_\theta(y_\theta - y_0)$.

Now, we introduce the concept of quasi-adjoint states that were first considered for linear problems by Serovajskiy [31].

Definition 24.7 We say that for given $\theta \in [0, 1]$ and $\widehat{u} \in \mathfrak{A}_{ad}$, a distribution ψ_θ is a quasi-adjoint state to $y_0 \in H_{u_0}$ if ψ_θ satisfies the following integral identity:

$$\int_\Omega |\nabla y_{\varepsilon_\theta,\theta}|^{p-2}\Big(\Big[I + (p-2)\frac{\nabla y_{\varepsilon_\theta,\theta}}{|\nabla y_{\varepsilon_\theta,\theta}|} \otimes \frac{\nabla y_{\varepsilon_\theta,\theta}}{|\nabla y_{\varepsilon_\theta,\theta}|}\Big]\nabla \psi_\theta, \nabla\varphi\Big)_{\mathbb{R}^N} u_\theta \, dx$$
$$+ p\int_\Omega |\nabla y_{\varepsilon_\theta,\theta}|^{p-2}\Big(\nabla y_{\varepsilon_\theta,\theta}, \nabla\varphi\Big)_{\mathbb{R}^N} u_\theta \, dx$$
$$+ p\int_\Omega |y_{\varepsilon_\theta,\theta} - y_d|^{p-2}\Big(y_{\varepsilon_\theta,\theta} - y_d\Big)\varphi \, dx = 0, \quad \forall \varphi \in H_{u_\theta}, \qquad (24.68)$$

or in terms of distributions, ψ_θ is a solution to the following degenerate boundary value problem

$$-\operatorname{div}(\rho_\theta A_\theta \nabla \psi_\theta) = g_\theta \quad \text{in } \Omega, \qquad \psi_\theta = 0 \quad \text{on } \partial\Omega. \qquad (24.69)$$

Here,

$$\rho_\theta = u_\theta |\nabla y_{\varepsilon_\theta,\theta}|^{p-2}, \qquad (24.70)$$
$$A_\theta = I + (p-2)\frac{\nabla y_{\varepsilon_\theta,\theta}}{|\nabla y_{\varepsilon_\theta,\theta}|} \otimes \frac{\nabla y_{\varepsilon_\theta,\theta}}{|\nabla y_{\varepsilon_\theta,\theta}|}, \qquad (24.71)$$
$$g_\theta = p \operatorname{div}\left(u_\theta |\nabla y_{\varepsilon_\theta,\theta}|^{p-2}\nabla y_{\varepsilon_\theta,\theta}\right) - p\,|y_{\varepsilon_\theta,\theta} - y_d|^{p-2}\left(y_{\varepsilon_\theta,\theta} - y_d\right), \qquad (24.72)$$

$I \in \mathscr{L}(\mathbb{R}^N; \mathbb{R}^N)$ is the identity matrix, $y_\theta := y\,(u_\theta) = y\,(u_0 + \theta(\widehat{u} - u_0))$ is the H_{u_θ}-solution of problem (24.34) and (24.35), $y_{\varepsilon_\theta,\theta} = y_0 + \varepsilon_\theta(y_\theta - y_0)$, and $\varepsilon_\theta = \varepsilon(u_\theta) \in (0, 1)$ is a constant coming from equality (24.66).

A crucial point of this definition is the choice of the class of test functions in integral identity (24.68) ($\varphi \in H_{u_\theta}$). At the end of this section, it will be shown that Definition 24.7 makes a sense, and moreover, under some additional assumptions, the quasi-adjoint states $\{\psi_\theta\}_{\theta\to 0}$ can be defined in a unique way for each $\theta \in [0, 1]$ in spite of the fact that boundary value problem (24.69) is degenerate in general.

Remark 24.9 If we assume that the quasi-adjoint state ψ_θ is defined by Definition 24.7 and the integral $\int_\Omega(\widehat{u} - u_0)\left(|\nabla y_0|^{p-2}\nabla y_0, \nabla\psi_\theta\right)_{\mathbb{R}^N} dx$ exists for all $\widehat{u} \in \mathfrak{A}_{ad}^{stab}$, then as $(y_\theta - y_0) \in H_{u_\theta}$ for each $\theta > 0$ (see (24.81) and Hypothesis (H2)), the element λ in (24.67) can be defined as the quasi-adjoint state. As a result, having put $\lambda = \psi_\theta$ in (24.67), the increment of the Lagrangian (24.67) can be simplified as

$$\int_\Omega (\widehat{u} - u_0)\Big[\, |\nabla y_0|^p + \left(|\nabla y_0|^{p-2}\nabla y_0, \nabla\psi_\theta\right)_{\mathbb{R}^N}\Big]\, dx \geq 0, \quad \forall \widehat{u} \in \mathfrak{A}_{ad}^{stab}. \qquad (24.73)$$

Thus, in order to derive the necessary optimality conditions, it remains to prove the existence and the compactness properties of the sequence of quasi-adjoint states

$\{\psi_\theta\}_{\theta\to 0}$ (with respect to some appropriate topology) and pass to the limit in (24.69)–(24.73) as $\theta \to +0$.

To begin with, we establish a few auxiliary results. The characteristic feature of the class of admissible controls $\mathfrak{A}_{ad}$ is the fact that strong convergence $u_k \to u$ in $L^1(\Omega)$ implies weak convergence in variable space $L^p(\Omega, u_k\,dx)^N$ of $\nabla y(u_k) \to \nabla y(u)$ as $k \to \infty$ (see, for instance, Theorem 24.3). However, we infer from Lemma 24.7 that the mapping $u_0 \mapsto y(u_0)$ enjoys stronger properties provided some "directional stability assumptions" on the space H_{u_0} hold. In particular, in this case, we have the following result (for the details, see the proof of Lemma 24.5).

Lemma 24.10 *Assume that for a given $\widehat{u} \in \mathfrak{A}_{ad}$, Hypotheses (H0)–(H3) are valid. Let $\theta \in [0,1]$, $u_\theta := u + \theta(\widehat{u} - u)$, and let $y_\theta = y(u_\theta)$ be the corresponding H_{u_θ}-solutions to the boundary value problem (24.26). Then, $u_\theta \to u_0$ in $L^1(\Omega)$,*

$$y_\theta \to y_0 \text{ in } L^p(\Omega), \quad \nabla y_\theta \to \nabla y_0 \text{ in the variable space } L^p(\Omega, u_\theta\,dx)^N \text{ as } \theta \to 0. \tag{24.74}$$

Taking this fact into account, we arrive at the following properties of the sequence

$$\left\{y_{\varepsilon_\theta,\theta} = y_0 + \varepsilon_\theta(y_\theta - y_0)\right\}_{\theta\to 0}. \tag{24.75}$$

Proposition 24.2 *Assume that for a given $\widehat{u} \in \mathfrak{A}_{ad}$, Hypothesis (H2) is valid. Then,*

$$y_{\varepsilon_\theta,\theta} \in H_{u_\theta}, \quad \forall \theta \in [0,1], \tag{24.76}$$

$$|\nabla y_{\varepsilon_\theta,\theta}|^p u_\theta \to |\nabla y_0|^p u_0 \text{ in } L^1(\Omega) \text{ as } \theta \to 0, \tag{24.77}$$

$$|y_{\varepsilon_\theta,\theta} - y_d|^{p-2}\left(y_{\varepsilon_\theta,\theta} - y_d\right) \to |y_0 - y_d|^{p-2}(y_0 - y_d) \text{ in } L^q(\Omega) \text{ as } \theta \to 0. \tag{24.78}$$

Proof By definition of the functions y_θ and y_0, we have

$$\|\nabla y_\theta\|_{L^p(\Omega,u_\theta\,dx)^N} < +\infty, \quad \text{and} \quad \|\nabla y_0\|_{L^p(\Omega,u_0\,dx)^N} < +\infty. \tag{24.79}$$

Using the convexity of the norm $\|\cdot\|_{L^p(\Omega,u_\theta\,dx)^N}$ and representation (24.75), we get

$$\|\nabla y_{\varepsilon_\theta,\theta}\|_{L^p(\Omega,u_\theta\,dx)^N} \le (1-\varepsilon_\theta)\|\nabla y_0\|_{L^p(\Omega,u_\theta\,dx)^N} + \varepsilon_\theta\|\nabla y_\theta\|_{L^p(\Omega,u_\theta\,dx)^N}. \tag{24.80}$$

Since

$$\|\nabla y_0\|^p_{L^p(\Omega,u_\theta\,dx)^N} = (1-\theta)\|\nabla y_0\|^p_{L^p(\Omega,u_0\,dx)^N} + \theta\|\nabla y_0\|^p_{L^p(\Omega,\widehat{u}\,dx)^N}, \tag{24.81}$$

by Hypothesis (H2) and $(24.79)_2$, it follows that $\|\nabla y_0\|_{L^p(\Omega,u_\theta\,dx)^N} < +\infty$. Thus, the inclusion $y_{\varepsilon_\theta,\theta} \in H_{u_\theta}$ is a direct consequence of the condition $y_\theta \in H_{u_\theta}$ and inequality (24.80). As for the property (24.77), in view of Lemma 24.10, we have, within a subsequence, that $u_\theta \to u_0$ a.e. in Ω, and hence (see (24.74)), $|\nabla y_\theta|^p(u_\theta - u_0) \to 0$ a.e. in Ω. Since $|\nabla y_\theta|^p u_\theta \geq 0$ a.e. in Ω and

$$\lim_{\theta\to 0}\int_\Omega |\nabla y_\theta|^p u_\theta\,dx = \int_\Omega |\nabla y_0|^p u_0\,dx \quad \text{by } (24.74)_2,$$

Scheffe's theorem implies strong convergence

$$|\nabla y_\theta|^p u_\theta \to |\nabla y_0|^p u_0 \quad \text{in } L^1(\Omega) \text{ as } \theta \to 0. \tag{24.82}$$

To conclude the proof, it remains to note that because of representation (24.75) and condition (24.76), the functions $y_{\varepsilon_\theta,\theta}$ inherit all limit properties of the sequence $\{y_\theta\}_{\theta\to 0}$, i.e., assertion (24.82) remains valid for the sequence $\{y_{\varepsilon_\theta,\theta}\}_{\theta\to 0}$ as well. Therefore, the property (24.78) is a direct consequence of the strong convergence $y_{\varepsilon_\theta,\theta} \to y_0$ in $L^p(\Omega)$ (see $(24.74)_1$).

We note that in view of property (24.77) (see Proposition 24.2), up to a subsequence we have $|\nabla y_{\varepsilon_\theta,\theta}|^{p-2}u_\theta \to |\nabla y_0|^{p-2}u_0$ a.e. in Ω. Since

$$\int_\Omega |\nabla y_{\varepsilon_\theta,\theta}|^{p-2}u_\theta\,dx = \int_\Omega |\nabla y_{\varepsilon_\theta,\theta}|^{p-2}u_\theta^{(p-2)/p}u_\theta^{2/p}\,dx = \left\{\begin{matrix}\widehat{p} = p/(p-2)\\ \widehat{q} = p/2\end{matrix}\right\}$$
$$\leq \|\nabla y_{\varepsilon_\theta,\theta}\|^{p-2}_{L^p(\Omega,u_\theta\,dx)^N}\|u_\theta\|^{2/p}_{L^1(\Omega)} \leq C\|f\|^{(p-2)/(p-1)}_{L^q(\Omega)}\|\xi_2\|^{2/p}_{L^1(\Omega)},$$

it follows that the sequence $\{|\nabla y_{\varepsilon_\theta,\theta}|^{p-2}u_\theta\}_{\theta\to 0}$ is equi-integrable. Hence, by Lebesgue's theorem, we arrive at the following property.

Corollary 24.3 *Under assumptions of Proposition 24.2,*

$$|\nabla y_{\varepsilon_\theta,\theta}|^{p-2}u_\theta \to |\nabla y_0|^{p-2}u_0 \quad \text{in } L^1(\Omega) \text{ as } \theta \to 0. \tag{24.83}$$

Proposition 24.3 *Let $(u_0, y_0) \in \Xi$ be an optimal pair for problem (24.26)–(24.28). Assume that for a given $\widehat{u} \in \mathfrak{A}_{ad}$, Hypotheses (H0)–(H3) are valid. Then, the matrix A_θ, given by (24.71), possesses the following properties:*

$$A_\theta \in L^\infty(\Omega; \mathbb{S}^N_{sym}), \quad \text{where } \mathbb{S}^N_{sym} \text{ is the set of all } N\times N \text{ symmetric matrices,} \tag{24.84}$$

$$A_\theta \to A_0 := I + (p-2)\frac{\nabla y_0}{|\nabla y_0|}\otimes\frac{\nabla y_0}{|\nabla y_0|} \quad \text{in } L^r(\Omega; \mathbb{S}^N_{sym})\ \forall\, r \in [1,\infty) \text{ as } \theta \to 0, \tag{24.85}$$

$$|\eta|^2 \leq (\eta, A_\theta\eta)_{\mathbb{R}^N} \leq \left[1 + (p-2)2^{N-1}\right]|\eta|^2 \quad \text{a.e. in } \Omega,\ \forall\, \eta \in \mathbb{R}^N,\ \forall\, \theta \in [0,1]. \tag{24.86}$$

Proof The proof is straightforward and, hence, omitted.

Our next intention is to study the variational problem (24.69). With that in mind, to each value $\theta \in [0, 1]$, we associate two weighted Sobolev spaces $\mathscr{H}_\theta$ and $\mathscr{W}_\theta$, where $\mathscr{W}_\theta$ is the set of functions $\psi \in W_0^{1,1}(\Omega)$ for which the norm

$$\|\psi\|_{\rho_\theta, A_\theta} = \left(\int_\Omega \left(\psi^2 + (\nabla\psi, A_\theta \nabla\psi)_{\mathbb{R}^N}\, \rho_\theta\right) dx\right)^{1/2} \tag{24.87}$$

is finite, and $\mathscr{H}_\theta$ is the closure of $C_0^\infty(\Omega)$ with respect to the norm (24.87). It is clear that $\mathscr{H}_\theta \subseteq \mathscr{W}_\theta$.

Remark 24.10 Some spaces of more or less similar type have been studied by Casas and Fernández [6], Murthy and Stampacchia [29], and Trudinger [32]. However, in contrast to the mentioned papers, we do not have a continuous embedding of $\mathscr{H}_\theta$ in the reflexive Banach space $W^{1,2}(\Omega)$.

For a fixed $\theta \in [0, 1]$, let us assume that

$$\left(y_{\varepsilon_\theta,\theta} - y_d\right) \in L^{2p-2}(\Omega) \ \text{ and } \ \rho_\theta^{-\sigma} := |\nabla y_{\varepsilon_\theta,\theta}|^{(2-p)\sigma} u_0^{-\sigma} \in L^1(\Omega) \tag{24.88}$$

for some $\sigma \in \left(\frac{N}{2}, +\infty\right)$. First of all, we note that due to Hypothesis (H3), the assumptions $(24.88)_2$ make a sense. As a result, the condition $(24.88)_2$ and Proposition 24.3 imply that the expression

$$\|\psi\|_{\mathscr{H}_\theta} = \left[\int_\Omega |\nabla\psi|^2 \rho_\theta\, dx\right]^{1/2} \tag{24.89}$$

can be considered as a norm on $\mathscr{H}_\theta$ and it is equivalent to the norm (24.87) (see [10, pp. 46] and Proposition 24.3). Moreover, in this case, the embedding $\mathscr{H}_\theta \hookrightarrow L^2(\Omega)$ is compact for a given $\theta > 0$ and due to Proposition 24.3 and estimates

$$\int_\Omega |\psi|\, dx \le \left(\int_\Omega \psi^2\, dx\right)^{1/2} |\Omega|^{1/2} \le C\|\psi\|_{\mathscr{H}_\theta}, \tag{24.90}$$

$$\int_\Omega |\nabla\psi|\, dx \le \left(\int_\Omega |\nabla\psi|^2 \rho_\theta\, dx\right)^{1/2} \left(\int_\Omega \rho_\theta^{-1}\, dx\right)^{1/2} \le C\|\psi\|_{\mathscr{H}_\theta}, \tag{24.91}$$

the space $\mathscr{H}_\theta$ is complete with respect to the norm $\|\cdot\|_{\mathscr{H}_\theta}$ with continuous embedding $\mathscr{H}_\theta \subset W_0^{1,1}(\Omega)$. Moreover, $\mathscr{H}_\theta$ is a Hilbert space with the inner product

$$(\psi_1, \psi_2)_{\mathscr{H}_\theta} = \int_\Omega (\nabla\psi_1, \nabla\psi_2)_{\mathbb{R}^N}\, \rho_\theta\, dx.$$

As a result, we can pass to the following variational formulation of the problem (24.69)

$$\begin{cases} \text{Find } \psi_\theta \in \mathscr{H}_\theta \text{ such that} \\ \displaystyle\int_\Omega (\nabla\varphi, A_\theta \nabla\psi_\theta)_{\mathbb{R}^N}\, \rho_\theta\, dx = \langle g_\theta, \varphi\rangle_{\mathscr{H}_\theta^*;\mathscr{H}_\theta}, \ \forall\, \varphi \in C_0^\infty(\Omega). \end{cases} \tag{24.92}$$

Since

$$\int_\Omega (\nabla\varphi, A_\theta \nabla\psi_\theta)_{\mathbb{R}^N}\, \rho_\theta\, dx \overset{\text{by Proposition 24.3}}{\leq} C\int_\Omega |\nabla\varphi||\nabla\psi_\theta|\rho_\theta\, dx \leq C\|\varphi\|_{\mathscr{H}_\theta}\|\psi_\theta\|_{\mathscr{H}_\theta}, \tag{24.93}$$

$$\int_\Omega (\nabla\psi_\theta, A_\theta \nabla\psi_\theta)_{\mathbb{R}^N}\, \rho_\theta\, dx \geq \int_\Omega |\nabla\psi_\theta|^2 \rho_\theta\, dx = \|\psi_\theta\|^2_{\mathscr{H}_\theta},$$

it follows that the bilinear form $a_\theta : \mathscr{H}_\theta \times \mathscr{H}_\theta \to \mathbb{R}$, where

$$a_\theta(\varphi, \psi) := \int_\Omega (\nabla\varphi, A_\theta\nabla\psi)_{\mathbb{R}^N}\, \rho_\theta\, dx,$$

is continuous and $\mathscr{H}_\theta$-coercive, whereas the right-hand side of (24.92) is a linear continuous functional on $\mathscr{H}_\theta$. Indeed, in view of condition $(24.88)_1$, we have

$$\begin{aligned} \langle g_\theta, \varphi\rangle_{\mathscr{H}_\theta^*;\mathscr{H}_\theta} &\leq p\int_\Omega |\nabla y_{\varepsilon_\theta,\theta}|^{p-1}|\nabla\varphi| u_\theta\, dx + p\int_\Omega |y_{\varepsilon_\theta,\theta} - y_d|^{p-1}|\varphi|\, dx \\ &= p(I_1 + I_2), \\ I_1 &\leq \left(\int_\Omega |\nabla y_{\varepsilon_\theta,\theta}|^2 |\nabla y_{\varepsilon_\theta,\theta}|^{p-2} u_\theta\, dx\right)^{1/2} \left(\int_\Omega |\nabla\varphi|^2 |\nabla y_{\varepsilon_\theta,\theta}|^{p-2} u_\theta\, dx\right)^{1/2} \\ &= \|y_{\varepsilon_\theta,\theta}\|^{p/2}_{H_{u_\theta}} \|\varphi\|_{\mathscr{H}_\theta}, \end{aligned} \tag{24.94}$$

$$I_2 \leq \|y_{\varepsilon_\theta,\theta} - y_d\|^{p-1}_{L^{2p-2}(\Omega)} \|\varphi\|_{L^2(\Omega)} \overset{\text{by (24.88)}}{\leq} C\|y_{\varepsilon_\theta,\theta} - y_d\|^{p-1}_{L^{2p-2}(\Omega)}\|\varphi\|_{\mathscr{H}_\theta}. \tag{24.95}$$

Therefore, by Lax–Milgram theorem, we immediately conclude that due to the assumptions (24.88), the variational problem (24.92) has a unique solution $\psi_\theta \in \mathscr{H}_\theta$ (denoted as $\mathscr{H}_\theta$-solution) with the a priori estimate

$$\begin{aligned} \|\nabla\psi_\theta\|_{\mathscr{H}_\theta} &\leq Cp\, \|y_{\varepsilon_\theta,\theta} - y_d\|^{p-1}_{L^{2p-2}(\Omega)} + p\, \|y_{\varepsilon_\theta,\theta}\|^{p/2}_{H_{u_\theta}} \\ &\leq Cp\left(\|y_{\varepsilon_\theta,\theta} - y_d\|^{p-1}_{L^{2p-2}(\Omega)} + \|f\|^{\frac{p}{2(p-1)}}_{L^q(\Omega)}\right). \end{aligned} \tag{24.96}$$

Remark 24.11 As obviously follows from the relation

$$\|\varphi\|^2_{\mathcal{H}_\theta} := \int_\Omega |\nabla \varphi|^2 \rho_\theta\,dx = \int_\Omega |\nabla \varphi|^2 |\nabla y_{\varepsilon_\theta,\theta}|^{p-2} u_\theta\,dx = \begin{Bmatrix} \widehat{p} = p/(p-2) \\ \widehat{q} = p/2 \end{Bmatrix}$$
$$\leq \left(\int_\Omega |\nabla y_{\varepsilon_\theta,\theta}|^p u_\theta\,dx\right)^{(p-2)/p} \left(\int_\Omega |\nabla \varphi|^p u_\theta\,dx\right)^{2/p} = \|y_{\varepsilon_\theta,\theta}\|^{p-2}_{H_{u_\theta}} \|\varphi\|^2_{H_{u_\theta}}, \quad (24.97)$$

which holds true for every $\varphi \in \mathcal{H}_\theta$, we have: $H_{u_\theta} \subset \mathcal{H}_\theta$ with continuous embedding. Hence, combining the inequalities (24.93)–(24.95) with estimate (24.97), we see that the integral identity (24.92) can be extended by continuity to any test functions $\varphi \in H_{u_\theta}$. Thus, the definition of quasi-adjoint states ψ_θ in the form of integral identity (24.68), where the test functions φ are considered as elements of the weighted Sobolev space H_{u_θ}, is correct provided assumptions (24.88) are valid for each $\theta \in [0, 1]$.

24.8 Substantiation of the Optimality Conditions for Optimal Control Problem (24.26)–(24.28) in the Framework of Weighted Sobolev Spaces

In view of Remark 24.11, we begin this section with the following hypothesis, which can be viewed as some supplement to the Hypotheses (H0)–(H3) (see (24.88))

(H4) For given $y_d \in L^p(\Omega)$, $\widehat{u} \in \mathfrak{A}^{stab}_{ad} \subset L^1(\Omega)$, $f \in L^q(\Omega)$ with $q = \frac{p}{p-1}$ and $p \geq 2$, there exist positive values $\widetilde{C} > 0$ and $\sigma \in \left(\frac{N}{2}, +\infty\right)$ such that

$$\sup_{\theta \in [0,1]} \|y_{\varepsilon_\theta,\theta} - y_d\|_{L^{2p-2}(\Omega)} \leq \widetilde{C}, \quad \rho_\theta^{-\sigma} \in L^1(\Omega). \quad (24.98)$$

Let $(u_0, y_0) \in \Xi$ be an optimal pair for problem (24.26)–(24.28). Let $\widehat{u} \in \mathfrak{A}^{stab}_{ad} \subset L^1(\Omega)$ be a fixed control function, and let $u_\theta := u_0 + \theta(\widehat{u} - u_0)$ for each $\theta \in [0, 1]$. Let, as before, $y_\theta := y(u_\theta) = y(u_0 + \theta(\widehat{u} - u_0))$ be an H_{u_θ}-solution of problem (24.34) and (24.35) and $y_{\varepsilon_\theta,\theta} = y_0 + \varepsilon_\theta(y_\theta - y_0)$, where the constant $\varepsilon_\theta = \varepsilon(u_\theta) \in (0, 1)$ is taken from equality (24.66). Having assumed that Hypotheses (H0)–(H4) are valid, we see that the sequence of quasi-adjoint states $\{\psi_\theta \in \mathcal{H}_\theta\}_{\theta \to 0}$ can be defined in a unique way for a special choice of the numerical sequence (see Hypothesis (H3)). Moreover, in view of (24.96), this sequence is bounded in the variable space $\mathcal{H}_\theta$, i.e.,

$$\sup_{\theta \to 0} \int_\Omega \left(\psi_\theta^2 + (\nabla \psi_\theta, A_\theta \nabla \psi_\theta)_{\mathbb{R}^N}\, \rho_\theta\right) dx$$
$$\leq C\left(\sup_{\theta \to 0} \|y_{\varepsilon_\theta,\theta} - y_d\|^{p-1}_{L^{2p-2}(\Omega)} + \|f\|^{\frac{p}{2(p-1)}}_{L^q(\Omega)}\right) < +\infty.$$

Therefore, in view of the property (24.83), we can extract a subsequence of sequence $\{\psi_\theta \in \mathscr{H}_\theta\}_{\theta\to 0}$, still denoted by the same index, such that (see the main properties of convergence in the variable L^p-spaces) $\psi_\theta \rightharpoonup \psi$ in $L^2(\Omega)$, $\nabla\psi_\theta \rightharpoonup v$ in the variable space $L^2(\Omega, \rho_\theta\, dx)^N$, where the last assertion means a fulfillment of the following conditions: $\{\nabla\psi_\theta\}_{\theta\to 0}$ is a bounded sequence in variable space $L^2(\Omega, |\nabla y_{\varepsilon_\theta,\theta}|^{p-2}u_\theta\, dx)^N$, $v \in L^2(\Omega, |\nabla y_0|^{p-2}u_0\, dx)^N$ and

$$\lim_{\theta\to 0}\int_\Omega (\nabla\psi_\theta, \nabla\varphi)_{\mathbb{R}^N}\, |\nabla y_{\varepsilon_\theta,\theta}|^{p-2}u_\theta dx = \int_\Omega (v, \nabla\varphi)_{\mathbb{R}^N}\, |\nabla y_0|^{p-2}u_0 dx,\ \forall\varphi \in C_0^\infty(\Omega).$$

Applying the arguments of Lemma 24.3 and estimates (24.90) and (24.91), we get: $v = \nabla\psi$ and

$$\begin{aligned}\|\psi\|^2_{\rho_0,A_0} &:= \int_\Omega \left(\psi^2 + (\nabla\psi, A_0\nabla\psi)_{\mathbb{R}^N}\,\rho_0\right) dx\\ &\overset{\text{by Proposition 24.3}}{\leq} \left(1+(p-2)2^{N-1}\right)\int_\Omega \left[\psi^2 + |\nabla\psi|^2|\nabla y_0|^{p-2}u_0\right] dx\\ &\leq C\left(1+(p-2)2^{N-1}\right)\left(\sup_{\theta\to 0}\|y_{\varepsilon_\theta,\theta} - y_d\|^{p-1}_{L^{2p-2}(\Omega)} + \|f\|^{\frac{p}{2(p-1)}}_{L^q(\Omega)}\right) < +\infty.\end{aligned}$$

As a result, we arrive at the following assertion.

Proposition 24.4 *If Hypotheses (H0)–(H4) are valid, then variational problem (24.92) has a unique $\mathscr{H}_\theta$-solution ψ_θ for every $\theta \in \{\theta_k\}_{k\in\mathbb{N}}$, where the sequence $\{\theta_k\}_{k\in\mathbb{N}}$ is given by Hypothesis (H3), and the sequence $\{\psi_\theta \in \mathscr{H}_\theta\}_{\theta\to 0}$ is relatively compact with respect to the following convergence:*

$$\psi_\theta \rightharpoonup \psi \text{ in } L^2(\Omega),\quad \nabla\psi_\theta \rightharpoonup \nabla\psi \text{ in the variable space } L^2(\Omega, \rho_\theta\, dx)^N. \quad (24.99)$$

Remark 24.12 It should be emphasized that in general, the limit function ψ to the sequence $\{\psi_\theta \in \mathscr{H}_\theta\}_{\theta\to 0}$, in the sense of (24.99), does not belong to the weighted space $\mathscr{H}_0 = \mathrm{cl}_{\|\cdot\|_{\rho_0,A_0}} C_0^\infty(\Omega)$, but it should be considered as some element of the weighted space $\mathscr{W}_0$, i.e., $\|\psi\|_{\rho_0,A_0} < +\infty$. So, the inclusion $\psi \in \mathscr{H}_0$ is an open question.

Thus, in order to finally characterize the limit function ψ, it remains to pass to the limit in (24.68) and (24.73) as $\theta \to 0$.

Lemma 24.11 *Let $(u_0, y_0) \in \Xi$ be an optimal pair to the problem (24.26)–(24.28). Assume that for a given $\widehat{u} \in \mathfrak{A}_{ad}$, Hypotheses (H0)–(H4) are valid. Let $\left\{y_\theta \in H_{u_\theta}\right\}_{\theta\to 0}$ and $\{\varphi_\theta \in \mathscr{H}_\theta\}_{\theta\to 0}$ be the given sequences such that*

$$y_\theta \to y_0 \text{ in } L^p(\Omega),\quad \nabla y_\theta \to \nabla y_0 \text{ in variable space } L^p(\Omega, u_\theta\, dx)^N, \quad (24.100)$$

$$\varphi_\theta \rightharpoonup \varphi \text{ in } L^2(\Omega),\quad \nabla\varphi_\theta \rightharpoonup \nabla\varphi \text{ in variable space } L^2(\Omega, |\nabla y_\theta|^{p-2}u_\theta\, dx)^N. \quad (24.101)$$

Then,

$$\lim_{\theta\to 0}\int_\Omega |\nabla y_\theta|^{p-2}\Big(\nabla y_\theta, \nabla\varphi_\theta\Big)_{\mathbb{R}^N} u_\theta\,dx = \int_\Omega |\nabla y_0|^{p-2}\Big(\nabla y_0, \nabla\varphi\Big)_{\mathbb{R}^N} u_0\,dx. \tag{24.102}$$

Proof By initial assumptions, the sequence $\left\{\nabla y_\theta \in L^p(\Omega, u_\theta\,dx)^N\right\}_{\theta\to 0}$ is bounded. Hence, there exists $C > 0$ such that

$$\|\nabla y_\theta\|^2_{L^2(\Omega,|\nabla y_\theta|^{p-2}u_\theta\,dx)^N} := \int_\Omega |\nabla y_\theta|^2|\nabla y_\theta|^{p-2}u_\theta\,dx = \|\nabla y_\theta\|^p_{L^p(\Omega,u_\theta\,dx)^N} \le C,$$

i.e., $\nabla y_\theta \in L^2(\Omega, |\nabla y_\theta|^{p-2}u_\theta\,dx)^N$ for all $\theta > 0$, and the sequence $\{\nabla y_\theta\}_{\theta\to 0}$ can be considered as a bounded sequence in variable Hilbert space $L^2(\Omega, |\nabla y_\theta|^{p-2}u_\theta\,dx)^N$. Taking into account the condition $(24.100)_2$, we have

$$\lim_{\theta\to 0}\int_\Omega |\nabla y_\theta|^2|\nabla y_\theta|^{p-2}u_\theta\,dx = \int_\Omega |\nabla y_0|^2|\nabla y_0|^{p-2}u_0\,dx.$$

Hence, the sequence $\left\{\nabla y_\theta \in L^2(\Omega, |\nabla y_\theta|^{p-2}u_\theta\,dx)^N\right\}_{\theta\to 0}$ strongly converges to ∇y_0 in variable space $L^2(\Omega, |\nabla y_\theta|^{p-2}u_\theta\,dx)^N$. As a result, in the left-hand side of (24.102), we have a product of weakly and strongly convergent sequences in variable space $L^2(\Omega, |\nabla y_\theta|^{p-2}u_\theta\,dx)^N$. Therefore, relation (24.102) is a direct consequence of the strong convergence definition in variable spaces (see (24.11)).

Lemma 24.12 *Let* $(u_0, y_0) \in \Xi$ *be an optimal pair for problem (24.26)–(24.28). Assume that for a given* $\widehat{u} \in \mathfrak{A}_{ad}$*, Hypotheses (H0)–(H4) are valid. Let* $\{\psi_\theta\}_{\theta\to 0}$ *be a bounded sequence in variable space* $\mathscr{H}_\theta$*. Assume that* ψ_θ *converges to* ψ *in the sense of (24.99). Then,*

$$\lim_{\theta\to 0}\int_\Omega (\nabla\varphi, A_\theta\nabla\psi_\theta)_{\mathbb{R}^N}\,\rho_\theta\,dx = \int_\Omega (\nabla\varphi, A_0\nabla\psi)_{\mathbb{R}^N}\,\rho_0\,dx \quad \forall\,\varphi \in C_0^\infty(\Omega), \tag{24.103}$$

$$\textit{i.e.,}\quad A_\theta\nabla\psi_\theta \rightharpoonup A_0\nabla\psi \;\; \textit{in } L^2(\Omega, \rho_\theta\,dx)^N. \tag{24.104}$$

Proof Following the Lemma 24.3 and Corollary 24.3, we can suppose that the sequence

$$\left\{A_\theta\rho_\theta := \left(I + (p-2)\frac{\nabla y_{\varepsilon_\theta,\theta}}{|\nabla y_{\varepsilon_\theta,\theta}|}\otimes\frac{\nabla y_{\varepsilon_\theta,\theta}}{|\nabla y_{\varepsilon_\theta,\theta}|}\right)|\nabla y_{\varepsilon_0,\theta}|^{p-2}u_\theta\right\}_{\theta\to 0}$$

is such that $\rho_\theta \to \rho_0$ in $L^1(\Omega)$, $A_\theta^{-1} \to A_0^{-1}$, and $(A_\theta\rho_\theta - A_0\rho_0) \to 0$ a.e. in Ω. Then, condition $(24.99)_2$ together with (24.12) implies that sequences $\{(A_\theta\rho_\theta - A_0\rho_0)\,\nabla\psi_\theta\}_{\theta\to 0}$ and $\{(\rho_\theta - \rho_0)\,A_0\nabla\psi_\theta\}_{\theta\to 0}$ are equi-integrable and converge to zero almost everywhere in Ω. Hence, by Lebesgue's theorem, we obtain

$$(A_\theta\rho_\theta - A_0\rho_0)\,\nabla\psi_\theta \to 0,\ \ (\rho_\theta-\rho_0)\,A_0\nabla\psi_\theta \to 0 \ \text{ in } L^1(\Omega)^N \ \text{ as } \theta\to 0. \tag{24.105}$$

As a result, we finally have

$$\begin{aligned}
&\left|(\nabla\varphi, A_\theta\nabla\psi_\theta)_{\mathbb{R}^N}\,\rho_\theta dx - \int_\Omega (\nabla\varphi, A_0\nabla\psi)_{\mathbb{R}^N}\,\rho_0 dx\right| \\
&\le \int_\Omega \left|\,(\nabla\varphi, (A_\theta\rho_\theta - A_0\rho_0)\,\nabla\psi_\theta)_{\mathbb{R}^N}\,\right| dx + \int_\Omega \left|\,(\nabla\varphi, A_0\,(\rho_\theta-\rho_0)\,\nabla\psi_\theta)_{\mathbb{R}^N}\,\right| dx \\
&\qquad + \left|\int_\Omega (\nabla\varphi, A_0\nabla\psi_\theta)_{\mathbb{R}^N}\,\rho_\theta\,dx - \int_\Omega (\nabla\varphi, A_0\nabla\psi)_{\mathbb{R}^N}\,\rho_0\,dx\right| \\
&\qquad\qquad = I_1 + I_2 + I_3 \quad \forall\,\varphi\in C_0^\infty(\Omega),
\end{aligned}$$

where $\lim_{\theta\to 0} I_i = 0$ for $i=1,2$ by (24.105), and $\lim_{\theta\to 0} I_3 = 0$ by $(24.99)_2$ and (24.13). The proof is complete.

We are now in a position to pass to the limit in variational problem (24.68) as $\theta\to 0$. Having assumed that Hypotheses (H0)–(H4) are valid for a given $\widehat{u}\in\mathfrak{A}_{ad}$, we get

$$\begin{aligned}
&\lim_{\theta\to 0}\int_\Omega |\nabla y_{\varepsilon_\theta,\theta}|^{p-2}\left(\left[I+(p-2)\frac{\nabla y_{\varepsilon_\theta,\theta}}{|\nabla y_{\varepsilon_\theta,\theta}|}\otimes\frac{\nabla y_{\varepsilon_\theta,\theta}}{|\nabla y_{\varepsilon_\theta,\theta}|}\right]\nabla\psi_\theta, \nabla\varphi\right)_{\mathbb{R}^N} u_\theta\,dx \\
&\quad\overset{\text{by Lemma 24.12}}{=} \int_\Omega |\nabla y_0|^{p-2}\left(\left[I+(p-2)\frac{\nabla y_0}{|\nabla y_0|}\otimes\frac{\nabla y_0}{|\nabla y_0|}\right]\nabla\psi, \nabla\varphi\right)_{\mathbb{R}^N} u_0\,dx,\,,
\end{aligned}$$

$$\begin{aligned}
&\lim_{\theta\to 0}\int_\Omega |\nabla y_{\varepsilon_\theta,\theta}|^{p-2}\Big(\nabla y_{\varepsilon_\theta,\theta}, \nabla\varphi\Big)_{\mathbb{R}^N} u_\theta\,dx \overset{\text{by Lemma 24.11}}{=} \int_\Omega |\nabla y_0|^{p-2}\Big(\nabla y_0, \nabla\varphi\Big)_{\mathbb{R}^N} u_0\,dx, \\
&\lim_{\theta\to 0}\int_\Omega |y_{\varepsilon_\theta,\theta}-y_d|^{p-2}\Big(y_{\varepsilon_\theta,\theta}-y_d\Big)\varphi\,dx \overset{\text{by (24.78)}}{=} \int_\Omega |y_0-y_d|^{p-2}\Big(y_0-y_d\Big)\varphi\,dx,
\end{aligned}$$

for all $\varphi\in C_0^\infty(\Omega)$. Thus, the weak limit of the sequence $\{\psi_\theta\}_{\theta\to 0}$ in the sense of (24.99) satisfies the following integral identity:

$$\begin{aligned}
&\int_\Omega |\nabla y_0|^{p-2}\left(\left[I+(p-2)\frac{\nabla y_0}{|\nabla y_0|}\otimes\frac{\nabla y_0}{|\nabla y_0|}\right]\nabla\psi, \nabla\varphi\right)_{\mathbb{R}^N} u_0\,dx \\
&\qquad + p\int_\Omega |\nabla y_0|^{p-2}\Big(\nabla y_0, \nabla\varphi\Big)_{\mathbb{R}^N} u_0\,dx \\
&\qquad + p\int_\Omega |y_0-y_d|^{p-2}\Big(y_0-y_d\Big)\varphi\,dx = 0, \quad \forall\,\varphi\in C_0^\infty(\Omega),
\end{aligned} \tag{24.106}$$

or, in other words, it is a weak solution to the degenerate Dirichlet elliptic problem

$$-\operatorname{div}(\rho_0 A_0\nabla\psi) = g_0 \ \text{ in } \Omega, \qquad \psi = 0 \ \text{ on } \partial\Omega, \tag{24.107}$$

where

$$\rho_0 = u_0 |\nabla y_0|^{p-2}, \tag{24.108}$$

$$A_0 = I + (p-2)\frac{\nabla y_0}{|\nabla y_0|} \otimes \frac{\nabla y_0}{|\nabla y_0|}, \tag{24.109}$$

$$g_0 = p \operatorname{div}\left(u_0 |\nabla y_0|^{p-2} \nabla y_0\right) - p\,|y_0 - y_d|^{p-2}(y_0 - y_d)\,. \tag{24.110}$$

In order to realize the limit passage in the inequality (24.73), we adopt the following "directional stability" property of the weak limit of the sequence of quasi-adjoint states $\{\psi_\theta \in \mathscr{H}_\theta\}_{\theta\to 0}$ (in the sense of (24.99)).

(H5) There exists a positive value $\delta > 0$ such that $\nabla\psi_\theta$ lies in "non-variable" weighted space $L^2(\Omega, |\nabla y_0|^{p-2} u_0\, dx)^N$ for all θ such that $0 < \theta \leq \delta$.

It is clear now that due to the Hypotheses (H2)–(H5), the inequality (24.73) becomes correctly defined for each $\widehat{u} \in \mathfrak{A}_{ad}^{stab}$. Indeed, in this case, we have

$$\int_\Omega \widehat{u}\left(|\nabla y_0|^{p-2}\nabla y_0, \nabla\psi_\theta\right)_{\mathbb{R}^N}$$
$$\leq \left\| \frac{\widehat{u}}{u_0} \right\|_{L^\infty(\Omega)} \left(\int_\Omega |\nabla \psi_\theta|^2 |\nabla y_0|^{p-2} u_0\, dx \right)^{1/2} \|y_0\|_{H_{u_0}}^{p/2} < +\infty \tag{24.111}$$

by (H2) and (H5) for all $\widehat{u} \in \mathfrak{A}_{ad}^{stab}$. To proceed further, we make use of the following property which is crucial for the substantiation of the limit passage in inequality (24.73).

Proposition 24.5 *Let $\{\psi_\theta \in \mathscr{H}_\theta\}_{\theta\to 0}$ be the sequence of quasi-adjoint states, and let $\psi \in \mathscr{W}_0$ be its limit in the sense of (24.99). Then, validity of Hypotheses (H0)–(H5) ensures the relation*

$$\lim_{\theta\to 0} \int_\Omega (\nabla y_0, \nabla\psi_\theta)_{\mathbb{R}^N} |\nabla y_0|^{p-2} \widehat{u}\, dx = \int_\Omega (\nabla y_0, \nabla\psi)_{\mathbb{R}^N} |\nabla y_0|^{p-2} \widehat{u}\, dx \quad \forall \widehat{u} \in \mathfrak{A}_{ad}^{stab}.$$

Proof Let $\widehat{u} \in \mathfrak{A}_{ad}^{stab}$ be a fixed control. Then, Corollary 24.3 implies that

$$\left(\frac{|\nabla y_0|^{p-2}}{|\nabla y_{\varepsilon_\theta,\theta}|^{p-2}} \frac{\widehat{u}}{u_\theta} - \frac{\widehat{u}}{u_0} \right) \to 0 \text{ a.e. in } \Omega \text{ as } \theta \to 0.$$

By weak convergence $\nabla\psi_\theta \rightharpoonup \nabla\psi$ in the variable space $L^2(\Omega, |\nabla y_{\varepsilon_\theta,\theta}|^{p-2} u_\theta\, dx)^N$ and property (24.12), we have

$$w_\theta := (\nabla\psi_\theta, \nabla\varphi)_{\mathbb{R}^N} |\nabla y_{\varepsilon_\theta,\theta}|^{p-2} u_\theta \left(\frac{|\nabla y_0|^{p-2}}{|\nabla y_{\varepsilon_\theta,\theta}|^{p-2}} \frac{\widehat{u}}{u_\theta} - \frac{\widehat{u}}{u_0} \right) \to 0 \text{ a.e. in } \Omega \text{ as } \theta \to 0$$

for every test function $\varphi \in C_0^\infty(\Omega)$. Since

$$
\begin{aligned}
w_\theta &= (\nabla\psi_\theta, \nabla\varphi)_{\mathbb{R}^N} |\nabla y_0|^{p-2}\widehat{u} - (\nabla\psi_\theta, \nabla\varphi)_{\mathbb{R}^N} |\nabla y_\theta|^{p-2} u_\theta \frac{\widehat{u}}{u_0} \\
&= (\nabla\psi_\theta, \nabla\varphi)_{\mathbb{R}^N} \left(|\nabla y_0|^{p-2} - |\nabla y_{\varepsilon_\theta,\theta}|^{p-2}\right)\widehat{u} \\
&+ \theta\, (\nabla\psi_\theta, \nabla\varphi)_{\mathbb{R}^N} |\nabla y_{\varepsilon_\theta,\theta}|^{p-2}\widehat{u}\left(1 - \frac{\widehat{u}}{u_0}\right),
\end{aligned}
$$

it follows from estimate (24.111) and definition of the set $\mathfrak{A}_{ad}^{stab}$ that the sequence $\{w_\theta\}_{\theta\to 0}$ is equi-integrable and $w_\theta \to 0$ a.e. in Ω as $\theta \to 0$. Hence, by Lebesgue theorem, we deduce:

$$
w_\varepsilon \to 0 \text{ in } L^1(\Omega) \text{ as } \theta \to 0. \tag{24.112}
$$

Taking this fact into account, we can provide the following estimation:

$$
\begin{aligned}
&\left|\int_\Omega (\nabla\varphi, \nabla\psi_\theta)_{\mathbb{R}^N} |\nabla y_0|^{p-2}\widehat{u}\,dx - \int_\Omega (\nabla\varphi, \nabla\psi)_{\mathbb{R}^N} |\nabla y_0|^{p-2}\widehat{u}\,dx\right| \\
&\le \int_\Omega \left|(\nabla\psi_\theta, \nabla\varphi)_{\mathbb{R}^N} |\nabla y_0|^{p-2}\widehat{u} - (\nabla\psi_\theta, \nabla\varphi)_{\mathbb{R}^N} |\nabla y_{\varepsilon_\theta,\theta}|^{p-2} u_\theta \frac{\widehat{u}}{u_0}\right| dx \\
&+ \left|\int_\Omega (\nabla\psi_\theta, \nabla\varphi)_{\mathbb{R}^N} |\nabla y_{\varepsilon_\theta,\theta}|^{p-2} u_\theta \frac{\widehat{u}}{u_0}\,dx - \int_\Omega (\nabla\varphi, \nabla\psi)_{\mathbb{R}^N} |\nabla y_0|^{p-2}\widehat{u}\,dx\right| \\
&= \underbrace{\int_\Omega |w_\theta|\,dx}_{I_1} \\
&+ \underbrace{\left|\int_\Omega (\nabla\psi_\theta, \nabla\varphi)_{\mathbb{R}^N} |\nabla y_{\varepsilon_\theta,\theta}|^{p-2} u_\theta \frac{\widehat{u}}{u_0}\,dx - \int_\Omega (\nabla\varphi, \nabla\psi)_{\mathbb{R}^N} |\nabla y_0|^{p-2} u_0 \frac{\widehat{u}}{u_0}\,dx\right|}_{I_2},
\end{aligned}
$$

where $\lim_{\theta\to 0} I_1 = 0$ by (24.112). As for the equality $\lim_{\theta\to 0} I_2 = 0$, it immediately follows from weak convergence $(24.99)_2$, condition $\widehat{u}/u_0 \in L^\infty(\Omega)$, and property (24.13). Thus, $\nabla\psi_\theta \rightharpoonup \nabla\psi$ weakly in $L^2(\Omega, |\nabla y_0|^{p-2}\widehat{u}\,dx)^N$. In order to complete the proof, it is enough to note that the function ∇y_0 belongs to $L^2(\Omega, |\nabla y_0|^{p-2}\widehat{u}\,dx)^N$ by condition $(24.63)_1$. So, by density of $C_0^\infty(\Omega)$ in $L^2(\Omega, |\nabla y_0|^{p-2}\widehat{u}\,dx)^N$, we can put $\varphi = y_0$ in the last inequality. The proof is complete.

As a result, the passage to the limit in (24.73) becomes evident by Proposition 24.5, and combining this fact with the relation (24.106), we arrive at the following final conclusion.

Theorem 24.6 *Let $y_d \in L^p(\Omega)$ and $f \in L^q(\Omega)$ be the given functions. Let $(u_0, y_0) \in \Xi$ be an optimal pair for problem (24.26)–(24.28). Then, the fulfillment of Hypotheses (H0)–(H5) implies the existence of an element $\psi \in L^2(\Omega)$ such that $\nabla\psi \in L^2(\Omega, |\nabla y_0|^{p-2}u_0\,dx)^N$ and*

$$\int_\Omega (\widehat{u} - u_0)\Big[|\nabla y_0|^p + \big(|\nabla y_0|^{p-2}\nabla y_0, \nabla\psi\big)_{\mathbb{R}^N} \Big] dx \geq 0, \quad \forall \widehat{u} \in \mathfrak{A}_{ad}^{stab}, \tag{24.113}$$

$$\int_\Omega |\nabla y_0|^{p-2} (\nabla y_0, \nabla w)_{\mathbb{R}^N}\, u_0(x)\, dx = \int_\Omega fw\, dx, \quad \forall w \in H_{u_0}, \tag{24.114}$$

$$\begin{aligned}
&\int_\Omega |\nabla y_0|^{p-2}\Big(\Big[I + (p-2)\frac{\nabla y_0}{|\nabla y_0|} \otimes \frac{\nabla y_0}{|\nabla y_0|}\Big] \nabla\psi, \nabla\varphi\Big)_{\mathbb{R}^N} u_0\, dx \\
&\quad + p\int_\Omega |\nabla y_0|^{p-2}\Big(\nabla y_0, \nabla\varphi\Big)_{\mathbb{R}^N} u_0\, dx \\
&\quad + p\int_\Omega |y_0 - y_d|^{p-2}\Big(y_0 - y_d\Big)\varphi\, dx = 0, \quad \forall \varphi \in C_0^\infty(\Omega).
\end{aligned} \tag{24.115}$$

Remark 24.13 Let us assume that condition (24.52) holds true with constant $C > 0$. Then, Theorem 24.5 implies that the weighted Sobolev spaces H_u and W_u coincide for each admissible control $u \in \mathfrak{A}_{ad}$. Then, it is easy to show that the space H_u is stable along every direction $\widehat{u} - u$, where $u, \widehat{u} \in \mathfrak{A}_{ad}$. Hence, Hypothesis (H0) can be omitted in Theorem 24.6. At the same time, if we assume that $\xi_2 \in L^\infty(\Omega)$ and $\xi_1^{-1} \in L^\infty(\Omega)$, i.e., we deal with an optimal control problem for non-degenerate p-harmonic equation, then $H_u = W_u = W_0^{1,p}(\Omega)$ for all $u \in \mathfrak{A}_{ad}$ and $\mathfrak{A}_{ad}^{stab} \equiv \mathfrak{A}_{ad}$. Hence, Hypotheses (H0) and (H4) become trivial.

Remark 24.14 Let us assume for a moment that

$$\int_\Omega |y_0 - y_d|^{p-2}\Big(y_0 - y_d\Big)\varphi\, dx = \int_{\Omega\setminus\overline{S}_1(y_0)} |y_0 - y_d|^{p-2}\Big(y_0 - y_d\Big)\varphi\, dx, \; \forall\varphi \in C_0^\infty(\Omega),$$

where the set $S_1(y_0)$ is defined by $(24.54)_2$ with the property (24.55). Then, the integral identity (24.115) can be represented as follows:

$$\int_{\Omega\setminus\overline{S}_1(y_0)} |\nabla y_0|^{p-2}\Big(\Big[I + (p-2)\frac{\nabla y_0}{|\nabla y_0|} \otimes \frac{\nabla y_0}{|\nabla y_0|}\Big] \nabla\psi, \nabla\varphi\Big)_{\mathbb{R}^N} u_0\, dx \tag{24.116}$$

$$\begin{aligned}
&+ p\int_{\Omega\setminus\overline{S}_1(y_0)} |\nabla y_0|^{p-2}\Big(\nabla y_0, \nabla\varphi\Big)_{\mathbb{R}^N} u_0\, dx \\
&+ p\int_{\Omega\setminus\overline{S}_1(y_0)} |y_0 - y_d|^{p-2}\Big(y_0 - y_d\Big)\varphi\, dx = 0, \quad \forall\varphi \in C_0^\infty(\mathbb{R}^N; \partial\Omega\setminus\partial S_1(y_0)).
\end{aligned} \tag{24.117}$$

Hence, formally, it can be associated with the following degenerate elliptic boundary value problem for the adjoint variable $\psi \in L^2(\Omega\setminus\overline{S}_1(y_0))$

$$-\operatorname{div}(\rho_0 A_0 \nabla\psi) = g_0 \quad \text{in } \Omega,$$

$$\psi = 0 \quad \text{on } \partial\Omega\setminus\partial S_1(y_0)), \qquad \rho_0\frac{\partial\psi}{\partial n_{A_0}} = 0 \quad \text{on } \partial S_1(y_0))\setminus\partial\Omega,$$

where ρ_0, A_0, and g_0 are defined in (24.108)–(24.110).

24.9 The Hardy–Poincaré Inequality and Uniqueness of the Adjoint State

The main goal of this section is to study the well-posedness of variational problem (24.115). With that in mind, we make use of the following version of the Hardy–Poincaré inequality: for a given internal point $x^* \in \Omega$, there exists a constant $\widehat{C}(\Omega) > 0$ such that for every $v \in H_0^1(\Omega)$

$$\int_\Omega \left[|\nabla v|^2_{\mathbb{R}^N} - \lambda_* \frac{v^2}{|x - x^*|^2_{\mathbb{R}^N}} \right] dx \geq \widehat{C}(\Omega) \int_\Omega v^2\, dx, \tag{24.118}$$

where $\lambda_* := (N-2)^2/4$ and $N \geq 2$. We begin with the following auxiliary results.

Lemma 24.13 *Let $(u_0, y_0) \in \Xi$ be an optimal pair to the problem (24.26)–(24.28). Assume the function $|\nabla y_0|^{p-2} u_0$ belongs to the class of Muckenhoupt weight A_2, and $\nabla \ln \left(|\nabla y_0|^{p-2} u_0 \right) \in L^2(\Omega)^N$. Then, each element*

$$\psi \in \mathscr{W}_0 := \left\{ \varphi \in W_0^{1,1}(\Omega) \ :\ \varphi \in L^2(\Omega),\ \nabla \varphi \in L^2(\Omega, |\nabla y_0|^{p-2} u_0\, dx)^N \right\}$$

can be represented in a unique way as follows:

$$\psi = |\nabla y_0|^{(2-p)/2} u_0^{-1/2} z_0, \quad \textit{where } z_0 \in W_0^{1,1}(\Omega) \cap L^2(\Omega). \tag{24.119}$$

Proof Since $|\nabla y_0|^{p-2} u_0$ belongs to the class of Muckenhoupt weights A_2, it follows that $\mathscr{H}_0 = \mathscr{W}_0 = \mathrm{cl}_{\|\cdot\|_{\mathscr{H}_0}} C_0^\infty(\Omega)$ and there exists a constant $C > 0$ such that (see [10, 12])

$$\int_\Omega |\psi|^2 |\nabla y_0|^{p-2} u_0\, dx \leq C \int_\Omega |\nabla \psi|^2 |\nabla y_0|^{p-2} u_0\, dx, \quad \text{for each element } \psi \text{ of } \mathscr{W}_0. \tag{24.120}$$

Let us fix an element $\psi \in \mathscr{W}_0$. Then, $\psi \in \mathscr{H}_0$ and for $z_0 := |\nabla y_0|^{(p-2)/2} u_0^{1/2} \psi$, we have

$$\|z_0\|^2_{L^2(\Omega)} = \int_\Omega \psi^2 |\nabla y_0|^{p-2} u_0\, dx \leq C \|\nabla \psi\|^2_{L^2(\Omega; |\nabla y_0|^{p-2} u_0\, dx)^N} = C \|\psi\|^2_{\mathscr{H}_0} < \infty,$$

where the constant $C > 0$ comes from the Poincaré inequality (24.120). Using the evident equality

$$\nabla z_0 = \left(\frac{1}{2} \sqrt{\rho_0} \psi \nabla \ln \rho_0 + \sqrt{\rho_0} \nabla \psi \right) \Big|_{\rho_0 = |\nabla y_0|^{p-2} u_0},$$

and applying the Hölder inequality with exponents $p' = q' = 2$, we get

$$\begin{aligned}\|\nabla z_0\|_{L^1(\Omega)^N} \le & \frac{1}{2}\int_\Omega |\nabla y_0|^{(p-2)/2} u_0^{1/2} |\psi| |\nabla \ln \left(|\nabla y_0|^{p-2} u_0\right)| \, dx \\ & + |\Omega|^{1/2} \left(\int_\Omega |\nabla y_0|^{p-2} u_0 |\nabla \psi|^2 \, dx\right)^{1/2} \\ \le & \frac{1}{2} \|\psi\|_{L^2(\Omega; |\nabla y_0|^{p-2} u_0 \, dx)} \left(\int_\Omega |\nabla \ln \left(|\nabla y_0|^{p-2} u_0\right)|^2 \, dx\right)^{1/2} + |\Omega|^{1/2} \|\psi\|_{\mathscr{H}_0} \\ \le & \left(\frac{C}{2} \|\nabla \ln \left(|\nabla y_0|^{p-2} u_0\right)\|_{L^2(\Omega)^N} + |\Omega|^{1/2}\right) \|\psi\|_{\mathscr{H}_0} < \infty.\end{aligned}$$

Thus, $z_0 \in W^{1,1}(\Omega) \cap L^2(\Omega)$. Since the element $z_0 := |\nabla y_0|^{(p-2)/2} u_0^{1/2} \psi$ inherits the trace properties along $\partial\Omega$ from its parent element ψ, we finally obtain $z_0 \in W_0^{1,1}(\Omega) \cap L^2(\Omega)$. The proof is complete.

As an obvious consequence of this result and continuity of the embedding of Sobolev spaces $H_0^1(\Omega) \hookrightarrow W_0^{1,1}(\Omega)$, we can give the following conclusion.

Corollary 24.4 *If* $\nabla \ln \left(|\nabla y_0|^{p-2} u_0\right) \in L^2(\Omega)^N$ *and* $|\nabla y_0|^{p-2} u_0 \in A_2$, *then there exists a non-empty dense subset* $D(y_0, u_0)$ *of* $H_0^1(\Omega)$ *such that*

$$|\nabla y_0|^{(2-p)/2} u_0^{-1/2} z \in \mathscr{H}_0, \quad \forall z \in D(y_0, u_0). \tag{24.121}$$

Remark 24.15 It is clear that Corollary 24.4 remains true if we relax the condition $|\nabla y_0|^{p-2} u_0 \in A_2$ to the following one: There exists a constant $C > 0$ such that inequality (24.120) holds true for every $\psi \in \mathscr{H}_0$. In this case, the function $|\nabla y_0|^{p-2} u_0$ is not obligatory of the class of Muckenhoupt weights A_2, and hence, we can not guarantee the fulfillment of the equality $\mathscr{H}_0 = \mathscr{W}_0$.

We introduce the following linear mapping:

$$\mathfrak{F} : D(y_0, u_0) \subset H_0^1(\Omega) \to \mathscr{H}_0, \quad \text{where } \mathfrak{F}z = |\nabla y_0|^{(2-p)/2} \sqrt{u_0^{-1}}\, z. \tag{24.122}$$

Since domain $D(y_0, u_0)$ of $\mathfrak{F}$ is dense in Banach space $H_0^1(\Omega)$, it follows that for $\mathfrak{F}$, as for a densely defined operator, there exists an adjoint operator

$$\mathfrak{F}^* : D\left(\mathfrak{F}^*\right) \subset \mathscr{H}_0^* \to H^{-1}(\Omega)$$

such that

$$\langle \mathfrak{F}^* v, z\rangle_{H^{-1}(\Omega), H_0^1(\Omega)} = \langle v, \mathfrak{F}z\rangle_{\mathscr{H}_0^*, \mathscr{H}_0}, \quad \forall z \in D(y_0, u_0) \text{ and } \forall v \in D\left(\mathfrak{F}^*\right),$$

where

$$D\left(\mathfrak{F}^*\right) = \left\{ v \in \mathscr{H}_0^* \;\middle|\; \begin{array}{c}\text{there exists } C > 0 \text{ such that for all } z \in D(y_0, u_0) \\ \left|\langle v, \mathfrak{F}z\rangle_{\mathscr{H}_0^*, \mathscr{H}_0}\right| \le C \|z\|_{H_0^1(\Omega)}\end{array} \right\}.$$

Notice that in general, the adjoint operator $\mathfrak{F}^*$ is not densely defined. For our further analysis, we need to introduce some preliminaries. Let λ be a positive constant such that $\lambda < \lambda_* := (N-2)^2/4$. Let $\{x_1, x_2, \dots, x_L\} \subset \Omega$ be a given collection of points. We define a subset $\mathfrak{M}(\Omega) \subset H_{u_0}$ as follows: $y \in \mathfrak{M}(\Omega)$ if and only if $y \in H_{u_0}$ and

$$-\widehat{C}(\Omega) \le V_y(x) \le \frac{2\lambda}{L}\sum_{i=1}^{L}\frac{1}{|x-x_i|^2} \quad \text{a.e. in } \Omega, \tag{24.123}$$

for some positive constant $\widehat{C}(\Omega) > 0$, where the symmetric matrix A_0 is defined by (24.109), and

$$V_y(x) = -\operatorname{div}\left(A_0\nabla \ln\left(|\nabla y|^{p-2}u_0\right)\right) - \frac{1}{2}\left(\nabla \ln\left(|\nabla y|^{p-2}u_0\right), A_0\nabla \ln\left(|\nabla y|^{p-2}u_0\right)\right)_{\mathbb{R}^N}.$$

Let us consider the following linear operator

$$\mathscr{A}_0\psi := -\operatorname{div}\left(\rho_0 A_0 \nabla\psi\right), \quad \forall\, \psi \in \mathscr{H}_0,$$

where ρ_0 and A_0 are defined by (24.108) and (24.109). As follows from Proposition 24.3, the operator $\mathscr{A}_0 : \mathscr{H}_0 \to \mathscr{H}_0^*$ is obviously strictly monotone

$$\begin{aligned}\langle \mathscr{A}_0(\psi-\phi), \psi-\phi\rangle_{\mathscr{H}_0^*;\mathscr{H}_0} &= \int_\Omega |\nabla y|^{p-2}u_0\left(A_0(\nabla\psi-\nabla\phi), \nabla\psi-\nabla\phi\right)_{\mathbb{R}^N}\\ &\ge \|\nabla\psi-\nabla\phi\|^2_{L^2(\Omega,|\nabla y|^{p-2}u_0\,dx)^N} = \|\psi-\phi\|^2_{\mathscr{H}_0},\end{aligned}$$

semicontinuous and $\mathscr{H}_0$-coercive

$$\begin{aligned}\left|\langle \mathscr{A}_0\psi, \phi\rangle_{\mathscr{H}_0^*;\mathscr{H}_0}\right| &\le \left[1+(p-2)2^{N-1}\right]\|\psi\|_{\mathscr{H}_0}\|\phi\|_{\mathscr{H}_0}, \forall\, \psi, \phi \in \mathscr{H}_0,\\ \langle \mathscr{A}_0\psi, \psi\rangle_{\mathscr{H}_0^*;\mathscr{H}_0} &\ge \|\psi\|^2_{\mathscr{H}_0}.\end{aligned}$$

We are now in a position to establish another important property of this operator.

Lemma 24.14 *Assume that an optimal pair $(u_0, y_0) \in \Xi$ to the problem (24.26)–(24.28) is such that*

$$(H6)\quad \begin{cases} \nabla \ln\left(|\nabla y_0|^{p-2}u_0\right) \in L^2(\Omega)^N, \quad y_0 \in \mathfrak{M}(\Omega) \subset H_{u_0},\\ \displaystyle\int_\Omega |\psi|^2|\nabla y_0|^{p-2}u_0\,dx \le C\int_\Omega |\nabla\psi|^2|\nabla y_0|^{p-2}u_0\,dx, \ \forall\, \psi \in \mathscr{H}_0 \text{ with some } C > 0.\end{cases}$$

Then,

$$\langle \mathscr{A}_0\left(\mathfrak{F}z\right), \mathfrak{F}v\rangle_{\mathscr{H}_0^*;\mathscr{H}_0} = \langle B_0(z), v\rangle_{H^{-1}(\Omega),H_0^1(\Omega)}, \tag{24.124}$$

where

$$B_0(z) = -\mathrm{div}\,(A_0 \nabla z) - \frac{1}{2} V(x) z, \tag{24.125}$$

$$V(x) = -\mathrm{div}\left(A_0 \nabla \ln\left(|\nabla y_0|^{p-2} u_0\right)\right) - \frac{1}{2}\left(\nabla \ln\left(|\nabla y_0|^{p-2} u_0\right), A_0 \nabla \ln\left(|\nabla y_0|^{p-2} u_0\right)\right)_{\mathbb{R}^N}, \tag{24.126}$$

and the linear operator B_0 defines an isomorphism from $H_0^1(\Omega)$ into its dual $H^{-1}(\Omega)$.

Proof Let v and z be arbitrary elements of $D(y_0, u_0) \subset H_0^1(\Omega)$. Then, by Corollary 24.4 (see also Remark 24.15), we have $\mathfrak{F}z, \mathfrak{F}v \in \mathscr{H}_0$. Therefore, following the definition of operator $\mathfrak{F}$ and taking into account that $\dfrac{\nabla \rho_0}{\rho_0} = \nabla \ln \rho_0$, we arrive at the following chain of transformations:

$$\begin{aligned}
\mathscr{A}_0(\mathfrak{F}z) &= -\mathrm{div}\left(\rho_0 A_0 \nabla\left(\frac{z}{\sqrt{\rho_0}}\right)\right) = -\mathrm{div}\left(\sqrt{\rho_0} A_0 \left(\nabla z - \frac{1}{2} z \nabla \ln \rho_0\right)\right)\\
&= -\frac{1}{2}\left(\frac{\nabla \rho_0}{\sqrt{\rho}}, A_0 \nabla z\right)_{\mathbb{R}^N} + \frac{1}{4} z \left(\frac{\nabla \rho_0}{\sqrt{\rho}}, A_0 \nabla \ln \rho_0\right)_{\mathbb{R}^N} - \sqrt{\rho_0}\,\mathrm{div}\,(A_0 \nabla z)\\
&\quad + \frac{\sqrt{\rho_0}}{2} (\nabla z, A_0 \nabla \ln \rho_0)_{\mathbb{R}^N} + \frac{\sqrt{\rho_0}}{2} z\,\mathrm{div}(A_0 \nabla \ln \rho_0)\\
&= \sqrt{\rho_0}\left[-\mathrm{div}\,(A_0 \nabla z) - \frac{z}{2}\left(-\frac{1}{2}(\nabla \ln \rho_0, A_0 \nabla \ln \rho_0)_{\mathbb{R}^N} - \mathrm{div}(A_0 \nabla \ln \rho_0)\right)\right]\\
&= \sqrt{\rho_0}(-\mathrm{div}(A_0 \nabla z) - \frac{1}{2} z\, V(x)).
\end{aligned}$$

Hence,

$$\begin{aligned}
\langle \mathscr{A}_0(\mathfrak{F}z), \mathfrak{F}v \rangle_{\mathscr{H}_0^*;\mathscr{H}_0} &= \langle \sqrt{\rho_0}\Big(-\mathrm{div}\,(A_0 \nabla z) - \frac{1}{2} V(x) z\Big), \frac{v}{\sqrt{\rho_0}} \rangle_{\mathscr{H}_0^*;\mathscr{H}_0}\\
&= \langle -\mathrm{div}\,(A_0 \nabla z) - \frac{1}{2} V(x) z, v \rangle_{H^{-1}(\Omega);H_0^1(\Omega)} = \langle B_0(z), v \rangle_{H^{-1}(\Omega),H_0^1(\Omega)}.
\end{aligned}$$

To conclude the proof, it remains to show that operator $B_0 := -\mathrm{div}\,(A_0 \nabla) - \frac{1}{2} V(x)$ defines an isomorphism from $H_0^1(\Omega)$ into its dual $H^{-1}(\Omega)$. With that in mind, we make use of the Hardy–Poincaré inequality (24.118), where $\lambda * := (N-2)^2/4$ and $N \geq 2$. According to this result and the fact that $y_0 \in \mathfrak{M}(\Omega)$, we have

$$-\widehat{C}(\Omega) \leq V(x) \leq \frac{2\lambda}{L} \sum_{i=1}^{L} \frac{1}{|x - x_i|^2} < \frac{(N-2)^2}{2L} \sum_{i=1}^{L} \frac{1}{|x - x_i|^2} \quad \text{a.e. in } \Omega \tag{24.127}$$

and therefore,

$$\left(1+\frac{\widehat{C}(\Omega)}{2C}\right)\|v\|^2_{H^1_0(\Omega)} \geq \int_\Omega \left[|\nabla v|^2 + \frac{\widehat{C}(\Omega)}{2}v^2\right] dx$$
$$\geq \int_\Omega \left[|\nabla v|^2 - \frac{\lambda}{L}\left(\sum_{i=1}^{L}\frac{1}{|x-x_i^*|}\right)v^2\right] dx = \left(1-\frac{\lambda}{\lambda_*}\right)\int_\Omega |\nabla v|^2\, dx$$
$$+\frac{\lambda}{\lambda_*}\int_\Omega \left[|\nabla v|^2 - \frac{\lambda_*}{L}\left(\sum_{i=1}^{L}\frac{1}{|x-x_i^*|}\right)v^2\right] dx$$
$$\geq \left(1-\frac{\lambda}{\lambda_*}\right)\int_\Omega |\nabla v|^2\, dx + \frac{\lambda\widehat{C}(\Omega)}{\lambda_*}\int_\Omega v^2\, dx \geq \left(1-\frac{\lambda}{\lambda_*}\right)\|v\|^2_{H^1_0(\Omega)}. \quad (24.128)$$

Thus, in view of (24.127) and (24.128),

$$\|[v]\|_0^2 := \int_\Omega \left[|\nabla v|^2 - \frac{1}{2}V(x)v^2\right] dx = \int_\Omega \left[(\nabla v, \nabla v)_{\mathbb{R}^N} - \frac{1}{2}V(x)v^2\right] dx$$

is equivalent to the standard norm of $H_0^1(\Omega)$, and therefore, the operator B_0 given by (24.125) defines an isomorphism from $H_0^1(\Omega)$ into its dual $H^{-1}(\Omega)$.

The last step of our analysis is to show that the adjoint state ψ to $y_0 \in H_{u_0}$ can be defined as a unique solution in $\mathscr{H}_0$ of degenerate variational problem (24.115) even if the weight $|\nabla y_0|^{p-2}u_0$ does not belong to the Muckenhoupt class A_2.

Lemma 24.15 *Assume that Hypotheses (H4) and (H6) (see Lemma 24.14) are valid. Then, variational problem (24.115) admits a unique solution $\psi \in \mathscr{H}_0$.*

Proof Since $y_0 \in H_{u_0}$, Hypothesis (H4) implies that (see (24.94) and (24.95))

$$\langle g_0, \varphi\rangle_{\mathscr{H}_0^*;\mathscr{H}_0} \leq p\int_\Omega |\nabla y_0|^{p-1}|\nabla\varphi|u_0\, dx + p\int_\Omega |y_0 - y_d|^{p-1}|\varphi|\, dx$$
$$\leq \|y_0\|^{p/2}_{H_{u_0}}\|\varphi\|_{\mathscr{H}_0} + C\|y_0 - y_d\|^{p-1}_{L^{2p-2}(\Omega)}\|\varphi\|_{\mathscr{H}_0}.$$

Hence, $g_0 := p\,\mathrm{div}\left(u_0|\nabla y_0|^{p-2}\nabla y_0\right) - p\,|y_0 - y_d|^{p-2}(y_0 - y_d)$ can be considered as a distribution on $\mathscr{H}_0$. Therefore, equality (24.115) can be represented as follows:

$$\langle -\mathrm{div}\left(|\nabla y_0|^{p-2}u_0 A_0\nabla\psi\right) - g_0, \phi\rangle_{\mathscr{H}_0^*;\mathscr{H}_0} = 0, \quad \forall\phi \in \mathfrak{F}(D(y_0, u_0)) = \mathscr{H}_0.$$

Then, Hypothesis (H6) and Lemma 24.14 lead to the following transformations:

$$\langle -\mathrm{div}\left(|\nabla y_0|^{p-2} u_0 A_0 \nabla \psi\right), \phi\rangle_{\mathscr{H}_0^*;\mathscr{H}_0} = \langle \mathscr{A}_0(\mathfrak{F}z), \mathfrak{F}v\rangle_{\mathscr{H}_0^*;\mathscr{H}_0}$$
$$= \langle -\mathrm{div}\,(A_0\nabla z) - \frac{1}{2}V(x)z, v\rangle_{H^{-1}(\Omega);H_0^1(\Omega)} = \langle B_0(z), v\rangle_{H^{-1}(\Omega),H_0^1(\Omega)},$$
$$\langle g_0, \phi\rangle_{\mathscr{H}_0^*;\mathscr{H}_0} = -p\int_\Omega |\nabla y_0|^{p-2}\left(\nabla y_0, \nabla\phi\right)_{\mathbb{R}^N} u_0\,dx$$
$$-p\int_\Omega |y_0 - y_d|^{p-2}\left(y_0 - y_d\right)\phi\,dx = \langle g_0, \mathfrak{F}v\rangle_{\mathscr{H}_0^*;\mathscr{H}_0} = \langle \mathfrak{F}^* g_0, v\rangle_{H^{-1}(\Omega),H_0^1(\Omega)}$$

provided $\phi \in \mathfrak{F}(D(y_0, u_0))$. By Hardy–Poincaré inequality (see (24.118)), the expression

$$\int_\Omega \left[(\nabla v, \nabla z)_{\mathbb{R}^N} - \frac{1}{2}V(x)vz\right] dx$$

can be considered as a scalar product in $H_0^1(\Omega)$. Then, by Riesz representation theorem, there exists a unique element $z_0 \in H_0^1(\Omega)$ such that

$$\langle B_0(z_0), v\rangle_{H^{-1}(\Omega),H_0^1(\Omega)} = \langle \mathfrak{F}^* g_0, v\rangle_{H^{-1}(\Omega),H_0^1(\Omega)}, \quad \forall v \in H_0^1(\Omega).$$

Thus, $\psi := \mathfrak{F}z_0 = |\nabla y_0|^{(2-p)/2}\sqrt{u_0^{-1}}\,z_0$ is a unique solution to the Dirichlet boundary value problem (24.107). Moreover, by Remark 24.15, we finally get $\psi \in \mathscr{H}_0$.

As a result, the optimality conditions to optimal control problem (24.26)–(24.28) can be reformulated as follows (see for comparison Theorem 24.6):

Theorem 24.7 *Let $y_d \in L^p(\Omega)$ and $f \in L^q(\Omega)$ be the given functions. Let $(u_0, y_0) \in \Xi$ be an optimal pair to the problem (24.26)–(24.28). Then, the fulfillment of Hypotheses (H0)–(H6) implies the existence of a unique element $z_0 \in H_0^1(\Omega)$ such that for all $\widehat{u} \in \mathfrak{A}_{ad}^{stab}$,*

$$\int_\Omega (\widehat{u} - u_0)\left[|\nabla y_0|^p + \sqrt{\frac{|\nabla y_0|^{p-2}}{u_0}}\left(\nabla y_0, \nabla z_0 - \frac{z_0}{2}\nabla \ln\left(|\nabla y_0|^{p-2}u_0\right)\right)_{\mathbb{R}^N}\right] dx \geq 0,$$
$$-\,\mathrm{div}\left(u_0(x)|\nabla y_0|^{p-2}\nabla y_0\right) = f \quad in\ \Omega,$$
$$y_0 = 0 \quad on\ \partial\Omega,$$
$$-\,\mathrm{div}\left(\left[I + (p-2)\frac{\nabla y_0}{|\nabla y_0|}\otimes\frac{\nabla y_0}{|\nabla y_0|}\right]\nabla z_0\right) - \frac{1}{2}V(x)z_0 = \mathfrak{F}^* g_0 \quad in\ \Omega,$$
$$z_0 = 0 \quad on\ \partial\Omega,$$

where potential term $V(x)$ and distribution $g_0 \in \mathscr{H}_0^*$ are defined by (24.126) and (24.110), respectively.

Acknowledgments This work was partially supported by the Ukrainian State Fund for Fundamental Researches under grant GP/F66/14921 and by the National Academy of Sciences of Ukraine under grant 2284.

References

1. Al'shin, A.B., Korpusov, M.O., Sveshnikov, A.G.: Blow-up in Nonlinear Sobolev Type Equations. De Gruyter, Berlin (2011)
2. Boccardo, L., Gallouët, T., Marcellini, P.: Anisotropic equations in L^1. Differ. Integral Equ. **9**, 209–212 (1996)
3. Buttazzo, G., Kogut, P.I.: Weak optimal controls in coefficients for linear elliptic problems. Revista Matematica Complutense **24**, 83–94 (2011)
4. Buttazzo, G., Varchon, N.: On the optimal reinforcement of an elastic membrane. Riv. Mat. Univ. Parma. **4**(7), 115–125 (2005)
5. Casas, E.: Optimal control in the coefficients of elliptic equations with state constraints. Appl. Math. Optim. **26**, 21–37 (1992)
6. Casas, E., Fernandez, L.A.: Optimal control of quasilinear elliptic equations with non differentiable coefficients at the origin. Rev. Matematica Univ. Compl. Madrid. **4**(2-3), 227–250 (1991)
7. Chiadó Piat V., Serra Cassano F.: Some remarks about the density of smooth functions in weighted Sobolev spaces. J. Convex Anal. **1**(2), 135–142 (1994)
8. D'Apice, C., De Maio, U., Kogut, O.P.: On shape stability of Dirichlet optimal control problems in coefficients for nonlinear elliptic equations. Adv. Differ. Equ. **15**(7–8), 689–720 (2010)
9. D'Apice, C., De Maio, U., Kogut, O.P.: Optimal control problems in coefficients for degenerate equations of monotone type: shape stability and attainability problems. SIAM J. Control Optim. **50**(3), 1174–1199 (2012)
10. Drabek P., Kufner A., Nicolosi F.: Non linear elliptic equations, singular and degenerate cases. University of West Bohemia (1996)
11. Drabek, P., Milota, Y.: Methods of Nonlinear Analysis. Applications to Differential Equations. Birkhäuser, Berlin (2007)
12. Fabes, E.B., Kenig, C.E., Serapioni, R.P.: The local regularity of solutions of degenerate elliptic equations. Commun. Partial Differ. Equ. **7**, 77–116 (1982)
13. Fursikov, A.V.: Optimal Control of Distributed Systems. Theory and Applications. AMS, Providence (2000)
14. Gajewski, H., Gröger, K., Zacharias, K.: Nichtlineare Operatorgleichungen und Operatordifferentialgleichungen. Academie-Varlar, Berlin (1974)
15. Giusti, E.: Minimal Surfaces and Functions of Bounded Variation. Birkhäuser, Boston (1984)
16. Ioffe, A., Tikhomirov, V.: Extremal Problems. North-Holland, Amsterdam (1979)
17. Kogut, P.I., Leugering, G.: Optimal control problems for partial differential equations on reticulated domains. Approximation and Asymptotic Analysis. Systems and Control. Birkhäuser Verlag, Boston (2011)
18. Kogut, P.I., Leugering, G.: Matrix-valued L^1-optimal controls in the coefficients of linear elliptic problems. Z. Anal. Anwend. **32**(4), 433–456 (2013)
19. Kogut P.I., Leugering G.: On existence of optimal solutions to boundary control problem for an elastic body with quasistatic evolution of damage. In: Continuous and Distributed Systems: Theory and Applications, Ch. 19. Solid Mechanics and Its Applications, vol. 211, pp. 265–286. Springer, Berlin (2014)
20. Kogut P.I., Kupenko O.P., Leugering G.: Optimal control problems in coefficients for nonlinear Dirichlet problems of monotone type: Optimality conditions. Part I. ZAA **34**(1), 85–108 (2015) doi:10.4171/ZAA/1530
21. Kogut P.I., Kupenko O.P., Leugering G.: Optimal control problems in coefficients for nonlinear Dirichlet problems of monotone type: Optimality conditions. Part II. ZAA **34**(1), 199–219 (2015) doi:10.4171/ZAA/1536
22. Kovalevsky A.A.: On L^1-functions with very singular behaviour. Nonlinear Anal. **85**(7), 66–77 (2013). arXiv:1010.0570v1
23. Kupenko, O.P.: Optimal control problems in coefficients for degenerate variational inequalities of monotone type. I. Existence of solutions. J. Comput. Appl. Math. **106**(3), 88–104 (2011)

24. Kupenko O.P., Manzo R.: On an optimal L^1-control problem in coefficients for linear elliptic variational inequality. Abstr. Appl. Anal. 1–13 (2013) doi:10.1155/2013/821964
25. Lions, J.-L.: Some Methods of Solving Non-Linear Boundary Value Problems. Dunod-Gauthier-Villars, Paris (1969)
26. Lurie, K.A.: Optimum control of conductivity of a fluid moving in a channel in a magnetic field. J. Appl. Math. Mech. **28**, 316–327 (1964)
27. Manfredi, J.J.: p-harmonic functions in the plane. Proc. Am. Math. Soc. **103**, 473–479 (1988)
28. Murat, F.: Un contre-exemple pour le probleme du controle dans les coefficients. C. R. Acad. Sci. Paris. Ser. A–B **273**, A708–A711 (1971) (in French)
29. Murthy, M.K.V., Stampacchia, G.: Boundary value problems for some degenerate elliptic operators. Ann. Mat. Pura Appl. **80**, 1–122 (1968)
30. Pastukhova, S.E.: Degenerate equations of monotone type: Lavrentev phenomenon and attainability problems. Sbornik: Math. **198**(10), 1465–1494 (2007)
31. Serovajskiy S.Ya.: Variational Inequalities in non-linear optimal control problems. Methods Means Math. Modell. 156–169 (1977)
32. Trudinger, N.S.: Linear elliptic equations with measurable coefficients. Ann. Scuola Norm. Sup Pisa. **27**, 265–308 (1973)
33. Zhikov, V.V.: On Lavrentiev phenomenon. Russ. J. Math. Phys. **3**(2), 249–269 (1994)
34. Zhikov, V.V.: Weighted Sobolev spaces. Sbornik: Math. **189**(8), 27–58 (1998)
35. Zhikov, V.V.: On an extension of the method of two-scale convergence and its applications. Sbornik: Math. **191**(7), 973–1014 (2000)
36. Zhikov, V.V., Pastukhova, S.E.: Homogenization of degenerate elliptic equations. Siberian Math. J. **49**(1), 80–101 (2006)

